www.wadsworth.com

wadsworth.com is the World Wide Web site for Wadsworth Publishing Company and is your direct source to dozens of online resources.

At *wadsworth.com* you can find out about supplements, demonstration software, and student resources. You can also send e-mail to many of our authors and preview new publications and exciting new technologies.

wadsworth.com
Changing the way the world learns®

Social Problems
of the Modern World

A Reader

FRANCES V. MOULDER

Three Rivers Community College

Wadsworth
Thomson Learning.

Australia • Canada • Mexico • Singapore • Spain • United Kingdom • United States

Publisher: Eve Howard
Marketing Manager: Diane McOscar
Marketing Assistant: Kelli Goslin
Project Editor: Jerilyn Emori
Print Buyer: Karen Hunt
Permissions Editor: Susan Walters
Production Service: Vicki Moran/
   Publishing Support Services

Copy Editor: Adrienne Armstrong
Illustrator: Lotus Art
Cover Designer: Yvo Riezebos
Cover Image: PhotoDisc
Compositor: Gustafson Graphics
Text and Cover Printer:
Webcom Limited

For permission to use material from this text, contact us by:
   Web: http://www.thomsonrights.com
   Fax: 1-800-730-2215
   Phone: 1-800-730-2214

For more information, contact
Wadsworth/Thomson Learning
10 Davis Drive
Belmont, CA 94002-3098
USA
http://www.wadsworth.com

International Headquarters
Thomson Learning
International Division
290 Harbor Drive, 2nd Floor
Stamford, CT 06902-7477
USA

UK/Europe/Middle East/South Africa
Thomson Learning
Berkshire House
168-173 High Holborn
London WC1V 7AA
United Kingdom

Asia
Thomson Learning
60 Albert Street, #15-01
Albert Complex
Singapore 189969

Canada
Nelson Thomson Learning
1120 Birchmount Road
Toronto, Ontario M1K 5G4
Canada

Library of Congress Cataloging-in-Publication Data

Moulder, Frances V. (Frances Valentine)
    Social problems of the modern world: a reader/ Frances Moulder.
        p. cm.
    ISBN 0-534-56682-0
    1. Social problems.   I. Title.
    HN17.5 .M67   2000
    361.1—dc21   99-058514

Contents

CHAPTER 9

The Cultural Survival of Indigenous Peoples 232

CHAPTER 10

Racial/Ethnic Conflicts and the Danger of Genocide 262

Preface to the Instructor

As the world system and modernization perspectives have emphasized, all societies are profoundly shaped by their participation in a larger global world and by world historical trends and processes. A few years ago, I began to rethink all the courses I teach with this fundamental insight in mind. I began an effort to "internationalize" or "globalize" my courses. For the Social Problems course, this meant looking at the social problems of individual nations as problems accompanying the transformation of the modern world as a whole.

As I searched for textbooks and readers with a global perspective, I discovered that few exist for the Social Problems course. Most textbooks for the course are still centered on the United States. If the global dimensions of problems are covered, they are often covered in an ad hoc way rather than systematically integrated into the book. Although there are several readers available on global social problems, these are not adequate for a 100- or 200-level Social Problems course. Some of these are collections of readings on current events from the popular press. These readers are confusing to students because their conceptual and theoretical frameworks are eclectic and drawn primarily from the fields of political science and economics, rather than sociology. Other readers are sociological, but the material is written for a scholarly audience. These readers are too academic or abstruse, and tend to put students to sleep.

Social Problems of the Modern World has been designed to fill this gap. A global perspective has been incorporated throughout the book. Each chapter of the reader covers a problem typically covered in the Social Problems course. Some of the readings on each problem deal with the United States; others focus on other nations or on the global dimensions of the problem. (On average, one-third to one-half of the articles in each chapter focus on the United States.) Because of the global perspective, a few chapters have been included that are not always found in the typical Social Problems texts: "The Global Media," "The Cultural Survival of Indigenous Peoples," and "Racial/Ethnic Conflicts and the Danger of Genocide." I believe instructors and students will agree that this material is an important enhancement.

In selecting material for the reader, I used the criteria listed below. These reflect my philosophy of teaching and learning, which is centered on *active learning, critical thinking, and multiculturalism*. They also reflect the idea that we study social problems *in order to work toward meeting human needs*.

- The writing should be accessible to a general audience of college students at the introductory level—clear, readable, interesting. Some of it should challenge students, stretching their abilities, without frustrating them.

- Although the authors do not necessarily need to be sociologists, the content should utilize sociological concepts and perspective, or at least be consistent with sociological concepts and perspective.

- Each chapter should cover the various dimensions of social problems analysis—that is, attention to how the problem is defined, the nature and scope of the problem, the human costs, causes, and solutions.

- Within the sociological perspective, there should be multiple perspectives on the problems, so as to encourage students to think critically. Order (for example, structural-functionalist, modernization), conflict (for example, Marxist, world system, feminist), and symbolic interactionist (for example, social constructionist) perspectives should be represented, as well as the contributions of people who seek to make the study of race/ethnicity, gender, and sexual orientation more central to the study of world issues.

- There should be a diversity of voices, including first-person accounts, from different nationalities and social backgrounds, to help students develop a more connected, or empathic, knowledge of world problems, and to appeal to students from a variety of backgrounds.

- Many of the writings should contain a message of hope—that is, focusing on *solutions* to the problems and *social change* rather than simply analyzing negative conditions and their causes.

- Solutions presented should contain material on grassroots or "bottom up" organizations, in keeping with the prominent place of these organizations today in working for global social change.

- For "current events" articles, preference should be given to those containing analysis that will still be relevant after the immediate situation has passed.

To my surprise, it proved to be quite a balancing act to meet all these criteria. For example, it was especially difficult to incorporate multiple theoretical perspectives without sacrificing the *global* dimension. Although the order, conflict, and symbolic interactionist perspectives have all influenced the sociological analysis of social problems, their influence on the analysis of *global* issues is uneven. The conflict perspective has been on the rise in the last twenty years, while the modernization perspective has been waning. Symbolic interactionists are more likely to write about problems in the United States than about global matters. To compound the difficulty, much of the discourse on world problems is carried on in a conceptual framework that is different from that of mainstream sociology—for example, the languages of development studies, environmentalism, political science, and economics. It was also necessary to reject a good deal of material, otherwise appropriate, because it required too high a level of academic preparation. Despite these difficulties, I believe I have successfully performed the balancing act, and that instructors will find that taken as a whole, the reader fulfills the criteria specified.

The format for the chapters is as follows: Each chapter begins with an Introduction to the readings. The Introduction frames the problem in global perspective, discusses key concepts and theoretical perspectives on the problem, and introduces each of the readings. I wrote the Introductions to take into account the needs of instructors who prefer to use a reader alone, as well as those who would use the book as a supplement to a standard textbook. I believe the content of the chapter introductions is sufficient to enable the reader to stand alone, without being so inclusive that it repeats the content of the standard textbook. Each chapter Introduction concludes with a section "For Additional Research." This section provides suggestions for students on additional readings, relevant community organizations, and projects for active learning. I decided to be selective, rather than encyclopedic, in the "For Additional Research" section. Each chapter lists three books, four organizations, and three "action projects." For the book suggestions, I chose a few good books that students at the first- and second-year level will find readable, or at least not impossibly challenging. For the list of organizations, I included some of the better-known organizations working for solutions to the issue covered in the chapter. I emphasized nongovernmental, grassroots organizations over government and corporate-sponsored organizations, in keeping with one of the themes of the reader. The "action projects" are aimed at inspiring active learning. I deliberately made them somewhat diverse, so that each will appeal to students with different interests, talents, time availability, or levels of academic preparation. In my experience, the best way to get students engaged in the subject matter of a course is to give them many opportunities for pursuing individual interests. Some of the "action projects" involve use of the Internet, others give students the opportunity to take part in social action and reflect on their experience, and others ask students to conduct exploratory research in the community. The chapter Introduction is followed by the readings, arranged in the order they are mentioned in the Introduction.

ACKNOWLEDGMENTS

I would never have started or finished *Social Problems of the Modern World* without the help of many people. I am grateful to the late Terence K. Hopkins and to Immanuel Wallerstein, for their intellectual inspiration over the years. My former colleagues at Fresno City College, Richard Valencia and Gerry Bill, encouraged and supported the project in an earlier incarnation. I appreciate the understanding I have received from my current institution, Three Rivers Community College, in particular from Booker T. DeVaughn, President, and Paul R. Susen, Dean of Instruction. I am also indebted to a number of colleagues for their encouragement, especially Irene Clampet, Mildred Hodge, David Holdridge, Madge Manfred, and Jim Wright. My special thanks go to Kathryn Brown-Tracy, Research Assistant, and Janice DeWolf, Library Circulation Assistant. I could never have completed this anthology without them: Kathryn Brown-Tracy brought her enthusiasm for learning about world issues, and helped search computer databases for appropriate material; Janice DeWolf patiently connected me with interlibrary resources, when our own library was lacking. I also received indispensable assistance from Paula Sherman, Teaching Intern from Eastern Connecticut State University, and Anne-Marie Jantsch, Research Intern from Southern Illinois State University. Paula Sherman, now a graduate student in history at the University of Connecticut, contributed material on indigenous peoples, and also educated me as to the virtues of computer assistance in a project such as this. I am very grateful to students in my classes in Current Social Problems and World Issues over the years who were willing to share with me their thoughts and reactions to the various kinds of readings I tried out on them. I am also indebted to Susan J. Ferguson, Grinnell College, for her good advice on the process of constructing an anthology. My special thanks also to my husband, Peter G. Anderheggen, who helped with this project in many ways, including proofreading the first draft, making countless photocopies, and above all, supporting me through both the bad and the good times with his enthusiasm and optimism.

Many reviewers contributed valuable suggestions. I am grateful to:

David Allen, University of New Orleans
Mary Blair-Loy, Washington State University
Walter Carroll, Bridgewater State College
William Cross, Illinois College
Michael Cruz, Texas Women's University
Phillip Davis, Georgia State University
Douglas Degher, Northern Arizona University
Julia Hall, Drexell University
Linda Nyce, Bluffton College
Stacy Ruth, Jones County Junior College
Paula Snyder, Columbus State Community College
Terry Timmins, Orange Coast College
Jeffry Will, University of North Florida
Mark Winton, University of Central Florida

At Wadsworth, I have benefited from the support of many people. My special thanks to Eve Howard, Publisher, for her vision and guidance; and to Ari Levenfeld, Assistant Editor, for his enthusiasm about the project and his hard work on the development of the book. Thanks also to Jerilyn Emori, Senior Production Editor;

Susan Walters, Permissions Editor; and Vicki Moran, Publishing Support Services, for their patient efforts in coordinating all the complex tasks that are involved in putting together a work of this type, and to Yvo Riezebos for his inspired work on the cover.

This book is dedicated to my mother, Rebecca H. Moulder, and to all the people everywhere who are working "from the bottom up" to make this world a better place.

SOCIAL PROBLEMS OF THE MODERN WORLD

1

Introduction to the Study of Global Social Problems

When, in a city of 100,000, only one man is unemployed, that is his personal trouble, and for its relief we properly look to the character of the man, his skills, and his immediate opportunities. But when in a nation of 50 million employees, 15 million men are unemployed, that is an issue, and we may not hope to find its solution within the range of opportunities open to any one individual. The very structure of opportunities has collapsed. Both the correct statement of the problem and the range of possible solutions require us to consider the economic and political institutions of the society, and not merely the personal situation and character of . . . individuals.

C. WRIGHT MILLS, 1959

Like it or not, our lives and our children's lives will be lived in the global economy. We'd better fix it.

JEREMY BRECHER, 1994

TOWARD A GLOBAL PERSPECTIVE ON SOCIAL PROBLEMS

We live in a world in which people are increasingly interconnected. Nowhere is this more obvious than in the consumer culture. People all over the world wear clothing made in China or Malaysia, drive cars made in Sweden or Mexico, and listen to American music on stereos made in Korea or Japan. It is perhaps less obvious, but the problems people experience around the world are also increasingly intertwined. In the United States, people lose their jobs when factories are moved to countries where wages are lower. In Brazil, people lose their homelands when the rain forests are cut down to make furniture for the world market.

The popular buzzword, *globalization*, reflects our growing consciousness of this interconnected world. However, most people have not yet fully understood just how profoundly our lives are shaped by this situation. Most of us have not clearly thought through what this means for the futures of ourselves and our children, or our efforts to solve the social problems confronting our communities. As a result, we are all too often casualties of global forces that appear to be beyond our control, and misled by powerful corporations, politicians, and media pundits who come forward to interpret global events for us. We are citizens in a global world we neither comprehend nor control. This book seeks to help its readers gain a deeper understanding of how social problems faced by people in the United States and around the world are increasingly shaped by our common fate in this interconnected world.

SOCIOLOGY AND THE STUDY OF SOCIAL PROBLEMS

What is a *social problem*? Sociologists, like the general public, have only recently come to recognize the need for a global perspective on social problems.

In the early twentieth century, sociologists typically associated social problems, such as unemployment, crime, or poverty, with *deviant individuals*. As a result, when they sought to solve social problems, they focused on changing individual behavior. Although this approach is still alive today, sociologists had by and large arrived at a different understanding by mid-century. At this time, sociologists turned away from an emphasis on individuals to a consideration of the *social structures* of nations—organizations and institutions such as corporations, governments, and the media. The first reading in this chapter, "The Promise of Sociology," by C. Wright Mills, is a classic statement of this perspective written in 1959. Mills forcefully argued that when a problem affects large numbers of people, we must look beyond individuals to *social structures*—the larger economic, political, and social patterns of a society. We cannot solve the problems of a society that is structurally flawed by changing individuals, one at a time. Sociologists came to define social problems as *problems that concern large numbers of people, have social-structural causes, and require social-structural solutions.* This approach remains fundamental to the sociological perspective. It was a step forward from the individualistic approach. By demonstrating that social problems have structural causes, sociologists helped steer the public away from unproductive scapegoating of individuals to an awareness of the need for social change.

In the latter part of the twentieth century, some sociologists began to attempt to understand the structure of the world as a whole. Two perspectives that were influential in this transition are the *modernization perspective* and the *sociology of the world system.* (These perspectives will be discussed more fully in Chapter 2, "The World Economy and World Poverty.") According to the *modernization perspective*, nations of the world are undergoing a process of modernization, encompassing changes such as industrialization, urbanization, and growing social complexity. *Traditional* societies, which are small scale and dominated by tradition, are evolving into *modern* societies, which are complex and constantly changing. From this perspective, in order to understand social problems, we need to

understand this process of global modernization. In the modernization perspective, social problems can be seen as failures in modernization. For example, societies that fail to industrialize will be poor and conflict-ridden. On an optimistic note, this perspective suggests that these problems can be solved through government intervention to lead society on the road to modernization.

According to the *world system perspective*, we need to see the world as a single *system*. In this perspective, pioneered by sociologists Immanuel Wallerstein and Terence K. Hopkins, the fundamental dynamic of the system is the growth of the *capitalist world economy*, an economy structured around private capital accumulation and a world division of labor extending beyond the boundaries of individual nations. This world economy originated in the sixteenth century and has been growing in scope and depth over the centuries, incorporating more and more of the world's peoples ever more closely. In the world system perspective, the nature of individual societies is greatly affected by the way in which they have become incorporated into the capitalist world economy. Some nations, such as the United States and Japan, have become wealthy *core* nations, with their powerful corporations exploiting the labor and resources of the rest of the world. Others have become impoverished *periphery* nations, such as Guatemala or Congo, their resources and people exploited by corporations from the core nations. *Semi-periphery* nations, such as South Korea or Taiwan, stand in between—exploited by the core, yet in turn exploiting the periphery. It follows from this perspective that to understand social problems in any nation, we must examine how the nation's economic, political, and other institutions have been shaped historically by its position in the world economy. The second reading in this chapter, "Open Veins of Latin America," by Eduardo Galeano, is a powerful statement by a Latin American author on the way in which Latin American societies and their problems have been shaped by their unequal relationships with the United States and other core nations.

Today's social problems are *global social problems*, problems *that concern large numbers of people world-*

wide, have global social-structural causes, and require global social-structural solutions. Because people everywhere have been drawn into the capitalist world economy, problems originating in individual nations, such as stock market declines, or a fall in the prices of agricultural products, can spread rapidly and affect massive numbers of people around the world. There are also global social-structural *causes* for many of the problems that concern people in the United States and other individual nations. For example, the activities of powerful multinational corporations, international financial institutions, and international trade agreements have contributed to the rapid destruction of the rain forests. The growth of global social problems is also reflected in the emergence of global organizations and social movements aimed at finding solutions. For example, women's rights movements in individual nations increasingly interact inside and outside of the United Nations as one global movement for "women's rights as human rights," aimed at ensuring the equality of women on a global basis. Because so many of today's social problems are global in nature, we cannot hope to understand and solve them by focusing on individual nations, one at a time. If the task for Mills' time was to begin to see the social, rather than simply the individual, causes of problems, the challenge today is to understand the global, rather than simply the national causes.

TOPICS ENCOMPASSED BY THE SOCIOLOGY OF SOCIAL PROBLEMS

The sociology of social problems encompasses a variety of interesting topics, in particular the study of *definitions, causes,* and *solutions.*

One of the more fascinating aspects of the study of social problems is that social problems are *socially defined (socially constructed).* That is, as people interact with one another in society, they decide whether they are going to identify something as a problem, and how they will define the nature of the problem. In today's complex soci-

eties, a chorus of conflicting voices is raised in the public discussion of social problems—corporations, the media, nonprofit organizations, government officials and agencies, political parties, and citizens' groups—all seeking to define the issue and shape how others will see it. Controversy over the definition of social problems is even more pronounced on a worldwide scale, as the world contains so many different cultures, each with its own set of values, beliefs, and social structures. This is also true of multi-ethnic societies such as the United States that have large immigrant populations from diverse cultures. "When in Rome," by Nina Schuyler, examines the clash between American legal norms regarding the age of marriage and the culture of recent immigrants to the United States from the Middle East.

A significant aspect of the sociology of social problems is the study of *causes,* or the underlying forces giving rise to social problems. An understanding of *power* is often important in making this analysis. Power refers to the ability of an individual or an organization to get its way, even though others may resist. Powerful corporations, governments, and other social organizations are often involved in generating social problems that persist despite the protest of those adversely affected by the problem. When an organization stands to benefit economically or politically from certain arrangements, it is said to have a *vested interest* in the situation. Powerful organizations with vested interests are often highly effective in getting the general public to overlook social problems or to accept the organization's definition of the situation. The above-mentioned reading, "Open Veins of Latin America," can also be read as an example of the vested interests of wealthy nations and social classes in relation to the problem of poverty.

In response to global social problems, an amazing variety of projects have been organized around the world today in an effort to find *solutions.* Sociologists classify these efforts in various ways. One classification is based on the nature of the goals. A *social organization* is a group that is focused on achieving specific goals. A *social movement* encompasses one or more groups focused on achieving the broader goal of *change*, the transformation of individuals or social structures. Responses to global social problems can also be classified based on the status in society of those initiating the response. Scarpitti and Cylke and others, for example, classify social movements into two broad categories: *"top down"* and *"bottom up"* (Scarpitti and Cylke, 1995, p. xiii). "Top down" efforts are initiated by those in charge of powerful organizations, such as corporations and governments. "Bottom up," or "grassroots," movements are initiated by ordinary citizens. "Top down" efforts to solve social problems are potentially highly effective, because elites have the power to make things happen. However, due to vested interests, elite solutions may be limited in scope and fail to uncover the root causes of problems. "Bottom up" efforts may be less effective (at least in the short run), due to ordinary citizens' lack of power. However, these may have a better vision of the comprehensive changes that may be needed in order to solve the problem. Marvin E. Olsen discusses these two types of movements (which he terms "social mobilization" and "social confrontation") in "Strategies for Changing Society." In this article, Olsen also identifies a third type of effort, "social involvement," in which elites and citizens may work together toward change.

SOCIOLOGICAL RESEARCH ON SOCIAL PROBLEMS

Sociology is a *science*, rather than an art or philosophy. This means that the ideas sociologists have about social problems and their solutions are based upon *research*—that is, systematic investigation of the world. Sociologists use a variety of different research methods to study social problems. The most typical are observation, interviews, and the use of existing data such as historical documents or statistics collected by government or private agencies.

Applied sociologists conduct research with the aim of using the knowledge obtained in a practical way to improve society. An interesting type of applied sociology is *participatory research*. Participatory research, used globally by sociologists affiliated with "bottom up" social movements, seeks participation in the research by the people who are affected by the problem in question. The people are invited to take part in designing the topics to be studied, developing the methods of research, and planning how the research findings will be utilized.

THEORETICAL PERSPECTIVES ON SOCIAL PROBLEMS

Sociological research is influenced by a variety of *theoretical perspectives* or *paradigms*. A theoretical perspective is a viewpoint shared by a community of scholars. Sociologists working within a particular theoretical perspective make similar assumptions about the definition of, causes of, and solutions to social problems. Three major theoretical perspectives have dominated sociological thinking during the last half of the twentieth century—the *order* perspective, the *conflict* perspective, and the *symbolic interactionist* perspective. The origins of these perspectives can be traced back to sociology's beginnings in the eighteenth and nineteenth centuries in Europe. Major social changes and problems rocked Europe at that time—industrialization, urbanization, the growth of capitalism, the conflict of classes, warfare, and the development of powerful nation-states. Early sociologists who were trying to understand society developed their ideas in this context. The perspectives were further developed in the twentieth century, as sociology spread to the United States and other parts of the world. The same social problem can be analyzed from multiple theoretical perspectives. As seen in the example of the family given below, we can choose any issue and analyze it from either the order, conflict, or symbolic interactionist perspectives.

The order perspective emphasizes social stability. A major example of the order perspective is *structural-functionalism*. Sociologists using the structural-functionalist perspective study the *functions* that various groups and organizations in a society are fulfilling for one another or for the society as a whole. For example, families function to socialize children into the society's cultural values, so that the society has continuity from one generation to the next. Social problems are viewed against this backdrop of order, and are seen as caused by some disorganization or *dysfunction* in society. For example, family breakdown may occur as a result of rapid social changes accompanying the mass migration of people from the countryside to the city. From this perspective, it seems that social problems can be taken care of through the orderly functioning of the existing social structures. For example, government programs may intervene to help dysfunctional families resume their typical functions. The structural-functionalist perspective is linked to the earlier twentieth-century idea mentioned above that social problems are caused by deviant individuals. For example, government programs may also punish or attempt to rehabilitate those who deviate, thus restoring order in society. The modernization perspective discussed above is also an example of the order perspective, with its emphasis on governments taking action to direct societies onto the path of modernization.

In contrast, sociologists employing the conflict perspective emphasize social conflict and instability. They focus on the inequities of wealth and power in society, such as the inequalities of social class, race, ethnicity, and gender. They see these inequalities as causing conflict, which in turn may result in basic social change, rather than restoration of an unjust order. Viewed against this backdrop of conflict, social problems are seen as caused by the vested interests of the powerful. From this perspective, it appears that social problems may not be solved, in the absence of fundamental social-structural changes in the society. For example, sociologists working in the conflict perspective see contemporary family problems as caused by the lack of economic opportunities afforded to the poor in a society with profound class inequalities. A solution will require basic changes in our economic and political institutions so that all citizens have access to jobs, education, and health care. The

world system perspective discussed above is an example of the conflict perspective, with its focus on the inequalities and conflicts among core, periphery, and semi-periphery nations.

Sociologists working in the symbolic interactionist perspective study the meanings that people in society give to social behavior. In contrast to the other two perspectives, symbolic interactionists are less interested in analyzing larger social structures, and more concerned with how individuals interact to create a world that is meaningful to them. They are also less interested in identifying the objective causes of and solutions for problems than they are in understanding the *social construction* of social problems. For example, sociologists working in the symbolic interactionist perspective might study the controversy over the definition of "family" in the United States in recent years, in which some have argued that the nuclear family with father as breadwinner is the most valid family form, while others have argued that there are a variety of acceptable family forms, including the single-parent family, the family in which both parents work, families headed by gay and lesbian parents, and so on. The symbolic interactionist might study the nature of the arguments made by each side of the debate, the vested interests on the part of those making the claims on each side, and the public response to the debate.

THE PLAN OF THIS BOOK

This collection of readings has a number of features that make it special. First, a global perspective is maintained throughout the book. Each chapter of this reader covers a social problem that affects people in the United States, but is simultaneously a problem that can be better comprehended if we deal with it in the global context.

- world economy and world poverty
- democracy and human rights
- gender and sexual orientation
- population growth
- environmental destruction

- war and militarism
- the global media
- the cultural survival of indigenous peoples
- racial/ethnic conflicts and the danger of genocide
- global crime
- world health

From one-third to one-half of the articles in each chapter are focused on the United States. Other articles frame the problem in global perspective, or discuss the nature of the problem in other nations of the world.

Second, each chapter of the reader covers the basic topics of concern to the study of social problems—definition, causes, and solutions—with a key aspect of this reader being its significant coverage of the search for solutions. Clearly, one of the key reasons for developing a better understanding of social problems is to be able to help in the struggle to alleviate or eliminate them. Each chapter contains readings dealing with solutions, as well as readings on definitions of the problem, and analysis of its causes. There is also a final chapter, "Prospects for the Future," which takes a more in-depth look at solutions. In the coverage of solutions, readings have been chosen that emphasize "bottom up," or grassroots, organizations and movements. In recent years, there has been a great proliferation of grassroots efforts aimed at solving social problems. Such movements have in fact led the way toward effective and practical solutions, dragging "top down" organizations reluctantly into more enlightened policies. The book's inclusion of "bottom up" responses will help students get a more realistic picture of the dynamics of social change in the world. In addition to emphasizing solutions, the reader also focuses on the *causal interconnections* among social problems. Social problems are often causally interrelated—that is, the existence of one problem leads to others. It is difficult to find solutions to a problem when we have not taken into consideration the way in which it is intertwined with others. The reader's coverage of interconnections will help students develop a more comprehensive and sophisticated picture of problems and solutions.

Third, this reader seeks to present students with a variety of perspectives on global social problems. As stated at the beginning of this chapter, the major goal of this reader is to help students learn to think more deeply about these problems—that is, to think about them *critically*, rather than simply accepting at face value the interpretation of problems given us by authorities such as politicians, corporations, or the media. One of the best ways to learn how to think critically is to compare and contrast multiple perspectives on an issue. Toward this end, readings are included that reflect the major theoretical perspectives in sociology. The authors have also been chosen to represent a variety of backgrounds, in terms of nationality, racial/ethnic identity, gender, and sexual orientation. The variety of voices found in this reader should help students develop a broader perspective. Unfortunately, many of us graduate from high school with distorted views about the world's problems. We have been taught little about the viewpoints of people in other nations, and have limited and often nationalistic views of world issues. High school history textbooks, for example, emphasize the heroic and democratic aspects of American history and foreign policy, and downplay injustices and oppression (Loewen, 1996).

Finally, an effort has been made to choose articles that are interesting and worthwhile to read. Unfortunately, many academic sociologists tend to write primarily for scholarly audiences in a style that does not appeal to the general reader (Ritzer, 1998, p. 449). Some academic sociologists *do* write in a readable style, and such work is included here. However, many of the articles included here have been written for a popular audience by journalists, activists, and others who have a sociological perspective on social problems. The sociological perspective is alive and well beyond the confines of academic circles. Many generations of sociology students have gone on to become writers, journalists, and activists. The sociological perspective has also entered the public imagination as a result of academic sociologists who have taken part in social movements, testified in court cases, lobbied national and international bodies, or worked in participatory research projects. As a result, it has been possible to find a good deal of sociological writing on global social problems that will not put the reader to sleep. This is not to say that all of the readings included are "easy" to read. Many are, but some are more challenging. It was impor-tant to include the more challenging material in order to make sure that students are exposed to important current ideas in the field and to prepare them for higher-level courses. Students should finish this text having developed a better ability to read and analyze college-level writing in the social sciences.

WORKS CITED

Loewen, James W. 1996. *Lies My Teacher Told Me.* New York: Simon and Schuster.

Ritzer, George. 1998. "Writing to be Read: Changing the Culture and Reward Structure of American Sociology." *Contemporary Sociology*, 27: 446–453.

Scarpitti, Frank R., and F. Kurt Cylke, Jr. 1995. *Social Problems: The Search for Solutions.* Los Angeles: Roxbury Publishing Company.

QUESTIONS FOR DISCUSSION

1.   In "The Promise of Sociology," written in the 1950s, Mills makes a distinction between "personal troubles" and "public (or social) issues" that are caused by social-structural factors. At the end of the reading, Mills gives some examples of public issues. What are these? Are these examples out of date? Why or why not?

2.   What does Galeano mean by the "open veins" of Latin America? Do you believe the United States could ever be in the position of having "open veins"? Why or why not?

3.   Imagine that you are the judge in the Nebraska case discussed by Schuyler in "When in Rome." What do *you* see as the more serious social problem—that is, the failure of U.S. law to respect cultural pluralism, or the family's deprivation of the rights of the girls under U.S. law? How will you decide the case?

4.   What is one social problem existing on your college campus or in your community? What approaches have been taken to solve this problem? Categorize them as "social mobilization" (or "top down"), "social involve-ment," or "social confrontation" (or "bottom up"). Which of the strategies pursued has been more effective, and why? (Consider Olsen's analysis of "typical problems" and "assessment.")

 INFOTRAC COLLEGE EDITION: EXERCISE

What is the nature of current research on social prob-lems? Using InfoTrac® College Edition, explore one of the major sociology journals devoted to publishing research on social problems—*Social Problems*. InfoTrac College Edition contains the last few years of this jour-nal, and you can view a list of all of the articles pub-lished in it. Click on the PowerTrac icon, and under Choose Search Index, select Journal Name (in) and enter the search term Social Problems.

Review the list of articles: What kinds of problems are currently being researched? Are there certain problems that seem to be receiving more attention than others? How many of the articles deal with problems in nations other than the United States? Try to identify an article that uses one of the three theoretical perspectives dis-cussed in this chapter. After looking over the entire list, select one or two articles that interest you, and read them. (Note: These are somewhat technical articles writ-ten for an audience of other researchers. You may find you have selected an article that is somewhat difficult to read; if so, simply look for the main points.) What is your reaction to what you read?

FOR ADDITIONAL RESEARCH

Books

Dresser, Norrine. 1996. *Multicultural Manners: New Rules of Etiquette for a Changing Society*. New York: John Wiley & Sons, Inc.

Mills, C. Wright. 1959. *The Sociological Imagination*. New York: Oxford University Press.

So, Alvin Y. 1990. *Social Change and Development*. Newbury Park, CA: Sage Publications.

Organizations

American Sociological Association
1722 N Street NW
Washington, DC 20036
(202) 833-3410
www.asanet.org

International Sociological Association
Secretariat: Faculdad C.C. Politicas y Sociologia
28223 Madrid
Spain
(34) 913 52 76 50
www.ucm.es/info/isa/index.html

The Society for the Study of Social Problems
906 McClung Tower
University of Tennessee
Knoxville, TN 37996-0490
(423) 974-3620
http://web.utk.edu/~sssp

Sociological Practice Association
Richard Bedea, SPA Administrative Officer
Department of Social Sciences
Anne Arundel Community College
Arnold, MD 21012-9989
(310) 541-2835

ACTION PROJECTS

1.   Make an inventory of your local community's connections to the world system. For example, which businesses buy and sell on the world market? Are there any foreign-owned corporations located in your community? Does your town or city have a partnership with a town in another country? Are local churches, schools, colleges, or other community groups involved in social justice work overseas? How comprehensive is your local newspaper's coverage of international news? Do local unions support union activities overseas?

2.   Interview a foreign student or recent immigrant to the United States to learn something about the culture and the social problems of their nation of origin. For example, how has globalization affected their country?

What is the educational system like there? What roles do men and women play? What are some of the important religious beliefs and values? Did social problems of the nations play any part in the person's decision to come to the United States? What did this person expect the United States would be like before they came here, and how has their actual experience contrasted with expectations?

3.   Visit the United Nations Web site (www.un.org), and go to the sections on economic and social development. Locate the *Report on the World Social Situation*. What social problems does the report highlight? Read one part of the report that interests you the most. What did you learn that was unexpected or surprising?

1

The Promise of Sociology

C. WRIGHT MILLS

Nowadays men often feel that their private lives are a series of traps. They sense that within their everyday worlds, they cannot overcome their troubles, and in this feeling, they are often quite correct: What ordinary men are directly aware of and what they try to do are bounded by the private orbits in which they live; their visions and their powers are limited to the close-up scenes of job, family, neighborhood; in other milieux, they move vicariously and remain spectators. And the more aware they become, however vaguely, of ambitions and of threats which transcend their immediate locales, the more trapped they seem to feel.

Underlying this sense of being trapped are seemingly impersonal changes in the very structure of continent-wide societies. The facts of contemporary history are also facts about the success and failure of individual men and women. When a society is industrialized, a peasant becomes a worker; a feudal lord is liquidated or becomes a businessman. When classes rise or fall, a man is employed or unemployed; when the rate of investment goes up or down, a man takes new heart or goes broke. When wars happen, an insurance salesman becomes a rocket launcher; a store clerk, a radar man; a wife lives alone; a child grows up without a father. Neither the life of an individual nor the history of a society can be understood without understanding both.

• • •

Perhaps the most fruitful distinction with which the sociological imagination works is between "the personal troubles of milieu" and "the public issues of social structure." This distinction is an essential tool of the sociological imagination and a feature of all classic work in social science.

Troubles occur within the character of the individual and within the range of his immediate relations with others; they have to do with his self and with those limited areas of social life of which he is directly and personally aware. Accordingly, the statement and the resolution of troubles properly lie within the individual as a biographical entity and within the scope of his immediate milieu—the social setting that is directly open to his personal experience and to some extent his willful activity. A trouble is a private matter. Values cherished by an individual are felt by him to be threatened.

Issues have to do with matters that transcend these local environments of the individual and the range of his inner life. They have to do with the organization of many such milieux into the institutions of a historical society as a whole, with the ways in which various milieux overlap and interpenetrate to form the larger structure of social and historical life. An issue is a public matter. Some value cherished by publics is felt to be threatened. Often there is a debate about what that value really is and about what it is that really threatens it. This debate is often without focus if only because it is the very nature of an issue, unlike even widespread trouble, that it cannot very well be defined in terms of the immediate and everyday environments of ordinary men. An issue, in fact, often involves a crisis in institutional arrangements, and often too it involves what Marxists call "contradictions" or "antagonisms."

In these terms, consider unemployment. When, in a city of 100,000, only one man is unemployed, that is his personal trouble, and for its relief we properly look to the character of the man, his skills, and his immediate opportunities. But when in a nation of 50 million employees, 15 million men are unemployed, that is an issue, and we may not hope to find its solution within the range of opportunities open to any one individual. The very structure of opportunities has collapsed. Both the correct statement of the problem and the range of possible solutions require us to consider the economic and political institutions of the society, and not merely the personal situation and character of a scatter of individuals.

Consider war. The personal problem of war, when it occurs, may be how to survive it or how to die in it with honor; how to make money out of it; how to climb into the higher safety of the military apparatus; or how to contribute to the war's termination. In short, according to one's values, to find a set of milieux and within it to survive the war or make one's death in it meaningful. But the structural issues of war have to do with its causes; with what types of men it throws up into command; with its effects upon economic and political, family and religious institutions, with the unorganized irresponsibility of a world of nation-states.

Consider marriage. Inside a marriage a man and a woman may experience personal troubles, but when the divorce rate during the first four years of marriage is 250 out of every 1,000 attempts, this is an indication of a structural issue having to do with the institutions of marriage and the family and other institutions that bear upon them.

Or consider the metropolis—the horrible, beautiful, ugly, magnificent sprawl of the great city. For many upper-class people, the personal solution to "the problem of the city" is to have an apartment with private garage under it in the heart of the city, and forty miles out, a house by Henry Hill, garden by Garrett Eckbo, on a hundred acres of private land. In these two controlled environments—with a small staff at each end and a private helicopter connection—most people could solve many of the problems of personal milieux caused by the facts of the city. But all this, however splendid, does not solve the public issues that the structural fact of the city poses. What should be done with this wonderful monstrosity? Break it all up into scattered units, combining residence and work? Refurbish it as it stands? Or, after evacuation, dynamite it and build new cities according to new plans in new places? What should those plans be? And who is to decide and to accomplish whatever choice is made? These are structural issues; to confront them and to solve them requires us to consider political and economic issues that affect innumerable milieux.

Insofar as an economy is so arranged that slumps occur, the problem of unemployment becomes incapable of personal solution. Insofar as war is inherent in the nation-state system and in the uneven industrialization of the world, the ordinary individual in his restricted milieu will be powerless—with or without psychiatric aid—to solve the troubles this system or lack of system imposes upon him. Insofar as the family as an institution turns women into darling little slaves and men into their chief providers and unweaned dependents, the problem of a satisfactory marriage remains incapable of purely private solution. Insofar as the overdeveloped megalopolis and the overdeveloped automobile are built-in features of the overdeveloped society, the issues of urban living will not be solved by personal ingenuity and private wealth.

What we experience in various and specific milieux, I have noted, is often caused by structural changes. Accordingly, to understand the changes of many personal milieux we are required to look beyond them. And the number and variety of such structural changes increase as the institutions within which we live become more embracing and more intricately connected with one another. To be aware of the idea of social structure and to use it with sensibility is to be capable of tracing such linkages among a great variety of milieux, To be able to do that is to possess the sociological imagination.

Open Veins of Latin America

EDUARDO GALEANO

The division of labor among nations is that some specialize in winning and others in losing. Our part of the world, known today as Latin America, was precocious: it has specialized in losing ever since those remote times when Renaissance Europeans ventured across the ocean and buried their teeth in the throats of the Indian civilizations. Centuries passed, and Latin America perfected its role. We are no longer in the era of marvels when fact surpassed fable and imagination was shamed by the trophies of conquest—the lodes of gold, the mountains of silver. But our region still works as a menial. It continues to exist at the service of others' needs, as a source and reserve of oil and iron, of copper and meat, of fruit and coffee, the raw materials and foods destined for rich countries which profit more from consuming them than Latin America does from producing them. The taxes collected by the buyers are much higher than the prices received by the sellers; and after all, as Alliance for Progress coordinator Covey T. Oliver said in July 1968, to speak of fair prices is a "medieval" concept, for we are in the era of free trade.

• • •

Latin America is the region of open veins. Everything, from the discovery until our times, has always been transmuted into European—or later United States—capital, and as such has accumulated in distant centers of power. Everything: the soil, its fruits and its mineral-rich depths, the people and their capacity to work and to consume, natural resources and human resources. Production methods and class structure have been successively determined from outside for each area by meshing it into the universal gearbox of capitalism. To each area has been assigned a function, always for the benefit of the foreign metropolis of the moment, and the endless chain of dependency has been endlessly extended. The chain has many more than two links. In Latin America it also includes the oppression of small countries by their larger neighbors and, within each country's frontiers, the exploitation by big cities and ports of their internal sources of food and labor. (Four centuries ago sixteen of today's twenty biggest Latin American cities already existed.)

For those who see history as a competition, Latin America's backwardness and poverty are merely the result of its failure. We lost; others won. But the winners happen to have won thanks to our losing: the history of Latin America's underdevelopment is, as someone has said, an integral part of the history of world capitalism's development. *Our defeat was always implicit in the victory of others; our wealth has always generated our poverty by nourishing the prosperity of others—the empires and their native overseers. In the colonial and neocolonial alchemy, gold changes into scrap metal and food into poison.* Potosí, Zacatecas, and Ouro Prêto became desolate warrens of deep, empty tunnels from which the precious metals had been taken; ruin was the fate of Chile's nitrate pampas and of Amazonia's rubber forests. Northeast Brazil's sugar and Argentina's quebracho belts, and communities around oil-rich Lake Maracaibo, have become painfully aware of the mortality of wealth which nature bestows and imperialism appropriates. The rain that irrigates the centers of imperialist power drowns the vast suburbs of the system. In the same way, and symmetrically, the well-being of our dominating classes—dominating inwardly, dominated from outside—is the curse of our multitudes condemned to exist as beasts of burden.

The gap widens. Around the middle of the last century the world's rich countries enjoyed a 50 percent higher living standard than the poor countries. Development develops inequality: in April 1969 Richard Nixon told the Organization of American States (OAS) that by the end of the twentieth century the United States' per capita income would be fifteen times higher than Latin America's. The strength of the imperialist system as a whole rests on the necessary inequality of its parts, and this inequality assumes ever more dramatic dimensions. The oppressor countries get steadily richer in absolute terms—and much more so in relative terms—through the dynamic of growing disparity. The capitalist "head office" can allow itself the luxury of creating and believing its own myths of opulence, but the poor countries on the capitalist periphery know that myths cannot be eaten. The United States citizen's average income is seven times that of a Latin American and grows ten times faster. And averages are deceptive in view of the abyss that yawns between the many poor and the rich few south of the Rio Grande. According to the United Nations, the amount shared by 6 million Latin Americans at the top of the social pyramid is the same as the amount shared by 140 million at the bottom. There are 60 million *campesinos* whose fortune amounts to $.25 a day. At the other extreme, the pimps of misery accumulate $5 billion in their private Swiss or U.S. bank accounts. Adding insult to injury, they squander in sterile ostentation and luxury, and in unproductive investments constituting no less than half the total investment, the capital that Latin America could devote to the replacement, extension, and generation of job-creating means of production. Harnessed as they have always been to the constellation of imperialist power, our ruling classes have no interest whatsoever in determining whether patriotism might not prove more profitable than treason, and whether begging is really the only formula for international politics. Sovereignty is mortgaged because "there's no other way." The oligarchies' cynical alibis confuse the impotence of a social class with the presumed empty destinies of their countries.

3

When in Rome

NINA SCHUYLER

Lincoln could not be more different from Iraq. But for the 700 Iraqi refugees who have moved there after years of persecution under Saddam Hussein and more time spent in Saudi Arabian refugee camps, the broad, flat plains of Nebraska are now home.

So no one in the Iraqi community thought twice about the November marriage of two Iraqi sisters, ages 13 and 14, to two Iraqi men, ages 28 and 34—no one, that is, except the state of Nebraska, which has charged the two grooms with

From Nina Schuyler, "When in Rome," *In These Times,* February 17, 1997, pp. 27–29. Copyright © 1997 by In These Times. Reprinted by permission.

statutory rape, a crime that carries a maximum sentence of 50 years. The father, who immigrated to the United States three years ago, faces two counts of child abuse, with a maximum sentence of one year in jail and a $1,000 fine. The girls' mother is charged with contributing to the delinquency of a minor, which carries the same punishment.

The case came to the state's attention when the older sister fled her new husband. The husband and the girl's father went to the police seeking help to find the girl, but were arrested instead. The legal marriage age in Nebraska is 17, and it is illegal for anyone 18 or older in the state to have sex with anyone who is under 18. The defendants are scheduled to go on trial in February.

At the crux of cases in which a foreign cultural practice collides with U.S. criminal law is the defendants' reliance on the so-called cultural defense. The defense attorneys in the Nebraska case plan to argue that the grooms and the parents were following the Iraqi custom of arranged marriages and did not think they were violating the law. According to this argument, someone raised in a foreign culture should not be held fully accountable for conduct that violates U.S. law if the act would be acceptable in his or her native culture. Foreign defendants have used this argument to win a reduced sentence or even exoneration for acts that U.S. law characterizes as assault, battery, rape, kidnap, mayhem and murder.

The cultural defense has generated debate among progressive legal scholars because it poses a difficult dilemma: Progressives support multiculturalism, but they also believe in the expansion of women's rights. The cultural defense pits these two values against each other since the victims in these cases are usually immigrant women and children.

Legal doctrine does not provide an easy answer because these cases encapsulate two competing legal traditions in U.S. criminal law. On the one hand, the law provides for individualized justice and aims to tailor punishment to a particular defendant's degree of culpability. For instance, for most crimes, if the defendant didn't think he was doing anything illegal, then society should not punish him to the full extent of the law. On the other hand, legal tradition also holds that the state must ensure equal protection from criminal conduct for all members of society. In keeping with that goal, the U.S. judicial system has never accepted the argument that ignorance of the law is an excuse.

Feminists rely on this latter legal principle to argue that cultural evidence has no place in a criminal trial. The United States offers women more civil rights than the countries from which most immigrants hail. When courts recognize cultural practices that condone crimes against women, they embrace patriarchal values about the subordinate status of women and children that feminists have worked hard to remove from the legal system. In essence, when a court allows such evidence into a criminal trial, it sets up a two-tiered legal system in which immigrant women and children receive less protection.

"The message gets sent that non-European immigrant women and children are not as worthy as European women and children," says Doriane Lambelet Coleman, a lecturer at Duke University School of Law in North Carolina and author of a 1996 *Columbia Law Review* article titled "Individualizing Justice Through Multiculturalism: The Liberals' Dilemma."

Margaret Fung, executive director of the Asian-American Defense and Education Fund in New York, made the same point in response to an immigrant spousal-killing case. "You don't want to import [immigrant] cultural values into our judicial system," she said. "We don't want women victimized by backward customs."

After receiving heavy criticism for her remarks from the Chinese-American community, however, Fung revised her position. To bar the cultural defense "would be to promote the idea that when people come to America, they have to give up their way of doing things," she now says. "That is an idea we cannot support."

Ironically, conservatives end up on the same side as feminists, but with a simpler argument: When in Rome, do as the Romans do.

No one knows how often cultural evidence gets introduced in criminal trials, since many of the cases are plea-bargained and therefore go unreported. Nevertheless, criminal defense attorneys agree that it's on the rise.

"We've seen these cases come up around the country, and they will continue because the

United States has allowed in a large number of refugees and has not given them any education about U.S. laws," says Sanford J. Pollack, the attorney representing the father of the two Iraqi girls.

Courts have let cultural evidence into all phases of the criminal trial—before indictment, during the trial to determine guilt, and in the sentencing phase. Of reported cases, most resulted in more lenient sentences than would have been expected otherwise.

- In 1988, a Hmong man from Laos kidnaped and raped an 11-year-old girl in Minnesota. The prosecutor interviewed members of the Hmong community, who explained the custom of marriage-by-capture. After these conversations, the prosecutor decided it would be difficult to convince a jury that the defendant was guilty. Instead of being charged with rape, the defendant was allowed to plead guilty to sexual intercourse with a child under 12, for which he paid a $1,000 fine and received no jail time.

- In 1987, Dong Lu Chen bludgeoned his wife to death with a claw hammer after she admitted to having been unfaithful. The New York district attorney's office charged Chen with murder in the second degree. Chen's lawyer argued that his conduct followed Chinese custom that allows husbands to dispel their shame in this way. The judge decided that the defendant "was driven to violence by traditional Chinese values about adultery and loss of manhood," and that these influences made Chen more "susceptible to cracking under the circumstances." He was found guilty of second-degree manslaughter and given five years on probation with no jail time.

- In 1986, the DeKalb, Ga., district attorney's office charged a Somali nurse with child abuse for allegedly severing her 2-year-old niece's clitoris. While cultural evidence was not decisive in the case—the nurse was acquitted because it was not clear who had done the actual cutting—the Somali community was angry that their traditional practice of female genital mutilation would be the subject of a

prosecution. "The Somali woman doesn't need an alien woman telling her how to treat her private parts," says Samme Warsame, founder of the Somali Relief Adjustment Organization Inc. in Atlanta. "The decision must come from the Somalis." More cases of this type are likely to come before the courts in the wake of decisions by the federal government and many states last year to make genital mutilation illegal.

- In 1985, Japanese immigrant Fomiko Kimura drowned her two children—an infant daughter and 4-year-old son—in the Pacific Ocean after learning of her husband's infidelity. She tried to kill herself as well. Charged with first-degree murder, she introduced evidence of the Japanese custom of *oyako-shinju,* or parent-child suicide, traditionally practiced by Japanese wives dishonored by a husband's adultery. She was allowed to plead guilty to involuntary manslaughter, and got one year in jail and five years on probation.

In the Nebraska case, Terrell R. Cannon, the grooms' lawyer, says he will argue that the defendants followed their ancient tradition of arranged marriages. He contends that the U.S. Constitution guarantees freedom of religion and that the state needs a compelling interest to justify any regulation of that freedom. "They should be allowed to practice their religion in this country," he says.

Pollack, the father's attorney, emphasizes that the two girls agreed to the marriages. "I have a hundred people who will testify that the girls consented to the marriages," Pollack says. "Under Iraqi custom, if the girls consent, these are legal marriages."

But is a 13-year-old capable of giving her full consent in such a situation? "I don't care what country you are in—a 13-year-old is a child, and children should be allowed to be children," says Nancy O'Malley, senior deputy district attorney at the Alameda district attorney's office in California.

"You live in our state, you live by our laws," Lancaster County Deputy Attorney Jodi Nelson told reporters. "I have yet to find a law that states: 'Oh, and by the way, if you immigrate here from another country, none of this applies.'"

Part of the problem with cultural evidence—and a reason to move cautiously in this area—is that the possibilities for its extension are limitless. O'Malley recently prosecuted a case against an African-American women whose infant died while in the care of her 9-year-old child. Charged with involuntary manslaughter, the mother argued that the childcare norms of the single black female parent are different from those of the middle-class white. Because of her race and gender, the woman argued, she does not earn enough to afford day care.

"She basically said, 'It's a cultural thing, and I shouldn't be held accountable for the baby's death,'" says O'Malley. "At some point, you've got to ask, 'What is culture?'"

O'Malley and others point out that a foreign country's custom or practice is not always uniform or clear-cut. A given practice may be acceptable to everyone or only to some segments of the population. Moreover, cultural norms change. For instance, when the U.S. Supreme Court outlawed segregation in public schools in *Brown vs. Board of Education* in 1954, it dramatically transformed American culture.

But Jacqueline Bhabha, associate director of the Center for International Studies at the University of Chicago and a lecturer at the law school, argues that cultural evidence is highly relevant in these cases. "Criminalizing people in this situation [the Nebraska case] has all sorts of drastic consequences, such as destroying the fabric of the family," she says. Rather than throwing people in jail, she says, the justice system should focus on educating refugees about the differences between their cultural practices and U.S. law.

Coleman, the law professor at Duke, says the answer in the Nebraska case and in the use of cultural evidence in general lies somewhere in the middle of this polarized debate. There is room in existing criminal-law doctrine for cultural evidence, says Coleman. If charged with murder, for instance, a foreign defendant can introduce cultural evidence as part of a conventional provocation or heat-of-passion defense.

But the most appropriate place for cultural evidence, she says, is in the sentencing phase, when criminal law focuses on the defendant's moral culpability.

"The defendants in the Nebraska case probably should be convicted," Coleman says. "They shouldn't be exonerated, as some courts have been doing by letting this evidence come in during the guilt phase of a trial, but they should not be sentenced as harshly as Americans born here."

• • •

4

Strategies for Changing Society

MARVIN E. OLSEN

• • •

Three sets of strategies for intentionally creating social change are examined in this paper. Social Mobilization strategies involve the exertion of influence "downward" by organizational, community, societal, or other leaders who are attempting to initiate social change from their positions of power. Social Involvement strategies involve reciprocal influence flows among

From Marvin E. Olsen, "Strategies for Changing Society," *Journal of Applied Sociology*, vol. 6, 1989, pp. 1–12. Copyright © 1989 by the Journal of Applied Sociology. Reprinted by permission.

leaders and citizens who are jointly attempting to change social life. Social Confrontation strategies involve the exercise of influence "upward" by citizens who are dissatisfied with existing conditions and want to change them despite resistance from system leaders. Each of these sets of change strategies is analyzed in terms of its major dynamics, guiding principles, possible courses of action, and typical problems. No one set of change strategies is inherently more desirable or effective than the others since they are intended to be used in quite different situations. Each of them is fraught with numerous potential difficulties, but each of them can be quite effective in producing intentional social change under particular circumstances.

INTRODUCTION

Social life is never static; it is constantly changing. "The actual world is a process," wrote the philosopher Alfred North Whitehead (1929). Expressed more broadly, "Process is reality, and reality is process." Nevertheless, we are often not satisfied with the course of undirected social change, and desire to direct it toward valued goals. How to guide and direct social change is the challenge of applied sociology. Using knowledge acquired through the study of social processes and conditions, applied sociology seeks to give humanity greater control over our collective social life.

In this endeavor, applied sociologists have given particular attention to developing strategies that social activists can use to direct the process of social change and attain desired goals in social life. Three basic sets of strategies for intentionally creating social change are examined in this paper, all of which have proven effective under particular conditions.

One set of change strategies is termed *Social Mobilization*. These strategies involve the exertion of influence "downward" by organizational, community, societal, or other leaders who are attempting to initiate social change from their positions of power. Their primary concern is to persuade citizens to adopt the intended changes in social life.

A second set of change strategies is termed *Social Involvement*. These strategies involve reciprocal influence flows among leaders and citizens who are jointly attempting to change social life. The primary concern in this case is to develop procedures for designing changes that are acceptable to all.

The third set of change strategies is termed *Social Confrontation*. These strategies involve the exercise of influence "upward" by citizens who are dissatisfied with existing conditions and want to change them despite resistance from leaders. The primary goal of these citizens is to overcome that resistance and force leaders to implement the desired changes.

Each of these three sets of change strategies can be analyzed in terms of its major dynamics, guiding principles, possible courses of action, and typical problems.

SOCIAL MOBILIZATION

Major Dynamics

The Social Mobilization approach to creating social change is leader-initiated and downward-oriented. Established system leaders identify an issue that they believe requires attention, formulate a policy concerning it and programs to enact that policy, and then seek to persuade citizens to accept the policy and to participate in the programs. In a highly autocratic system, the leaders might be able to enact their changes by fiat, provided they could exercise enough power to force others to comply. In most situations within modern societies, however, leaders rarely command that kind of absolute power. If their policies and programs are to be successful, they must convince the public that the intended changes will benefit them in some way.

While citizens do not participate directly in policy formation or program development in these situations, they may eventually be able to replace the current leaders if their policies and programs are widely unpopular. More immediately, citizens can express their views toward the proposed changes by either adopting or ignoring them.

Guiding Principles

If the majority of citizens are to be persuaded to adopt the proposed changes, they must feel that they will receive some kind of benefit from their cooperation. Social exchange theory offers several general principles that can facilitate this process. For example, leaders should initially offer some desired services to the public, so that people will feel indebted to them. They should strive to create a climate of trust between themselves and citizens. And they should be as explicit as possible about the kinds of benefits that people will realize through the proposed changes. This process of social exchange need not be strictly equal on both sides—that rarely happens in social life—but people must view the exchanges as fair and worthwhile.

Courses of Action

The possible courses of action constituting this mobilization set can be classified into the three categories of communicative, financial, and regulatory strategies (Olsen and Joerges 1983).

Communicative Strategies These strategies include providing information, making emotional appeals, and exerting social pressures. With an Information Strategy, leaders disseminate information about the current problem and the expected benefits of the proposed change. It is very widely used because of its ease and simplicity. Its effectiveness is seriously limited, however, by the fact that most of the information disseminated tends to be received only by people who are already convinced of the desirability of the change. Other people simply ignore the messages.

An Appeals Strategy involves sending messages with strong psychological or emotional appeals concerning the problem and the proposed change. This strategy is often more effective in reaching uncommitted individuals, especially if it touches on their fundamental beliefs and values. Nevertheless, altering individuals' personal attitudes can be very difficult. This approach is most effective when it is aimed at rather specific attitudes pertaining directly to the proposed change. It is much less effective when aimed at very broad attitudes.

A Pressures Strategy seeks to involve people in social situations where they will be influenced by others who already support the change. These situations may be informal neighborhood gatherings, meetings of voluntary associations, or activities of community organizations. Because people tend to be most receptive to new ideas when they are received from individuals they trust or within organizations to which they belong, this strategy is generally the most effective communicative technique. Creating social situations in which the desired pressures will be exerted can be demanding, however.

Financial Strategies These strategies include price adjustments, financial incentives, and tax disincentives. The Pricing Strategy is entirely passive and unintrusive, which makes it popular with political leaders. They simply allow the price of relevant goods or services to rise or fall in the marketplace according to supply and demand. If an item becomes scarce, its price will rise sharply; whereas, if it becomes overabundant, its price will drop sharply. There are numerous problems with this technique, however, especially when the goods or services are either vital necessities or complete luxuries, and hence display little price elasticity. For this and other reasons, the pricing strategy is very unpopular with the public.

An Incentives Strategy offers financial benefits to people who adopt the proposed change. These incentives may take such forms as grants, loans, tax credits, tax deductions, bonuses, or rebates. Needless to say, this strategy tends to be quite popular with most people. Nevertheless, it is not always successful because the incentives are frequently seen as insufficient to warrant altering established practices, or because people do not trust the offer.

A Disincentives Strategy, which is the converse of the incentives technique, penalizes people who do not adopt the intended change. Disincentives may be applied through taxes, fines, tariffs, or surcharges. Although this strategy does not require trust, it often fails because people find ways of circumventing or avoiding the financial penalty.

Regulatory Strategies Whereas communicative and financial strategies are aimed at individuals, regulatory techniques attempt to manipulate broader social conditions. This may be done by establishing performance standards, allocating goods or services, and restructuring communities or organizations. A Standards Strategy requires a government agency to establish performance or quality standards for goods or services, and to enforce adherence to them by those who produce the goods or offer the services. Because this technique is generally viewed as equitable and is relatively unintrusive, it tends to be rather popular. It does, however, have considerable administrative, surveillance, and enforcement costs.

An Allocation Strategy brings a government agency directly into the process through which goods or services are distributed to the public. Tactics can range from simple limits on the amount of items available through complex rationing schemes. This strategy entails all of the administrative, surveillance, and enforcement costs associated with standards, and in addition is normally very unpopular with the public. Consequently, it is usually a last-resort technique.

A Restructuring Strategy is by far the most complex means of promoting social change, and therefore, is generally the least used of all these strategies. Nevertheless, it can be the most effective in the long run. It involves changing the social (and sometimes also the physical) structure of an organization or community, so that it operates in a different way. When the social setting changes, people's actions almost invariably change in response to it. But because those action changes are voluntary, people tend to feel that they are in control of their lives and are not being manipulated. The obvious limitation of this strategy is that it can entail considerable planning and implementation costs.

Typical Problems

No matter how strongly leaders may desire to initiate social changes that they believe are necessary and desirable, most of the time they are not likely to be successful unless they can mobilize the public to adopt those changes. This can be a demanding task. Among the countless potential problems that may derail the efforts of system leaders, three are especially noteworthy. First, large numbers of citizens are likely to remain ignorant of the leaders' efforts, be apathetic or indifferent toward them, or flatly reject them. Second, the exchange transaction being proposed by the leaders may appear to be unfair or even exploitive to many people. And third, many of the mobilization strategies described above can be quite costly in terms of time, effort, and resources, without any guarantee of success. Through evaluation studies of these various mobilization strategies, however, we are gradually learning how to use them more effectively.

SOCIAL INVOLVEMENT

Major Dynamics

Because of the many problems inherent in the Social Mobilization approach to promoting social change, a quite different approach has been developed during the last 15 or so years. Instead of taking all the initiative in social change efforts, many leaders have begun to work *with* citizens rather than *on* them. The Social Involvement approach is a joint process in which system leaders and citizens work together to create social change. They establish and conduct a cooperative citizen involvement program to resolve a pressing problem, formulate a policy concerning it, and carry out an action program to deal with it.

Although the leaders usually retain final decision-making authority, an effective citizen participation program will involve interested citizens in meaningful ways at every step of the process. Therefore, the recommendations that result from the program are a shared result of collaborative effects. When this occurs, citizens do not need to be persuaded to adopt the proposed change. They do so willingly because they have participated in shaping it.

Guiding Principles

The basics of this approach lies in participatory democratic theory. Very briefly, this theory asserts that all people have a right to participate in making collective decisions which affect them within all spheres of their lives. Formulated during the political upheavals of the 1960s and 1970s, participatory democratic theory goes far beyond the idea of representative democracy as presently practiced in Western societies. Instead of merely selecting leaders to represent them and make decisions for them, all people should take part in actually making collective decisions. Moreover, that mandate is not restricted to the political sphere. It extends to work situations, community affairs, voluntary associations, and even the family. If participatory democracy is actually to occur, however, all decision-making procedures must be organized to facilitate and encourage as widespread and meaningful citizen involvement as possible.

Courses of Action

Most citizen involvement programs conducted thus far have been within communities, and the present discussion is limited to that realm. There is no inherent reason, however, why the process could not be utilized in complex organizations or many other settings.

Initially, many of these programs consisted of little more than one or two public hearings at which citizens expressed their views and feelings toward policies and programs being proposed by community leaders. These hearings were merely a formal ritual, and gave citizens little or no opportunity to make significant inputs to the decision-making process. Gradually, however, citizen involvement programs have been expanded and improved, so that today we know how to design and conduct participation programs which fully involve citizens.

A few years ago, several colleagues and I decided to compile and apply this knowledge. We examined all that was known about such programs, analyzed a considerable number of previous programs to discover their strengths and weaknesses, designed a comprehensive citizen involvement program based on that investigation, and tested it in a community that was struggling with an important policy decision. Since our program is described in detail in a recent book (Howell, Olsen and Olsen 1987), I shall not attempt to go through all its 33 steps here. Several features of our program can be noted, however.

First, it is divided into five phases, beginning at the time when a new policy or action is first being contemplated. It includes laying a foundation for a broad citizen involvement program within the community, organizing the program, conducting the program, reaching a final decision, and implementing the agreed-upon change. At the completion of each phase, the activities conducted thus far are evaluated and the program is modified as necessary to take account of any difficulties that have been encountered.

Second, it is a complex, long-term process. Involving large numbers of citizens in a community decision in a meaningful way cannot be accomplished in a few days or even a few weeks. In our trial community, the entire program lasted for about nine months; although with hindsight, it might have been completed in about six months. This process requires at least one professional leader working more or less full-time to coordinate the program, plus additional part-time staff persons. And it demands a considerable amount of time and effort by numerous citizens if it is to be successful.

Third, the process builds on and operates through the network of organizations presently existing in a community as extensively as possible. This is not intended to be a community organization process. Rather, it uses those existing organizations to ensure that the program is solidly grounded in the social structure of the community. This approach is vital if the program is to be taken seriously by both community leaders and citizens.

Fourth, it employs a number of interlinked techniques to encourage as many people as possible to participate in the program in at least one of several different ways. These techniques include a relatively large Citizen Task Force, a smaller Planning Committee, a central Steering Committee, numerous Working Groups that focus on specific aspects of the problem, a community survey, a workshop, all kinds of media publicity and information activities, written and telephone inputs from many citizens, and a formal public hearing.

Fifth, this is entirely a voluntary program and consequently tends to attract people who are concerned about and active in community affairs. While the program seeks to involve as many different kinds of people as possible, it is not a public referendum. Therefore, it is not representative of the entire community, but that is not its purpose. Its goal, rather, is to give all interested citizens meaningful opportunities to participate directly in making a community decision, if they so desire.

Sixth, this kind of program can be quite effective. In our trial community, almost everyone who participated in the program expressed strong satisfaction with it in a follow-up survey. More importantly, the process resulted in a final policy recommendation that was acceptable to virtually everyone involved. Although it was quite different from a provisional decision on the issue that had previously been made by the legal authorities, they agreed to adopt and implement the recommended action.

Typical Problems

There are several obvious and some less than obvious problems in conducting a citizen involvement program such as this. Some obvious problems are the time, effort, and resources needed to conduct a meaningful program. In addition, most citizens have little experience in participating in such a program, so that the process must be presented and taught to them in a manner that gains their trust rather than suspicion. Less obvious, but absolutely critical, is the requirement that the legal decision-makers must agree in advance to support and participate in the program, and to give serious consideration to its recommendations. If they make their final decision prior to or outside of the program, it becomes merely a sham exercise to legitimate that decision. If this happens, the public will sooner or later become aware of the situation, and will be unlikely to have any further interest in participating in citizen involvement programs.

SOCIAL CONFRONTATION

Major Dynamics

The Social Involvement approach to creating social change rests on the assumption that both leaders and citizens agree on the necessity of taking action to deal with a social issue or problem, even though they may not initially agree on what actions should be taken. In contrast, the Social Confrontation approach applies to situations in which only a set of citizens define an existing situation as a problem that requires action. In these situations, the relevant system leaders deny the existence of any problem or insist that existing practices are adequate to deal with it. In short, the citizens want change but the leaders do not.

The Social Confrontation approach is therefore a citizen-initiated, upward-oriented process. The concerned citizens must identify the nature of the problem, organize themselves for collective action, bring pressures to bear on leaders to accept a new policy and new programs despite their opposition, and ensure that the change is implemented. This is clearly a challenging process since the citizens are normally operating from a relatively weak power base (at least initially), but nevertheless must overcome the resistance of powerful opponents if they are to be successful. The central dynamics in this process are, therefore, power generation and exertion, coupled with negotiation and compromise.

Guiding Principles

The theoretical perspective that is most relevant to this process is social conflict theory. More specifically, the process draws on principles of intentional conflict creation, purposeful power exertion, and conflict resolution. They pertain to the questions of how to create conflict directed toward a specific objective without allowing it to become mass social disruption or rioting. These principles pertain to the matter of how to use social conflict to generate social power and how to focus sufficient power exertion on designated targets to force authorities to consider changing their policies and practices. They also deal with ways of resolving conflicts that are acceptable to all the involved parties.

This last concern is particularly crucial since the process of resolving conflict in a manner that produces desired social change calls for strategies which are quite different from those required to create conflict and exert power. Consequently, this approach demands a basic shift in philosophy and tactics as it progresses, which can be extremely difficult to achieve.

Courses of Action

Throughout human history, relatively powerless citizens have sought to change social conditions despite opposition from powerful system leaders. They have used strategies ranging from passive noncompliance to violent revolution, but a large proportion of those efforts have failed. However, one set of confrontation strategies has proved to be quite effective in such situations. Originating in Henry David Thoreau's concept of "civil disobedience" and demonstrated by Gandhi in his struggle for an Indian independence from Great Britain, this approach had been widely used during the twentieth century by labor unions and in the Civil Rights Movement of the 1960s. Analytically, it can be divided into five overlapping stages, each of which involves different kinds of activities.

Organization In this initial stage, the concerned citizens must create a social movement organization or transform an existing body such as a labor union into such an organization. This involves specifying goals for the movement and developing rhetoric that expresses the problem and those goals in clear and simple language. The movement must find dynamic, charismatic leaders who can attract a core of dedicated and active members and a larger number of less involved supporters. And it must conduct a public relations campaign to gain sympathy with its objectives, attract financial contributions, and legitimize the movement in the eyes of the public.

Conflict The next step in this stage is to identify key leverage points in the existing system where it and its authorities are especially vulnerable. In the economy, these are likely to be major production activities or consumption practices, while in the political system they may be enforcement of regulations or public images of officials. The movement then carefully designs and carries out actions aimed at those key leverage points that hinder, disrupt, block, or prevent normal activities. Very often this involves strikes, boycotts, sit-ins, demonstrations, or other public protest activities. All such actions must be carefully aimed at specific key targets and not get out of hand. The purpose of these actions is to exert pressures on the system and its authorities that are too costly for them to ignore.

Since the initial reaction of system authorities to the movement's actions is likely to be indifference or repression, the movement must be prepared to continue its conflict activities—often with gradual escalation—until the authorities are willing to negotiate. To gain the sympathy and support of the public in this situation, movement activists must be willing to suffer, without resistance, any punishments or reprisals that may be inflicted on them by the authorities. The moral justice of their cause and their willingness to suffer for it thus become their primary power resource in the struggle against the existing system. Mass media coverage is quite essential in this effort. In addition, the authorities must be sensitive to public opinion. If their stance is simply to "shoot all demonstrators on sight, and the public be damned," this strategy will obviously fail.

Negotiation While continuing to exert pressures on the system, the activists must constantly reiterate their desire to negotiate with system authorities. And no matter what kinds of threats or punishments are directed toward them because of their actions, they must not abandon the struggle until the authorities agree to negotiate with them.

When that time comes, the movement must be able to make a crucial transformation. Up to then, its goals will have been lofty and far-reaching, its leaders will have been ideological and charismatic, and its tactics will have been strident and conflictual. Now, however, it is time to sit down at the bargaining table with their enemy and participate in calm, rational bargaining. Some leaders of social movements can make this turnaround in tactics, but many cannot because they are too committed to the revolutionary rhetoric of "Give me liberty or give me death." In that case, they must designate or the movement must select new leaders who can operate more pragmatically.

The members of the movement's negotiating team must be willing to participate in good faith in the bargaining process with system authorities for whatever length of time is needed to reach a settlement. More importantly, they must be able to be flexible and to compromise gradually, letting go on minor points while continuing to stand firm on their most critical demands. In other words, they must be willing to eventually accept partial victory rather than insisting on "everything or nothing." A successful bargaining process must arrive at a compromise that is seen by both sides as a "win-win" resolution of the conflict, so that both feel they have gained some worthwhile outcomes.

Alteration Once a resolution of the conflict has been agreed upon by both sides, the movement must be willing to cooperate fully with the authorities in developing new programs and practices to implement the intended change. This process is most likely to be effective if the proposed alterations are aimed at the structure of the entire system rather than specific actions of particular individuals, if they emphasize alterations that will have minimal immediate disruptive effects on the system while maximizing long-term beneficial consequences, and if at least some of the system alterations can be implemented immediately so that supporters on both sides of the conflict can see that something is being accomplished. In addition, the movement leaders must insist that the entire process of change be constantly monitored to ensure that the agreed-upon alterations are actually being carried out.

Acceptance This final stage in the change process calls for movement leaders with a third set of organizational skills. Within the movement, they must be able to convince most of its members and supporters that the agreed-upon plan of action will eventually satisfy the most critical of their desires for change. If the current movement leaders cannot do this, they are likely to be overthrown by ideological radicals who claim that the present leaders have "sold out" to the system. When the radicals are successful in this challenge, the resulting "Thermadorian" reaction usually destroys whatever agreements were reached with system authorities. This throws the movement

back to the confrontation stage, and invites severe repressions from the authorities.

Outside the movement, its leaders must support the system authorities for their acceptance of the agreed-upon changes so that the authorities no longer feel threatened by the movement. This must be done both privately and in public, so as to encourage public support and legitimacy for the alterations that are being implemented. If these efforts are successful, system authorities who initially opposed all change are very likely to become increasingly accepting of the change as will also the larger public. People's attitudes, values, and beliefs will gradually shift in response to the new conditions. Eventually, the changes will be fully accepted and incorporated into the system so that in the future people may likely wonder, "why was all that controversy necessary?"

Typical Problems

This Social Confrontation approach to creating social change can easily break down at several critical points. The movement organization may not be able to find strong leaders, attract enough members and supporters, or obtain sufficient financial resources. There may be no leverage points where the existing system is vulnerable, or the movement may not be able to identify them or gain access to them. The movement's conflict tactics may not exert sufficient pressures on the system to force its authorities to bargain with the movement, or those authorities may be able to diffuse or squelch the conflict. The movement leaders or the system authorities may be unwilling or unable to negotiate in a spirit of compromise. The negotiations may end in an impasse that neither side can overcome. If agreement is reached on new policies, the more ideologically committed members of the movement may not accept them, or system authorities may fail to implement the programs and practices to which they have agreed. This is indeed a highly problematic way of creating social change. But it can be made to work for the benefit of citizens who are committed to achieving change.

ASSESSMENT

Each of the three sets of strategies for creating social change outlined in this paper—Social Mobilization, Social Involvement, and Social Confrontation—is fraught with numerous potential problems. Nevertheless, each of them can be effective under particular circumstances. No one approach is inherently more desirable or effective that the others, since they are intended to be used in quite different situations.

There are many situations in which system leaders see a need for social change that is not initially accepted by most of the public. A typical example with which I have worked extensively is promoting energy conservation. Beginning in the late 1970s, an increasing number of public officials came to understand that conserving energy can be much more beneficial—both financially and environmentally—than supplying ever-growing amounts of new energy. Yet much of the public was very slow and reluctant to adopt this perspective. Many national and local officials therefore began experimenting with all of the mobilization strategies discussed above, seeking the most effective way of encouraging citizens to conserve energy.

In quite a number of other situations during the past 15 years, especially within local communities, public officials and citizens have agreed that something had to be done to deal with a community problem, but did not initially agree on the best course of action. In several of those communities, citizen involvement programs such as the one described in this paper have enabled them to find a solution to the problem that was generally acceptable to everyone. In our trial community, public officials agreed to abandon their initial plan to build a new dam on a nearby river, and instead to use those funds to weatherize all the houses in the community. As we learn more about designing and conducting such programs,

they should become an even more widely used way of achieving desired social change.

Meanwhile, there will undoubtedly continue to be many situations in which citizens seek change but system authorities are indifferent or opposed to their demands. In such situations, the concerned citizens must be able and willing to organize themselves for action if their demands are to be accepted. As we learn more about using confrontation strategies in a purposeful and effective manner, citizens will discover that they are not powerless to change social systems despite strong opposition. The Civil Rights Movement provided dramatic evidence that this approach can be extremely effective when skillfully conducted.

In sum, it is possible to direct the process of social change toward desired goals. To do this effectively, however, we must select a broad approach that is appropriate to the particular situation, and then develop and implement action strategies that will enable us to achieve our goals. Social science—and especially applied sociology—can provide the knowledge necessary to make these strategic decisions in an informed manner, and thus facilitate the process of intentionally creating social change. Are we willing to undertake this challenge?

REFERENCES

Howell, Robert E., Marvin E. Olsen and Darryll Olsen. 1987. *Designing a Citizen Involvement Program: A Guidebook for Involving Citizens in the Resolution of Environmental Issues.* Corvallis, Oregon: Western Rural Development Center.

Olsen, Marvin E. and Bernward Joerges. 1983. "Consumer Energy Conservation Programs." Pp. 561–591 in *Handbook of Social Intervention*, edited by Edward Seidman. Beverly Hills, California: Sage Publications.

Whitehead, Alfred North, 1929. *Process and Reality.* New York: Macmillan Co.

2

The World Economy
and World Poverty

*The division of labor among nations is that some
specialize in winning, and others in losing.*

EDUARDO GALEANO

This collection of readings begins with the
topic of "the world economy and world
poverty," for several reasons. First, the world
economy is a central factor shaping all facets of
life in the contemporary world. We cannot
understand many of the world's social problems
without understanding world economic structures and processes. This is especially the case with
the problem of poverty, which is inextricably tied
to the dynamics of the global economy. Second,
poverty is a central problem because it causes or
exacerbates so many *other* world problems—
hunger, migration, homelessness, environmental
destruction, and violent conflict, to name just a
few. If humanity can find solutions to the problem of poverty, we will be closer to solving a variety of related problems. Third, world poverty is an
especially *striking* problem, considering the
unprecedented abundance that exists in the world
today. How can we justify, for example, that in the
poorest nations, 166 children of every 1,000 die
before the age of 5, when a plenitude of resources
exist to save their lives? One United Nations
study concluded that a concerted effort to reduce
infant mortality worldwide would cost only $25
billion a year, less than what people in the United
States spend each year on cigarettes (Bradshaw
and Wallace, 1996, p. 18)! Finally, some of the
most influential theories about the social, political, and economic dynamics of the modern world
originated in efforts to explain world poverty. For

example, important theoretical debates between
modernization theorists and world system theorists emerged as sociologists tried to explain the
gap between the richer and poorer nations of the
world, and make recommendations as to paths
out of poverty. In short, this topic provides a good
foundation from which to begin to deal with a
number of important problems and questions.

Poverty can be defined in an *absolute* and in a
relative sense. *Absolute poverty* refers to conditions
that are life threatening, such as a severe shortage
of food or inadequate sanitation leading to malnutrition and disease. *Relative poverty* refers to
conditions of deprivation relative to others who
are better off—that is, conditions that may seem
unjust, but are not necessarily life threatening. Vast
numbers of people in the world live in poverty
today, no matter which of the two definitions we
use. To get a picture of the scope of relative
poverty in the world, nations can be divided into
three categories based on their levels of per capita
income. *Most* of the world's peoples live in relative poverty as compared to those in the wealthy
nations, such as the United States, Japan, or the
nations of Western Europe. Only about 15 percent of the world's population live in wealthy
nations such as these having a per capita income
ranging from $8,000 to $20,000. Another third
live in the somewhat less affluent nations, such as
Korea, Greece, or Thailand, having a per capita
income of about $5,000. *Fully half* of the world's
population live in the poorest nations, such as
Haiti, India, or Congo, having a per capita income
of just a few hundred to a couple of thousand
dollars (Macionis, 1996, pp. 189–191). Absolute

24

poverty is also a problem affecting more than a billion people. Fifty-three percent of the people in Africa south of the Sahara, 42.9 percent of the people in South Asia, and 41.5 percent of the people in Latin America and the Caribbean live in absolute poverty (Isbister, 1995, p. 17)! The human costs of absolute poverty are depicted in the article by Luis Urrea, "Tortilla Curtain." Life-threatening poverty in Mexico leads large numbers of Mexicans each year to attempt to cross the border into the United States. Sadly, the very journey to cross the border is often as perilous to the migrants as staying at home.

How can we explain global poverty and its persistence in the face of affluence? What is the relationship between the global economy and global poverty? What are some solutions to the problem of poverty? Contrasting viewpoints are provided by the *structural-functionalist, modernization, Marxist,* and *world system* perspectives. The structural-functionalist, modernization, and world system perspectives were discussed briefly in Chapter 1. Here they will be discussed in more detail, with emphasis on how each perspective analyzes world poverty.

The *structural-functionalist perspective* is an example of the *order* perspective, as discussed briefly in Chapter 1. From this perspective, the best way to understand any phenomenon in society is to look for the positive, or negative, *functions* (or consequences) it has for other parts of society, or for society as a whole. As functionalist Robert Merton noted, functions may be *manifest*, or intended, or *latent*, unanticipated and unintended. Poverty might seem to be a phenomenon that is wholly dysfunctional—what could be positive about poverty? However, Herbert Gans, in "The Uses of Poverty: The Poor Pay All," outlines a number of (mostly latent) functions that poverty serves in the United States. When the poor are stigmatized as undeserving, argues Gans, this works to the advantage of many groups of the nonpoor.

The *modernization perspective* is an order perspective on world issues that originated in the wealthy nations of the world following World War II (the late 1940s and early 1950s). This perspective continues to be a vital tradition in sociological thinking about the world, and is very similar to the perspective that dominates the thinking of governments and corporations in the wealthy nations of the world. In this perspective, the world can be divided into nations that are "modern" and those that are "traditional." Modern nations are characterized by industrialization, urbanization, complex social organizations, a diversity of occupations, and high levels of education and literacy (Huntington, 1996, p. 68). The modern nations have become wealthy because they have a "modern" culture and society that favor the growth of advanced technology. For example, cultural values such as the work ethic, innovation, and individualism favor technological change, as do social structures such as universal education and democratic political institutions. Traditional nations remain poor because they have a "traditional" culture and society that devalue change. For example, values that emphasize family ties over individual merit retard technological change, as do political systems based on family or clan rather than a democratic, national government. Modernization theory sees the wealthy nations as having a special role in helping the poor nations, and is optimistic about the prospects of reducing poverty in the world. For example, when capitalists in the modern nations invest in the traditional nations, this brings them advanced technology and teaches their people the job skills and work habits they are lacking. When the governments of modern nations provide foreign aid to the traditional nations, they help them with the funds they need to modernize—to purchase advanced technology, to create new educational institutions, and to build strong, democratic governments. Because of its focus on programs created and led by the elites in the wealthy nations, modernization theory can be seen as an example of the "top down" approach to solutions discussed in Chapter 1.

Modernization theory has often been accused of being ethnocentric, because it seems to say that non-Western nations must adopt Western beliefs and values and give up their own cultures. However, modernization theory today does not simply equate modern values with "Western" or "American" values. A number of recent studies have focused on non-Western nations that have undergone fairly dramatic technological change, such as Japan or Hong Kong. They have attempted to prove that the cultures of these areas, while historically very different from those of the West, did have features that encouraged modernization (So, 1990, pp. 60–87). In *The Clash of Civilizations*, Samuel P. Huntington, an influential modernization theorist, argues that modernization does not necessarily mean Westernization, and discusses the variety of approaches that non-Western societies have taken to the process of modernization (Huntington, 1996, pp. 56–78).

Both the *Marxist* and the *world system perspective* contrast sharply with the functionalist and modernization perspectives and are examples of the *conflict perspective* discussed in Chapter 1. The Marxist perspective originated with the ideas of Karl Marx, the nineteenth-century philosopher and activist. Although Marx lived 150 years ago, his thinking continues to have a major influence on sociology and many other disciplines, as well as on the ideas of activists in many nations working to find solutions to world problems. Marx argued that economic activities, rather than cultural values, play the leading role in shaping human behavior. He analyzed the dynamics of *capitalism*, and saw poverty as inherent in a capitalist economy. Marx defined capitalism as an economic system based on two social classes—the *capitalist class* of property owners (*bourgeoisie*), and the *working class* of wage laborers (*proletariat*). These classes are always in conflict, as the capitalists are engaged in a ceaseless quest to accumulate more and more capital (wealth that can be invested to produce more wealth), and seek to reduce workers' wages to the poverty level, or below, to do so. Marx saw capitalism as having a contradictory nature. On the one hand, he argued that capitalism has brought the world unparalleled technological change resulting in great material progress. On the other hand, he saw capitalism as creating social problems, because it prioritizes the private capital accumulation of the capitalist class over the human needs of the majority. Marx saw

"bottom up" revolutions as the solutions to the problem of poverty, given the vested interests of the capitalist class in perpetuating poverty. He placed little faith in government efforts to alleviate poverty, and criticized the governments in capitalist nations as being little more than representatives of the capitalist class. Marx believed that one day the impoverished workers around the world would undertake revolutions against their governments that would replace capitalism with socialism, and ultimately communism, which he envisioned as a classless society, in which material abundance would be shared by all—"from each according to his ability, to each according to his need."

Interestingly, Marx identified what we today call "globalization" as one of the fundamental features of capitalism. He was one of the first thinkers to recognize that capitalism was uniting formerly isolated or independent nations and communities into an interdependent whole. In contrast to the modernization perspective, which sees the impact of the wealthier nations on the poorer nations as basically positive, Marx saw capitalism's effect on the poor nations as a mixture of progressive and destructive elements. On the one hand, Marx believed it was inevitable that many of these nations would become capitalist in order to experience technological change. He saw European capitalism as bringing advanced technology to nations that were trapped in backward ideas and institutions. He believed that native capitalist classes would emerge and join Europeans in the further development of the productive forces. On the other hand, Marx wrote about the human misery involved in the Europeans' enslavement of indigenous peoples in the Caribbean, South America, and Africa, and the vast loss of resources as the Europeans established colonies and pillaged them for gold, silver, and other valuables. Marx, in fact, saw the wealth of the European nations as directly related to the poverty of the colonies they conquered. In his theory of "primitive accumulation," he argued that the wealth plundered from Latin America played a direct role in the origins of capitalism, giving the nations of Europe a head start in accumulating capital they have enjoyed ever since. Marx pointed out that the native capitalist classes that were emerging in the colonial areas would also play a part in exploiting the indigenous poor.

The *world system perspective* originated in the United States in the 1960s and 1970s among American sociologists who were influenced by Marxism. The world system perspective has elaborated on the ideas of Marx to create a more in-depth analysis of how capitalism has operated on a global level over the last 500 years. Capitalism is envisioned as a *world system*, which Immanuel Wallerstein has defined as "a unit with a single division of labor and multiple cultural systems" (Wallerstein, 1979, p. 5). This means that there is one world economy that goes beyond the borders of each individual nation or culture. Governments or other organizations within a single nation may attempt to control or influence the impact of the world economy on the peoples within the nation's borders, but they will meet with varying degrees of success, because the world economy is a structure with a life of its own. The key feature of the world system is the global *division of labor* in which different areas of the world specialize in producing certain kinds of products and trade with others to get what they need. This exchange is highly unequal, and leads to three distinct and unequal areas of the world, the *core, periphery,* and *semi-periphery.*

Periphery areas today include the poor, formerly colonial nations of the world. Since the European colonial expansion of the sixteenth century, these nations were increasingly forced to specialize in production of low-value products utilizing low levels of technology, such as agricultural crops, minerals, and other primary products. The *core* nations today are the formerly colonial conquerors that have come to specialize in production of high-value products utilizing advanced technology, such as manufactured items. The poverty of the periphery nations is due to the *unequal exchange* relationship involved between core and periphery. The periphery must sell its products for a low price, and buy manufactured products at a high price. This inequality causes a continual drain of capital from the periphery and

accumulation of capital at the core (Shannon, 1996, pp. 34–37). Nations in the *semi-periphery* include those few nations that are somewhere in between the core and periphery and have some of the characteristics of each. These are formerly colonial nations that were once in the periphery, but have industrialized in recent years. Like the periphery, they are less developed technologically than the core, which continues to trade with the semi-periphery on an unequal basis. At the same time, they have some higher-tech industry, like the core, and gain by trading with the periphery on an unequal basis.

Like Marxism, and in contrast with the modernization perspective, the world system perspective is an example of the "bottom up" approach to solutions discussed in Chapter 1. The poor of the world need to revolt against the inequality inherent within the global division of labor, in order to garner a larger share of the wealth being produced. However, in contrast to the Marxist perspective, the world system perspective does not see fundamental solutions to world poverty through revolutions occurring *within individual nations.* Because the system is a *world* system, fundamental solutions require a transformation of the entire world system *as a whole.* In this perspective, socialist revolutions in individual nations, such as the former Soviet Union, China, Cuba, or Vietnam, were ultimately doomed to failure, because these nations were trapped within and subject to the influences of the larger and more powerful capitalist world system. In the world system perspective, *antisystemic movements,* social movements that seek to transform the entire world system, are humanity's best hope for ultimately eliminating world poverty.

Frances Moore Lappé and Joseph Collins's "Why Can't People Feed Themselves?" discusses the impact of the world system on hunger. This article shows the importance of understanding the world as a *historical* system and argues that world hunger today originates in processes first established centuries ago under colonialism. Medea Benjamin and Andrea Freeman's "Fair Trade: Buying and Selling for Justice" discusses the emergence of the fair trade movement, a "bottom up" movement that seeks to give poor farmers and workers greater control over the processes of the world economy. This movement is building *Alternative Trade Organizations,* or ATOs. These organizations bypass the large corporations and directly link producers in the periphery with consumers in the core nations. They pay the people in the poor nations a fairer price for their products than the large corporations, enabling them to have a better chance at climbing out of poverty.

WORKS CITED

Bradshaw, York W., and Michael Wallace. 1996. *Global Inequalities.* Thousand Oaks, CA: Pine Forge Press.

Huntington, Samuel P. 1996. *The Clash of Civilizations.* New York: Simon and Schuster.

Isbister, John. 1995. *Promises Not Kept.* 3rd ed. West Hartford, CT: Kumarian Press.

Macionis, John J. 1996. *Society: The Basics.* 3rd ed. Upper Saddle River, NJ: Prentice Hall.

Shannon, Thomas R. 1996. *An Introduction to the World System Perspective.* 2nd ed. Boulder, CO: Westview Press.

So, Alvin Y. 1990. *Social Change and Development.* Newbury Park, CA: Sage Publications.

Wallerstein, Immanuel. 1979. *The Capitalist World-Economy.* New York: Cambridge University Press.

QUESTIONS FOR DISCUSSION

1. In "Tortilla Curtain," Urrea describes the horrible conditions experienced by people seeking to cross the border from Mexico into the United States. Why do you believe so many people are willing to leave their homes in Central America to come to the United States, even in the face of such conditions?

2. What are some differences between the order and conflict perspectives on world poverty? What are some similarities in the two perspectives? Which perspective do *you* find most useful in explaining the conditions described in "Tortilla Curtain"?

3. In "Why Can't People Feed Themselves?" Lappé and Collins discuss why they have rejected the modernization perspective on world poverty. Briefly state their reasoning. Why do you think it might have taken them "decades," as they say, to figure this out? What are some of the historical strategies, according to Lappé and Collins, that created the problem of world hunger?

4. What is the goal of the alternative trade movement? Do you believe this movement is an effective way to solve the problem of world poverty? Why or why not? In answering this, what criteria did you use to judge "effectiveness"?

INFOTRAC COLLEGE EDITION: EXERCISE

This chapter discussed contrasting perspectives on the causes of world poverty. These sociological perspectives can be used to better understand why so many people migrate illegally to the United States (the scene so vividly described in "Tortilla Curtain"), despite efforts of the Immigration and Naturalization Service to prevent this. Using InfoTrac College Edition, look up the following article, which analyzes U.S. immigration policies:

"Masters of the Game: How the U.S. Protects the Traffic in Cheap Mexican Labor." Wade Graham. *Harper's Magazine.* July, 1996v293n1754p35. (*Hint*: Enter the search terms Wade Graham, using the Subject Guide.)

What are some of the *functions* that illegal immigration serves? What *vested interests* in the continuation of illegal immigration exist in the *world system*? Why has the *modernization* of Mexico failed to result in a decline in immigration to the United States? Do you agree with the author that U.S. immigration policy actually preserves, rather than limits, illegal immigration from Mexico?

FOR ADDITIONAL RESEARCH

Books

Bradshaw, York W., and Michael Wallace. 1996. *Global Inequalities*. Thousand Oaks, CA: Pine Forge Press.

Brecher, Jeremy, and Tim Costello. 1994. *Global Village or Global Pillage*. Boston: South End Press.

Isbister, John. 1998. *Promises Not Kept*. 4th ed. West Hartford, CT: Kumarian Press.

Organizations

Food First (Institute for Food and Development Policy)
398 60th Street
Oakland, CA 94618
www.foodfirst.org

Habitat for Humanity International
121 Habitat Street
Americus, GA 31709
(912) 924-6935
www.habitat.org/

Oxfam America
115 Broadway
Boston, MA 02116
(617) 482-1211
www.oxfam.org

Sweatshop Watch
310 Eighth Street, Suite 309
Oakland, CA 94607
www.sweatshopwatch.org

ACTION PROJECTS

1. Go to the United Nations Web site (www.un.org) and locate the United Nations' programs on economic and social development. From there, go to the sections on social development and sustainable development; also locate the *Bulletin on the Eradication of Poverty* and the *Report on the World Social Situation*. How does the United Nations define poverty? According to the United Nations, how many people in the world are poor? Is world poverty increasing or decreasing? What are some of the United Nations' programs aimed at reducing poverty?

2. Contact one of the organizations listed above in the "For Additional Research" section. Find out what you can do to help the organization, and get together with other students to organize a campus project. Meet together to reflect on what you learned from the project, in relation to the topic of world poverty.

3. Using the Internet, search for additional organizations working to alleviate global poverty. Make a list of those you find and share it with other students and your instructor. Go to the Web site of those you find most interesting. How do they define the issue? What do they see as causes? Solutions? Are they "top down" or "bottom up" organizations? Would you consider giving some time or money to these organizations? Why or why not?

5

Tortilla Curtain

LUIS URREA

When I was younger, I went to war. The Mexican border was the battlefield. There are many Mexicos; there are also many Mexican borders, any one of which could fill its own book. I, and the people with me, fought on a specific front. We sustained injuries and witnessed deaths. There were machine guns pointed at us, knives, pistols, clubs, even skyrockets. I caught a street-gang member trying to stuff a lit cherry bomb into our gas tank. On the same night, a drunk mariachi opened fire on the missionaries through the wall of his house.

We drove five beat-up vans. We were armed with water, medicine, shampoo, food, clothes, milk, and doughnuts. At the end of a day, like returning veterans from other battles, we carried secrets in our hearts that kept some of us awake at night, gave others dreams and fits of crying. Our faith sustained us—if not in God or "good," then in our work.

Others of us had no room for or interest in such drama, and came away unscathed—and unmoved. Some of us sank into the mindless joy of fundamentalism, some of us drank, some of us married impoverished Mexicans. Most of us took it personally. Poverty *is* personal: it smells and it shocks and it invades your space. You come home dirty when you get too close to the poor. Sometimes you bring back vermin: they hide in your hair, in your underpants, in your intestines. These unpleasant possibilities are a given. They are the price you occasionally have to pay.

In Tijuana and environs, we met the many ambassadors of poverty: lice, scabies, tapeworm, pinworm, ringworm, fleas, crab lice. We met diphtheria, meningitis, typhoid, polio, *turista* (diarrhea), tuberculosis, hepatitis, VD, impetigo, measles, chronic hernia, malaria, whooping cough. We met madness and "demon possession."

These were the products of dirt and disregard—bad things afflicting good people. Their world was far from our world. Still, it would take you only about twenty minutes to get there from the center of San Diego.

For me, the worst part was the lack of a specific enemy. We were fighting a nebulous, all-pervasive *It*. Call it hunger. Call it despair. Call it the Devil, the System, Capitalism, the Cycle of Poverty, the Fruits of the Mexican Malaise. It was a seemingly endless circle of disasters. Long after I'd left, the wheel kept on grinding.

At night, the Border Patrol helicopters swoop and churn in the air all along the line. You can sit in the Mexican hills and watch them herd humans on the dusty slopes across the valley. They look like science fiction crafts, their hard-focused lights raking the ground as they fly.

Borderlands locals are so jaded by the sight of nightly people-hunting that it doesn't even register in their minds. But take a stranger to the border, and she will *see* the spectacle: monstrous Dodge trucks speeding into and out of the landscape; uniformed men patrolling with flashlights, guns, and dogs; spotlights; running figures; lines of people hurried onto buses by armed guards; and the endless clatter of the helicopters with their harsh white beams. A Dutch woman once told me it seemed altogether "un-American."

But the Mexicans keep on coming—and the Guatemalans, the Salvadorans, the Panamanians, the Colombians. The seven-mile stretch of Interstate 5 nearest the Mexican border is, at times, so congested with Latin American pedestrians that it resembles a town square.

They stick to the center island. Running down the length of the island is a cement wall. If the "illegals" (currently, "undocumented workers"; formerly, "wetbacks") are walking north and a Border Patrol vehicle happens along, they simply hop over the wall and trot south. The officer will have to drive up to the 805 interchange, or Dairy Mart Road, swing over the overpasses, then drive south. Depending on where this pursuit begins, his detour could entail five to ten miles of driving. When the officer finally reaches the group, they hop over the wall and trot north. Furthermore, because freeway arrests would endanger traffic, the Border Patrol has effectively thrown up its hands in surrender.

It seems jolly on the page. But imagine poverty, violence, natural disasters, or political fear driving you away from everything you know. Imagine how bad things get to make you leave behind your family, your friends, your lovers; your home, as humble as it might be; your church, say. Let's take it further—you've said good-bye to the graveyard, the dog, the goat, the mountains where you first hunted, your grade school, your state, your favorite spot on the river where you fished and took time to think.

Then you come hundreds—or thousands—of miles across territory utterly unknown to you. (Chances are, you have never traveled farther than a hundred miles in your life.) You have walked, run, hidden in the backs of trucks, spent part of your precious money on bus fare. There is no AAA or Travelers Aid Society available to you. Various features of your journey north might include police corruption; violence in the forms of beatings, rape, murder, torture, road accidents; theft; incarceration. Additionally, you might experience loneliness, fear, exhaustion, sorrow, cold, heat, diarrhea, thirst, hunger. There is no medical attention available to you. There isn't even Kotex.

Weeks or months later, you arrive in Tijuana. Along with other immigrants, you gravitate to the bad parts of town because there is nowhere for you to go in the glittery sections where the *gringos* flock. You stay in a run-down little hotel in the red-light district, or behind the bus terminal. Or you find your way to the garbage dumps, where you throw together a small cardboard nest and claim a few feet of dirt for yourself. The garbage-pickers working this dump might allow you to squat, or they might come and rob you or burn you out for breaking some local rule you cannot possibly know beforehand. Sometimes the dump is controlled by a syndicate, and goon squads might come to you within a day. They want money, and if you can't pay, you must leave or suffer the consequences.

In town, you face endless victimization if you aren't street-wise. The police come after you, street thugs come after you, petty criminals come after you; strangers try your door at night as you sleep. Many shady men offer to guide you across the border, and each one wants all your money now, and promises to meet you at a prearranged spot. Some of your fellow travelers end their journeys right here—relieved of their savings and left to wait on a dark corner until they realize they are going nowhere.

If you are not Mexican, and can't pass as *tijuanense,* a local, the tough guys find you out. Salvadorans and Guatemalans are routinely beaten up and robbed. Sometimes they are disfigured. Indians—Chinantecas, Mixtecas, Guasaves, Zapotecas, Mayas—are insulted and pushed around; often they are lucky—they are merely ignored. They use this to their advantage. Often they don't dream of crossing into the United States: a Mexican tribal person would never be able to blend in, and they know it. To them, the garbage dumps and street vending and begging in Tijuana are a vast improvement over their former lives. As Doña Paula, a Chinanteca friend of mine who lives at the Tijuana garbage dump, told me, "This is the garbage dump. Take all you need. There's plenty here for *everyone!*"

If you are a woman, the men come after you. You lock yourself in your room, and when you must leave it to use the pestilential public bathroom at the end of your floor, you hurry, and you check every corner. Sometimes the lights are out in the toilet room. Sometimes men listen at the door. They call you "good-looking" and "bitch" and "*mamacita*," and they make kissing sounds at you when you pass.

You're in the worst part of town, but you can comfort yourself—at least there are no death squads here. There are no torturers here, or bandit land barons riding into your house. This is the last barrier, you think, between you and the United States—*Los Yunaites Estaites.*

You still face police corruption, violence, jail. You now also have a wide variety of new options available to you: drugs, prostitution, white slavery, crime. Tijuana is not easy on newcomers. It is a city that has always thrived on taking advantage of a sucker. And the innocent are the ultimate suckers in the Borderlands.

If you have saved up enough money, you go to one of the *coyotes* (people-smugglers), who guide travelers through the violent canyons immediately north of the border. Lately, these men are also called *polleros*, or "chicken-wranglers." Some of them are straight, some are land pirates. Negotiations are tense and strange: *polleros* speak a Spanish you don't quite understand—like the word *polleros*. Linguists call the new border-speak "Spanglish," but in Tijuana, Spanglish is mixed with slang and *pochismos* (the polyglot hip talk of Mexicans infected with *gringoismo*; the *cholos* in Mexico, or Chicanos on the American side).

Suddenly, the word for "yes," *sí,* can be *simón* or *siról.* "No" is *chale.* "Bike" (*bicicleta*) is *biaca.* "Wife" (*esposa*) is *waifa.* "The police" (*la policía*) are *la chota.* "Women" are *rucas* or *morras.* You don't know what they're talking about.

You pay them all your money—sometimes it's your family's lifelong savings. Five hundred dol-

lars should do it. "*Orale,*" the dude tells you, which means "right on." You must wait in Colonia Libertad, the most notorious *barrio* in town, ironically named "Liberty."

The scene here is baffling. Music blares from radios. Jolly women at smoky taco stands cook food for the journeys, sell jugs of water. You can see the Border Patrol agents cruising the other side of the fence; they trade insults with the locals.

When the appointed hour comes, you join a group of *pollos* (chickens) who scuttle along behind the *coyote.* You crawl under the wires, or, if you go a mile east, you might be amazed to find that the famous American Border Fence simply stops. To enter the United States, you merely step around the end of it. And you follow your guide into the canyons. You might be startled to find groups of individuals crossing the line without *coyotes* leading them at all. You might wonder how they mastered the canyons, and you might begin to regret the loss of your money.

If you have your daughters or mothers or wives with you—or if you are a woman—you become watchful and tense, because rape and gang rape are so common in this darkness as to be utterly unremarkable. If you have any valuables left after your various negotiations, you try to find a sly place to hide them in case you meet *pandilleros* (gang members) or *rateros* (thieves—rat-men). But, really, where can you put anything? Thousands have come before you, and the hiding places are pathetically obvious to robbers: in shoulder bags or clothing rolls, pinned inside clothes, hidden in underwear, inserted in body orifices.

If the *coyote* does not turn on you suddenly with a gun and take everything from you himself, you might still be attacked by the *rateros.* If the *rateros* don't get you, there are roving zombies that you can smell from fifty yards downwind—these are the junkies who hunt in shambling packs. If the junkies somehow miss you, there are the *pandilleros*—gang-bangers from either side of the border who are looking for some bloody fun. They adore "taking off" illegals because it's the perfect crime: there is no way they can ever be caught. They are Tijuana *cholos*, or Chicano *vatos*, or Anglo head-bangers.

Their sense of fun relies heavily on violence. Gang beatings are their preferred sport, though rape in all its forms is common, as always. Often the *coyote* will turn tail and run at the first sight of *pandilleros*. What's another load of desperate chickens to him? He's just making a living, taking care of business.

If he doesn't run, there is a good chance he will be the first to be assaulted. The most basic punishment these young toughs mete out is a good beating, but they might kill him in front of the *pollos* if they feel the immigrants need a lesson in obedience. For good measure, these boys— they are mostly *boys,* aged twelve to nineteen, bored with Super Nintendo and MTV—beat people and slash people and trash the women they have just finished raping.

Their most memorable tactic is to hamstring the *coyote* or anyone who dares speak out against them. This entails slicing the muscles in the victim's legs and leaving him to flop around in the dirt, crippled. If you are in a group of *pollos* that happens to be visited by these furies, you are learning border etiquette.

Now, say you are lucky enough to evade all these dangers on your journey. Hazards still await you and your family. You might meet white racists, complimenting themselves with the tag "Aryans"; they "patrol" the scrub in combat gear, carrying radios, high-powered flashlights, rifles, and bats. Rattlesnakes hide in bushes—you didn't count on that complication. Scorpions, tarantulas, black widows. And, of course, there is the Border Patrol (*la migra*).

They come over the hills on motorcycles, on horses, in huge Dodge Ramcharger four-wheel drives. They yell, wear frightening goggles, have guns. Sometimes they are surprisingly decent; sometimes they are too tired or too bored to put much effort into dealing with you. They collect you in a large group of fellow *pollos*, and a guard (a Mexican Border Patrol agent!) jokes with your group in Spanish. Some cry, some sulk, most laugh. Mexicans hate to be rude. You don't know what to think—some of your fellow travelers take their arrest with aplomb. Sometimes the officers know their names. But you have been told repeatedly that the Border Patrol sometimes beats or kills people. Everyone talks about the Mexican girl molested inside its building.

The Border Patrol puts you into trucks that take you to buses that take you to compounds that load you onto other buses that transport you back to Tijuana and put you out. Your *coyote* isn't bothered in the least. Some of the regulars who were with you go across and get brought back a couple of times a night. But for you, things are different. You have been brought back with no place to sleep. You have already spent all your money. You might have been robbed; so you have only your clothes—maybe not all of them. The robbers may have taken your shoes. You might be bloodied from a beating by *pandilleros*, or an "accident" in the Immigration and Naturalization Service compound. You can't get proper medical attention. You can't eat, or afford to feed your family. Some of your compatriots have been separated from their wives or their children. Now their loved ones are in the hands of strangers, in the vast and unknown United States. The Salvadorans are put on planes and flown back to the waiting arms of the military. As you walk through the cyclone fence, back into Tijuana, the locals taunt you and laugh at your misfortune.

If you were killed, you have nothing to worry about.

6

The Uses of Poverty:
The Poor Pay All

HERBERT J. GANS

Some twenty years ago Robert K. Merton applied the notion of functional analysis[1] to explain the continuing though maligned existence of the urban political machine: if it continued to exist, perhaps it fulfilled latent—unintended or unrecognized—positive functions. Clearly it did. Merton pointed out how the political machine provided central authority to get things done when a decentralized local government could not act, humanized the services of the impersonal bureaucracy for fearful citizens, offered concrete help (rather than abstract law or justice) to the poor, and otherwise performed services needed or demanded by many people but considered unconventional or even illegal by formal public agencies.

Today, poverty is more maligned than the political machine ever was; yet it, too, is a persistent social phenomenon. Consequently, there may be some merit in applying functional analysis to poverty, in asking whether it also has positive functions that explain its persistence.

Merton defined functions as "those observed consequences [of a phenomenon] which make for the adaptation or adjustment of a given [social] system." I shall use a slightly different definition; instead of identifying functions for an entire social system, I shall identify them for the interest groups, socioeconomic classes, and other population aggregates with shared values that "inhabit" a social system. I suspect that in a modern heterogeneous society, few phenomena are functional or dysfunctional for the society as a whole, and that most result in benefits to some groups and costs to others. Nor are any phenomena indispensable; in most instances, one can suggest what Merton calls "functional alternatives" or equivalents for them, i.e., other social patterns or policies that achieve the same positive functions but avoid the dysfunctions.[2]

Associating poverty with positive functions seems at first glance to be unimaginable. Of course, the slumlord and the loan shark are commonly known to profit from the existence of poverty, but they are viewed as evil men, so their activities are classified among the dysfunctions of poverty. However, what is less often recognized, at least by the conventional wisdom, is that poverty also makes possible the existence or expansion of respectable professions and occupations, for example, penology, criminology, social work, and public health. More recently, the poor have provided jobs for professional and paraprofessional "poverty warriors," and for journalists and social scientists, this author included, who have supplied the information demanded by the revival of public interest in poverty.

Clearly, then, poverty and the poor may well satisfy a number of positive functions for many non-poor groups in American society. I shall describe thirteen such functions—economic, social, and political—that seem to me most significant.

---

[1]"Manifest and Latent Functions," in *Social Theory and Social Structure* (Glencoe, Ill: The Free Press, 1949), p. 71.

[2]I shall henceforth abbreviate positive functions as functions and negative functions as dysfunctions. I shall also describe functions and dysfunctions, in the planner's terminology, as benefits and costs.

From Herbert J. Gans, "The Uses of Poverty: The Poor Can Pay All," *Social Policy,* July/August, 1971, pp. 20–24. Copyright © 1971 by Social Policy Corporation. Reprinted by permission.

THE FUNCTIONS OF POVERTY

First, the existence of poverty ensures that society's "dirty work" will be done. Every society has such work: physically dirty or dangerous, temporary, dead-end and underpaid, undignified and menial jobs. Society can fill these jobs by paying higher wages than for "clean" work, or it can force people who have no other choice to do the dirty work—and at low wages. In America, poverty functions to provide a low-wage labor pool that is willing—or, rather, unable to be *un*willing—to perform dirty work at low cost. Indeed, this function of the poor is so important that in some Southern states, welfare payments have been cut off during the summer months when the poor are needed to work in the fields. Moreover, much of the debate about the Negative Income Tax and the Family Assistance Plan has concerned their impact on the work incentive, by which is actually meant the incentive of the poor to do the needed dirty work if the wages therefrom are no larger than the income grant. Many economic activities that involve dirty work depend on the poor for their existence: restaurants, hospitals, parts of the garment industry, and "truck farming," among others, could not persist in their present form without the poor.

Second, because the poor are required to work at low wages, they subsidize a variety of economic activities that benefit the affluent. For example, domestics subsidize the upper middle and upper classes, making life easier for their employers and freeing affluent women for a variety of professional, cultural, civic, and partying activities. Similarly, because the poor pay a higher proportion of their income in property and sales taxes, among others, they subsidize many state and local governmental services that benefit more affluent groups. In addition, the poor support innovation in medical practice as patients in teaching and research hospitals and as guinea pigs in medical experiments.

Third, poverty creates jobs for a number of occupations and professions that serve or "service" the poor, or protect the rest of society from them. As already noted, penology would be minuscule without the poor, as would the police. Other activities and groups that flourish because of the existence of poverty are the numbers game, the sale of heroin and cheap wines and liquors, pentecostal ministers, faith healers, prostitutes, pawn shops, and the peacetime army, which recruits its enlisted men mainly from among the poor.

Fourth, the poor buy goods others do not want and thus prolong the economic usefulness of such goods—day-old bread, fruit and vegetables that would otherwise have to be thrown out, secondhand clothes, and deteriorating automobiles and buildings. They also provide incomes for doctors, lawyers, teachers, and others who are too old, poorly trained, or incompetent to attract more affluent clients.

In addition to economic functions, the poor perform a number of social functions.

Fifth, the poor can be identified and punished as alleged or real deviants in order to uphold the legitimacy of conventional norms. To justify the desirability of hard work, thrift, honesty, and monogamy, for example, the defenders of these norms must be able to find people who can be accused of being lazy, spendthrift, dishonest, and promiscuous. Although there is some evidence that the poor are about as moral and law-abiding as anyone else, they are more likely than middle-class transgressors to be caught and punished when they participate in deviant acts. Moreover, they lack the political and cultural power to correct the stereotypes that other people hold of them and thus continue to be thought of as lazy, spendthrift, etc., by those who need living proof that moral deviance does not pay.

Sixth, and conversely, the poor offer vicarious participation to the rest of the population in the uninhibited sexual, alcoholic, and narcotic behavior in which they are alleged to participate and which, being freed from the constraints of affluence, they are often thought to enjoy more than the middle classes. Thus many people, some social scientists included, believe that the poor not only are more given to uninhibited behavior (which may be true, although it is often motivated by despair more than by lack of inhibition) but derive more pleasure from it than affluent people (which research by Lee Rainwater, Walter Miller, and others shows to be patently untrue). However, whether the poor actually have more sex and enjoy it more is irrelevant; so long as middle-class people believe this to be true, they can participate in it vicariously when instances are reported in factual or fictional form.

Seventh, the poor also serve a direct cultural function when culture created by or for them is adopted by the more affluent. The rich often collect artifacts from extinct folk cultures of poor people; and almost all Americans listen to the blues, Negro spirituals, and country music, which originated among the Southern poor. Recently they have enjoyed the rock styles that were born, like the Beatles, in the slums; and in the last year, poetry written by ghetto children has become popular in literary circles. The poor also serve as culture heroes, particularly, of course, to the left; but the hobo, the cowboy, the hipster, and the mythical prostitute with a heart of gold have performed this function for a variety of groups.

Eighth, poverty helps to guarantee the status of those who are not poor. In every hierarchical society someone has to be at the bottom; but in American society, in which social mobility is an important goal for many and people need to know where they stand, the poor function as a reliable and relatively permanent measuring rod for status comparisons. This is particularly true for the working class, whose politics is influenced by the need to maintain status distinctions between themselves and the poor, much as the aristocracy must find ways of distinguishing itself from the *nouveaux riches.*

Ninth, the poor also aid the upward mobility of groups just above them in the class hierarchy. Thus a goodly number of Americans have entered the middle class through the profits earned from the provision of goods and services in the slums, including illegal or nonrespectable ones that upper-class and upper-middle-class businessmen shun because of their low prestige. As a result, members of almost every immigrant group have financed their upward mobility by providing slum housing, entertainment, gambling, narcotics, etc., to later arrivals—most recently to Blacks and Puerto Ricans.

Tenth, the poor help to keep the aristocracy busy, thus justifying its continued existence. "Society" uses the poor as clients of settlement houses and beneficiaries of charity affairs; indeed,

the aristocracy must have the poor to demonstrate its superiority over other elites who devote themselves to earning money.

Eleventh, the poor, being powerless, can be made to absorb the costs of change and growth in American society. During the nineteenth century, they did the backbreaking work that built the cities; today, they are pushed out of their neighborhoods to make room for "progress." Urban renewal projects to hold middle-class taxpayers in the city and expressways to enable suburbanites to commute downtown have typically been located in poor neighborhoods, since no other group will allow itself to be displaced. For the same reason, universities, hospitals, and civic centers also expand into land occupied by the poor. The major costs of the industrialization of agriculture have been borne by the poor, who are pushed off the land without recompense; and they have paid a large share of the human cost of the growth of American power overseas, for they have provided many of the foot soldiers for Vietnam and other wars.

Twelfth, the poor facilitate and stabilize the American political process. Because they vote and participate in politics less than other groups, the political system is often free to ignore them. Moreover, since they can rarely support Republicans, they often provide the Democrats with a captive constituency that has no other place to go. As a result, the Democrats can count on their votes, and be more responsive to voters—for example, the white working class—who might otherwise switch to the Republicans.

Thirteen, the role of the poor in upholding conventional norms (see the *fifth* point, above) also has a significant political function. An economy based on the ideology of laissez faire requires a deprived population that is allegedly unwilling to work or that can be considered inferior because it must accept charity or welfare in order to survive. Not only does the alleged moral deviancy of the poor reduce the moral pressure on the present political economy to eliminate poverty but socialist alternatives can be made to look quite unattractive if those who will benefit most from them can be described as lazy, spendthrift, dishonest, and promiscuous.

ALTERNATIVES

I have described thirteen of the more important functions poverty and the poor satisfy in American society, enough to support the functionalist thesis that poverty, like any other social phenomenon, survives in part because it is useful to society or some of its parts. This analysis is not intended to suggest that because it is often functional, poverty *should* exist, or that it *must* exist. For one thing, poverty has many more dysfunctions than functions; for another, it is possible to suggest functional alternatives.

For example, society's dirty work could be done without poverty, either by automation or by paying "dirty workers" decent wages. Nor is it necessary for the poor to subsidize the many activities they support through their low-wage jobs. This would, however, drive up the costs of these activities, which would result in higher prices to their customers and clients. Similarly, many of the professionals who flourish because of the poor could be given other roles. Social workers could provide counseling to the affluent, as they prefer to do anyway; and the police could devote themselves to traffic and organized crime. Other roles would have to be found for badly trained or incompetent professionals now relegated to serving the poor, and someone else would have to pay their salaries. Fewer penologists would be employable, however. And pentecostal religion could probably not survive without the poor—nor would parts of the second- and third-hand-goods market. And in many cities, "used" housing that no one else wants would then have to be torn down at public expense.

Alternatives for the cultural functions of the poor could be found more easily and cheaply. Indeed, entertainers, hippies, and adolescents are already serving as the deviants needed to uphold traditional morality and as devotees of orgies to "staff" the fantasies of vicarious participation.

The status functions of the poor are another matter. In a hierarchical society, some people must be defined as inferior to everyone else with respect to a variety of attributes, but they need not be poor in the absolute sense. One could conceive of a society in which the "lower class," though last in the pecking order, received 75 percent of the median income, rather than 15–40 percent, as is now the case. Needless to say, this would require considerable income redistribution.

The contribution the poor make to the upward mobility of the groups that provide them with goods and services could also be maintained without the poor's having such low incomes. However, it is true that if the poor were more affluent, they would have access to enough capital to take over the provider role, thus competing with, and perhaps rejecting, the "outsiders." (Indeed, owing in part to antipoverty programs, this is already happening in a number of ghettos, where white storeowners are being replaced by Blacks.) Similarly, if the poor were more affluent, they would make less willing clients for upper-class philanthropy, although some would still use settlement houses to achieve upward mobility, as they do now. Thus "Society" could continue to run its philanthropic activities.

The political functions of the poor would be more difficult to replace. With increased affluence the poor would probably obtain more political power and be more active politically. With higher incomes and more political power, the poor would be likely to resist paying the costs of growth and change. Of course, it is possible to imagine urban renewal and highway projects that properly reimbursed the displaced people, but such projects would then become considerably more expensive, and many might never be built. This, in turn, would reduce the comfort and convenience of those who now benefit from urban renewal and expressways. Finally, hippies could serve also as more deviants to justify the existing political economy—as they already do. Presumably, however, if poverty were eliminated, there would be fewer attacks on that economy.

In sum, then, many of the functions served by the poor could be replaced if poverty were eliminated, but almost always at higher costs to others, particularly more affluent others. Consequently, a functional analysis must conclude that poverty persists not only because it fulfills a number of positive functions but also because many of the functional alternatives to poverty would be quite dysfunctional for the affluent members of society. A functional analysis thus ultimately arrives at much the same conclusion as radical sociology, except that radical thinkers treat as manifest what I describe as latent: that social phenomena that are functional for affluent or powerful groups and dysfunctional for poor or powerless ones persist; that when the elimination of such phenomena through functional alternatives would generate dysfunctions for the affluent, or powerful, they will continue to persist; and that phenomena like poverty can be eliminated only when they become dysfunctional for the affluent or powerful, or when the powerless can obtain enough power to change society.

7

Why Can't People Feed Themselves?

FRANCES MOORE LAPPÉ AND
JOSEPH COLLINS

Question: You have said that the hunger problem is not the result of overpopulation. But you have not answered the most basic and simple question of all: Why can't people feed themselves? As Senator Daniel P. Moynihan put it bluntly, when addressing himself to the Third World, "Food growing is the first thing you do when you come down out of the trees. The question is, how come the United States can grow food and you can't?"

Our Response: In the very first speech I, Frances, ever gave after writing Diet for a Small Planet, *I tried to take my audience along the path that I had taken in attempting to understand why so many are hungry in this world. Here is the gist of that talk that was, in truth, a turning point in my life:*

When I started I saw a world divided into two parts: a *minority* of nations that had "taken off" through their agricultural and industrial revolutions to reach a level of unparalleled material abundance and a *majority* that remained behind in a primitive, traditional, undeveloped state. This lagging behind of the majority of the world's peoples must be due, I thought, to some internal deficiency or even to several of them. It seemed obvious that the underdeveloped countries must be deficient in natural resources—particularly good land and climate—and in cultural development, including modern attitudes conducive to work and progress.

But when looking for the historical roots of the predicament, I learned that my picture of these two separate worlds was quite false. My "two separate worlds" were really just different sides of the same coin. One side was on top largely because the other side was on the bottom. Could this be true? How were these separate worlds related?

Colonialism appeared to me to be the link. Colonialism destroyed the cultural patterns of production and exchange by which traditional societies in "underdeveloped" countries previously had met the needs of the people. Many precolonial social structures, while dominated by exploitative elites, had evolved a system of mutual obligations among the classes that helped to ensure at least a minimal diet for all. A friend of mine once said: "Precolonial village existence in subsistence agriculture was a limited life indeed, but it's certainly not Calcutta." The misery of starvation in the streets of Calcutta can only be understood as the end-point of a long historical process—one that has destroyed a traditional social system.

"Underdeveloped," instead of being an adjective that evokes the picture of a static society, became for me a verb (to "underdevelop") meaning the *process* by which the minority of the world has transformed—indeed often robbed and degraded—the majority.

That was in 1972. I clearly recall my thoughts on my return home. I had stated publicly for the first time a world view that had taken me years of study to grasp. The sense of relief was tremendous. For me the breakthrough lay in realizing that today's "hunger crisis" could not be described in static, descriptive terms. Hunger and underdevelopment must always be thought of as a *process*.

To answer the question "why hunger?" it is counterproductive to simply *describe* the conditions in an underdeveloped country today. For these conditions, whether they be the degree of malnutrition, the levels of agricultural production, or even the country's ecological endowment, are not static factors—they are not "givens." They are rather the *results* of an ongoing historical process. As we dug ever deeper into that historical process for the preparation of this book, we began to discover the existence of scarcity-creating mechanisms that we had only vaguely intuited before.

We have gotten great satisfaction from probing into the past since we recognized it is the only way to approach a solution to hunger today. We have come to see that it is the *force* creating the condition, not the condition itself, that must be the target of change. Otherwise we might change the condition today, only to find tomorrow that it has been recreated—with a vengeance.

Asking the question "Why can't people feed themselves?" carries a sense of bewilderment that there are so many people in the world not able to feed themselves adequately. What astonished us, however, is that there are not *more* people in the world who are hungry—considering the weight of the centuries of effort by the few to undermine the capacity of the majority to feed themselves. No, we are not crying "conspiracy!" If these forces were entirely conspiratorial, they would be easier to detect and many more people would by now have risen up to resist. We are talking about something more subtle and insidious; a heritage of colonial order in which people with the advantage of considerable power sought their own self-interest, often arrogantly believing they were acting in the interest of the people whose lives they were destroying.

THE COLONIAL MIND

The colonizer viewed agriculture in the subjugated lands as primitive and backward. Yet such a view contrasts sharply with the documents from the colonial period now coming to light. For example, A. J. Voelker, a British agricultural scientist assigned to India during the 1890s, wrote:

> Nowhere would one find better instances of keeping land scrupulously clean from weeds, of ingenuity in device of water-raising appliances, of knowledge of soils and their capabilities, as well as of the exact time to sow and reap, as one would find in Indian agriculture. It is wonderful, too, how much is known of rotation, the system of "mixed crops" and of fallowing. . . . I, at least, have never seen a more perfect picture of cultivation."[1]

None the less, viewing the agriculture of the vanquished as primitive and backward reinforced the colonizer's rationale for destroying it. To the colonizers of Africa, Asia, and Latin America, agriculture became merely a means to extract wealth—much as gold from a mine—on behalf of the colonizing power. Agriculture was no longer seen as a source of food for the local population, nor even as their livelihood. Indeed the English economist John Stuart Mill reasoned that colonies should not be thought of as civilizations or countries at all but as "agricultural establishments" whose sole purpose was to supply the "larger community to which they belong." The colonized society's agriculture was only a subdivision of the agricultural system of the metropolitan country. As Mill acknowledged, "Our West India colonies, for example, cannot be regarded as countries. . . . The West Indies are the place where England *finds it convenient* to carry on the production of sugar, coffee and a few other tropical commodities."[2]

Prior to European intervention, Africans practiced a diversified agriculture that included the introduction of new food plants of Asian or American origin. But colonial rule simplified this diversified production to single cash crops—often to the exclusion of staple foods—and in the process sowed the seeds of famine.[3] Rice farming once had been common in Gambia. But with colonial rule so much of the best land was taken over by peanuts (grown for the European market) that rice had to be imported to counter the mounting prospect of famine. Northern Ghana, once famous for its yams and other foodstuffs, was forced to concentrate solely on cocoa. Most of the Gold Coast thus became dependent on cocoa. Liberia was turned into a virtual plantation subsidiary of Firestone Tire and Rubber. Food production in Dahomey and southeast Nigeria was all but abandoned in favor of palm oil; Tanganyika (now Tanzania) was forced to focus on sisal and Uganda on cotton.

The same happened in Indochina. About the time of the American Civil War the French decided that the Mekong Delta in Vietnam would be ideal for producing rice for export. Through a production system based on enriching the large landowners, Vietnam became the world's third largest exporter of rice by the 1930s; yet many landless Vietnamese went hungry.[4]

Rather than helping the peasants, colonialism's public works programs only reinforced export crop production. British irrigation works built in nineteenth-century India did help increase production, but the expansion was for spring export crops at the expense of millets and legumes grown in the fall as the basic local food crops.

Because people living on the land do not easily go against their natural and adaptive drive to grow food for themselves, colonial powers had to force the production of cash crops. The first strategy was to use physical or economic force to get the local population to grow cash crops instead of food on their own plots and then turn them over to the colonizer for export. The second strategy was the direct takeover of the land by large-scale plantations growing crops for export.

FORCED PEASANT PRODUCTION

As Walter Rodney recounts in *How Europe Underdeveloped Africa*, cash crops were often grown literally under threat of guns and whips.[5] One visitor to the Sahel commented in 1928: "Cotton is an artificial crop and one the value of which is not entirely clear to the natives. . ." He wryly noted the "enforced enthusiasm with which the natives . . . have thrown themselves into . . . planting cotton."[6] The forced cultivation of cotton was a major grievance leading to the Maji Maji wars in Tanzania (then Tanganyika) and behind the nationalist revolt in Angola as late as 1960.[7]

Although raw force was used, taxation was the preferred colonial technique to force Africans to grow cash crops. The colonial administrations simply levied taxes on cattle, land, houses, and even the people themselves. Since the tax had to be paid in the coin of the realm, the peasants had either to grow crops to sell or to work on the plantations or in the mines of the Europeans.[8] Taxation was both an effective tool to "stimulate" cash cropping and a source of revenue that the colonial bureaucracy needed to enforce the system. To expand their production of export crops to pay the mounting taxes, peasant producers were forced to neglect the farming of food crops. In 1830, the Dutch administration in Java made the peasants an offer they could not refuse; if they would grow government-owned export crops on one fifth of their land, the Dutch would remit their land taxes.[9] If they refused and thus could not pay the taxes, they lost their land.

Marketing boards emerged in Africa in the 1930s as another technique for getting the profit from cash crop production by native producers into the hands of the colonial government and international firms. Purchases by the marketing boards were well below the world market price. Peanuts bought by the boards from peasant cultivators in West Africa were sold in Britain for more than *seven times* what the peasants received.[10]

The marketing board concept was born with the "cocoa hold-up" in the Gold Coast in 1937. Small cocoa farmers refused to sell to the large cocoa concerns like United Africa Company (a subsidiary Anglo-Dutch firm, Unilever—which we know as Lever Brothers) and Cadbury until they got a higher price. When the British government stepped in and agreed to buy the cocoa directly in place of the big business concerns, the smallholders must have thought they had scored at least a minor victory. But had they really? The following year the British formally set up the West African Cocoa Control Board. Theoretically, its purpose was to pay the peasants a reasonable price for their crops. In practice, however, the board, as sole purchaser, was able hold down the prices paid the peasants for their crops when the world prices were rising. Rodney sums up the real "victory":

> None of the benefits went to Africans, but rather to the British government itself and to the private companies. . . Big companies like the United African Company and John Holt were given . . . quotas to fulfill on behalf of the boards. As agents of the government, they were no longer exposed to direct attack, and their profits were secure.[11]

These marketing boards, set up for most export crops, were actually controlled by the companies. The chairman of the Cocoa Board was none other than John Cadbury of Cadbury Brothers (ever had a Cadbury chocolate bar?) who was part of a buying pool exploiting West African cocoa farmers.

The marketing boards funneled part of the profits from the exploitation of peasant producers indirectly into the royal treasury. While the Cocoa Board sold to the British Food Ministry at low prices, the ministry upped the price for British manufacturers, thus netting a profit as high as 11 million pounds in some years.[12]

These marketing boards of Africa were only the institutionalized rendition of what is the essence of colonialism—the extraction of wealth. While profits continued to accrue to foreign interests and local elites, prices received by those actually growing the commodities remained low.

PLANTATIONS

A second approach was direct takeover of the land either by the colonizing government or by private foreign interests. Previously self-provisioning farmers were forced to cultivate the plantation fields through either enslavement or economic coercion.

After the conquest of the Kandyan Kingdom (in present day Sri Lanka), in 1815, the British designated all the vast central part of the island as crown land. When it was determined that coffee, a profitable export crop, could be grown there, the Kandyan lands were sold off to British investors and planters at a mere five shillings per acre, the government even defraying the cost of surveying and road building.[13]

Java is also a prime example of a colonial government seizing territory and then putting it into private foreign hands. In 1870, the Dutch declared all uncultivated land—called waste land—property of the state for lease to Dutch plantation enterprises. In addition, the Agrarian Land Law of 1870 authorized foreign companies to lease village-owned land. The peasants, in chronic need of ready cash for taxes and foreign consumer goods, were only too willing to lease their land to the foreign companies for very modest sums and under terms dictated by the firms. Where land was still held communally, the village headman was tempted by high cash commissions offered by plantation companies. He would lease the village land even more cheaply than would the individual peasant or, as was frequently the case, sell out the entire village to the company.[14]

The introduction of the plantation meant the divorce of agriculture from nourishment, as the notion of food value was lost to the overriding claim of "market value" in international trade. Crops such as sugar, tobacco, and coffee were selected, not on the basis of how well they feed people, but for their high price value relative to their weight and bulk so that profit margins could be maintained even after the costs of shipping to Europe.

SUPPRESSING PEASANT FARMING

The stagnation and impoverishment of the peasant food-producing sector was not the mere by-product of benign neglect, that is, the unintended consequence of an overemphasis on export production. Plantations—just like modern "agro-industrial complexes"—needed an abundant and readily available supply of low-wage agricultural workers. Colonial administrations thus devised a variety of tactics, all to undercut self-provisioning agriculture and thus make rural populations dependent on plantation wages. Government services and even the most minimal infrastructure (access to water, roads, seeds, credit, pest and disease control information, and so on) were systematically denied. Plantations usurped most of the good land, either making much of the rural population landless or pushing them onto marginal soils. (Yet the plantations have often held much of their land idle simply to prevent the peasants from using it—even to this day. Del Monte owns 57,000 acres of Guatemala but plants only 9000. The rest lies idle except for a few thousand head of grazing cattle.)[15]

In some cases a colonial administration would go even further to guarantee itself a labor supply. In at least twelve countries in the eastern and southern parts of Africa the exploitation of mineral wealth

(gold, diamonds, and copper) and the establishment of cash-crop plantations demanded a continuous supply of low-cost labor. To assure this labor supply, colonial administrations simply expropriated the land of the African communities by violence and drove the people into small reserves.[16] With neither adequate land for their traditional slash-and-burn methods nor access to the means—tools, water, and fertilizer—to make continuous farming of such limited areas viable, the indigenous population could scarcely meet subsistence needs, much less produce surplus to sell in order to cover the colonial taxes. Hundreds of thousands of Africans were forced to become the cheap labor source so "needed" by the colonial plantations. Only by laboring on plantations and in the mines could they hope to pay the colonial taxes.

The tax scheme to produce reserves of cheap plantation and mining labor was particularly effective when the Great Depression hit and the bottom dropped out of cash crop economies. In 1929 the cotton market collapsed, leaving peasant cotton producers, such as those in Upper Volta, unable to pay their colonial taxes. More and more young people, in some years as many as 80,000, were thus forced to migrate to the Gold Coast to compete with each other for low-wage jobs on cocoa plantations.[17]

The forced migration of Africa's most able-bodied workers—stripping village food farming of needed hands—was a recurring feature of colonialism. As late as 1973 the Portuguese "exported" 400,000 Mozambican peasants to work in South Africa in exchange for gold deposited in the Lisbon treasury.

The many techniques of colonialism to undercut self-provisioning agriculture in order to ensure a cheap labor supply are no better illustrated than by the story of how, in the mid-nineteenth century, sugar plantation owners in British Guiana coped with the double blow of the emancipation of slaves and the crash in the world sugar market. The story is graphically recounted by Alan Adamson in *Sugar Without Slaves*.[18]

Would the ex-slaves be allowed to take over the plantation land and grow the food they needed? The planters, many ruined by the sugar slump, were determined they would not. The planter-dominated government devised several schemes for thwarting food self-sufficiency. The price of crown land was kept artificially high, and the purchase of land in parcels smaller than 100 acres was outlawed—two measures guaranteeing that newly organized ex-slave cooperatives could not hope to gain access to much land. The government also prohibited cultivation on as much as 400,000 acres—on the grounds of "uncertain property titles." Moreover, although many planters held part of their land out of sugar production due to the depressed world price, they would not allow any alternative production on them. They feared that once the ex-slaves started growing food it would be difficult to return them to sugar production when world market prices began to recover. In addition, the government taxed peasant production, then turned around and used the funds to subsidize the immigration of laborers from India and Malaysia to replace the freed slaves, thereby making sugar production again profitable for the planters. Finally, the government neglected the infrastructure for subsistence agriculture and denied credit for small farmers.

Perhaps the most insidious tactic to "lure" the peasant away from food production—and the one with profound historical consequences—was a policy of keeping the price of imported food low through the removal of tariffs and subsidies. The policy was double-edged: first, peasants were told they need not grow food because they could always buy it cheaply with their plantation wages; second, cheap food imports destroyed the market for domestic food and thereby impoverished local food producers.

Adamson relates how both the Governor of British Guiana and the Secretary for the Colonies Earl Grey favored low duties on imports in order to erode local food production and thereby release labor for the plantations. In 1851 the governor rushed through a reduction of the duty on cereals in order to "divert" labor to the sugar estates. As Adamson comments, "Without realizing it, he [the governor] had put his finger on the most mordant feature of monoculture: . . . its convulsive need to destroy any other sector of the economy which might compete for 'its' labor."[19]

Many colonial governments succeeded in establishing dependence on imported foodstuffs. In 1647 an observer in the West Indies wrote to Governor Winthrop of Massachusetts: "Men are so intent upon planting sugar that they had rather buy foode at very deare rates than produce it by labour, so infinite is the profitt of sugar workes . . ."[20] By 1770, the West Indies were importing most of the continental colonies' exports of dried fish, grain, beans, and vegetables. A dependence on imported food made the West Indian colonies vulnerable to any disruption in supply. This dependence on imported food stuffs spelled disaster when the thirteen continental colonies gained independence and food exports from the continent to the West Indies were interrupted. With no diversified food system to fall back on, 15,000 plantation workers died of famine between 1780 and 1787 in Jamaica alone.[21] The dependence of the West Indies on imported food persists to this day.

SUPPRESSING PEASANT COMPETITION

We have talked about the techniques by which indigenous populations were forced to cultivate cash crops. In some countries with large plantations, however, colonial governments found it necessary to *prevent* peasants from independently growing cash crops not out of concern for their welfare, but so that they would not compete with colonial interests growing the same crop. For peasant farmers, given a modicum of opportunity, proved themselves capable of outproducing the large plantations not only in terms of output per unit of land but, more important, in terms of capital cost per unit produced.

In the Dutch East Indies (Indonesia and Dutch New Guinea) colonial policy in the middle of the nineteenth century forbade the sugar refineries to buy sugar cane from indigenous growers and imposed a discriminatory tax on rubber produced by native smallholders.[22] A recent unpublished United Nations study of agricultural development in Africa concluded that large-scale agricultural operations owned and controlled by foreign commercial interests (such as the rubber plantations of Liberia, the sisal estates of Tanganyika [Tanzania], and the coffee estates of Angola) only survived the competition of peasant producers because "the authorities actively supported them by suppressing indigenous rural development."[23]

The suppression of indigenous agricultural development served the interests of the colonizing powers in two ways. Not only did it prevent direct competition from more efficient native producers of the same crops, but it also guaranteed a labor force to work on foreign-owned estates. Planters and foreign investors were not unaware that peasants who could survive economically by their own production would be under less pressure to sell their labor cheaply to the large estates.

The answer to the question, then, "Why can't people feed themselves?" must begin with an understanding of how colonialism actively prevented people from doing just that.

Colonialism

- forced peasants to replace food crops with cash crops that were then expropriated at very low rates;

- took over the best agricultural land for export crop plantations and then forced the most able-bodied workers to leave the village fields to work as slaves or for very low wages on plantations;

- encouraged a dependence on imported food;

- blocked native peasant cash crop production from competing with cash crops produced by settlers or foreign firms.

These are concrete examples of the development of underdevelopment that we should have perceived as such even as we read our history schoolbooks. Why didn't we? Somehow our schoolbooks always seemed to make the flow of history appear to have its own logic—as if it could not have been any other way. I, Frances, recall, in particular, a grade-school, social studies

pamphlet on the idyllic life of Pedro, a nine-year-old boy on a coffee plantation in South America. The drawings of lush vegetation and "exotic" huts made his life seem romantic indeed. Wasn't it natural and proper that South America should have plantations to supply my mother and father with coffee? Isn't that the way it was *meant* to be?

NOTES

[1] Radha Sinha, *Food and Poverty* (New York: Holmes and Meier, 1976), p. 26.

[2] John Stuart Mill, *Political Economy*, Book 3, Chapter 25 (emphasis added).

[3] Peter Feldman and David Lawrence, "Social and Economic Implications of the Large-Scale Introduction of New Varieties of Foodgrains," Africa Report, preliminary draft (Geneva: UNRISD, 1975), pp.107–108.

[4] Edgar Owens, *The Right Side of History*, unpublished manuscript, 1976.

[5] Walter Rodney, *How Europe Underdeveloped Africa* (London: Bogle-L'Ouverture Publications, 1972), pp. 171–172.

[6] Ferdinand Ossendowski, *Slaves of the Sun* (New York: Dutton, 1928), p. 276.

[7] Rodney, *How Europe Underdeveloped Africa*, pp. 171–172.

[8] Ibid., p. 181.

[9] Clifford Geertz, *Agricultural Involution* (Berkeley and Los Angeles: University of California Press, 1963), pp. 52–53.

[10] Rodney, *How Europe Underdeveloped Africa*, p. 185.

[11] Ibid., p. 184.

[12] Ibid., p. 186.

[13] George L. Beckford, *Persistent Poverty: Underdevelopment in Plantation Economies of the Third World* (New York: Oxford University Press, 1972), p. 99.

[14] Ibid., p. 99, quoting from Erich Jacoby, *Agrarian Unrest in Southeast Asia* (New York: Asia Publishing House, 1961), p. 66.

[15] Pat Flynn and Roger Burbach, North American Congress on Latin America, Berkeley, California, recent investigation.

[16] Feldman and Lawrence, "Social and Economic Implications," p. 103.

[17] Special Sahelian Office Report, Food and Agriculture Organization, March 28, 1974, pp. 88–89.

[18] Alan Adamson, *Sugar Without Slaves: The Political Economy of British Guiana, 1838–1904* (New Haven and London: Yale University Press, 1972).

[19]Ibid., p. 41.

[20]Eric Williams, *Capitalism and Slavery* (New York: Putnam, 1966), p. 110.

[21]Ibid., p. 121.

[22]Gunnar Myrdal, *Asian Drama*, vol. 1 (New York: Pantheon, 1966), pp. 448–449.

[23]Feldman and Lawrence, "Social and Economic Implications," p. 189.

8

Fair Trade: Buying and Selling for Justice

MEDEA BENJAMIN AND
ANDREA FREEMAN

I was in graduate school studying music in Wisconsin. During vacation my husband and I went on a trip to Peru. I'd only been out of the country once before, and what shocked me most about what we saw was how hard the women work. That's not to say the men don't work, but suddenly the old adage "Men work from sun to sun but women's work is never done" took on new meaning.

One day in Cuzco—which is this breathtaking site of the old Incan ruins—I suddenly had what I thought was the most brilliant idea in the world. I'd import some of the beautiful clothing and crafts the Peruvian women made, sell them in the United States, and send the money back down to the women. At the same time, the crafts would be an entree for me to talk to Americans about the lives of Third World women.

When I returned to the United States and did some homework, I discovered that there were already a number of groups doing the same thing in other countries. Tapping into their expertise and experience, in 1986 I set up the People's Exchange International.

JESSICA LINDNER

One day I heard about a woman who was importing Tanzanian coffee and sending the profits back to the

producers in Tanzania. It sounded like a terrific way to go beyond the charity approach to hunger by supporting fair trade. So I started going around to some of the larger churches here in Canada and encouraging them to do this sort of thing. They all said, "Good idea, but . . ." or "We'd love to Peter, but. . . ." After getting a heavy dose of "buts," my wife—who is herself a pastor of a small church—finally turned to me and said, "I guess that leaves us."

So in 1981 we turned our spare bedroom into an office, our garage into a warehouse, and our old beat-up station wagon into a delivery truck. By 1984 we were selling $500,000 worth of coffee, tea, cashews, and vanilla beans, and sending most of the money right back down to the producers. That's how Bridge-head Trading was born.

PETER DAVIES

I grew up in a small rural town in southern Indiana. I had a religious upbringing and was always sensitive to people's needs, but my first real exposure to poverty was during the college semester I spent in Kenya. That's where I learned not only the outrageous extent of human suffering but also the systemic injustice that perpetuates that suffering. The other thing I learned was the incredible power my country, the United States, has in the world. And this knowledge compelled me to come home and do something about it.

At first it never occurred to me that selling crafts could be a vehicle for venting my outrage. I thought of trade as exploitative and crafts as apolitical and indirect. I was more fired up about direct political actions like divestment and boycotts—actions I still wholeheartedly support. But I wanted to do something positive that would give Americans a greater understanding of struggles in the Third World. Working with Jubilee Crafts, I realized that for the people I wanted to give a voice to, crafts could be a powerful mode of speaking.

MELISSA MOYE

As a Catholic priest in Holland, I had a lot of ties with the Third World through the missionaries in our order. But I later left the priesthood and went to work with the Eduardo Mondlane Foundation, a Dutch group supporting the independence struggle of the Portuguese colonies of Mozambique, Angola, and Guinea Bissau in Africa.

When these groups received their independence in 1975, they turned to their former supporters for help in their even more daunting struggle—the struggle for economic independence. I didn't know the first thing about importing and exporting. I certainly was no businessman. But I also knew these countries were desperately poor and had few places to turn. Commercial enterprises were only interested in profits, not in helping these newly independent nations build up their war torn economies.

I was living in a houseboat in one of the Amsterdam canals at the time, and turned it into the headquarters of a non-profit import company called Stichting Ideele. Ten years later, with a staff of twenty, we were doing about $7 million worth of trade with Third World countries."

CARL GRASVELD

Jessica Lindner, Peter Davies, Melissa Moye, and Carl Grasveld are part of a growing movement for fair trade with the Third World. Their groups are known as Alternative Trade Organizations, or ATOs, and their goal is to create a more equitable system of trade with the Third World. For Third World producers, ATOs offer a rare opportunity to export in a non-exploitative fashion. It also offers them the confidence and sense of self-worth that comes from earning their money instead of relying on gifts or handouts. For consumers, ATOs provide an opportunity to support the poor in a way that goes far beyond charity, and to generate respect and understanding for people of other cultures.

ORIGINS OF THE ALTERNATIVE TRADE MOVEMENT

The international alternative trade movement is an amalgam of three distinct trends. One is church-related, another is an outgrowth of development efforts, and the third is based on political criteria.

In the United States, the church-related groups are the oldest and largest. Two of the major groups are SELFHELP Crafts, which is affiliated with the Mennonite Church, and SERRV Self-Help Handcrafts, which is run by the Church of the Brethren. Both began buying crafts from European war victims after World War II and later turned their attention to the Third World. In 1987 SELFHELP operated 110 retail stores and SERRV was selling its goods to some fifteen hundred churches around the country.

ATOs with a development focus began in Europe in the 1960s. The first Third World shops—stores selling goods produced by Third World cooperatives—were started by the Dutch branch of the development group Oxfam. They later spread throughout Europe, Australia, and New Zealand. By 1987, Oxfam-UK, for example, was selling $14 million worth of Third World goods in five hundred retail stores. And Switzerland, a country with only six million inhabitants, has more than five hundred Third World shops.

Jim Goetsch discovered these Third World shops while visiting Europe and brought the idea back to the United States. "It seemed like a great way to not only support Third World producers," Jim explains, "but also to begin creating a clearer understanding among the American people of the need for basic structural change in the world's marketplace." In 1970 Jim founded Friends of the Third World in Fort Wayne, Indiana and later helped start other stores around the country.

The third trend in ATOs, the political support groups, also have strong roots in Europe. One of the main groups in this category is Stichting Ideele in Holland. Stichting differs from both the development and church groups in that it does not work directly with producers but with governments. "We work with some of the poorest countries in the world," director Carl Grasveld explains. "They are poor for many reasons, but one is that trade has tended to work against them. For many, trying to make gains from trade is like trying to walk up a 'down' escalator. There is, however, usually no alternative to trade—apart from permanent dependence on loans and aid, and thus permanent vulnerability. Therefore, a number of countries are trying to put their commercial relations on a new footing by seeking new partners and a better deal."

Before working with a government, Stichting looks carefully at its policies to see if they are aimed at lessening inequalities within its society and providing its citizens with decent education, health care, and jobs. Based on these criteria, Stichting works with such countries as Zimbabwe, Mozambique, Angola, Cuba, Nicaragua, and Vietnam. Stichting has worked closely with a number of parallel groups in North America, including Bridgehead Trading in Canada and Equal Exchange in Cambridge, Massachusetts.

These three distinct trends in the ATO movement are by no means mutually exclusive. Pueblo to People, a Houston-based group selling food and crafts from Central America, focuses on both development and politics. It works with cooperatives to help them develop technical skills, but it also sees itself as part of a broader movement for social change. Jubilee Crafts, based in Philadelphia, combines the religious and political aspects. It is firmly faith-based, but it is also one of the most politically outspoken trade groups in the United States.

Compared to Europe, the alternative trade movement in the United States is in its infancy.

"The European movement is much more developed, but it isn't necessarily a model for us," Rink Dickinson of Equal Exchange explains. "Europe has a long social democratic history, and ATOs can often get substantial institutional support. Stichting Ideele in Holland started with a big loan from a union. Traidcraft in Great Britain, one of the largest ATOs with a staff of over one hundred, benefits from government subsidies. Unfortunately, we haven't been able to get that kind of government or union support here in the United States.

"What's different about the United States, and ironically this works to our advantage, is that our government is directly involved in so many places around the world—but not usually on the side of bringing greater economic and political justice to the poor. More and more Americans are learning about that. Thousands of people have been to Nicaragua, for example, and now drink Nicaraguan coffee as a statement. There are a lot of people who are looking for creative ways to support Third World economies. That's a big challenge, but we're excited about being able to take up our small part of that challenge."

WHAT DISTINGUISHES ATOS FROM COMMERCIAL IMPORTERS?

While the philosophies and strategies of the alternative trade groups are quite varied, they share a number of characteristics which distinguish them from commercial importers:

- Unlike other businesses, their goal is to benefit the poor, not to maximize profits.

    Most ATOs struggle to keep expenses and salaries as low as possible, so they can return the bulk of the money to the Third World. Staff members of Friends of the Third World go so far as to live on incomes of under $2,000 a year so they don't have to pay taxes that would support the military. Members of Pueblo to People work as a collective, earning less than $1,000 a month. "While we hope to increase our salaries as the business grows, we certainly didn't go into this work to get rich," said Jim McClure, who took a $15,000 pay cut to join Pueblo to People. Groups such as SELFHELP Crafts and SERRV keep their costs low by relying heavily on volunteer staff.

By keeping expenses low, most ATOs are able to pay their producers substantially more than commercial importers pay. Pueblo to People returns an average of 40 percent of the retail price to the producers, compared to about 10 percent for commercial retailers. Some ATOs buy from their producers at the going market rate, but then channel their profits into development projects in the producers' country or into educational activities in their own country.

ATOs also try to ensure that much of the final processing of the product is done in the Third World itself, so more money will remain in the producer country. Commercial importers usually reserve the labor-intensive activities for the Third World where labor is cheap, while the processing and packaging are done in the industrialized countries. In the case of cashews, for example, many U.S. companies import raw cashews from India, and then roast, process, and package them in the United States. In contrast, Pueblo to People helps villagers learn to do the processing and packaging themselves so they can double the value of the cashews. New Wind in Finland developed a simple tea bag machine so the Tanzanians it works with can export the higher-priced tea bags instead of loose tea.

■ ATOs see the education of consumers as an essential part of their work.

Groups that sell through mail order often use their catalogues to explain where and under what conditions the products are made. Some groups, like Pueblo to People, explain the political conditions within the countries from which it buys. Jubilee Crafts goes even further, using its catalogue to call for a change in U.S. government and corporate policies that hurt the poor. "We want to present the crafts in such a way," the catalogue states, "that they might become a stimulus for Americans to begin thinking about the role that our life-styles and government play in keeping many people poor and powerless."

Some wholesale distributors put a tag on the product itself, explaining where and how it is made. The Boston-based group, Equal Exchange, which is a wholesale distributor of Nicaraguan coffee, sends out educational material on Nicaragua with every order.

Many of the ATOs sell at fairs, churches, and schools, and use these as prime opportunities to do educational work. "We see every sale as a chance to talk to people about what we're doing and why," says Melissa Moye of Jubilee Crafts. "Say I'm selling crafts to a women's group at a small rural church. The women are usually fascinated by the Guatemalan weavings, since lots of them do embroidering and quilting themselves and appreciate the manual dexterity. They're often amazed that people in these 'uncivilized' countries can do such fine work! From there I try to explain the conditions under which the Guatemalan women work. I explain how many of them are the sole supporters of their children because their husbands have been killed by the military. I explain how many women can't do this kind of weaving anymore because they have been pushed into 'model villages' where traditional activities are considered subversive. For us, educating the U.S. public is certainly as important as selling the craft."

Not all groups, however, take their educational role this far. "For some of the larger church-based ATOs," Melissa explains, "it is difficult to have an educational component that goes beyond an explanation of who makes the products and under what conditions. They have a broad-based constituency and can't afford to alienate anyone. While we are church-based, we are not church-funded, which gives us a lot more latitude in our educational work. We can—

and do—discuss the political situation in the producer countries, the role of the U.S. government there, and what we should be doing to make things better."

■ ATOS often work with producers shunned by commercial distributors.

"The people we work with are often the poorest, the least educated, the least skilled in things like keeping books and quality control," explains Jim McClure of Pueblo to People. "On top of that, we work in areas of tremendous social upheaval. The people we're working with in El Salvador and Guatemala, for example, could be killed tomorrow by their militaries. Who would want to operate under such conditions? It's certainly not the best atmosphere for maximizing profits, but it's under such conditions that you find the people in most need."

ATOs also tend to be more sensitive to the needs and problems of producer groups they work with. Randy Gibson of SERRV says that unlike commercial importers, ATOs work with producers to iron out problems rather than cutting them off if they make mistakes. "Say we start working with a new group in India that doesn't have experience in packing and shipping their goods. Well, suppose they did a lousy packing job and when the first shipment arrives, half the crafts are broken. A commercial buyer wouldn't speak to them again. We'd subtract the losses from their inventory, but we'd give them another chance. That's the difference."

Carl Grasveld of Stichting Ideele gives another example of how ATOs are willing to work through problems with the producers. "Before independence from Portugal, Mozambique was a large exporter of sesame seeds. But when the Portuguese owners fled the country, the machinery began to break down and none of the Mozambican workers had been trained to fix it. The seeds were not cleaned properly, and the commercial distributors refused to buy them anymore. So we began to buy the seeds and clean them in

Europe. At the same time we helped the Mozambicans revamp the machines and train the workers to maintain them."

ATOs also work to strengthen the marketing capacity of their client groups, and they urge the groups to market through as many channels as possible. "Unlike commercial distributors," says Paul Leatherman of SELFHELP Crafts, "we don't care who else the groups work with. In fact, we encourage and help them to find other marketing outlets—other nonprofits or even commercial outlets. We also encourage them to sell as much of their products domestically as they can. We certainly don't want the groups to be dependent on us as their sole outlet."

• • •

WHAT IMPACT DO ATOS HAVE?

Total sales of the international ATO movement amounted to about $75 million in 1987, some $8 million of which were U.S. sales. Given that this represents only a tiny portion of Third World imports of food and crafts, do ATOs really have much of an impact?

The impact of ATOs goes beyond the dollar figures. "First of all," explained SELFHELP Director, Paul Leatherman, "we work with groups that are too poor and unsophisticated to have access to commercial buyers. So while we may not be reaching large numbers of people, we're reaching some of the neediest. Secondly, by buying directly from the villagers and eliminating the local middleman, we often have a ripple effect by raising the price the middlemen pay. I've seen this happen in Saharanpur, in northern India, where our purchases of carved wood at 10 to 20 percent above the local price forced the middlemen to up their prices, too. So our influence is greater than the figures suggest."

Pueblo to People has witnessed the same effects in Honduras. "We're now buying up 30 percent of Honduras' cashew crop," says Jim

McClure. "So the remaining 70 percent also goes up in value. And when we go into a Guatemalan village and buy a few hundred pieces of clothing, the producers are in a better position to demand more from other buyers."

ATOs have also seen that increasing people's incomes by marginal amounts can have a dramatic effect on their lives. In Guatemala, Pueblo to People is starting to foster the production of coffee and herbal teas—hibiscus, lemon grass, and chamomile. It's working in an area where farmers can't make enough money on their own small plots, so are forced to migrate to the coast every year to pick cotton and coffee, for low pay and under horrendous conditions. It is hoped that the extra income the farmers will gain from producing herbal teas for the ATOs will provide an alternative to migrant labor.

While ATOs are certain of having helped thousands upon thousands of Third World producers, they are also painfully aware of how much more they could be doing. In the past, many ATOs have been plagued by an identity crisis. While they were small businesses, they shunned the business world. But to really grow and help the greatest number of producers, they had to become more professional and "hard-nosed" in their dealings.

Equal Exchange, an example of one of the up and coming ATOs in the United States, feels strongly about the need to be more businesslike. "The three of us who founded Equal Exchange previously worked at a cooperative food warehouse that distributed food to over 450 co-ops in the Northeast," Jonathan Rosenthal explains. "Experience taught us the dangers of being lax about the business end. We may not be out to get rich, but we are out to run a successful business. And the only way to do that is to be aggressive about getting capital, about marketing, about advertising."

Stichting Ideele also sees the need for expanding and   professionalizing the ATO movement.

Stichting itself is anxiously probing the possibility of entering into joint ventures with progressive governments. To get access to new technologies and markets, many Third World governments have to rely on multi-national corporations, or, in rare cases, joint ventures with multi-nationals. But ATOs could forge a model for a new type of joint venture that would look out for the best interests of the producer country. Such ventures would enormously increase the impact of ATOs.

Stichting has begun such an arrangement with the government of Cape Verde in West Africa. Although it is an island, Cape Verde had no fishing industry. Stichting and the government of Cape Verde have started a tuna business in which the government owns 51 percent and Stichting 49 percent. Unlike relationships in which the multi-nationals call all the shots, this type of joint venture is based on equality. The contract states that in ten to fifteen years Stichting will sell its shares to Cape Verde at their present value plus inflation. It also states that all profits are to be reinvested in the business. The three-person board of directors is comprised of one government representative, one representative from Stichting, and one representative of the Cape Verde workers. It is a truly unique experiment that may open up an entirely new role for ATOs.

Stichting also feels that ATOs could significantly aid Third World countries in developing and marketing new products. "In Cape Verde and Angola we are helping to develop the fishing industry," says Carl. "In Cuba we are helping to link some products—from rum to candy to lobsters—with commercial distributors in Europe. In Mozambique we are designing new products like sesame and cashew butter. If poor Third World countries are ever to break out of the dependency of a few unstable products, we must assist them in finding new ones."

3

Democracy and Human Rights

Every day in those dark times, we asked ourselves,
what is democracy, if people are starving?
How can you trust a vote when a man
will vote for whomever gives him the money
to feed his children that night?

JEAN BERTRAND ARISTIDE

The history of the modern world is filled with movements for democracy and human rights. From the historic French Revolution to today's Chinese Pro-Democracy movement, people have sought to establish systems of self-government, and to be free from the threat of unlawful imprisonment, torture, murder, and other deprivations of human rights. Do democracy and human rights always accompany capitalist economic growth? Many advocates of the modernization perspective have argued that this is the case. These hopes were buoyed by the downfall of socialism in the Soviet Union and Eastern Europe. In the early 1990s, many scholars and politicians predicted a new world order of increasingly affluent capitalist democracies and the reduction of human rights abuses. At the time of this writing, this does not appear to be the way things are headed. Research indicates that economic growth and globalization have not appreciably diminished human rights abuses (Smith, Bolyard, and Ippolito, 1998). The chapter begins with one example of the many violations of human rights that occur regularly around the world. In "376 Killed. Did World Bank Know?" the *Earth Island Journal* reports on a little-known series of murders of peasants in Guatemala from 1980 to 1982. It is argued that the murders were carried out by the military and paramilitary death squads when the peasants protested being reset-

tled to make way for an economic development project funded by the World Bank.

The terms *democracy* and *human rights* have been defined in various ways. In general, there are two different visions of democracy. Democracy has been defined as *representative democracy* or as *participatory democracy*. Representative democracy refers to a form of government in which citizens vote in free elections for representatives to govern the nation. This notion of democracy is the one crafted in the eighteenth-century democratic revolutions, and is usually what spokespeople for contemporary governments mean when they say democracy. Participatory democracy refers to a more active form of democracy in which there is ongoing citizen participation in governing. The idea of participatory democracy took shape in grassroots movements that emerged alongside the representative democratic states that developed in the eighteenth and nineteenth centuries. It has flourished in the twentieth century, and is often what the spokespeople for "bottom up" movements mean when they say democracy. There are also two different conceptions of human rights. Human rights have been defined as *civil and political rights* or as *social and economic rights*. The idea of human rights as *civil and political rights* encompasses rights such as freedom of speech, freedom of association, freedom of the press, the right to vote, to be free of slavery, and so on. These are the type of rights familiar to most Americans as the rights guaranteed in the U.S. Constitution. This idea of human rights accompanied the development of representative democracies. The idea of human rights as *social and economic rights* includes such rights as the right to education, employment, or health care, a conception of rights that goes far beyond the more narrow idea of civil and political rights. This idea of human rights accompanied the rise of "bottom up" movements of workers and consumers, and has been championed by liberal and socialist parties and governments. Two articles in this chapter examine the meaning and implication of diverse conceptions of democracy and human rights. The first is the 1948 "United Nations Universal Declaration of Human Rights." This visionary document encompasses both the narrower and broader conceptions of human rights, reflecting the clash of views within the United Nations. This reading is followed by "The Domestic Scene" by Noam Chomsky. In this article, written from the conflict perspective, Chomsky examines the state of democracy in the United States and distinguishes between *political democracy* and *economic democracy*.

What accounts for the lack of democracy and deprivation of human rights in so much of the world today, more than 50 years since the United Nations Universal Declaration of Human Rights? Many governments around the world are simply opposed to democracy, and do not accept the idea that there are universal human rights. For example, as this book was being prepared for publication, the world witnessed the vicious "ethnic cleansing" of ethnic Albanian citizens of Kosovo by the Yugoslavian government. And although the world has seen a rise in democracies and a decline in dictatorships during the 1990s, many of the new democratic governments are democratic in name only. In some of the so-called democracies, there is widespread violence by government and private bodies against the poor and powerless. From the conflict perspective, the progress of democracy and human rights appears shaky. It is difficult for democracy to flourish, given the extremes of

economic and social inequality that are created by global capitalism. Extreme inequalities militate against effective citizen participation, and globalization is exacerbating such inequalities. It can be argued that even the United States and other representative democracies are not fully democratic. This is because of the enormous economic power exercised by corporations and the rich, as Chomsky argues in his article mentioned above, "The Domestic Scene." From the modernization perspective, it appears that there is a brighter future for democracy and human rights. Democracy will continue to develop, as societies everywhere undergo cultural change in the wake of economic change. Ideas opposed to democracy and human rights will give way to a more modern system of values. On an optimistic note, in "Who Rules Now?" Stanley Rothman and Amy E. Black review recent research into power in the United States, and conclude that power is increasingly widely shared. Benjamin Barber draws on *both* modernization and conflict perspectives in "Jihad vs. McWorld." He argues that both the capitalist world market and ethnocentric cultural values are an ongoing threat to democracy and human rights. Barber concludes his article with some interesting speculations on the vision of democracy that activists need to have in order to effectively combat the undemocratic forces in the contemporary world.

Grassroots organizations and the United Nations have been especially active in focusing attention on the problem of the lack of human rights and democracy. The last article in this chapter provides a glimpse into the variety of activities underway in relation to various dimensions of the problem. In "States of Bondage," Manfred Nowak reports on progress being made in United Nations efforts to eradicate torture.

WORKS CITED

Smith, Jackie, Melissa Bolyard, and Anna Ippolito. 1998. "World System vs. World Culture: Globalization and Human Rights." Paper presented at Annual Meeting of the American Sociological Association, 21–25 August, 1998. San Francisco, CA.

QUESTIONS FOR DISCUSSION

1.   Which rights listed in the United Nations Universal Declaration of Human Rights do you believe we have in the United States? Which rights do we lack or have only partially? Why?

2.   What does Chomsky mean by "political democracy" and "economic democracy"? How does Chomsky characterize the media's role in relation to democracy in the United States? Do you agree with his analysis? Why or why not?

3.   Rothman and Black argue that power in America is increasingly widely shared. What are some of their reasons? Do you agree? Explain.

4.   What does Barber mean by "Jihad" and "McWorld"? How do each of these patterns undermine democracy, in his opinion? What does Barber think should be done to solve the problem? Do you agree with him? Why or why not?

5.   What are some of the controversies making it difficult to arrive at an international definition of torture, according to Nowak in "States of Bondage"? What steps has the United Nations taken in order to prevent and punish torture? How effective do these seem, in your opinion?

 INFOTRAC COLLEGE EDITION: EXERCISE

One of the readings in this chapter is the United Nations Universal Declaration of Human Rights. Do United Nations declarations such as this have a practical impact? That is, has this document been helpful to people around the world who are struggling to prevent human rights abuses? Or is it just words on paper? Using InfoTrac College Edition, look up the following article:

"An Available Instrument of Subversion." Archbishop Desmond Tutu. *UNChronicle*. Winter, 1998 v35i4p16.

(*Hint*: Enter the search terms human rights and Africa, using the Subject Guide.)

How did the United Nations Declaration affect the struggles of Africans in South Africa against *apartheid*? Do you believe this document can be of assistance to citizens of representative democracies, such as the United States? If not, why not? If so, which aspects of the document do you see as most relevant to the United States?

FOR ADDITIONAL RESEARCH

Books

An-Na'im, Abdullahi Ahmed, ed. 1992. *Human Rights in Cross-Cultural Perspectives: A Quest for Consensus*. Philadelphia: University of Pennsylvania Press.

Human Rights Watch. 1997. *Human Rights Watch World Report 1998*. New York: Human Rights Watch.

Mandela, Nelson. 1994. *Long Walk to Freedom*. New York: Little Brown and Company.

Organizations

American Civil Liberties Union
125 Broad Street, 18th Floor
New York, NY 10004
(212) 549-2500
www.aclu.org

Amnesty International USA
322 8th Avenue
New York, NY 10001
(212) 807-8400
www.amnesty.org

Human Rights Documentation Exchange
P.O. Box 2327
Austin, TX 78768
(512) 476-9841

Human Rights Watch
485 Fifth Avenue
New York, NY 10017
(212) 972-8400
www.hrw.org

ACTION PROJECTS

1.   Learn more about human rights violations around the world by contacting one of the first three organizations on the list above. What are its goals, activities, and accomplishments? Also identify a specific project of the organization that interests you and get more in-depth information on it. (For example, Human Rights Watch has an "Africa Watch" project, and under that, projects for specific African nations; the organization also has a prison project, a women's rights project, and so on.)

2.   The human rights organizations on the list above usually sponsor campaigns against human rights abuses. For example, Amnesty International often organizes campaigns to publicize the situation of human rights activists and others who are subjected to abuses in prison in various nations. Contact one of the organizations and identify a campaign you feel is worth supporting. Organize a group on campus to take part in the cam-

paign. Reflect on what you learned from this experience about the issues of democracy and human rights.

3.   If you attend a large university, there may be foreign or immigrant students or faculty who were active in a democracy movement or sympathetic to it in their nation of origin. Locate such a person and interview them to learn about their views on democracy and human rights. Why did they join or sympathize with the movement? What activities did they take part in? How did this involvement affect their decision to emigrate to the United States? Have they modified their views since coming to the United States? Are they still involved in the movement? What, in their view, are the prospects for the movement? What did you expect to learn *before* you did the interview? Did the interview surprise you in any way—was it what you expected?

9

376 Killed. Did the World Bank Know?

EARTH ISLAND JOURNAL

An investigation by Witness for Peace reveals some 376 Guatemalan people—primarily women and children—were killed in a series of massacres, between 1980 and 1982, after they resisted eviction from the Rio Negro village to make way for the Chixoy Reservoir.

The World Bank and Inter-American Development Bank (IDB) both provided loans for the Chixoy Dam. The World Bank made its second loan in 1985—after the murders took place. Internal reports from the World Bank and IDB refer to resettlement problems at Chixoy, but nowhere mention that more than one-in-ten of the villagers slated for relocation were killed shortly before the reservoir filled.

Fearful survivors first began to speak out in 1993. In November 1993, forensic experts began exhuming bodies.

The Witness for Peace report called the World Bank's role "at best a calculated cover up, and at worst an act of complicity in the violence."

The campaign of terror against Rio Negro's indigenous Maya Achi community began in early 1980, when villagers refused to move to the cramped housing and poor land at the resettlement site provided by the Guatemalan power utility INDE.

In March 1980, military police from the dam site shot seven people. In July, two Rio Negro representatives agreed to an INDE request to visit the dam site to present their resettlement documents. The villagers' mutilated bodies were discovered one week later.

In February 1982, the local military commander ordered 73 Rio Negro men and women to report to Xococ, a village upstream from the reservoir. Only one woman returned. According to Witness for Peace, the rest were raped, tortured and then murdered by Xococ's Civil Defense Patrol or PAC—one of the state's notorious paramilitary death squads.

On March 13, 1982, ten soldiers and 25 patrollers arrived in Rio Negro, rounded up the remaining women and children and marched them to a hill above the village. The women were then beaten with clubs and rifles and strangled with ropes and sticks. Survivors told Witness for Peace how patrollers "killed the children by tying ropes around their ankles and swinging them, smashing their heads and bodies into rocks and trees."

The death toll amounted to 70 women and 107 children. Only two women managed to escape. Patrollers took 18 children to Xococ as slaves.

In May, 82 more Rio Negro villagers were slaughtered. In September, soldiers machine-gunned and burned-to-death 92 people—including 35 orphaned Rio Negro children—in another village near the dam. INDE began to fill the reservoir soon after this final massacre.

Many villagers believe INDE encouraged the violence to allow its officials to pocket compensation payments due relocated villagers.

The IDB lent Guatemala $105 million to build Chixoy in 1975 and another $70 million in 1981. The World Bank lent $72 million for the dam in 1978 and another $45 million in 1985. The local people claim that everyone at the dam site knew about the massacres.

The closest the World Bank's confidential 1991 "Project Completion Report" on Chixoy came to mentioning the slaughter is a reference to the resettlement plans as "conceptually . . . seriously flawed."

Chixoy's geological problems, combined with internal corruption, has caused the dam's total cost to soar to some $1.2 billion—521 percent higher than forecast in 1974. The dam began official operation in 1983, but closed for repairs after only five months. It did not restart operation for two years and continues to suffer technical problems. The 1991 project report on the dam concluded: "With hindsight, it has proved to be an unwise and uneconomic disaster."

10

The United Nations Universal Declaration of Human Rights

Whereas recognition of the inherent dignity and of the equal and inalienable rights of all members of the human family is the foundation of freedom, justice and peace in the world,

Whereas disregard and contempt for human rights have resulted in barbarous acts which have outraged the conscience of mankind, and the advent of a world in which human beings shall enjoy freedom of speech and belief and freedom from fear and want has been proclaimed as the highest aspiration of the common people,

Whereas it is essential, if man is not to be compelled to have recourse, as a last resort, to rebellion against tyranny and oppression, that human rights should be protected by the rule of law,

Whereas it is essential to promote the development of friendly relations between nations,

Whereas the peoples of the United Nations have in the Charter reaffirmed their faith in fundamental human rights, in the dignity and worth of the human person and in the equal rights of men and women and have determined to promote social progress and better standards of life in larger freedom,

Whereas Member States have pledged themselves to achieve, in co-operation with the United Nations, the promotion of universal respect for and observance of human rights and fundamental freedoms,

Whereas a common understanding of these rights and freedoms is of the greatest importance for the full realization of this pledge,

Now, therefore, the General Assembly:

Proclaims this Universal Declaration of Human Rights as a common standard of achievement for all peoples and all nations, to the end that every individual and every organ of society, keeping this Declaration constantly in mind, shall strive by teaching and education to promote respect for these rights and freedoms and by progressive measures, national and international, to secure their universal and effective recognition and observance, both among the peoples of Member States themselves and among the peoples of territories under their jurisdiction.

Article 1: All human beings are born free and equal in dignity and rights. They are endowed with reason and conscience and should act towards one another in a spirit of brotherhood.

Article 2: Everyone is entitled to all the rights and freedoms set forth in this Declaration, without distinction of any kind, such as race, color, sex, language, religion, political or other opinion, national and social origin, property, birth or other status.

Furthermore, no distinction shall be made on the basis of the political, jurisdictional or international status of the country or territory to which a person belongs, whether it be independent, trust, non-self-governing or under any other limitation of sovereignty.

Article 3: Everyone has the right to life, liberty and security of person.

Article 4: No one shall be held in slavery or servitude; slavery and the slave trade shall be prohibited in all their forms.

Article 5: No one shall be subjected to torture or to cruel, inhuman or degrading treatment or punishment.

Article 6: Everyone has the right to recognition everywhere as a person before the law.

Article 7: All are equal before the law and are entitled without any discrimination to equal protection of the law. All are entitled to equal protection against any discrimination in violation of this Declaration and against any incitement to such discrimination.

Article 8: Everyone has the right to an effective remedy by the competent national tribunals for acts violating the fundamental rights granted him by the constitution or by law.

Article 9: No one shall be subjected to arbitrary arrest, detention or exile.

Article 10: Everyone is entitled in full equality to a fair and public hearing by an independent and impartial tribunal, in the determination of his rights and obligations and of any criminal charge against him.

Article 11: Everyone charged with a penal offence has the right to be presumed innocent until proved guilty according to law in a public trial at which he has had all the guarantees necessary for his defence. No one shall be held guilty of any penal offence on account of any act or omission which did not constitute a penal offence, under national or international law, at the time when it was committed. Nor shall a heavier penalty be imposed than the one that was applicable at the time the penal offence was committed.

Article 12: No one shall be subjected to arbitrary interference with his privacy, family, home or correspondence, nor to attacks upon his honors and reputation. Everyone has the right to the protection of the law against such interference or attacks.

Article 13: Everyone has the right to freedom of movement and residence within the borders of each State. Everyone has the right to leave any country, including his own, and to return to his country.

Article 14: Everyone has the right to seek and enjoy in other countries asylum from prosecution. This right may not be invoked in the case of prosecutions genuinely arising from non-political crimes or from acts contrary to the purposes and principles of the United Nations.

Article 15: Everyone has the right to a nationality. No one shall be arbitrarily deprived of his nationality nor denied the right to change his nationality.

Article 16: Men and women of full age, without any limitation due to race, nationality or religion, have the right to marry and to found a family. They are entitled to equal rights as to marriage, during marriage and at its dissolution. Marriage shall be entered into only with the free and full consent of the intending spouses. The family is the natural and fundamental group unit of society and is entitled to protection by society and the State.

Article 17: Everyone has the right to own property alone as well as in association with others. No one shall be arbitrarily deprived of his property.

Article 18: Everyone has the right to freedom of thought, conscience and religion; this right includes freedom to change his religion or belief, either alone or in community with others and in public or private, to manifest his religion or belief in teaching, practice, worship and observance.

Article 19: Everyone has the right to freedom of opinion and expression; this right includes freedom to hold opinions without interference and to seek, receive and impart information and ideas through any media and regardless of frontiers.

Article 20: Everyone has the right to freedom of peaceful assembly and association. No one may be compelled to belong to an association.

Article 21: Everyone has the right to take part in the government of his country, directly or through freely chosen representatives. Everyone has the right to equal access to public service in his country. The will of the people shall be the basis of the authority of government; this will shall be expressed in periodic and genuine elections which shall be by universal and equal suffrage and shall be held by secret vote or by equivalent free voting procedures.

Article 22: Everyone, as a member of society, has the right to social security and is entitled to realization, through national effort and international co-operation and in accordance with the organization and resources of each State, of the economic, social and cultural rights indispensable for his dignity and the free development of his personality.

Article 23: Everyone has the right to work, to free choice of employment, to just and favorable conditions of work and to protection against unemployment. Everyone, without any discrimination, has the right to equal pay for equal work. Everyone who works has the right to just and favorable remuneration ensuring for himself and his family an existence worthy of human dignity, and supplemented, if necessary, by other means of social protection. Everyone has the right to form and to join trade unions for the protection of his interests.

Article 24: Everyone has the right to rest and leisure, including reasonable limitation of working hours and periodic holidays with pay.

Article 25: Everyone has the right to a standard of living adequate for the health and well-being of himself and of his family, including food, clothing, housing and medical care and necessary social services, and the right to security in the event of unemployment, sickness, disability, widowhood, old age or other lack of livelihood in circumstances beyond his control. Motherhood and childhood are entitled to special care and assistance. All children, whether born in or out of wedlock, shall enjoy the same social protection.

Article 26: Everyone has the right to education. Education shall be free, at least in the elementary and fundamental stages. Elementary education shall be compulsory. Technical and professional education shall be made generally available and higher education shall be equally accessible to all on the basis of merit. Education shall be directed to the full development of the human personality and to the strengthening of respect for human rights and fundamental freedoms. It shall promote understanding, tolerance and friendship among all nations, racial or religious groups, and shall further the activities of the United Nations for the maintenance of peace. Parents have a prior right to choose the kind of education that shall be given to their children.

Article 27: Everyone has the right freely to participate in the cultural life of the community, to enjoy the arts and to share in scientific advancement and its benefits. Everyone has the right to the protection of the moral and material interests resulting from any scientific, literary or artistic production of which he is the author.

Article 28: Everyone is entitled to a social and international order in which the rights and freedoms set forth in this Declaration can be fully realized.

Article 29: Everyone has duties to the community in which alone the free and full development of his personality is possible. In the exercise of his rights and freedoms, everyone shall be subject only to such limitations as are determined by law solely for the purpose of securing due recognition and respect for the rights and freedoms of others and of meeting the just requirements of morality, public order and the general welfare in a democratic society. These rights and freedoms may in no case be exercised contrary to the purposes and principles of the United Nations.

Article 30: Nothing in this Declaration may be interpreted as implying for any State, group or person any right to engage in any activity or to perform any act aimed at the destruction of any of the rights and freedoms set forth herein.

11

The Domestic Scene

NOAM CHOMSKY

I would like to conclude these lectures with some comments about U.S. society itself, asking how state policies are fashioned and what possibilities there are of modifying them. The basic question reduces to this: To what extent is the United States a democratic society, in which the general population is able to influence public policy? There is no simple answer to this question. It is one that has many dimensions. Let us consider a few of these.

One crucial dimension in terms of which one can evaluate the democratic credentials of some political system has to do with the power of the state to coerce its citizens and protect itself from their scrutiny and control, its power to prevent free expression and free association, to maintain state secrets and conduct its affairs without public awareness and influence. Such questions were vigorously debated in the early years of the Republic after the U.S. War of Independence. If the people are sovereign, libertarians argued, then the state must be subordinated to them, not conversely. If, for example, legislators have the constitutional right of free expression with immunity from prosecution, then citizens should have no less a right: specifically, they should be free to condemn the government and its practices without fear of prosecution for "seditious libel," the doctrine that the state can be criminally assaulted by mere speech and writing, short of action, and that the state has the right to punish this crime through the courts or the Parliament. It is a remarkable fact, worth remembering, that through the 18th century there was virtually no challenge to this doctrine of the common law, which was accepted as legitimate by leading advocates of libertarian ideals: John Milton, John Locke, Benjamin Franklin, Thomas Jefferson, and others. Few even went so far as to declare truth to be a defense against libel; in fact true charges were regarded as even more culpable, since they brought authority into disrepute and threatened civil order. The struggle is far from over, even in the Western industrial democracies, where it is most advanced. This is, of course, only one of the many aspects of the question of the locus of sovereignty in the political system and the rights accorded to the people and to the state authorities.

Along this dimension, the United States is near the libertarian extreme in the spectrum of existing societies. Relatively speaking, the United States is a free and open society, in which the state has limited means of coercion to exercise against its own citizens. This is a very important fact. It means that an aroused public can influence policy in many ways, ranging from political action to civil disobedience and resistance. It is also possible to learn a good deal about the government, its plans and its practices. In these respects, the U.S. is probably more free and open than any other society in the world. Despite flaws in practice, the protection granted to citizens by the Bill of Rights, and in more recent years, the rights afforded by the Freedom of Information Act (which permits wide access to state documents), are unusual if not unique among existing political systems. It is not surprising that statist reactionaries of the Reaganite variety are seeking to abridge these rights as part of their project of aggrandizement of the state and expansion of its power.

In these respects, the United States is at the opposite extreme in the world spectrum from the

From Noam Chomsky, "The Domestic Scene," in *The Managua Lectures,* 1987, pp. 113–119, 123–126.

second superpower, a closed society in which the state is protected from scrutiny and has ample means to coerce the population.

Along this crucial dimension, then, the United States is among the most free societies in the world, and it will remain so despite Reaganite assaults on individual freedom and efforts to enhance state power. This is, again, a critical and important fact. We may note, however, that there is little reason to expect a correlation between the internal freedom of some society and its external violence and repression, and history reveals no such correlation. A society that is relatively free and open at home may be brutal and murderous abroad.

Why may we be fairly confident that despite the efforts of reactionary elements of the Reaganite variety, the state will remain limited in its power to coerce and control? The answer can be found in broader aspects of U.S. society. The United States is a capitalist democracy, to the extent that such a concept is meaningful (the extent is limited, since capitalism poses severe barriers to meaningful democracy, a matter to which I will turn in a moment). Of course, the U.S. is not truly a capitalist society; no such system could long survive, for reasons that have been well understood, most clearly within business circles, for a century. Business demands that the state intervene in the economy to regulate markets and otherwise support business interests, and also that it employ its means of violence in the international arena in the manner described by Woodrow Wilson in the private papers I cited in the first lecture, among other services the state must provide for the wealthy and privileged. On the other hand, business does not want the state to be a powerful competitor, interfering with the prerogatives of the businessman or organizing popular forces that might act in the parliamentary arena or elsewhere to counter business dominance of the society. Thus, business has long had a love-hate relation to the state; it wants a strong state to serve its needs, a state capable of intervening in domestic affairs and the international system; it wants a weak state that will not interfere with private

privilege, but will enhance it. To a large extent, political debate in a capitalist democracy such as the United States reduces to efforts on the part of various segments of the business community to resolve this problem in a way that will suit their sometimes conflicting interests within a shared consensus.

Though remote from the ideal, the U.S. is closer in many respects to a capitalist order than other leading industrial democracies. In a capitalist system, everything tends to become a commodity, including freedom: you can have as much of it as you can buy. The wealthy and privileged therefore have an interest in maintaining personal freedom and limiting the coercive power of the state, since they are the prime beneficiaries. For Black teen-agers in the ghetto, the system of formal liberties has little significance, since they have only limited access to it—and again, reactionaries of the Reaganite variety attempt to limit this access still further by undermining legal aid to the poor, reducing legal constraints on police power, and so on. The wealthy and privileged will defend personal freedom from state encroachment, though in times of rising class struggle and domestic challenge to their effective rule, this may change. Given the interest of dominant elites in limiting state power, we can be fairly confident that individual rights will withstand the onslaughts of statist reactionaries. One consequence is that dissident minorities also benefit from the freedom defended by the privileged, roughly to the extent that they share in existing privilege. And in a wealthy society like the United States, that includes a substantial part of the population, in greater or lesser degree.

A second crucial dimension along which democratic credentials can be evaluated is simply this: Who makes the basic decisions about what happens within the society and how it acts in the international arena? Here we may distinguish two major categories of decisions: investment decisions and political decisions. The former have to do with what is produced, how it is produced, what work is done, how production and profits are distributed and to whom, how the conditions of work are managed and controlled, and so on. The second category has to do with state policy: which groups actually participate in shaping it?

As far as investment decisions are concerned, in law and in practice they are excluded from popular control in the United States, which does nor aspire to democracy in the full sense but only to *capitalist* democracy, something rather different. To 18th century libertarians, the prime enemies of freedom were the feudal system, slavery and two powerful institutions: the Church and the State. They could envision a social order in which individuals (more accurately, white male property owners) would be more or less equal and free, once these barriers to liberty were removed. They could not foresee the centralization of effective power in the industrial and financial system of corporate capitalism. To apply their libertarian ideals to the modern world, one must go far beyond a concern for the coercive role of the Church and the State. The true inheritors of classical liberalism, in my view, are the libertarian socialists and anarchists, who oppose hierarchic structures and authoritarian institutions in a far broader realm.

In a capitalist democracy, the primary concern of everyone must be to ensure that the wealthy are satisfied; all else is secondary. Unless the wants of investors are satisfied, there is no production, no work, no resources available for welfare, in short, no possibility of survival. It is not a matter of "all or none," but "more or less." Only to the extent that the demands of the wealthy—those who control investment decisions—are satisfied can the population at large hope for a decent existence in their role as servants of private power, who rent themselves to those who own and manage the private economy. This too is a factor of fundamental importance.

Another feature of a capitalist democracy such as the United States is the inequity in distribution of resources, which translates into vast differences in the ability to participate in a meaningful way even in the narrow margin of decisions that remain within the political system. Furthermore,

the political system, like every other aspect of capitalist democracy, must be dedicated to ensuring that the demands of the wealthy are satisfied, or the society will decline and collapse. The threat to withhold investment, or capital flight, can suffice to set very narrow limits for decisions within the political system, a fact of which Latin Americans are well aware.

In the real world, state policy is largely determined by those groups that command resources, ultimately by virtue of their ownership and management of the private economy or their status as wealthy professionals. The major decision-making positions in the Executive branch of the government are typically filled by representatives of major corporations, banks and investment firms, a few law firms that cater primarily to corporate interests and thus represent the broad interests of owners and managers rather than some parochial interest, and selected intellectuals who become "experts," as Henry Kissinger once explained without irony, by virtue of their ability to articulate the consensus of the powerful and to manage their affairs for them. The Legislative branch is more varied, but overwhelmingly, it is drawn from the business and professional classes. This has, in fact, been true since the 1780s, when for a brief period, before the Republic was fully formed, legislators were drawn from a wide range of social strata. If a Senator or Representative leaves Congress, he (or occasionally, she) will not return to a position as industrial worker, small farmer, truck driver, clerk, etc., but, typically, to a business or law firm. Accordingly, in their commitments, associations and perceptions of social reality, legislators represent the business and wealthy professional classes.

Furthermore, the external conditions of policy formation are set by the same narrow elite of privileged groups. They carry out the planning studies, finance the political parties, dominate Washington lobbying, and in a variety of other ways, determine the conditions within which the political system functions.

In short, a capitalist democracy is, at best, a very limited form of democracy.

All of this has long been understood. John Jay, the President of the Continental Congress and the first Chief Justice of the US. Supreme Court, held that "the people who own the country ought to govern it." The political system as well as the social system was designed to serve the needs of the propertied classes; others might benefit incidentally, as conditions allowed. And so affairs have proceeded since. The United States, while unusual among industrial democracies in the relative inability of the state to coerce its citizens and protect itself from their scrutiny, is also unusual in the narrowness of choice afforded within the political system. There is no political party based on labor and the poor, responsive to some extent to their needs and interests and committed to limited reforms of the capitalist system, such as the socialist, labor, or Communist parties in Europe. To a large degree, the U.S. is a one-party state, where the ruling party has two factions that compete for control of the government. U.S. political history is, to a significant extent, a history of conflict among those in a position to make investment decisions; where few major issues divide them, there is a period of political harmony, and where such issues do arise, there is political conflict over them. The general public is afforded an opportunity to ratify elite decisions, but the option of participating in making them is limited, very largely, to privileged elites.

Much of the public is aware of its marginalization and of the essential irrelevance of the political system to its concerns. Close to half the electorate does not even take the trouble to go to the polls in Presidential elections, and of those who vote, many do so independently of the public stand of the candidates on crucial issues. Take the most recent (1984) Presidential election, for example. This is almost invariably described as a landslide victory for Ronald Reagan and his "conservatism"—actually, a form of reactionary jingoism that would be anathema to true conservatives. In fact, there was no such "landslide." Reagan received less than 30% of the potential vote. Of those who voted for Reagan, about 60% felt that his legislative program would harm

the country, while about 1% of the electorate voted for him because they considered him a "real conservative."

Polls taken after the election showed that half the public believe that the government is run "by a few big interests looking out for themselves." As always, voting was highly skewed towards privileged sectors, much higher among white collar than blue collar workers, and very low among the poor and unemployed, who evidently do not consider themselves to be represented within the political system. These facts are particularly noteworthy in the light of the extraordinary efforts "to bring out the vote" and the unremitting patriotic propaganda about the magnificence of American democracy. The rather accurate conceptions of half the population would be castigated as "extremist" or "Marxist" if they were to receive articulate expression. But much of the population understands the accuracy of John Jay's dictum, despite the dedicated efforts undertaken within the doctrinal system to convince them otherwise.

Although I know of no direct study of the question, it is a fair guess that as level of education increases, the level of understanding of these social realities will decline. We see evidence for this conclusion in that these topics can barely be discussed within the ideological institutions managed by the educated classes: the media, the schools, the universities, the journals of opinion. In a rare moment of candor, the Trilateral Commission study on the "Crisis of Democracy," which I mentioned earlier, described the schools and universities as among the institutions responsible for "the indoctrination of the young." Those who are more subject to indoctrination, which continues in later life through the media, journals, popular and often scholarly books, are more likely to be subject to its illusions. Furthermore, the educated classes are not only the main targets of the system of indoctrination but also its practitioners; their self-interest dictates that they adopt and believe its dictates, if they are to be able to fulfill their role as educators, journalists, or "responsible intellectuals" with access to privilege, influence, and respect. What is more, the victims of the system of exploitation develop an intuitive understanding of reality through their own lives. The banality, superficiality and often sheer silliness of cultivated discourse therefore comes as little surprise.

• • •

I have mentioned two central dimensions along which the democratic credentials of some sociopolitical system may be evaluated: the power of the state to coerce its citizens and to protect itself from their control; the locus of decision-making in the social, economic and political systems. A third crucial dimension has to do with the ideological system. To what extent are ordinary people able to become informed, a prerequisite to democratic participation? I have addressed this question repeatedly throughout these lectures. The right of free expression is vigorously maintained in the United States, in that state controls are very weak by comparative standards. On the other hand, the ideological system operates within very narrow constraints and those who do not accept them are effectively excluded. Debate is permitted, even encouraged, as long as it adopts the fundamental principles of the ideological system. In the case of the Vietnam war, for example, when it was clear that the costs to the U.S. were mounting severely, it became possible to debate the issue of the war in the national press, but only within certain limits. One could take the position of the hawks, who held that with sufficient dedication the U.S. could win, or the position of the doves, who held that success was unlikely though "we all pray" that the hawks are right and we will "all be saluting the wisdom and statesmanship of the American government" in conducting a war that was turning Vietnam into "a land of ruin and wreck" if the hawks prove to be right in their judgment, as explained by historian Arthur Schlesinger, regarded as an "antiwar leader" in the establishment media. Those who held that aggression was wrong even if it could succeed were systematically excluded

from the discussion. To this day, as I have already mentioned, there is no such event as the U.S. attack against South Vietnam in official history, though this was clearly the central element in the Indochina war.

Much the same is true in other cases, some already discussed. The debate—such as it is—over Nicaragua today in the mainstream is a revealing example. As I pointed out in the third lecture in reviewing the national press, debate is tolerated, but within very narrow limits. Recall that in the crucial first three months of 1986, as debate was heating up over the impending vote on contra aid in Congress, the two major national newspapers assured 100 percent uniformity on the central issue, permitting nothing sympathetic to the Sandinista government. There was no mention at all of the not-insignificant fact that in sharp contrast to U.S. clients in the region, the Sandinistas do not slaughter their own population; Sandinista social reforms, the prime reason for the U.S. attack, merited two passing phrases. Editorial commentary since 1980 is similar, as I noted. While the imposition of a State of Siege in Nicaragua in October 1985 elicited outraged denunciations, the *renewal* of the Salvadoran State of Siege two days later passed without comment; indeed, it has never been mentioned in a *New York Times* editorial. All of this is particularly instructive in the light of the unquestionable fact that the Salvadoran State of Siege has been applied with incomparably greater harshness since it was instituted in 1980, and that unlike Nicaragua, El Salvador is not under attack by the regional superpower.

Elsewhere, I have examined press coverage on these and other issues in greater detail, as have others. The results are quite regularly the same: suppression or apologetics with regard to crimes of the United States and its clients; anguish and outrage, often based on the kind of flimsy evidence that would be dismissed with contempt if adduced in connection with the U.S. and its clients, or on outright fabrication, with regard to the crimes of official enemies. One expects to find such behavior in the official press of a totalitarian state. The extent to which much the same is true in a press that operates without overt state controls will come as a surprise and a shock to those who choose to inquire into the matter honestly. Documentation of this matter is quite extensive, but invariably ignored as much too inconvenient in discussion of the nature of the media, which are—the ultimate irony—regularly condemned for their "adversarial" stance with regard to state and private power.

The reasons for the systematic deference of the media towards external power are not difficult to discern. The media represent the same interests that control the state and private economy, and it is therefore not very surprising to discover that they generally act to confine public discussion and understanding to the needs of the powerful and privileged. The media are, in the first place, major corporations. Their primary market is business (advertisers), and like other corporations, they must bend to the needs of the community of investors. In the unlikely event that they might seek to pursue an independent path, they would quickly be called to account, and could not survive. Their top management (editors, etc.) is drawn from the ranks of wealthy professionals who tend naturally to share the perceptions of the privileged and powerful, and who have achieved their position, and maintain it, by having demonstrated their efficiency in the task of serving the needs of dominant elites. Furthermore, by virtue of their associations, class status, aspirations, and so on, they tend to share the perceptions and commitments of those who hold effective power. Thus it is only to be expected that the framework of interpretation, selection of what counts as "news," permitted opinion, etc., will fall well within the range that conforms to the needs of the nexus of state-private power that controls the economy and the political system.

Journalists and columnists have the choice of conforming or being excluded, and in a wealthy society, the rewards for conformity can be substantial. Those who choose to conform, hence to remain within the system, will soon find that they internalize the beliefs and attitudes that they express and that shape their work; it is a very rare individual who can believe one thing and say another on a regular basis. A certain range of opinion is tolerated, generally on narrow tactical questions within a shared consensus as to "the national interest," and one should not discount the professional integrity of the better and more honest journalists. But the institutional structure of the system is in its essence hostile to independence of mind, and it is hardly surprising that it is so rarely exhibited. The point is not that the journalists or commentators are dishonest; rather, unless they happen to conform to the institutional requirements, they will find no place in the corporate media. At the margins of the system— e.g., in the listener-supported local radio—one can find deviation from the prevailing norms, and there are deviations amounting to "statistical error" even within the mainstream on occasion, but these norms, rooted in the institutional structure, are very rarely violated. With some variations, much the same is true in the schools and universities, for similar reasons.

As in the case of the political system, the United States is unusual among the capitalist democracies in the ideological constraints observed by the media. One would be hard put to find even a mild democratic socialist in the mass media, and a genuine opposition press is difficult to imagine. In these respects, the United States departs from the norm among capitalist democracies, for a variety of reasons that I cannot pursue here—one of them being its power and importance in the global system.

• • •

12

Who Rules Now?

American Elites in the 1990s

STANLEY ROTHMAN AND
AMY E. BLACK

Arguments about the structure of power in the United States have a long history and range from Marxist and Marxist-influenced populist interpretations such as those of C. Wright Mills and G. William Domhoff through various "new class" interpretations fathered by Irving Kristol and supported by many neo-conservatives, to pluralist theories of the kind advanced by Robert Dahl, Arnold Rose, Suzanne Keller and Etzioni-Halevy among many others. Despite their differing ideological perspectives, Mills, Domhoff and Kristol share a similar perception of American politics. American society is dominated by one or two unified elites dedicated to maintaining or increasing their power. To Mills it is the military-industrial complex; to Domhoff a ruling class and to Kristol a "new class" competing to seize power from the traditional bourgeoisie. More recently Michael Lind has popularized the notion of an "overclass" consisting of affluent white males who, despite minor disagreements, share the common values which enable them to dominate the society. Lind's writing resembles, to some extent, that of some radical feminists who see domination as located entirely at the intersection of race, class and gender. According to this view, America is ruled by an affluent white patriarchy.

Various pluralist theorists, on the other hand, tend to see power as relatively dispersed (though not equally) among a number of leadership groups which strive to implement policies which they support as right, just, or simply advantageous to them. In the United States, such elites consist of high-level business leaders, journalists, political figures, and bureaucrats, among a number of other groups, who, while sharing some values, also disagree about many issues.

Given the extensive debate about these matters, scholars have completed surprisingly few empirical studies of elites. In fact, only two major surveys have been conducted in recent decades, those by Sidney Verba and his colleagues and by Lerner, Nagai and Rothman. The groups chosen in each of these studies differ somewhat, and the purpose of the research also varies, but, overall, one finding remains constant. Although consisting primarily (but not entirely) of relatively affluent white males, the views of the members of various leadership groups in the United States in the early 1980s differed sharply in a number of key areas.

In 1995, Rothman and Black replicated the Lerner, Nagai, Rothman survey conducting computer-assisted telephone interviews with members of nine strategic elite groups. The following analysis examines the points of intersection between the two studies, comparing only replicated or similar questions for the elite groups included in both studies. For the purposes of simplification, all data reported herein will refer to the *American Elites* study as the 1980 study and the more recent study as the 1995 study. Our goal was to examine what changes, if any, had taken place among various leadership in the 15 years or so since the first study had been completed.

The term elite is one of art among sociologists, referring to institutions or groups which have a larger share of power or influence (or sometimes just excellence, depending on the scale) than does the average group or institution. Our sample of elites (read leadership groups) included random samples of businessmen, federal judges, top level federal bureaucrats, journalists on leading newspapers and television news programs, the writers, producers and directors of top-grossing motion pictures or television shows, etc.

WHO ARE AMERICA'S LEADERS?

Key studies demonstrate that American leaders are still overwhelmingly white. Except for elective office and the military, African Americans have not made much progress since 1980. They remain a relatively small portion of top business executives, lawyers, government bureaucrats, religious leaders, journalists and the movers and shakers in Hollywood. Asian and Hispanics are hardly represented at all. White women are another matter. In some leadership groups, for example the bureaucracy, the legal profession and the mass media, their representation has risen sharply in the past 10 years. Today, for example, over a third of top-level journalists are women. This fact would seem to weaken feminist analysis of the American "patriarchy" but so does a comparison of male members of the American elite, for, despite being male and white, they differ widely from each other on a number of important issues.

A plurality of American leaders are nominally Protestants and some American leaders seem to have become more religious in the past 15 years. The proportion answering "none" to the question of religious preference dropped from 50 percent to 21 percent among journalists during that period. Yet about 40 percent never attend church as compared to about 20 percent of business men and women. However, the two groups are moving in opposite directions. More journalists claim to attend church today than made that claim in the 1980s. On the other hand, somewhat fewer business leaders do. The same pattern of increased church attendance is also to be found among the Hollywood elite, though Hollywood leaders are still far less religious than the population as a whole.

Among the groups mentioned, top-level business executives seem to have come from the most modest backgrounds as measured by the education of their parents. In the 1980 study we found that 30 percent of business leaders were raised in blue collar families as compared to 17 percent of journalists. The pattern seems to be similar today. It is to be expected. Those who are poor want to move up the economic ladder. Those whose parents have made it, want to engage in interesting work and/or to perform direct social service.

Lind has argued that the members of the overclass tend to have graduated from Ivy League colleges and universities at which they derived their common outlook. More recently Nathan Glazer, arguing the case for a moderate form of affirmative action, maintains that leaders in American society tend to come from elite schools, including not only those of the Ivy League, but also such institutions as the University of California at Berkeley. The data we have uncovered would not seem to support either claim.

In our 1980 study of American leadership groups my colleagues and I rated colleges and universities on a scale based on the work of two noted political scientists, Seymour Martin Lipset and C. Everett Ladd. Our classification of top colleges and universities was far broader than Ivy League schools. Nevertheless, except for lawyers and journalists, no more than 25 percent of the members of any of the groups we studied attended these elite institutions. Thus, while 69 percent of elite lawyers graduated from elite schools no more than 28 percent of business leaders had similar backgrounds.

When I repeated the elite study in 1995 with Professor Amy Black we found that the proportion of those members of various elites who had graduated from Ivy League schools was less than two out of ten. Much like the 1980 study, the percentages were highest among lawyers and journalists and quite low among business people, Hollywood motion picture and television leaders, federal judges or bureaucrats.

One group does seem to have benefited from an Ivy League education to an astonishing degree. If the shift in the Ivy League from admissions based on character and parental alumni status to a more meritocratic policy assisted anyone from the late 1940s to the mid-1960s it was Jewish Americans. Some 55 percent of Americans of Jewish background who are members of elite groups graduated from an Ivy League college or university.

WHAT DO THEY BELIEVE?

Lind claims that the "overclass" is economically conservative and socially liberal. We found a rather different picture. In both 1980 and 1995 business leaders and elite journalists were far apart on economic issues, with the former (as one might expect) being far more conservative than the latter on such issues as to whether or not the government should guarantee jobs for all, or on the need to reduce the income gap between the rich and the poor. Top religious leaders are almost as liberal as journalists on some of these questions. Thus, 71 percent of leading journalists and 7 out of 10 religious leaders want government to guarantee jobs for all. Only 34 percent of business leaders agree. Interestingly enough, only 63 percent of TV and motion picture elites support job guarantees, more than business executives but fewer than journalists.

If business leaders are economically conservative, they are not socially liberal, at least as compared to journalists and the mass entertainment elite. They do share a "pro-choice" attitude with other groups, but they are less accepting of adultery than motion picture and television leaders as well as of homosexuality, though on that issue, they, as well as the general public, seem to be changing. This is a change, one suspects, which the more accepting mass entertainment industry has probably played some role in bringing about, for, while the Hollywood elite is somewhat more conservative than journalists on economic issues, it is slightly more liberal on social issues. Thus on the question of whether homosexuality is as acceptable a lifestyle as heterosexuality, more than six out of 10 entrepreneurs answer negatively, as compared to 27 percent of journalists, but only 21 percent of the Hollywood elite.

Corporate executives also share what seems to be a growing disenchantment with affirmative action and a growing belief that "minority gains in recent years have often come at the expense of white males." Here again, journalists as well as the mass entertainment elite are well to the left of them, if one wishes to describe that issue in left-right terms. Fewer than three out of 10 business executives support "special preferences" for minorities as compared to just over half of the Hollywood elite.

The voting patterns of various elites are congruent with their attitudes. The journalists, Hollywood leaders, and bureaucrats are consistently Democrats while business leaders and judges are consistently Republicans. In the 1992 elections, slightly more than three-quarters of business elites voted Republican as compared to eight percent of the leaders of the entertainment industry and seven percent of journalists.

AND SO?

White males still dominate the major leadership groups in the United States, though blacks and women, especially the latter, have made gains since our initial study. However, white males are not a monolithic group. They continue to differ sharply on a variety of issues, even though the data provide some evidence of a moderation of the culture wars. Those elites concerned with the creation and dissemination of ideas and entertainment remain to the left of most leadership groups on economic policy. But the responses from all of the elite groups reveal a general drift in a more conservative direction, though the picture is mixed and depends on the issues.

Comparisons of responses on social and cultural issues reveal both continuity and change. Cultural elites remain well to the left of more traditional groups on such issues as homosexuality and adultery as well as on race relations. Indeed, on social issues there is some evidence that more traditional leadership groups are following the example of the cultural elites, opening up the society to perspectives (i.e. homosexuality as an alternative lifestyle) which break sharply with traditional outlooks. Furthermore, despite a greater degree of support for policies which stress free markets, cultural elites remain quite hostile to business. When asked, cultural elites express the belief that business leaders have too much power. Of course, except for journalists, other leadership groups also believe that journalists have too much power. Except for journalists again, most of the groups we studied believe that journalists are at least as powerful as business executives in our society.

In short, it appears that, despite some retreats, those who would remake the culture are bringing about significant change. The same may be said of various marginal groups like women and especially blacks, despite their lack of representation among elite groups. They have produced important changes in American life. At this point, however, after many gains by African Americans, the area of race relations seems to have become somewhat more polarized in some respects. Our study appears to show a sharpening of the differences between white and black elites on the issue of affirmative action though many whites seem unwilling to press their anti-affirmative action views publicly in the face of sharp resistance by African

Americans. The wording of the questions upon which we relied differed slightly in our two studies so our very preliminary results must be used with caution. Nevertheless, whereas 53 percent of whites were opposed to special preferences for blacks in hiring in 1980, the proportion rose to more than 65 percent of our 1995 sample. Similarly, while in 1980, very few whites (less than one in twenty) expressed the view that minority gains came at the expense of whites, almost half our white respondents support that view now. Perhaps more disturbing, almost 60 percent of black elites (as compared to about four out of ten whites) agree that we will never achieve racial equality in the United States. These, of course, are very preliminary findings. Nevertheless they are sobering.

All in all, whatever changes have occurred, and whatever tensions exist, it is clear that the United States is not governed by a monolithic "overclass" or "white patriarchy." As we pointed out earlier, the old quasi-Marxist and populist notions put forward by a previous generation of intellectuals seem rather dated at this point, and feminist characterizations of power in our society are not supported by the evidence.

White males, many from relatively humble backgrounds, still dominate various leadership groups, but white Protestant dominion is a thing of the past. In the not too distant future, we expect that the key role still played by white males will be replaced by a greater ethnic and gender pluralism among the leaders of American social and political life. As usual, those on the top will continue to be characterized by more or less sharp ideological disagreements as they mostly have been in the past. Even now it is clear for those who wish to look that power is widely distributed in the United States. As a nation, we are far from being dominated by a homogeneous overclass. Indeed it seems to us that we are increasingly characterized, for good or evil, by a sort of public anarchy in which the center of power is not at all clear, perhaps because there is no one fixed power center.

• • •

13

Jihad vs. McWorld

BENJAMIN R. BARBER

Just beyond the horizon of current events lie two possible political figures—both bleak, neither democratic. The first is a retribalization of large swaths of humankind by war and bloodshed: a threatened Lebanonization of national states in which culture is pitted against culture, people against people, tribe against tribe—a Jihad in the name of a hundred narrowly conceived faiths against every kind of interdependence, every kind of artificial social cooperation and civic mutuality. The second is being borne in on us by the onrush of economic and ecological forces that demand integration and uniformity and that mesmerize the world with fast music, fast computers, and fast food—with MTV, Macintosh, and McDonald's, pressing nations into one commercially homogenous global network: one McWorld tied together by technology, ecology, communications, and commerce. The planet is falling precipitantly apart *and* coming reluctantly together at the very same moment.

These two tendencies are sometimes visible in the same countries at the same instant: thus Yugoslavia, clamoring just recently to join the New Europe, is exploding into fragments; India is trying to live up to its reputation as the world's largest integral democracy while powerful new fundamentalist parties like the Hindu nationalist Bharatiya Janata Party, along with nationalist assassins, are imperiling its hard-won unity. States are breaking up or joining up: the Soviet Union has disappeared almost overnight, its parts forming new unions with one another or with like-minded nationalities in neighboring states. The old interwar national state based on territory and political sovereignty looks to be a mere transitional development.

The tendencies of what I am here calling the forces of Jihad and the forces of McWorld operate with equal strength in opposite directions, the one driven by parochial hatreds, the other by universalizing markets, the one re-creating ancient subnational and ethnic borders from within, the other making national borders porous from without. They have one thing in common: neither offers much hope to citizens looking for practical ways to govern themselves democratically. If the global future is to put Jihad's centrifugal whirlwind against McWorld's centripetal black hole, the outcome is unlikely to be democratic—or so I will argue.

MCWORLD, OR THE GLOBALIZATION OF POLITICS

Four imperatives make up the dynamic of McWorld: a market imperative, a resource imperative, an information-technology imperative, and an ecological imperative. By shrinking the world and diminishing the salience of national borders, these imperatives have in combination achieved a considerable victory over factiousness and particularism, and not least of all over their most virulent traditional form—nationalism. It is the realists who are now Europeans, the utopians who dream nostalgically of a resurgent England or Germany, perhaps even a resurgent Wales or Saxony. Yesterday's wishful cry for one world has yielded to the reality of McWorld.

From Benjamin R. Barber, "Jihad vs. McWorld." Originally published in *The Atlantic Monthly,* March, 1992. Reprinted by permission of Benjamin R. Barber who is Whitman Professor of Political Science and Director of the Walt Whitman Center at Rutgers University and the author of many books including *Strong Democracy* (1984), *Jihad versus McWorld* (Times Books, 1995) and *A Place for Us* (Farrar, Straus & Giroux, 1998).

The market imperative. Marxist and Leninist theories of imperialism assumed that the quest for ever-expanding markets would in time compel nation-based capitalist economies to push against national boundaries in search of an international economic imperium. Whatever else has happened to the scientist predictions of Marxism, in this domain they have proved farsighted. All national economies are now vulnerable to the inroads of larger, transnational markets within which trade is free, currencies are convertible, access to banking is open, and contracts are enforceable under law. In Europe, Asia, Africa, the South Pacific, and the Americas such markets are eroding national sovereignty and giving rise to entities—international banks, trade associations, transnational lobbies like OPEC and Greenpeace, world news services like CNN and the BBC, and multinational corporations that increasingly lack a meaningful national identity—that neither reflect nor respect nationhood as an organizing or regulative principle.

The market imperative has also reinforced the quest for international peace and stability, requisites of an efficient international economy. Markets are enemies of parochialism, isolation, fractiousness, war. Market psychology attenuates the psychology of ideological and religious cleavages and assumes a concord among producers and consumers—categories that ill fit narrowly conceived national or religious cultures. Shopping has little tolerance for blue laws, whether dictated by pub-closing British paternalism, Sabbath-observing Jewish Orthodox fundamentalism, or no-Sunday-liquor-sales Massachusetts puritanism. In the context of common markets, international law ceases to be a vision of justice and becomes a workaday framework for getting things done—enforcing contracts, ensuring that governments abide by deals, regulating trade and currency relations, and so forth.

Common markets demand a common language, as well as a common currency, and they produce common behaviors of the kind bred by cosmopolitan city life everywhere. Commercial pilots, computer programmers, international bankers, media specialists, oil riggers, entertainment celebrities, ecology experts, demographers, accountants, professors, athletes—these compose a new breed of men and women for whom religion, culture, and nationality can seem only marginal elements in a working identity. Although sociologists of everyday life will no doubt continue to distinguish a Japanese from an American mode, shopping has a common signature throughout the world. Cynics might even say that some of the recent revolutions in Eastern Europe have had as their true goal not liberty and the right to vote but well-paying jobs and the right to shop (although the vote is proving easier to acquire than consumer goods). The market imperative is, then, plenty powerful; but, notwithstanding some of the claims made for "democratic capitalism," it is not identical with the democratic imperative.

The resource imperative. Democrats once dreamed of societies whose political autonomy rested firmly on economic independence. The Athenians idealized what they called autarky, and tried for a while to create a way of life simple and austere enough to make the polis genuinely self-sufficient. To be free meant to be independent of any other community or polis. Not even the Athenians were able to achieve autarky, however: human nature, it turns out, is dependency. By the time of Pericles, Athenian politics was inextricably bound up with a flowering empire held together by naval power and commerce—an empire that, even as it appeared to enhance Athenian might, ate away at Athenian independence and autarky. Master and slave, it turned out, were bound together by mutual insufficiency.

The dream of autarky briefly engrossed nineteenth-century America as well, for the underpopulated, endlessly bountiful land, the cornucopia of natural resources, and the natural barriers of a continent walled in by two great seas led many to believe that America could be a world unto itself. Given this past, it has been harder for Americans than for most to accept the inevitability of interdependence. But the rapid depletion of resources even in a country like ours, where they once seemed inexhaustible, and the maldistribution of arable soil and mineral resources on the planet, leave even the wealthiest societies ever more resource-dependent and many other nations in permanently desperate straits.

Every nation, it turns out, needs something another nation has; some nations have almost nothing they need.

The information-technology imperative. Enlightenment science and the technologies derived from it are inherently universalizing. They entail a quest for descriptive principles of general application, a search for universal solutions to particular problems, and an unswerving embrace of objectivity and impartiality.

Scientific progress embodies and depends on open communication, a common discourse rooted in rationality, collaboration, and an easy and regular flow and exchange of information. Such ideals can be hypocritical covers for power-mongering by elites, and they may be shown to be wanting in many other ways, but they are entailed by the very idea of science and they make science and globalization practical allies.

Business, banking, and commerce all depend on information flow and are facilitated by new communication technologies. The hardware of these technologies tends to be systemic and integrated—computer, television, cable, satellite, laser, fiber-optic, and microchip technologies combining to create a vast interactive communications and information network that can potentially give every person on earth access to every other person, and make every datum, every byte, available to every set of eyes. If the automobile was, as George Ball once said (when he gave his blessing to a Fiat factory in the Soviet Union during the Cold War), "an ideology on four wheels," then electronic telecommunication and information systems are an ideology at 186,000 miles per second—which makes for a very small planet in a very big hurry. Individual cultures speak particular languages; commerce and science increasingly speak English; the whole world speaks logarithms and binary mathematics.

Moreover, the pursuit of science and technology asks for, even compels, open societies. Satellite footprints do not respect national borders; telephone wires penetrate the most closed societies. With photocopying and then fax machines having infiltrated Soviet universities and *samizdat* literary circles in the eighties, and computer modems having multiplied like rabbits in communism's bureaucratic warrens thereafter, *glasnost* could not be far behind. In their social requisites, secrecy and science are enemies.

The new technology's software is perhaps even more globalizing than its hardware. The information arm of international commerce's sprawling body reaches out and touches distinct nations and parochial cultures, and gives them a common face chiseled in Hollywood, on Madison Avenue, and in Silicon Valley. Throughout the 1980s one of the most-watched television programs in South Africa was *The Cosby Show*. The demise of apartheid was already in production. Exhibitors at the 1991 Cannes film festival expressed growing anxiety over the "homogenization" and "Americanization" of the global film industry when, for the third year running, American films dominated the awards ceremonies. America has dominated the world's popular culture for much longer, and much more decisively. In November of 1991 Switzerland's once insular culture boasted best-seller lists featuring *Terminator 2* as the No. 1 movie, *Scarlett* as the No. 1 book, and Prince's *Diamonds and Pearls* as the No. 1 record album. No wonder the Japanese are buying Hollywood film studios even faster than Americans are buying Japanese television sets. This kind of software supremacy may in the long term be far more important than hardware superiority, because culture has become more potent than armaments. What is the power of the Pentagon compared with Disneyland? Can the Sixth Fleet keep up with CNN? McDonald's in Moscow and Coke in China will do more to create a global culture than military colonization ever could. It is less the goods than the brand names that do the work, for they convey life-style images that alter perception and challenge behavior. They make up the seductive software of McWorld's common (at times much too common) soul.

Yet in all this high-tech commercial world there is nothing that looks particularly democratic. It lends itself to surveillance as well as liberty, to new forms of manipulation and covert control as well as new kinds of participation, to skewed, unjust market outcomes as well as greater productivity. The consumer society and the open society are not quite synonymous. Capitalism and democracy have a relationship, but it is something less than a marriage. An efficient free market after all requires that consumers be free to vote their dollars on competing goods, not that citizens be free to vote their values and beliefs on competing political candidates and programs. The free market flourished in junta-run Chile, in military-governed Taiwan and Korea, and, earlier, in a variety of autocratic European empires as well as their colonial possessions.

The ecological imperative. The impact of globalization on ecology is a cliché even to world leaders who ignore it. We know well enough that the German forests can be destroyed by Swiss and Italians driving gas-guzzlers fueled by leaded gas. We also know that the planet can be asphyxiated by greenhouse gases because Brazilian farmers want to be part of the twentieth century and are burning down tropical rain forests to clear a little land to plough, and because Indonesians make a living out of converting their lush jungle into toothpicks for fastidious Japanese diners, upsetting the delicate oxygen balance and in effect puncturing our global lungs. Yet this ecological consciousness has meant not only greater awareness but also greater inequality, as modernized nations try to slam the door behind them, saying to developing nations, "The world cannot afford *your* modernization; ours has wrung it dry!"

Each of the four imperatives just cited is transnational, transideological, and transcultural. Each applies impartially to Catholics, Jews, Muslims, Hindus, and Buddhists; to democrats and totalitarians; to capitalists and socialists. The Enlightenment dream of a universal rational society has to a remarkable degree been realized—but in a form that is commercialized, homogenized, depoliticized, bureaucratized, and, of course, radically incomplete, for the movement toward McWorld is in competition with forces of global breakdown, national dissolution, and centrifugal corruption. These forces, working in the opposite direction, are the essence of what I call Jihad.

JIHAD, OR THE LEBANONIZATION OF THE WORLD

OPEC, the World Bank, the United Nations, the International Red Cross, the multinational corporation . . . there are scores of institutions that reflect globalization. But they often appear as ineffective reactors to the world's real actors: national states and, to an ever greater degree, subnational factions in permanent rebellion against uniformity and integration—even the kind represented by universal law and justice. The headlines feature these players regularly: they are cultures, not countries; parts, not wholes; sects, not religions; rebellious factions and dissenting minorities at war not just with globalism but with the traditional nation-state. Kurds, Basques, Puerto Ricans, Ossetians, East Timoreans, Quebecois, the Catholics of Northern Ireland, Abkhasians, Kurile Islander Japanese, the Zulus of Inkatha, Catalonians, Tamils, and, of course Palestinians—people without countries, inhabiting nations not their own, seeking smaller worlds within borders that will seal them off from modernity.

A powerful irony is at work here. Nationalism was once a force of integration and unification, a movement aimed at bringing together disparate clans, tribes, and cultural fragments under new, assimilationist flags. But as Ortega y Gasset noted more than sixty years ago, having won its victories, nationalism changed its strategy. In the 1920s, and again today, it is more often a reactionary and divisive force, pulverizing the very nations it once helped cement together. The force that creates nations is "inclusive," Ortega wrote in *The Revolt of the Masses.* "In periods of consolidation, nationalism has a positive value, and is a lofty standard. But in Europe everything is more than consolidated, and nationalism is nothing but a mania. . . ."

This mania has left the post–Cold War world smoldering with hot wars; the international scene is little more unified than it was at the end of the Great War, in Ortega's own time. There were more than thirty wars in progress last year, most of them ethnic, racial, tribal, or religious in character, and the list of unsafe regions doesn't seem to be getting any shorter. Some new world order!

The aim of many of these small-scale wars is to redraw boundaries, to implode states and re-secure parochial identities: to escape McWorld's dully insistent imperatives. The mood is that of Jihad: war not as an instrument of policy but as an emblem of identity, an expression of community, an end in itself. Even where there is no shooting war, there is fractiousness, secession, and the quest for ever smaller communities. Add to the list of dangerous countries those at risk: In Switzerland and Spain, Jurassian and Basque separatists still argue the virtues of ancient identities, sometimes in the language of bombs. Hyper-disintegration in the former Soviet Union may well continue unabated—not just a Ukraine independent from the Soviet Union but a Bessarabian Ukraine independent from the Ukrainian republic; not just Russia severed from the defunct union but Tatarstan severed from Russia. Yugoslavia makes even the disunited, ex-Soviet, nonsocialist republics that were once the Soviet Union look integrated, its sectarian fatherlands springing up within factional motherlands like weeds within weeds within weeds. Kurdish independence would threaten the territorial integrity of four Middle Eastern nations. Well before the current cataclysm Soviet Georgia made a claim for autonomy from the Soviet Union, only to be faced with its Ossetians (164,000 in a republic of 5.5 million) demanding their own self-determination within Georgia. The Abkhasian minority in Georgia has followed suit. Even the good will established by Canada's once promising Meech Lake protocols is in danger, with Francophone Quebec again threatening the dissolution of the federation. In South Africa the emergence from apartheid was hardly achieved when friction between Inkatha's Zulus and the African National Congress's tribally identified members threatened to replace Europeans' racism with an indigenous tribal war.

After thirty years of attempted integration using the colonial language (English) as a unifier, Nigeria is now playing with the idea of linguistic multiculturalism—which could mean the cultural breakup of the nation into hundreds of tribal fragments. Even Saddam Hussein has benefited from the threat of internal Jihad, having used renewed tribal and religious warfare to turn last season's mortal enemies into reluctant allies of an Iraqi nationhood that he nearly destroyed.

The passing of communism has torn away the thin veneer of internationalism (workers of the world unite!) to reveal ethnic prejudices that are not only ugly and deep-seated but increasingly murderous. Europe's old scourge, anti-Semitism, is back with a vengeance, but it is only one of many antagonisms. It appears all too easy to throw the historical gears into reverse and pass from a Communist dictatorship back into a tribal state.

Among the tribes, religion is also a battlefield. ("Jihad" is a rich word whose generic meaning is "struggle"—usually the struggle of the soul to avert evil. Strictly applied to religious war, it is used only in reference to battles where the faith is under assault, or battles against a government that denies the practice of Islam. My use here is rhetorical, but does follow both journalistic practice and history.) Remember the Thirty Years War? Whatever forms of Enlightenment universalism might once have come to grace such historically related forms of monotheism as Judaism, Christianity, and Islam, in many of their modern incarnations they are parochial rather than cosmopolitan, angry rather than loving, proselytizing rather than ecumenical, zealous rather than rationalist, sectarian rather than deistic, ethnocentric rather than universalizing. As a result, like the new forms of hypernationalism, the new expressions of religious fundamentalism are fractious and pulverizing, never integrating. This is religion as the Crusaders knew it: a battle to the death for souls that if not saved will be forever lost.

The atmospherics of Jihad have resulted in a breakdown of civility in the name of identity, of comity in the name of community. International relations have sometimes taken on the aspect of gang war—cultural turf battles featuring tribal factions that were supposed to be sublimated as integral parts of large national, economic, post-colonial, and constitutional entities.

THE DARKENING FUTURE
OF DEMOCRACY

These rather melodramatic tableaux vivants do not tell the whole story, however. For all their defects, Jihad and McWorld have their attractions. Yet, to repeat and insist, the attractions are unrelated to democracy. Neither McWorld nor Jihad is remotely democratic in impulse. Neither needs democracy; neither promotes democracy.

McWorld does manage to look pretty seductive in a world obsessed with Jihad. It delivers peace, prosperity, and relative unity—if at the cost of independence, community, and identity (which is generally based on difference). The primary political values required by the global market are order and tranquillity, and freedom—as in the phrases "free trade," "free press," and "free love." Human rights are needed to a degree, but not citizenship or participation—and no more social justice and equality than are necessary to promote efficient economic production and consumption. Multinational corporations sometimes seem to prefer doing business with local oligarchs, inasmuch as they can take confidence from dealing with the boss on all crucial matters. Despots who slaughter their own populations are no problem, so long as they leave markets in place and refrain from making war on their neighbors (Saddam Hussein's fatal mistake). In trading partners, predictability is of more value than justice.

The Eastern European revolutions that seemed to arise out of concern for global democratic values quickly deteriorated into a stampede in the general direction of free markets and their ubiquitous, television-promoted shopping malls. East Germany's Neues Forum, that courageous gathering of intellectuals, students, and workers which overturned the Stalinist regime in Berlin in 1989, lasted only six months in Germany's

mini-version of McWorld. Then it gave way to money and markets and monopolies from the West. By the time of the first all-German elections, it could scarcely manage to secure three percent of the vote. Elsewhere there is growing evidence that *glasnost* will go and *perestroika*—defined as privatization and an opening of markets to Western bidders—will stay. So understandably anxious are the new rulers of Eastern Europe and whatever entities are forged from the residues of the Soviet Union to gain access to credit and markets and technology—McWorld's flourishing new currencies—that they have shown themselves willing to trade away democratic prospects in pursuit of them: not just old totalitarian ideologies and command-economy production models but some possible indigenous experiments with a third way between capitalism and socialism, such as economic cooperatives and employee stock-ownership plans, both of which have their ardent supporters in the East.

Jihad delivers a different set of virtues: a vibrant local identity, a sense of community, solidarity among kinsmen, neighbors, and countrymen, narrowly conceived. But it also guarantees parochialism and is grounded in exclusion. Solidarity is secured through war against outsiders. And solidarity often means obedience to a hierarchy in governance, fanaticism in beliefs, and the obliteration of individual selves in the name of the group. Deference to leaders and intolerance toward outsiders (and toward "enemies within") are hallmarks of tribalism—hardly the attitudes required for the cultivation of new democratic women and men capable of governing themselves. Where new democratic experiments have been conducted in retribalizing societies, in both Europe and the Third World, the result has often been anarchy, repression, persecution, and the coming of new, noncommunist forms of very old kinds of despotism. During the past year, Havel's velvet revolution in Czechoslovakia was imperiled by partisans of "Czechland" and of Slovakia as independent entities. India seemed little less rent by Sikh, Hindu, Muslim, and Tamil infighting than it was immediately after the British pulled out, more than forty years ago.

To the extent that either McWorld or Jihad has a *natural* politics, it has turned out to be more of an antipolitics. For McWorld, it is the antipolitics of globalism: bureaucratic, technocratic, and meritocratic, focused (as Marx predicted it would be) on the administration of things—with people, however, among the chief things to be administered. In its politico-economic imperatives McWorld has been guided by laissez-faire market principles that privilege efficiency, productivity, and beneficence at the expense of civic liberty and self-government.

For Jihad, the antipolitics of tribalization has been explicitly antidemocratic: one-party dictatorship, government by military junta, theocratic fundamentalism—often associated with a version of the *Führerprinzip* that empowers an individual to rule on behalf of a people. Even the government of India, struggling for decades to model democracy for a people who will soon number a billion, longs for great leaders; and for every Mahatma Gandhi, Indira Gandhi, or Rajiv Gandhi taken from them by zealous assassins, the Indians appear to seek a replacement who will deliver them from the lengthy travail of their freedom.

THE CONFEDERAL OPTION

How can democracy be secured and spread in a world whose primary tendencies are at best indifferent to it (McWorld) and at worst deeply antithetical to it (Jihad)? My guess is that globalization will eventually vanquish retribalization. The ethos of material "civilization" has not yet encountered an obstacle it has been unable to thrust aside. Ortega may have grasped in the 1920s a clue to our own future in the coming millennium.

> Everyone sees the need of a new principle of life. But as always happens in similar crises—some people attempt to save the situation by an artificial intensification of the very principle which has led to decay. This is the meaning of the "nationalist" outburst

of recent years. . . . things have always gone that way. The last flare, the longest; the last sign, the deepest. On the very eve of their disappearance there is an intensification of frontiers—military and economic.

Jihad may be a last deep sigh before the eternal yawn of McWorld. On the other hand, Ortega was not exactly prescient; his prophecy of peace and internationalism came just before blitzkrieg, world war, and the Holocaust tore the old order to bits. Yet democracy is how we remonstrate with reality, the rebuke our aspirations offer to history. And if retribalization is inhospitable to democracy, there is nonetheless a form of democratic government that can accommodate parochialism and communitarianism, one that can even save them from their defects and make them more tolerant and participatory: decentralized participatory democracy. And if McWorld is indifferent to democracy, there is nonetheless a form of democratic government that suits global markets passably well—representative government in its federal or, better still, confederal variation.

With its concern for accountability, the protection of minorities, and the universal rule of law, a confederalized representative system would serve the political needs of McWorld as well as oligarchic bureaucratism or meritocratic elitism is currently doing. As we are already beginning to see, many nations may survive in the long term only as confederations that afford local regions smaller than "nations" extensive jurisdiction. Recommended reading for democrats of the twenty-first century is not the U.S. Constitution or the French Declaration of Rights of Man and Citizen but the Articles of Confederation, that suddenly pertinent document that stitched together the thirteen American colonies into what then seemed a too loose confederation of independent states but now appears a new form of political realism, as veterans of Yeltsin's new Russia and the new Europe created at Maastricht will attest.

By the same token, the participatory and direct form of democracy that engages citizens in civic activity and civic judgment and goes well beyond just voting and accountability—the system I have called "strong democracy"—suits the political needs of decentralized communities as well as theocratic and nationalist party dictatorships have done. Local neighborhoods need not be democratic, but they can be. Real democracy has flourished in diminutive settings: the spirit of liberty, Tocqueville said, is local. Participatory democracy, if not naturally apposite to tribalism, has an undeniable attractiveness under conditions of parochialism.

Democracy in any of these variations will, however, continue to be obstructed by the undemocratic and antidemocratic trends toward uniformitarian globalism and intolerant retribalization which I have portrayed here. For democracy to persist in our brave new McWorld, we will have to commit acts of conscious political will—a possibility, but hardly a probability, under these conditions. Political will requires much more than the quick fix of the transfer of institutions. Like technology transfer, institution transfer rests on foolish assumptions about a uniform world of the kind that once fired the imagination of colonial administrators. Spread English justice to the colonies by exporting wigs. Let an East Indian trading company act as vanguard to Britain's free parliamentary institutions. Today's well-intentioned quick-fixers in the National Endowment for Democracy and the Kennedy School of Government, in the unions and foundations and universities zealously nurturing contacts in Eastern Europe and the Third World, are hoping to democratize by long distance. Post Bulgaria a parliament by first-class mail. Fed Ex the Bill of Rights to Sri Lanka. Cable Cambodia some common law.

Yet Eastern Europe has already demonstrated that importing free political parties, parliaments, and presses cannot establish a democratic civil society; imposing a free market may even have the opposite effect. Democracy grows from the bottom up and cannot be imposed from the top down. Civil society has to be built from the inside out. The institutional superstructure comes last.

Poland may become democratic, but then again it may heed the Pope, and prefer to found its politics on its Catholicism, with uncertain consequences for democracy. Bulgaria may become democratic, but it may prefer tribal war. The former Soviet Union may become a democratic confederation, or it may just grow into an anarchic and weak conglomeration of markets for other nations' goods and services.

Democrats need to seek out indigenous democratic impulses. There is always a desire for self-government, always some expression of participation, accountability, consent, and representation, even in traditional hierarchical societies. These need to be identified, tapped, modified, and incorporated into new democratic practices with an indigenous flavor. The tortoises among the democratizers may ultimately outlive or outpace the hares, for they will have the time and patience to explore conditions along the way, and to adapt their gait to changing circumstances. Tragically, democracy in a hurry often looks something like France in 1794 or China in 1989.

It certainly seems possible that the most attractive democratic ideal in the face of the brutal realities of Jihad and the dull realities of McWorld will be a confederal union of semiautonomous communities smaller than nation-stares, tied together into regional economic associations and markets larger than nation-states—participatory and self-determining in local matters at the bottom, representative and accountable at the top. The nation-state would play a diminished role, and sovereignty would lose some of its political potency. The Green movement adage "Think globally, act locally" would actually come to describe the conduct of politics.

This vision reflects only an ideal, however—one that is not terribly likely ro be realized. Freedom, Jean-Jacques Rousseau once wrote, is a food easy to eat but hard to digest. Still, democracy has always played itself out against the odds. And democracy remains both a form of coherence as binding as McWorld and a secular faith potentially as inspiriting as Jihad.

14

States of Bondage

MANFRED NOWAK

Torture is one of the most barbaric acts of state repression, and it constitutes a direct and deliberate attack on the core of the human personality. Like slavery, it is an expression of the almost unlimited power of one individual over another. In the case of slavery, the human being is degraded to the condition of a non-human object deprived of legal personality. Torture aims to destroy human dignity and reduce the victim to the status of a passive tool in the hands of the torturer.

In ancient and medieval times in Europe, torture was employed to aggravate criminal punishments—usually the death penalty—and to extort confessions. Its use was an officially accepted and legally regulated aspect of the criminal justice system.

Torture was officially abolished in all European countries between 1750 and 1830. Like the abolition of slavery, its suppression was the fruit of the humanism and rationalism of the Enlightenment. Although torture continued to be applied behind prison walls, there were comparably few allegations of its systematic use in the late nineteenth and early twentieth centuries. Whereas slavery and the slave trade were explicitly prohibited by a number of bilateral and multilateral treaties culminating in the 1926 Slavery Convention, torture was so much regarded as a phenomenon of the past that neither international human rights law nor even most domestic bills of rights of this period contained explicit prohibitions.

Under National Socialism in Germany and Stalinism in the U.S.S.R. torture was again practised in a systematic, albeit clandestine manner. It acquired a new rationale as a means of state repression against political, ethnic and religious "enemies of the people." As well as being used as a punishment and in obtaining confessions, torture now served additional functions: the extortion of denunciations and other information about third persons, intimidation and discouragement of the victim, and the creation of a general atmosphere of fear as a violent means of deterrence and repression.

Since these modern functions proved to be fairly efficient in combatting criminality and terrorism as well as "subversion" and other opposition tendencies, torture unfortunately began to spread again after the Second World War, to such an extent that it is now sometimes referred to as "the plague of the twentieth century." It was, for instance, systematically practised by European colonial powers against decolonization movements in the 1950s and 1960s (well-documented cases include those under French rule in Algeria and under Portuguese rule in that nation's former African colonies), under British anti-terrorist legislation in Northern Ireland, by the Greek colonels in the late 1960s, by many Latin American military dictatorships based on the ideology of "national security," by African dictators, by communist regimes, and by many other governments in many other regions.

WHAT IS TORTURE?

The international movement against torture can only rest on humanitarian grounds and on the moral, political and legal force of human rights. In contrast to their practice with regard to torture in earlier centuries and to most other human rights violations today, governments now practically never admit that they order or even tolerate its use. Condemnation of these barbaric acts is universal, which might make the "mobilization of shame" somewhat easier than, for instance, in the case of the campaign against capital punishment. On the other hand, the same fact leads governments to apply torture primarily in remote and clandestine detention centres on prisoners who are held incommunicado, which makes the gathering of evidence about its use usually very difficult. It also increases the risk of victims being killed after having being subjected to torture.

Torture and other forms of cruel, inhuman and degrading treatment or punishment are prohibited in various international treaties and agreements generally considered to have the force of law, among them the Geneva Conventions of 1949, the International Covenant on Civil and Political Rights of 1966 (CCPR), the 1984 UN Convention against Torture (CAT), the 1950 European Convention on Human Rights, the American Convention on Human Rights of 1969, and the African Charter on Human and Peoples' Rights of 1981. In addition article 5 of the Universal Declaration of Human Rights of 1948 is generally regarded as part of customary international law. The Vienna Declaration and Programme of Action again confirmed in the strongest terms that freedom from torture is a right that must be protected under all circumstances.

Despite this impressive evidence regarding the universality of the right to freedom from torture, doubts remain as to whether this universal standard can be applied equally to all political systems, religious and cultural groups. The governments of certain states claim that amputations and similar forms of corporal punishment provided for by Islamic law do not contravene the right to freedom from torture. While feminist groups around the world denounce female circumcision as a form of torture prohibited under international law, many Africans view the practice as an expression of their traditional culture. Similarly, the admissible minimum standard of prison conditions is said to depend on the cultural background

as well as the level of socio-economic development of the respective country.

These examples show that even an absolute right to freedom from torture may be relativized as a result of different interpretations given to the term in different cultures. Even in a single cultural environment, opinions as to what constitutes torture can differ considerably. In the case of Northern Ireland, the European Commission of Human Rights considered the so-called five techniques used by British security forces during interrogation (hooding detainees, subjecting them to constant and intense noise, depriving them of sleep and sufficient food and drink, and making them stand for long periods on their toes against a wall in painful posture) as torture; but the European Court of Human Rights viewed them only as inhuman treatment.

The drafters of the 1984 UN Convention Against Torture (CAT) aimed at solving the problem of defining a universal minimum standard of unacceptable behavior by elaborating a precise definition of torture. According to the Convention, torture is an act of public officials that intentionally inflicts severe physical or mental pain in order to fulfil a certain purpose, such as the extortion of information or confessions. The concept of severe pain is itself of course partly subjective.

The CAT definition further excludes "pain or suffering arising only from, inherent in or incidental to lawful sanctions." This limitation has been heavily criticized as an unacceptable loophole. Who is to decide whether a sanction such as corporal punishment is lawful? If cultural diversity is to be taken into account, the obvious answer would be the domestic legislature involved. Yet to leave the question exclusively with national authorities would strip the right to freedom from torture of its regulatory function. In the current circumstances, it is hard not to agree with the observation of the Special Rapporteur on Torture, whose post was created by the UN Commission on Human Rights in 1985, that "it is international law, and not domestic law, which ultimately determines whether a certain practice may be regarded as lawful."

In defining torture and other forms of inhuman treatment or punishment, one therefore has to strike a careful balance between the need for common universal minimum standards and the requirement to take political, social, religious and cultural particularities into account. This can only be done on a case-by-case basis by the competent international bodies. In cases involving certain dictatorships such as that which formerly existed in Uruguay, the Human Rights Committee of the United Nations found that interrogation techniques that included systematic beatings, electric shocks, burns, mock executions, hanging prisoners for extended periods by their arms or forcing their heads under water constituted torture and violated Article 7 of the International Covenant on Civil and Political Rights.

By these universal standards, there can be no doubt that the more severe forms of corporal punishment practised under certain religious laws, including amputation and stoning, would constitute inhuman or cruel punishment or even torture. Whether states under present international law are under a positive obligation to prohibit female circumcision is more difficult to judge. Since these traditional practices inflict severe physical and mental pain on the girls and women concerned and are, moreover, discriminatory on the ground of gender, states should at least try to prevent them by educational and similar means.

TORTURE AND HUMAN RIGHTS

Since most cases of torture take place in detention and/or are related to criminal proceedings, an indivisible relationship exists between freedom from torture, personal liberty and the right to a fair trial. To strengthen the guarantees of these other human rights may, therefore, have the indirect effect of preventing or reducing the occurrence of torture. Consequently, states should ensure the following minimum guarantees, which are not dependent on the political, social or cultural background of the country concerned:

- The absolute prohibition of incommunicado detention;
- The right of arrested persons to immediately contact relatives, a lawyer and a doctor;
- The right of arrested persons to be brought promptly (within forty-eight hours of arrest) to a judge and to have a medical examination by an independent doctor;
- Supervision of interrogations by an independent authority;
- Prohibition of the use of statements obtained by torture as evidence before courts.

Some of these standards can be derived from present international law; others should be added as a means to prevent torture. Another provision that links freedom from torture with personal liberty is the right of detained persons under Article 10 of the Covenant on Civil and Political Rights to be treated with humanity and dignity. This right provides a guarantee against harsh prison conditions that nonetheless do not amount to inhuman or degrading treatment as defined by Article 7. Again, what is considered inhuman will reflect a country's cultural background.

Nevertheless, the United Nations has adopted standard minimum rules for the treatment of prisoners. In one recent case, the UN Human Rights Committee even found that allowances of just five minutes a day for personal hygiene and for open-air exercise constitute inhuman prison conditions. Since it had already earlier expressed the view that states must establish a minimum standard of conditions of detention regardless of economic difficulties, this comparatively high standard will have to be applied equally in all States Parties to the Covenant.

In view of the fact that, despite its unanimous, absolute and universal condemnation, torture was practised on such an unprecedented scale in the 1970s, Amnesty International, the International Commission of Jurists and other NGOs called for more efficient and innovative implementation measures. In fact the traditional monitoring mechanisms under the terms of the Covenant on Civil and Political Rights and regional conventions only apply retrospectively, usually years after the actual violation has taken place, and offer little relief to the victim. A person under imminent threat of torture needs either immediate intervention or else international measures that effectively deter potential perpetrators.

CURRENT TRENDS

The main thrust of the Convention against Torture is the punishment of torturers by means of the criminal law. States Parties are obliged to treat all acts of torture as criminal offences, and to assign appropriate penalties for them. The CAT provides for universal jurisdiction, which means that torturers can be detained, prosecuted and punished in all States Parties regardless of the nationality of the torturer or his victims or the territory in which the act of torture was perpetrated. In addition to the traditional monitoring procedures, Article 20 of the CAT authorizes the Committee against Torture in cases of alleged systematic torture to carry out confidential enquiries, including visits to the territory of the state concerned if it gives its consent.

The experiences of the first six years of CAT's implementation have not been particularly encouraging. States are extremely hesitant about applying universal criminal jurisdiction against foreign nationals, and the Committee against Torture seems to be more occupied with its traditional task of examining state reports than with adopting a new and more efficient approach. However, in November 1993 the Committee issued for the first time a public statement in which it confirmed the practice of torture in Turkey.

The Special Rapporteur on Torture has a worldwide mandate to examine questions relevant to torture and to report on its occurrence and on the extent to which it was practised in all countries. His activities consist of communicating with governments on questions relating to torture, launching urgent appeals when he receives information that a person is at risk, paying visits to countries which invite him for consultation, analysing the phenomenon of torture and the

root causes conducive to it, and recommending measures for improvement.

To permit more efficient action to prevent torture, the Special Rapporteur time and again stressed the need for a universal system of preventive visits to places of detention. This system, originally proposed by Jean-Jacques Gautier and the Swiss Committee against Torture, is based on a very simple idea: states should give an international body the right to carry out visits to places within their jurisdiction where persons are deprived of their liberty by a public authority. This international body would then draw up a report on its findings and make the necessary recommendations. The mere fact of an international body having the right to inspect places of detention without giving previous notice would presumably have a deterrent effect on those responsible for torture.

Thus far, the fear that such a system would interfere too much with states' internal affairs has prevented its adoption at the universal level. That is why the Council of Europe took up the idea and adopted, in 1987, the European Convention for the Prevention of Torture, which follows the lines described above, and entered into force in 1989. The European Committee for the Prevention of Torture has made a very dynamic start by carrying out regular visits to all States Parties and by submitting very thorough and critical reports with detailed recommendations, which are taken seriously by governments.

On the basis of these encouraging experiences, efforts to establish a similar non-juridical and non-bureaucratic system at the universal level also have gained force in recent times. In 1992, the Commission on Human Rights set up an open-ended intersessional working group in order to elaborate a draft Optional Protocol to the UN Convention against Torture. In June 1993 the Vienna World Conference on Human Rights reaffirmed that efforts to eradicate torture should, first and foremost, be concentrated on prevention and therefore called for the early adoption of this draft.

• • •

4

Gender and Sexual Orientation

*Is it to be understood that the principles of the
Declaration of Independence bear no relation to half
of the human race?*

HARRIET MARTINEAU

Gender is one of the most sharply marked social divisions in the world. In every society in the world, the biological differences between the male and female *sex* are used as the basis for assigning human beings into separate *gender roles*. In many human societies, males and females have historically been placed into such different roles that we still have difficulty envisioning our common humanity. Julius Lester's "Being a Boy" illustrates the way in which people are shaped by gender roles, and makes the point that being assigned a gender role is hardly a painless process, for either boys or girls: "No, it wasn't easy for any of us, girls and boys, as we forced our beautiful, free-flowing child-selves into those narrow, constricting cubicles labeled *female* and *male*." Jamaica Kincaid's "Girl" reminds us of the tremendous effort parents in most societies put into teaching children their respective gender roles.

The contemporary study of gender distinguishes among *sex, gender, gender identity,* and *sexual orientation*. *Sex* refers to the biological differences between males and females. *Gender* refers to the social roles played by males and females in a given society. *Gender identity* refers to an individual's identification with one gender or the other. *Sexual orientation* refers to an individual's sexual attraction, which may be to one sex, or the other, or to both. In American culture, and many others, all of these phenomena are equated. Thus, for example, it is believed that a person born biologically male will grow up to play what we consider typically masculine roles, will identify with masculinity, and will be sexually attracted to females. A good deal of social control is exercised to prevent and punish deviance from this model. *Heterosexism* refers to prejudice and discrimination against nonheterosexuals. *Homophobia* refers to the irrational hatred and fear of homosexuals. Research reveals that this model is only one way for gender to be socially constructed. Gender roles vary from society to society, such that it is not possible to say that certain behaviors are "naturally" either masculine or feminine. Some societies do not identify people as "homosexual," or stigmatize behavior that would be labeled "homosexual" in the United States today. In a number of non-European cultures, a "third gender" is recognized, in which people born as biological males or females, or as intersexuals (people whose biological sex is ambiguous), play a gender role that is neither typically masculine nor feminine, and are not exclusively heterosexual in their sexual orientation (Weinrich and Williams, 1991). And in societies characterized by heterosexism, there have always been many people who have disobeyed the heterosexual model to pursue alternative gender identities and homosexual or bisexual sexual orientations. In "He Defies You Still: The Memoirs of a Sissy," Tommi Avicolli writes about surviving the painful experiences of growing up gay in the United States.

Gender inequality is a worldwide social problem. Although in the very earliest human societies men and women played different, but relatively equal roles, most societies in the world today are *patriarchies*, or societies in which men have more power than women. In patriarchal societies, gender inequality is structured into society's various organizations and institutions, such as the workplace, politics, education, and the family. These institutions are *sexist*—that is, they contain processes of *prejudice, stereotyping,* and *discrimination* against women, which ensure that men are privileged by the way the institutions operate. This results in *gender stratification*, a situation in which men have greater access to social resources of all kinds—property, money, jobs, personal liberties, education, health care—and tend to hold on to such advantages over time. In "The Burden of Womanhood," John Ward Anderson and Molly Moore present a chilling overview of gender inequality in the poor nations of the world.

Why do gender roles and gender inequality exist? The structural-functionalist perspective argues that gender roles were beneficial to societies in preindustrial, prescientific times. Because preindustrial production required heavy physical labor, males, with their greater body strength, worked as hunters and tillers of the land. Because women's bodies were inextricably linked to childbirth and breast-feeding, females raised children and took care of the home. Because it was essential to the survival of early societies that the tasks of production and reproduction were properly done, human beings took great care to build gender roles into their cultures and social structures. Today, however, modern technology has rendered rigid gender roles obsolete. Women's strength is more than sufficient to operate computer technology, for example, and women's bodies have been freed from biological necessity by contraception, bottle-feeding, and new reproductive technologies. In the structural-functionalist perspective, gender roles are rapidly changing due to modernization. If they hang on, it is because institutions tend to change slowly, but change they will. Some structural-functionalists are concerned that gender roles are changing too rapidly, and call for reforms to mitigate dysfunctional change. In "American Family Decline, 1960–1990," David Popenoe examines the weakening of the family and argues that this is a cause for alarm, especially because of the negative effects on children.

The order perspective has been strongly criticized by sociologists utilizing the social conflict perspective, including *feminist* sociologists who have been influenced by the social movements for women's rights. These sociologists have pointed

out that early human societies contained different gender roles without having gender inequality. Thus there does not seem to be an essential connection between biological differences and gender inequality. Although no one knows exactly the origins of gender inequality, it seems to have arisen thousands of years ago when societies evolved from a reliance on hunting and gathering to agriculture and trade. In the course of this evolution, men took control over property in land, tools, and commodities, and societies began to define women as male property, to be utilized to produce male heirs and confined to a life of work inside the home. When capitalist industrial societies evolved out of these agrarian patriarchies, they perpetuated gender inequality, resulting in *capitalist patriarchies*. In capitalist patriarchal societies, women are no longer confined to the "private sphere" of the home, and they go out to work for wages in the market economy. However, they do not enjoy equality with men at work, or in other aspects of social life. From this perspective, gender inequality persists in the world today, not because it is a relic of backward cultures and societies, but because men, particularly those men who are in control of corporations and other powerful organizations and institutions, have a vested interest in it. The connections between global capitalism and gender inequality are explored in "The Globetrotting Sneaker" by Cynthia Enloe.

Women everywhere, together with male allies, are joining together in movements to reduce and eliminate gender inequality. Likewise, movements of gay/lesbian/transsexual/transgendered/intersexual people are actively working to restructure the model of gender in heterosexist societies. These movements are by no means confined to the affluent. As "Widows Banding Together," by Margaret Owen, makes clear, even the poorest and most powerless women in impoverished nations such as Bangladesh and India are making their voices heard. At the 1995 United Nations Fourth World Conference on Women, held in Beijing, many women from grassroots organizations around the world came to attend the parallel "NGO Forum on Women." This event demonstrated the growing power of grassroots women's organizing, and highlighted the concept that *"women's rights are human rights."* Movements to transform our conceptions of gender often emphasize the need for unity among all those who suffer under various forms of domination in the world. This idea has never been better expressed than by Pastor Martin Niemuller in his reflection on how the Nazis came to power in Europe during the 1930s. His famous remarks, which begin with "First they came after the homosexuals," bring this chapter to a close.

As a result of the global women's movement, it is increasingly recognized that women's inequality contributes to other urgent world problems. International agencies now identify women's rights as an indispensable prerequisite for solving such problems as overpopulation, economic underdevelopment, and environmental destruction. These links will be explored in Chapter 5 on Population Growth and Chapter 6 on Environmental Destruction.

WORKS CITED

Weinrich, James D., and Walter L. Williams. 1991. "Strange Customs, Familiar Lives: Homosexualities in Other Cultures," in John C. Gonsiorek and James D.

Weinrich, eds., 1991. *Homosexuality: Research Implications for Public Policy.* Newbury Park, CA: Sage Publications, pp. 44–59.

QUESTIONS FOR DISCUSSION

1.   Using the readings by Lester, Kincaid, and Avicolli, explain the idea that "gender is *socially constructed.*"

2.   What do Anderson and Moore mean by "the burden of womanhood"? Do any of the conditions discussed in their article also exist in wealthy nations such as the United States? Discuss.

3.   After reading "American Family Decline, 1960–1990" and "The Globetrotting Sneaker," imagine a conversation between David Popenoe and Cynthia

Enloe. You have just asked them to answer the following question: "What is the most important problem facing families today?" How would each of them answer? How would each respond to the other? Which person do you agree with in this imaginary conversation?

4.   "Widows Banding Together" discusses a "bottom up" movement for the rights of widows in India. How does this movement illustrate the idea that "women's rights are *human* rights"?

 ## INFOTRAC COLLEGE EDITION: EXERCISE

Choose a specific aspect of gender inequality that interests you, and learn more about it, using Infotrac College Edition. For example, enter search terms such as female genital mutilation, sweatshops, child labor, or infanticide (go to Subdivisions: International Aspects).

Then click on the Periodical References and find and read several articles that look interesting on your topic. What did you learn about the definition of, causes of, and solutions to this problem?

FOR ADDITIONAL RESEARCH

Books

Basu, Amrita. 1995. *The Challenge of Local Feminisms.* Boulder, CO: Westview Press.

Blumberg, Rae Lesser, Cathy A. Rakowski, Irene Tinker, and Michael Monteon, eds. *EnGENDERing Wealth and Well-Being.* Boulder, CO: Westview Press.

Enloe, Cynthia. 1989. *Bananas Beaches and Bases.* Berkeley, CA: University of California Press.

Organizations

Boston Women's Health Book Collective
P.O. Box 192
West Somerville, MA 02144
(617) 625–0277

National Organization for Women
1000 16th Street NW, Suite 700
Washington, DC 20036-5705
(202) 331–0066
www.now.org

Women's International Information Project
c/o beyondmedia
59 East Van Buren Street, 14th Floor
Chicago, IL 60605
(312) 922-7780
www.beyondmedia.org

Women's International League for Peace and Freedom
1213 Race Street
Philadelphia, PA 19107
(215) 563-7110
Chris@wilpf.org

ACTION PROJECTS

1.   Go to the United Nations Web site (www.un.org) and locate the sections on women. Find the United Nations Convention on the Elimination of All Forms of Discrimination Against Women. Note the various forms of discrimination targeted in this document. Find out how many nations have ratified—agreed to be legally bound by the convention. Also locate the Platform for Action document that came out of the 1995 Fourth United Nations World Conference on Women held in Beijing. How does this plan differ from the Convention? Finally, find the National Plan of Action section, and read the plan the United States has made for implementation. Compare the U.S. plan with that of another country that interests you.

2.   Get together with other students and watch a video that deals with gender issues in global perspective. Some videos that may be available in your local video store include *Mississippi Massala, The Story of Qiu Ju,* and *Tilal (The Law).* (Look in the foreign films section.) Your col-

lege library may have some documentaries, such as *Small Happiness, Miss India Georgia, Fast Food Women, Dream Worlds,* or *Global Assembly Line.* Discuss how the video depicted gender issues. What problems did the video deal with? How were these defined, by the characters and/or filmmaker? Did the video frame the causes of the problem, and if so, how? Were any solutions depicted? How realistic was the video, given what you have learned in class about the problem of women's rights?

3.   Contact one of the organizations on the list above, or your local Women's Center, to learn about their goals, programs, and activities. If the organization is conducting a campaign and is looking for volunteers, get together with other students and organize a campus project. Or perhaps you can volunteer some time to help the organization on an individual basis. Reflect on what you learned from this project in relation to the topic of women's rights as human rights.

15

Three Voices on Gender

JULIUS LESTER
JAMAICA KINCAID
TOMMI AVICOLLI

JULIUS LESTER

Being a Boy

As boys go. I wasn't much. I mean, I tried to be a boy and spent many childhood hours pummeling my hardly formed ego with failure at cowboys and Indians, baseball, football, lying, and sneaking out of the house. When our neighborhood gang raided a neighbor's pear tree, I was the only one who got sick from the purloined fruit. I also failed at setting fire to our garage, an art at which any five-year-old boy should be adept. I was, however, the neighborhood champion at getting beat up. "That Julius can take it, man," the boys used to say, almost in admiration, after I emerged from another battle, tears brimming in my eyes but refusing to fall.

My efforts at being a boy earned me a pair of scarred knees that are a record of a childhood spent falling from bicycles, trees, the tops of fences, and porch steps; of tripping as I ran (generally from a fight), walked, or simply tried to remain upright on windy days.

I tried to believe my parents when they told me I was a boy, but I could find no objective proof for such an assertion. Each morning during the summer, as I cuddled up in the quiet of a corner with a book, my mother would push me out the back door and into the yard. And throughout the day as my blood was let as if I were a patient of 17th-century medicine, I thought of the girls sitting in the shade of porches, playing with their dolls, toy refrigerators and stoves.

There was the life, I thought! No constant pressure to prove oneself. No necessity always to be competing. While I humiliated myself on football and baseball fields, the girls stood on the sidelines laughing at me, because they didn't have to do anything except be girls. The rising of each sun brought me to the starting line of yet another day's Olympic decathlon, with no hope of ever winning even a bronze medal.

Through no fault of my own I reached adolescence. While the pressure to prove myself on the athletic field lessened, the overall situation got worse—because now I had to prove myself with girls. Just how I was supposed to go about doing this was beyond me, especially because, at the age of 14, I was four foot nine and weighed 78 pounds. (I think there may have been one 10-year-old girl in the neighborhood smaller than I.) Nonetheless, duty called, and with my ninth-grade gym-class jockstrap flapping between my legs, off I went.

To get a girlfriend, though, a boy had to have some asset beyond the fact that he was alive. I wasn't handsome like Bill McCord, who had girls after him like a cop killer has policemen. I wasn't ugly like Romeo Jones, but at least the girls noticed him: "That ol' ugly boy better stay 'way from me!" I was just there, like a vase your grandmother gives you at Christmas that you don't like or dislike, can't get rid of, and don't know what to do with. More than ever

I wished I were a girl. Boys were the ones who had to take the initiative and all the responsibility. (I hate responsibility so much that if my heart didn't beat of itself, I would now be a dim memory.)

It was the boy who had to ask the girl for a date, a frightening enough prospect until it occurred to me that she might say no! That meant risking my ego, which was about as substantial as a toilet-paper raincoat in the African rainy season. But I had to thrust that ego forward to be judged, accepted, or rejected by some girl. It wasn't fair! Who was she to sit back like a queen with the power to create joy by her consent or destruction by her denial? It wasn't fair—but that's the way it was.

But if (God forbid!) she should say Yes, then my problem would begin in earnest, because I was the one who said where we would go (and waited in terror for her approval of my choice). I was the one who picked her up at her house where I was inspected by her parents as if I were a possible carrier of syphilis (which I didn't think one could get from masturbating, but then again, Jesus was born of a virgin, so what did I know?). Once we were on our way, it was I who had to pay the bus fare, the price of the movie tickets, and whatever she decided to stuff her stomach with afterward. (And the smallest girls are all stomach.) Finally, the girl was taken home where once again I was inspected (the father looking covertly at my fly and the mother examining the girl's hair). The evening was over and the girl had done nothing except honor me with her presence. All the work had been mine.

Imagining this procedure over and over was more than enough: I was a sophomore in college before I had my first date.

I wasn't a total failure in high school, though, for occasionally I would go to a party, determined to salvage my self-esteem. The parties usually took place in somebody's darkened basement. There was generally a surreptitious wine bottle or two being passed furtively among the boys, and a record player with an insatiable appetite for Johnny Mathis records. Boys gathered on one side of the room and girls on the other. There were always a few boys and girls who'd come to the party for the sole purpose of grinding away their sexual frustrations to Johnny Mathis's falsetto, and they would begin dancing to their own music before the record player was plugged in. It took a little longer for others to get started, but no one matched my talent for standing by the punch bowl. For hours, I would try to make my legs do what they had been doing without effort since I was nine months old, but for some reason they would show all the symptoms of paralysis on those evenings.

After several hours of wondering whether I was going to die ("Julius Lester, a sixteen-year-old, died at a party last night, a half-eaten Ritz cracker in one hand and a potato chip dipped in pimiento-cheese spread in the other. Cause of death: failure to be a boy"), I would push my way to the other side of the room where the girls sat like a hanging jury. I would pass by the girl I wanted to dance with. If I was going to be refused, let it be by someone I didn't particularly like. Unfortunately, there weren't many in that category. I had more crushes than I had pimples.

Finally, through what surely could only have been the direct intervention of the Almighty, I would find myself on the dance floor with a girl. And none of my prior agony could compare to the thought of actually dancing. But there I was and I had to dance with her. Social custom decreed that I was supposed to lead, because I was the boy. Why? I'd wonder. Let her lead. Girls were better dancers anyway. It didn't matter. She stood there waiting for me to take charge. She wouldn't have been worse off if she'd waited for me to turn white.

But, reciting "Invictus" to myself, I placed my arms around her, being careful to keep my armpits closed because, somehow, I had managed to overwhelm a half jar of deodorant and a good-size bottle of cologne. With sweaty armpits, "Invictus," and legs afflicted again with polio, I took her in my arms, careful not to hold her so far away that she would think I didn't like her, but equally careful not to hold her so close that she could feel the catastrophe which had befallen me the instant I touched her hand. My penis, totally disobeying the lecture I'd given it before we left home, was as rigid as Governor Wallace's jaw would be if I asked for his daughter's hand in marriage.

God, how I envied girls at that moment. Wherever *it* was on them, it didn't dangle between their legs like an elephant's trunk. No wonder boys talked about nothing but sex. That thing was always there. Every time we went to the john, there *it* was, twitching around like a fat little worm on a fishing hook. When we took baths, it floated in the water like a lazy fish and God forbid we should touch it! It sprang to life like lightning leaping from a cloud. I wished I could cut it off, or at least keep it tucked between my legs, as if it were a tail that had been mistakenly attached to the wrong end. But I was helpless. It was there, with a life and mind of its own, having no other function than to embarrass me.

Fortunately, the girls I danced with were discreet and pretended that they felt nothing unusual rubbing against them as we danced. But I was always convinced that the next day they were all calling up their friends to exclaim: "Guess what, girl? Julius Lester got one! I ain't lyin'!"

Now, of course, I know that it was as difficult being a girl as it was a boy, if not more so. While I stood paralyzed at one end of a dance floor trying to find the courage to ask a girl for a dance, most of the girls waited in terror at the other, afraid that no one, not even I, would ask them. And while I resented having to ask a girl for a date, wasn't it also horrible to be the one who waited for the phone to ring? And how many of those girls who laughed at me making a fool of myself on the baseball diamond would have gladly given up their places on the sidelines for mine on the field?

No, it wasn't easy for any of us, girls and boys, as we forced our beautiful, free-flowing child-selves into those narrow, constricting cubicles labeled *female* and *male*. I tried, but I wasn't good at being a boy. Now, I'm glad, knowing that a man is nothing but the figment of a penis's imagination, and any man should want to be something more than that.

JAMAICA KINCAID

Girl

• • •

Wash the white clothes on Monday and put them on the stone heap; wash the color clothes on Tuesday and put them on the clothesline to dry;

don't walk barehead in the hot sun; cook pumpkin fritters in very hot sweet oil; soak your little clothes right after you take them off; when buying cotton to make yourself a nice blouse, be sure that it doesn't have gum on it, because that way it won't hold up well after a wash; soak salt fish overnight before you cook it; is it true that you sing benna in Sunday School?; always eat your food in such a way that it won't turn someone else's stomach; on Sundays try to walk like a lady and not like the slut you are so bent on becoming; don't sing benna in Sunday School; you mustn't speak to wharf-rat boys, not even to give directions; don't eat fruits on the street—flies will follow you; *but I don't sing benna on Sundays at all and never in Sunday School;* this is how to sew on a button; this is how to make a buttonhole for the button you have just sewed on; this is how to hem a dress when you see the hem coming down and so to prevent yourself from looking like the slut I know you are so bent on becoming; this is how you iron your father's khaki shirt so that it doesn't have a crease; this is how you iron your father's khaki pants so that they don't have a crease; this is how you grow okra—far from the house, because okra tree harbors red ants; when you are growing dasheen, make sure it gets plenty of water or else it makes your throat itch when you are eating it; this is how you sweep a corner; this is how you sweep a whole house; this is how you sweep a yard; this is how you smile to someone you don't like too much; this is how you smile to someone you don't like at all; this is how you smile to someone you like completely; this is how you set a table for tea; this is how you set a table for dinner; this is how you set a table for dinner with an important guest; this is how you set a table for lunch; this is how you set a table for breakfast; this is how to behave in the presence of men who don't know you very well, and this way they won't recognize immediately the slut I have warned you against becoming; be sure to wash every day, even if it is with your own spit; don't squat down to play marbles—you are not a boy, you know; don't pick people's flowers—you might catch something; don't throw stones at blackbirds, because it might not be a blackbird at all; this is how to make a bread pudding; this is how to make doukona; this is how to make pepper pot; this is how to make a good medicine for a cold; this is how to make a good medicine to throw away a child before it even becomes a child; this is how to catch a fish; this is how to throw back a fish you don't like, and that way something bad won't fall on you; this is how to bully a man; this is how a man bullies you; this is how to love a man, and if this doesn't work there are other ways, and if they don't work don't feel too bad about giving up; this is how to spit up in the air if you feel like it, and this is how to move quick so that it doesn't fall on you; this is how to make ends meet; always squeeze bread to make sure it's fresh; *but what if the baker won't let me feel the bread?;* you mean to say that after all you are really going to be the kind of woman who the baker won't let near the bread?

TOMMI AVICOLLI

He Defies You Still:
The Memoirs of a Sissy

You're just a faggot
No history faces you this morning
A faggot's dreams are scarlet
Bad blood bled from words that scarred

SCENE ONE

A homeroom in a Catholic high school in South Philadelphia. The boy sits quietly in the first aisle, third desk, reading a book. He does not look up, not even for a moment. He is hoping no one will remember he is sitting there. He wishes he were invisible. The teacher is not yet in the classroom so the other boys are talking and laughing loudly.

Suddenly, a voice from beside him:

"Hey, you're a faggot, ain't you?"

The boy does not answer. He goes on reading his book, or rather pretending he is reading his book. It is impossible to actually read the book now.

"Hey, I'm talking to you!"

The boy still does not look up. He is so scared his heart is thumping madly; it feels like it is leaping out of his chest and into his throat. But he can't look up.

"Faggot, I'm talking to you!"

To look up is to meet the eyes of the tormentor.

Suddenly, a sharpened pencil point is thrust into the boy's arm. He jolts, shaking off the pencil, aware that there is blood seeping from the wound.

"What did you do that for?" he asks timidly.

"Cause I hate faggots," the other boy says, laughing. Some other boys begin to laugh, too. A symphony of laughter. The boy feels as if he's going to cry. But he must not cry. Must not cry. So he holds back the tears and tries to read the book again. He must read the book. Read the book.

When the teacher arrives a few minutes later, the class quiets down. The boy does not tell the teacher what has happened. He spits on the wound to clean it, dabbing it with a tissue until the bleeding stops. For weeks he fears some dreadful infection from the lead in the pencil point.

SCENE TWO

The boy is walking home from school. A group of boys (two, maybe three, he is not certain) grab him from behind, drag him into an alley and beat him up. When he gets home, he races up to his room, refusing dinner ("I don't feel well," he tells his mother through the locked door) and spends the night alone in the dark wishing he would die. . . .

These are not fictitious accounts—I *was* that boy. Having been branded a sissy by neighborhood children because I preferred jump rope to baseball and dolls to playing soldiers, I was often taunted with "hey sissy" or "hey faggot" or "yoo hoo honey" (in a mocking voice) when I left the house.

To avoid harassment, I spent many summers alone in my room. I went out on rainy days when the street was empty.

I came to like being alone. I didn't need anyone, I told myself over and over again. I was an island. Contact with others meant pain. Alone, I was protected. I began writing poems, then short stories. There was no reason to go outside anymore. I had a world of my own.

> In the schoolyard today
> they'll single you out
> Their laughter will leave your ears ringing
> like the church bells
> which once awed you. . . .

School was one of the more painful experiences of my youth. The neighborhood bullies could be avoided. The taunts of the children living in those endless repetitive row houses could be evaded by staying in my room. But school was something I had to face day after day for some two hundred mornings a year.

I had few friends in school. I was a pariah. Some kids would talk to me, but few wanted to be known as my close friend. Afraid of labels. If I was a sissy, then he had to be a sissy, too. I was condemned to loneliness.

Fortunately, a new boy moved into our neighborhood and befriended me; he wasn't afraid of the labels. He protected me when the other guys threatened to beat me up. He walked me home from school; he broke through the terrible loneliness. We were in third or fourth grade at the time.

We spent a summer or two together. Then his parents sent him to camp and I was once again confined to my room.

SCENE THREE

High school lunchroom. The boy sits at a table near the back of the room. Without warning, his lunch bag is grabbed and tossed to another table. Someone opens it and confiscates a package of Tastykakes; another boy takes the sandwich. The empty bag is tossed back to the boy who stares at it, dumbfounded. He should be used to this; it has happened before.

Someone screams, "faggot," laughing. There is always laughter. It does not annoy him anymore.

There is no teacher nearby. There is never a teacher around. And what would he say if there were? Could he report the crime? He would be jumped after school if he did. Besides, it would be his word against theirs. Teachers never noticed anything. They never heard the taunts. Never heard the word, "faggot." They were the great deaf mutes, pillars of indifference; a sissy's pain was not relevant to history and geography and god made me to love honor and obey him, amen.

SCENE FOUR

High school Religion class. Someone has a copy of *Playboy*. Father N. is not in the room yet; he's late, as usual. Someone taps the boy roughly on the shoulder. He turns. A finger points to the centerfold model, pink fleshy body, thin and sleek. Almost painted. Not real. The other asks, mocking voice, "Hey, does she turn you on? Look at those tits!"

The boy smiles, nodding meekly; turns away.

The other jabs him harder on the shoulder, "Hey, whatsamatter, don't you like girls?"

Laughter. Thousands of mouths; unbearable din of laughter. In the Arena: thumbs down. Don't spare the queer.

"Wanna suck my dick? Huh? That turn you on, faggot!"

The laughter seems to go on forever . . .

> Behind you, the sound of their laughter
> echoes a million times
> in a soundless place
> They watch you walk/sit/stand/breathe. . . .

What did being a sissy really mean? It was a way of walking (from the hips rather than the shoulders); it was a way of talking (often with a lisp or in a high-pitched voice); it was a way of relating to others (gently, not wanting to fight, or hurt anyone's feelings). It was being intelligent ("an egghead" they called it sometimes); getting good grades. It means not being interested in sports, not playing football in the street after school; not discussing teams and scores and play-offs. And it involved not showing fervent interest in girls, not talking about scoring with tits or *Playboy* centerfolds. Not concealing naked women in your history book; or porno books in your locker.

On the other hand, anyone could be a "faggot." It was a catch-all. If you did something that didn't conform to what was the acceptable behavior of the group, then you risked being called a faggot. If you didn't get along with the "in" crowd, you were a faggot. It was the most commonly used put-down. It kept guys in line. They became angry when somebody called them a faggot. More fights started over someone calling someone else a faggot than anything else. The word had power. It toppled the male ego, shattered his delicate facade, violated the image he projected. He was tough. Without feeling. Faggot cut through all this. It made him vulnerable. Feminine. And feminine was the worst thing he could possibly be. Girls were fine for fucking, but no boy in his right mind wanted to be like them. A boy was the opposite of a girl. He was not feminine. He was not feeling. He was not weak.

Just look at the gym teacher who growled like a dog; or the priest with the black belt who threw kids against the wall in rage when they didn't know their Latin. They were men, they got respect.

But not the physics teacher who preached pacifism during lectures on the nature of atoms. Everybody knew what he was—and why he believed in the anti-war movement.

My parents only knew that the neighborhood kids called me names. They begged me to act more like the other boys. My brothers were ashamed of me. They never said it, but I knew. Just as I knew that my parents were embarrassed by my behavior.

At times, they tried to get me to act differently. Once my father lectured me on how to walk right. I'm still not clear on what that means. Not from the hips, I guess, don't "swish" like faggots do.

A nun in elementary school told my mother at Open House that there was "something wrong with me." I had draped my sweater over my shoulders like a girl, she said. I was a smart kid, but I should know better than to wear my sweater like a girl!

My mother stood there, mute. I wanted her to say something, to chastise the nun; to defend me. But how could she? This was a nun talking—representative of Jesus, protector of all that was good and decent.

An uncle once told me I should start "acting like a boy" instead of like a girl. Everybody seemed ashamed of me. And I guess I was ashamed of myself, too. It was hard not to be.

SCENE FIVE

Priest: Do you like girls, Mark?

Mark: Uh-huh.

Priest: I mean *really* like them?

Mark: Yeah—they're okay.

Priest: There's a role they play in your salvation. Do you understand it, Mark?

Mark: Yeah.

Priest: You've got to like girls. Even if you should decide to enter the seminary, it's important to keep in mind God's plan for a man and a woman. . . .

Catholicism of course condemned homosexuality. Effeminacy was tolerated as long as the effeminate person did not admit to being gay. Thus, priests could be effeminate because they weren't gay.

As a sissy, I could count on no support from the church. A male's sole purpose in life was to father children—souls for the church to save. The only hope a homosexual had of attaining salvation was by remaining totally celibate. Don't even think of touching another boy. To think of a sin was a sin. And to sin was to put a mark upon the soul. Sin—if it was a serious offense against god—led to hell. There was no way around it. If you sinned, you were doomed.

Realizing I was gay was not an easy task. Although I knew I was attracted to boys by the time I was about eleven, I didn't connect this attraction to homosexuality. I was not queer. Not I. I was merely appreciating a boy's good looks, his fine features, his proportions. It didn't seem to matter that I didn't appreciate a girl's looks in the same way. There was no twitching in my thighs when I gazed upon a beautiful girl. But I wasn't queer.

I resisted that label—queer—for the longest time. Even when everything pointed to it, I refused to see it. I was certainly not queer. Not I.

We sat through endless English classes, and History courses about the wars between men who were not allowed to love each other. No gay history was ever taught. No history faces you this morning. You're just a faggot. Homosexuals had never contributed to the human race. God destroyed the queers in Sodom and Gomorrah.

We learned about Michelangelo, Oscar Wilde, Gertrude Stein—but never that they were queer. They were not queer. Walt Whitman, the "father of American poetry," was not queer. No one was queer. I was alone, totally unique. One of a kind. Were there others like me somewhere? Another planet, perhaps?

In school, they never talked of the queers. They did not exist. The only hint we got of this other species was in religion class. And even then it was clouded in mystery—never spelled out. It was sin. Like masturbation. Like looking at *Playboy* and getting a hard-on. A sin.

Once a progressive priest in senior year religion class actually mentioned homosexuals—he said the word—but was into Erich Fromm, into homosexuals as pathetic and sick. Fixated at some early stage; penis, anal, whatever. Only heterosexuals passed on to the nirvana of sexual development.

No other images from the halls of the Catholic high school except those the other boys knew: swishy faggot sucking cock in an alley somewhere, grabbing asses in the bathroom. Never mentioning how much straight boys craved blow jobs, it was part of the secret.

It was all a secret. You were not supposed to talk about the queers. Whisper maybe. Laugh about them, yes. But don't be open, honest; don't try to understand. Don't cite their accomplishments. No history faces you this morning. You're just a faggot faggot no history just a faggot.

EPILOGUE

The boy marching down the Parkway. Hundreds of queers. Signs proclaiming gay pride. Speakers. Tables with literature from gay groups. A miracle, he is thinking. Tears are coming loose now. Someone hugs him.

> You could not control
> the sissy in me
> nor could you exorcise him
> nor electrocute him
> You declared him illegal illegitimate
> insane and immature
> But he defies you still.

16

The Burden of Womanhood

JOHN WARD ANDERSON AND
MOLLY MOORE

GANDHI NAGAR, INDIA

When Rani returned home from the hospital cradling her newborn daughter, the men in the family slipped out of her mud hut while she and her mother-in-law mashed poisonous oleander seeds into a dollop of oil and forced it down the infant's throat. As soon as darkness fell, Rani crept into a nearby field and buried her baby girl in a shallow, unmarked grave next to a small stream.

"I never felt any sorrow," Rani, a farm laborer with a weather-beaten face, said through an interpreter. "There was a lot of bitterness in my heart toward the baby because the gods should have given me a son."

Each year hundreds and perhaps thousands of new-born girls in India are murdered by their mothers simply because they are female. Some women believe that sacrificing a daughter guarantees a son in the next pregnancy. In other cases, the family cannot afford the dowry that would eventually be demanded for a girl's marriage.

And for many mothers, sentencing a daughter to death is better than condemning her to life as a woman in the Third World, with cradle-to-grave discrimination, poverty, sickness and drudgery.

"In a culture that idolizes sons and dreads the birth of a daughter, to be born female comes perilously close to being born less than human," the Indian government conceded in a recent report by its Department of Women and Child Development.

While women in the United States and Europe—after decades of struggling for equal rights—often measure sex discrimination by pay scales and seats in corporate board rooms, women in the Third World gauge discrimination by mortality rates and poverty levels.

"Women are the most exploited among the oppressed," says Karuna Chanana Ahmed, a New Delhi anthropologist who has studied the role of women in developing countries. "I don't think it's even possible to eradicate discrimination, it's so deeply ingrained."

This is the first in a series that will examine the lives of women in developing countries around the globe where culture, religion and the law often deprive women of basic human rights and sometimes relegate them to almost subhuman status. From South America to South Asia, women are often subjected to a lifetime of discrimination with little or no hope of relief.

As children, they are fed less, denied education and refused hospitalization. As teenagers, many are forced into marriage, sometimes bought and sold like animals for prostitution and slave labor. As wives and mothers, they are often treated little better than farmhands and baby machines. Should they outlive their husbands, they frequently are denied inheritance, banished from their homes and forced to live as beggars on the streets.

The scores of women interviewed for this series—from destitute villagers in Brazil and Bangladesh, to young professionals in Cairo, to factory workers in China—blamed centuries-old cultural and religious traditions for institutionalizing and giving legitimacy to gender discrimination.

Although the forms of discrimination vary tremendously among regions, ethnic groups and age levels in the developing world, Shahla Zia, an attorney and women's activist in Islamabad,

Pakistan, says there is a theme: "Overall, there is a social and cultural attitude where women are inferior—and discrimination tends to start at birth."

In many countries, a woman's greatest challenge is an elemental one: simply surviving through a normal life cycle. In South Asia and China, the perils begin at birth, with the threat of infanticide.

Like many rural Indian women, Rani, now 31, believed that killing her daughter 3½ years ago would guarantee that her next baby would be a boy. Instead, she had another daughter.

"I wanted to kill this child also," she says, brushing strands of hair from the face of the 2-year-old girl she named Asha, or Hope. "But my husband got scared because all these social workers came and said, 'Give us the child.'" Ultimately, Rani was allowed to keep her. She pauses. "Now I have killed, and I still haven't had any sons."

Amravati, who lives in a village near Rani in the Indian state of Tamil Nadu, says she killed two of her own day-old daughters by pouring scalding chicken soup down their throats, one of the most widely practiced methods of infanticide in southern India. She showed where she buried their bodies—under piles of cow dung in the tiny courtyard of her home.

"My mother-in-law and father-in-law are bedridden," says Amravati, who has two living daughters. "I have no land and no salary, and my husband met with an accident and can't work. Of course it was the right decision. I need a boy. Even though I have to buy clothes and food for a son, he will grow on his own and take care of himself. I don't have to buy him jewelry or give him a 10,000-rupee [$350] dowry."

Sociologists and government officials began documenting sporadic examples of female infanticide in India about 10 years ago. The practice of killing newborn girls is largely a rural phenomenon in India; although its extent has not been documented, one indication came in a recent survey by the Community Services Guild of Madras, a city in Tamil Nadu. Of the 1,250 women questioned, the survey concluded that more than half had killed baby daughters.

In urban areas, easier access to modern medical technology enables women to act before birth. Through amniocentesis, women can learn the sex of a fetus and undergo sex-selective abortions. At one clinic in Bombay, of 8,000 abortions performed after amniocentesis, 7,999 were of female fetuses, according to a recent report by the Indian government. To be sure, female infanticide and sex-selective abortion are not unique to India. Social workers in other South Asian states believe that some communities also condone the practice. In China, one province has had so many cases of female infanticide that a half-million bachelors cannot find wives because they outnumber women their age by 10 to 1, according to the official New China News Agency.

The root problems, according to village women, sociologists and other experts, are cultural and economic. In India, a young woman is regarded as a temporary member of her natural family and a drain on its wealth. Her parents are considered caretakers whose main responsibility is to deliver a chaste daughter, along with a sizable dowry, to her husband's family.

"They say bringing up a girl is like watering a neighbor's plant," says R. Venkatachalam, director of the Community Services Guild of Madras. "From birth to death, the expenditure is there." The dowry, he says, often wipes out a family's life savings but is necessary to arrange a proper marriage and maintain the honor of the bride's family.

After giving birth to a daughter, village women "immediately start thinking, 'Do we have the money to support her through life?' and if they don't, they kill her," according to Vasanthai, 20, the mother of an 18-month-old girl and a resident of the village where Rani lives. "You definitely do it after two or three daughters. Why would you want more?"

Few activists or government officials in India see female infanticide as a law-and-order issue, viewing it instead as a social problem that should be eradicated through better education, family planning and job programs. Police officials say few cases are reported and witnesses seldom cooperate.

"There are more pressing issues," says a top police official in Madras. "Very few cases come to our attention. Very few people care."

Surviving childbirth is itself an achievement in South Asia for both mother and baby. One of every 18 women dies of a pregnancy-related cause, and more than one of every 10 babies dies during delivery.

For female children, the survival odds are even worse. Almost one in every five girls born in Nepal and Bangladesh dies before age 5. In India, about one-fourth of the 12 million girls born each year die by age 15.

The high death rates are not coincidental. Across the developing world, female children are fed less, pulled out of school earlier, forced into hard labor sooner and given less medical care than boys. According to numerous studies, girls are handicapped not only by the perception that they are temporary members of a family, but also by the belief that males are the chief breadwinners and therefore more deserving of scarce resources.

Boys are generally breast-fed longer. In many cultures, women and girls eat leftovers after the men and boys have finished their meals. According to a joint report the United Nations Children's Fund and the government of Pakistan, some tribal groups do not feed high-protein foods such as eggs and meat to girls because of the fear it will lead to early puberty.

Women are often hospitalized only when they have reached a critical stage of illness, which is one reason so many mothers die in childbirth. Female children, on the other hand, often are not hospitalized at all. A 1990 study of patient records at Islamabad Children's Hospital in Pakistan found that 71 percent of the babies admitted under age 2 were boys. For all age groups, twice as many boys as girls were admitted to the hospital's surgery, pediatric intensive care and diarrhea units.

Mary Okumu, an official with the African Medical and Research Foundation in Nairobi, says that when a worker in drought-ravaged northern Kenya asked why only boys were lined up at a clinic, the worker was told that in times of drought, many families let their daughters die.

"Nobody will even take them to a clinic," Okumu says. "They prefer the boy to survive."

For most girls, however, the biggest barrier—and the one that locks generations of women into a cycle of discrimination—is lack of education.

Across the developing world, girls are withdrawn from school years before boys so they can remain at home and lug water, work the fields, raise younger siblings and help with other domestic chores. By the time girls are 10 or 12 years old, they may put in as much as an eight-hour work day, studies show. One survey found that a young girl in rural India spends 30 percent of her waking hours doing household work, 29 percent gathering fuel and 20 percent fetching water.

Statistics from Pakistan demonstrate the low priority given to female education: Only one-third of the country's schools—which are sexually segregated—are for women, and one-third of those have no building. Almost 90 percent of the women over age 25 are illiterate. In the predominantly rural state of Baluchistan, less than 2 percent of women can read and write.

In Islamic countries such as Pakistan and Bangladesh, religious concern about interaction with males adds further restrictions to females' mobility. Frequently, girls are taken out of school when they reach puberty to limit their contact with males—though there exists a strong impetus for early marriages. In Bangladesh, according to the United Nations, 73 percent of girls are married by age 15, and 21 percent have had at least one child.

Across South Asia, arranged marriages are the norm and can sometimes be the most demeaning rite of passage a woman endures. Two types are common—bride wealth, in which the bride's family essentially gives her to the highest bidder, and dowry, in which the bride's family pays exorbitant amounts to the husband's family.

In India, many men resort to killing their wives—often by setting them afire—if they are unhappy with the dowry. According to the country's Ministry of Human Resource Development, there were 5,157 dowry murders in 1991—one every hour and 42 minutes.

After being bartered off to a new family, with little education, limited access to health care and no knowledge of birth control, young brides soon become young mothers. A woman's adulthood is often spent in a near constant state of pregnancy, hoping for sons.

According to a 1988 report by India's Department of Women and Child Development: "The Indian woman on an average has eight to nine pregnancies, resulting in a little over six live births, of which four or five survive. She is estimated to spend 80 percent of her reproductive years in pregnancy and lactation." Because of poor nutrition and a hard workload, she puts on about nine pounds during pregnancy, compared with 22 pounds for a typical pregnant woman in a developed country.

A recent study of the small Himalayan village of Bemru by the New Delhi-based Center for Science and the Environment found that "birth in most cases takes place in the cattle shed," where villagers believe that holy cows protect the mother and newborn from evil spirits. Childbirth is considered unclean, and the mother and their newborn are treated as "untouchables" for about two weeks after delivery.

"It does not matter if the woman is young, old or pregnant, she has no rest, Sunday or otherwise," the study said, noting that women in the village did 59 percent of the work, often laboring 14 hours a day and lugging loads 1½ times their body weight. "After two or three . . . pregnancies, their stamina gives up, they get weaker, and by the late thirties are spent out, old and tired, and soon die."

Studies show that in developing countries, women in remote areas can spend more than two hours a day carrying water for cooking, drinking, cleaning and bathing, and in some rural areas they spend the equivalent of more than 200 days a year gathering firewood. That presents an additional hazard: The International Labor Organization found that women using wood fuels in India inhaled carcinogenic pollutants that are the equivalent of smoking 20 packs of cigarettes a day.

Because of laws relegating them to a secondary status, women have few outlets for relaxation or recreation. In many Islamic countries, they are not allowed to drive cars, and their appearance in public is so restricted that they are banned from such recreational and athletic activities as swimming and gymnastics.

In Kenya and Tanzania, laws prohibit women from owning houses. In Pakistan, a daughter legally is entitled to half the inheritance a son gets when their parents die. In some criminal cases, testimony by women is legally given half the weight of a man's testimony, and compensation for the wrongful death of a woman is half that for the wrongful death of a man.

After a lifetime of brutal physical labor, multiple births, discrimination and sheer tedium, what should be a woman's golden years often hold the worst indignities. In India, a woman's identity is so intertwined and subservient to her husband's that if she outlives him, her years as a widow are spent as a virtual nonentity. In previous generations, many women were tied to their husband's funeral pyres and burned to death, a practice called *suttee* that now rarely occurs.

Today, some widows voluntarily shave their heads and withdraw from society, but more often a spartan lifestyle is forced upon them by families and a society that place no value on old, single women. Widowhood carries such a stigma that remarriage is extremely rare, even for women who are widowed as teenagers.

In some areas of the country, women are forced to marry their dead husband's brother to ensure that any property remains in the family. Often they cannot wear jewelry or a *bindi*—the beauty spot women put on their foreheads—or they must shave their heads and wear a white sari. Frequently, they cannot eat fish or meat, garlic or onions.

"The life of a widow is miserable," says Aparna Basu, general secretary of the All India Women's Conference, citing a recent study showing that more than half the women in India age 60 and older are widows, and their mortality rate is three times higher than that of married women of the same age.

In South Asia, women have few property or inheritance rights, and a husband's belongings usually go to sons and occasionally daughters. A widow must rely on the largess of her children, who often cast their mothers on the streets.

Thousands of destitute Indian widows make the pilgrimage to Vrindaban, a town on the outskirts of Agra where they hope to achieve salvation by praying to the god Krishna. About 1,500 widows show up each day at the Shri Bhagwan prayer house, where in exchange for singing

"Hare Rama, Hare Krishna" for eight hours, they are given a handful of rice and beans and 1.5 rupees, or about 5 cents.

Some widows claim that when they stop singing, they are poked with sticks by monitors, and social workers allege that younger widows have been sexually assaulted by temple custodians and priests.

On a street there, an elderly woman with a *tilak* on her forehead—white chalk lines signifying that she is a devout Hindu widow—waves a begging cup at passing strangers.

"I have nobody," says Paddo Chowdhury, 65, who became a widow at 18 and has been in Vrindaban for 30 years. "I sit here, shed my tears and get enough food one way or another."

17

American Family Decline, 1960–1990

DAVID POPENOE

Family decline in America continues to be a debatable issue, especially in academia. Several scholars have recently written widely-distributed trade books reinforcing what has become the establishment position of many family researchers—that family decline is a "myth," and that "the family is not declining, it is just changing" (Coontz, 1992; Skolnick, 1991; Stacey, 1990). . . .

My view is just the opposite. Like the majority of Americans, I see the family as an institution in decline and believe that this should be a cause for alarm—especially as regards the consequences for children. . . .

Families have lost functions, social power, and authority over their members. They have grown smaller in size, less stable, and shorter in life span. People have become less willing to invest time, money, and energy in family life, turning instead to investments in themselves.

Moreover, there has been a weakening of child-centeredness in American society and culture. Familism as a cultural value has diminished. . . .

INSTITUTIONAL

There are three key dimensions to the strength of an institution: the institution's cohesion or the hold which it has over its members, how well it performs its functions, and the power it has in society relative to other institutions. The evidence suggests that the family as an institution has weakened in each of these respects.

First, individual family members have become more autonomous and less bound by the group; the group as a whole, therefore, has become less cohesive. . . .

With more women in the labor market, for example, the economic interdependence between husbands and wives has been declining. Wives are less dependent on husbands for economic support; more are able, if they so desire, to go it alone. This means that wives are less likely to stay in bad marriages for economic reasons. . . . By the same token, if a wife has economic independence (for example, through state welfare support), it is easier for a husband to abandon her if he so chooses.

From David Popenoe, "American Family Decline, 1960–1990: A Review and Appraisal," *Marriage and Family*, 55(3), Copyright © by the National Council on Family Relations, 3989 Central Ave., NE, Suite 550, Minneapolis, MN 55421. Reprinted by permission.

However one looks at it, and unfortunate though it may be, the decline of economic interdependence between husband and wife (primarily the economic dependence of the wife) appears to have led, in the aggregate, to weaker marital units as measured by higher rates of divorce and separation (for a contradictory view, see Greenstein, 1990).

As the marital tie has weakened in many families, so also has the tie between parents and children. A large part of the history of childhood and adolescence in the twentieth century is the decline of parental influence and authority and the growth in importance of both the peer group and the mass media (Hawes & Hiner, 1985; Modell, 1989). . . . Similarly, there is much less influence today of the elderly over their own children. . . .

The second dimension of family institutional decline is that the family is less able—and/or less willing—to carry out its traditional social functions. This is, in part, because it has become a less cohesive unit. The main family functions in recent times have been the procreation and socialization of children, the provision to its members of affection and companionship, sexual regulation, and economic cooperation. With a birthrate that is below the replacement level, it is demonstrably the case that the family has weakened in carrying out the function of procreation. A strong case can also be made that the family has weakened in conducting the function of child socialization. . . .

By almost everyone's reckoning, marriage today is a more fragile institution than ever before precisely because it is based mainly on the provision of affection and companionship. When these attributes are not provided, the marriage often dissolves. The chances of that happening today are near a record high.

A decline of the family regulation of sexual behavior is one of the hallmarks of the past 30 years (D'Emilio & Freedman, 1988). Against most parents' wishes, young people have increasingly engaged in premarital sex, at ever younger ages. And against virtually all spousal wishes, the amount of sexual infidelity among married couples has seemingly increased. . . .

Finally, the function of the family in economic cooperation has diminished substantially. . . . The family is less a pooled bundle of economic resources, and more a business partnership between two adults (and one which, in most states, can unilaterally be broken at any time). Witness, for example, the decline of joint checking accounts and the rise of prenuptial agreements. . . .

The third dimension of family institutional decline is the loss of power to other institutional groups. In recent centuries, with the decline of agriculture and the rise of industry, the family has lost power to the workplace and, with the rise of mandatory formal education, it has lost power to the school. The largest beneficiary of the transfer of power out of the family in recent years has been the state. State agencies increasingly have the family under surveillance, seeking compliance for increasingly restrictive state laws covering such issues as child abuse and neglect, wife abuse, tax payments, and property maintenance (Lasch, 1977; Peden & Glahe, 1986). . . .

CULTURAL

Family decline has also occurred in the sense that familism as a cultural value has weakened in favor of such values as self-fulfillment and egalitarianism (Bellah, Madsen, Sullivan, Swidler, & Tipton, 1985; Lasch, 1978; Veroff et al., 1981). . . . Familism refers to the belief in a strong sense of family identification and loyalty, mutual assistance among family members, a concern for the perpetuation of the family unit, and the subordination of the interests and personality of individual family members to the interests and welfare of the family group.

It is true that most Americans still loudly proclaim family values, and there is no reason to question their sincerity about this. The family ideal is still out there. Yet apart from the ideal, the value of family has steadily been chipped away. The percentage of Americans who believe that "the family should stay together for the sake of the children" has declined precipitously, for example, as noted above. And fewer Americans believe that it is important to have children, to be married if you do, or even to be married, period. . . .

EVALUATING FAMILY DECLINE

The net result—or bottom line—of each of these trends is, I submit, that Americans today are less willing than ever before to invest time, money, and energy in family life (Goode, 1984). Most still want to marry and most still want children, but they are turning more to other groups and activities, and are investing much more in themselves. . . .

The increase in individual rights and opportunities is, of course, one of the great achievements of the modern era. No one wants to go back to the days of the stronger family when the husband owned his wife and could do virtually anything he wanted to her short of murder, when the parents were the sole custodians of their children and could treat them as they wished, when the social status of the family you were born into heavily determined your social status for life, and when the psychosocial interior of the family was often so intense that it was like living in a cocoon. . . .

Many scholars have noted that the institution of the family could be said to have been in decline since the beginning of mankind. And people of almost every era seem to have bemoaned the loss of the family, even suggesting its imminent demise (Popenoe, 1988). Yet we, as human beings, have made some progress over the centuries. Why, therefore, should we be unduly alarmed about the family decline of our generation? This question is a good one and demands an answer.

. . . Once the only social institution in existence, the family over time has lost functions to such institutions as organized religion, education, work, and government (Lenski & Lenski, 1987). . . . Education and work are the latest functions to be split off from the family unit, the split having occurred for the most part over the past two centuries. . . .

From its earliest incarnation as a multifunctional unit, the streamlined family of today is left with just two principal functions: childrearing, and the provision to its members of affection and companionship. Both family functions have become greatly magnified over the years. . . .

[W]hat kind of family decline is underway today that we should be concerned about? There are two dimensions of today's family decline that make it both unique and alarming. The first is that it is not the extended family that is breaking up but the nuclear family. The nuclear family can be thought of as the last vestige of the traditional family unit; all other adult members have been stripped away, leaving but two—the husband and wife. The nuclear unit—man, woman, and child—is called that for good reason: It is the fundamental and most basic unit of the family. Breaking up the nucleus of anything is a serious matter.

The second dimension of real concern regards what has been happening to the two principal functions—childrearing, and the provision to its members of affection and companionship—with which the family has been left. It is not difficult to argue that the functions that have already been taken from the family—government, formal education, and so on—can in fact be better performed by other institutions. It is far more debatable, however, whether the same applies to childrearing and the provision of affection and companionship. There is strong reason to believe, in fact, that the family is by far the best institution to carry out these functions, and that insofar as these functions are shifted to other institutions, they will not be carried out as well. . . .

CONCLUSION

My argument, in summary, is that the family decline of the past three decades is something special—very special. It is "end-of-the-line" family decline. Historically, the family has been stripped down to its bare essentials—just two adults and two main functions. The weakening of this unit is much more problematic than any prior family change. People today, most of all children, dearly want families in their lives. They long for that special, and hopefully life-long, social and emotional bond that family membership brings. Adults can perhaps live much of their lives, with some success, apart from families. The problem is that children, if we wish them to become successful adults, cannot.

REFERENCES

Bellah, R. N., Madsen, R., Sullivan, W. M., Swidler, A., & Tipton, S. M. (1985). *Habits of the heart: Individualism and commitment in American Life.* Berkeley: University of California Press.

Coontz, S. (1992). *The way we never were.* New York: Basic Books.

D'Emilio, J., & Freedman, E. B. (1988). *Intimate matters: A history of sexuality in America.* New York: Harper & Row.

Goode, W. J. (1984). Individual investments in family relationships over the coming decades. *The Toqueville Review, 6,* 51–83.

Greenstein, T. N. (1990). Marital disruption and the employment of married women. *Journal of Marriage and the Family, 52,* 657–676.

Hawes, J. M., & Hiner, N. R. (Eds.). (1985). *American Childhood.* Westport, CT: Greenwood Press.

Lasch, C. (1978). *The culture of narcissism.* New York: W. W. Norton.

Lenski, G., & Lenski, J. (1987). *Human societies.* New York: McGraw Hill.

Modell, J. (1989). *Into one's own: From youth to adulthood in the United States. 1920–1975.* Berkeley, CA: University of California Press.

Peden, J. R., & Glahe, F. R. (Eds.). (1986). *The American family and the state.* San Francisco: Pacific Research Institute for Public Policy.

Popenoe, D. (1988). *Disturbing the nest: Family change and decline in modern societies.* New York: Aldine de Gruyter.

Skolnick, A. (1991). *Embattled paradise: The American family in an age of uncertainty.* New York: Basic Books.

Stacey, J. (1990). *Brave new families.* New York: Basic Books.

Veroff, J., Douven, E., & Kulka, R. A. (1981). *The inner American: A self-portrait from 1957 to 1976.* New York: Basic Books.

18

The Globetrotting Sneaker

CYNTHIA ENLOE

Four years after the fall of the Berlin Wall marked the end of the Cold War, Reebok, one of the fastest growing companies in United States history, decided that the time had come to make its mark in Russia. Thus it was with considerable fanfare that Reebok's executives opened their first store in downtown Moscow in July 1993. A week after the grand opening, store managers described sales as well above expectations.

Reebok's opening in Moscow was the perfect post-Cold War scenario: commercial rivalry replacing military posturing; consumerist tastes homogenizing heretofore hostile peoples; capital and managerial expertise flowing freely across newly porous state borders. Russians suddenly had the "freedom" to spend money on U.S. cultural icons like athletic footwear, items priced above and beyond daily subsistence: at the end of 1993, the average Russian earned the equivalent of $40 a month. Shoes on display were in the $100 range. Almost 60 percent of single parents, most of whom were women, were living in poverty. Yet in Moscow and Kiev, shoe promoters

From Cynthia Enloe, "The Globetrotting Sneaker," *Ms.,* March/April, 1995, pp. 10–15. Reprinted by permission of Ms. Magazine, © 1995.

Hourly Wages in Athletic Footwear Factories

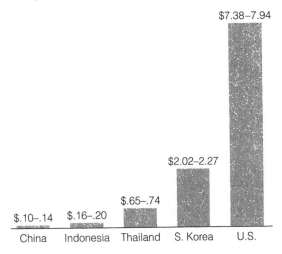

Source: Figures are estimates based on 1993 data from the International Textile, Garment, and Leather Workers Federation; International Labor Organization; and the U.S. Bureau of Labor Statistics.

had begun targeting children, persuading them to pressure their mothers to spend money on stylish, Western sneakers. And as far as strategy goes, athletic shoe giants have, you might say, a good track record. In the U.S. many inner-city boys who see basketball as a "ticket out of the ghetto" have become convinced that certain brand-name shoes will give them an edge.

But no matter where sneakers are bought or sold, the potency of their advertising imagery has made it easy to ignore this mundane fact: Shaquille O'Neal's Reeboks are stitched by someone; Michael Jordan's Nikes are stitched by someone; so are your roommate's, so are your grandmother's. Those someones are women, mostly Asian women who are supposed to believe that their "opportunity" to make sneakers for U.S. companies is a sign of their country's progress—just as a Russian woman's chance to spend two month's salary on a pair of shoes for her child allegedly symbolizes the new Russia.

As the global economy expands, sneaker executives are looking to pay women workers less and less, even though the shoes that they produce are capturing an ever-growing share of the footwear market. By the end of 1993, sales in the U.S.

alone had reached $11.6 billion. Nike, the largest supplier of athletic footwear in the world, posted a record $298 million profit for 1993—earnings that had nearly tripled in five years. And sneaker companies continue to refine their strategies for "global competitiveness"—hiring supposedly docile women to make their shoes, changing designs as quickly as we fickle customers change our tastes, and shifting factories from country to country as trade barriers rise and fall.

The logic of it all is really quite simple; yet trade agreements such as the North American Free Trade Agreement (NAFTA) and the General Agreement on Tariffs and Trade (GATT) are, of course, talked about in a jargon that alienates us, as if they were technical matters fit only for economists and diplomats. The bottom line is that all companies operating overseas depend on trade agreements made between their own governments and the regimes ruling the countries in which they want to make or sell their products. Korean, Indonesian, and other women workers around the world know this better than anyone. They are tackling trade politics because they have learned from hard experience that the trade deals their governments sign do little to improve the lives of workers. Guarantees of fair, healthy labor practices, of the rights to speak freely and to organize independently, will usually be left out of trade pacts—and women will suffer. The recent passage of both NAFTA and GATT ensures that a growing number of private companies will now be competing across borders without restriction. The result? Big business will step up efforts to pit working women in industrialized countries against much lower-paid working women in "developing" countries, perpetuating the misleading notion that they are inevitable rivals in the global job market.

All the "New World Order" really means to corporate giants like athletic shoemakers is that they now have the green light to accelerate long-standing industry practices. In the early 1980s, the field marshals commanding Reebok and Nike, which are both U.S.-based, decided to manufacture most of their sneakers in South Korea and Taiwan, hiring local women. L.A. Gear, Adidas,

A $70 Pair of Nike Pegasus: Where the Money Goes

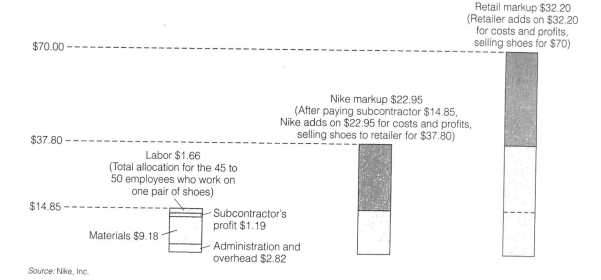

Source: Nike, Inc.

Fila, and Asics quickly followed their lead. In short time, the coastal city of Pusan, South Korea, became the "sneaker capital of the world." Between 1982 and 1989 the U.S. lost 58,500 footwear jobs to cities like Pusan, which attracted sneaker executives because its location facilitated international transport. More to the point, South Korea's military government had an interest in suppressing labor organizing, and it had a comfortable military alliance with the U.S. Korean women also seemed accepting of Confucian philosophy, which measured a woman's morality by her willingness to work hard for her family's well-being and to acquiesce to her father's and husband's dictates. With their sense of patriotic duty, Korean women seemed the ideal labor force for export-oriented factories.

U.S. and European sneaker company executives were also attracted by the ready supply of eager Korean male entrepreneurs with whom they could make profitable arrangements. This fact was central to Nike's strategy in particular. When they moved their production sites to Asia to lower labor costs, the executives of the Oregon-based company decided to reduce their corporate responsibilities further. Instead of owning factories outright, a more efficient strategy would be to subcontract the manufacturing to wholly foreign-owned—in this case, South Korean—companies. Let them be responsible for the workers' health and safety. Let them negotiate with newly emergent unions. Nike would retain control over those parts of sneaker production that gave its officials the greatest professional satisfaction and the ultimate word on the product: design and marketing. Although Nike was following in the footsteps of garment and textile manufacturers, it set the trend for the rest of the athletic footwear industry.

But at the same time, women workers were developing their own strategies. As the South Korean pro-democracy movement grew throughout the 1980s, increasing numbers of women rejected traditional notions of feminine duty. Women began organizing in response to the dangerous working conditions, daily humiliations and low pay built into their work. Such resistance was profoundly threatening to the government, given the fact that South Korea's emergence as an industrialized "tiger" had depended on women accepting their "role" in growing industries like sneaker manufacture. If women reimagined their lives as daughters, as wives, as workers, as citizens, it wouldn't just rattle their employers; it would shake the very foundations of the whole political system.

At the first sign of trouble, factory managers called in government riot police to break up employees' meetings. Troops sexually assaulted women workers, stripping, fondling, and raping them "as a control mechanism for suppressing women's engagement in the labor movement," reported Jeong-Lim Nam of Hyosung Women's University in Taegu. It didn't work. It didn't work because the feminist activists in groups like the Korean Women Workers Association (KWWA) helped women understand and deal with the assaults. The KWWA held consciousness-raising sessions in which notions of feminine duty and respectability were tackled along with wages and benefits. They organized independently of the male-led labor unions to ensure that their issues would be taken seriously, in labor negotiations and in the pro-democracy movement as a whole.

The result was that women were at meetings with management, making sure that in addition to issues like long hours and low pay, sexual assault at the hands of managers and health care were on the table. Their activism paid off: in addition to winning the right to organize women's unions, their earnings grew. In 1980, South Korean women in manufacturing jobs earned 45 percent of the wages of their male counterparts; by 1990, they were earning more than 50 percent. Modest though it was, the pay increase was concrete progress, given that the gap between women's and men's manufacturing wages in Japan, Singapore, and Sri Lanka actually *widened* during the 1980s. Last but certainly not least, women's organizing was credited with playing a major role in toppling the country's military regime and forcing open elections in 1987.

Without that special kind of workplace control that only an authoritarian government could offer, sneaker executives knew that it was time to move. In Nike's case, its famous advertising slogan—"Just Do It"—proved truer to its corporate philosophy than its women's "empowerment" ad campaign, designed to rally women's athletic (and consumer) spirit. In response to South Korean women workers' newfound activist self-confidence, the sneaker company and its subcontractors began shutting down a number of their South Korean factories in the late 1980s and early 1990s. After bargaining with government officials in nearby China and Indonesia, many Nike subcontractors set up shop in those countries, while some went to Thailand. China's government remains nominally Communist; Indonesia's ruling generals are staunchly anti-Communist. But both are governed by authoritarian regimes who share the belief that if women can be kept hard at work, low paid, and unorganized, they can serve as a magnet for foreign investors.

Where does all this leave South Korean women—or any woman who is threatened with a factory closure if she demands decent working conditions and a fair wage? They face the dilemma confronted by thousands of women from dozens of countries. The risk of job loss is especially acute in relatively mobile industries; it's easier for a sneaker, garment, or electronics manufacturer to pick up and move than it is for an automaker or a steel producer. In the case of South Korea, poor women had moved from rural villages into the cities searching for jobs to support not only themselves, but parents and siblings. The exodus of manufacturing jobs has forced more women into the growing "entertainment" industry. The kinds of bars and massage parlors offering sexual services that had mushroomed around U.S. military bases during the Cold War have been opening up across the country.

But the reality is that women throughout Asia are organizing, knowing full well the risks involved. Theirs is a long-term view; they are taking direct aim at companies' nomadic advantage, by building links among workers in countries targeted for "development" by multinational corporations. Through sustained grassroots efforts, women are developing the skills and confidence that will make it increasingly difficult to keep their labor cheap. Many are looking to the United Nations conference on women in Beijing, China, this September, as a rare opportunity to expand their cross-border strategizing.

The Beijing conference will also provide an important opportunity to call world attention to the hypocrisy of the governments and corporations doing business in China. Numerous athletic shoe companies followed Nike in setting up manufacturing sites throughout the country. This included Reebok—a company claiming its share of responsibilities for ridding the world of "injustice, poverty, and other ills that gnaw away at the social fabric," according to a statement of corporate principles.

Since 1988, Reebok has been giving out annual human rights awards to dissidents from around the world. But it wasn't until 1992 that the company adopted its own "human rights production standards"—after labor advocates made it known that the quality of life in factories run by its subcontractors was just as dismal as that at most other athletic shoe suppliers in Asia. Reebok's code of conduct, for example, includes a pledge to "seek" those subcontractors who respect workers' rights to organize. The only problem is that independent trade unions are banned in China. Reebok has chosen to ignore that fact, even though Chinese dissidents have been the recipients of the company's own human rights award. As for working conditions, Reebok now says it sends its own inspectors to production sites a couple of times a year. But they have easily "missed" what subcontractors are trying to hide—like 400 young women workers locked at night into an overcrowded dormitory near a Reebok-contracted factory in the town of Zhuhai, as reported last August in the *Asian Wall Street Journal Weekly.*

Nike's cofounder and CEO Philip Knight has said that he would like the world to think of Nike as "a company with a soul that recognizes the value of human beings." Nike, like Reebok, says it sends in inspectors from time to time to check up on work conditions at its factories; in Indonesia, those factories are run largely by South Korean subcontractors. But according to Donald Katz in a recent book on the company, Nike spokesman Dave Taylor told an in-house newsletter that the factories are "[the subcontractors'] business to run." For the most part, the company relies on regular reports from subcontractors regarding its "Memorandum of Understanding," which managers must sign, promising to impose "local government standards" for wages, working conditions, treatment of workers, and benefits.

In April, the minimum wage in the Indonesian capital of Jakarta will be $1.89 *a day*—among the highest in a country where the minimum wage varies by region. And managers are required to pay only 75 percent of the wage directly; the remainder can be withheld for "benefits." By now, Nike has a well-honed response to growing criticism of its low-cost labor strategy. Such wages should not be seen as exploitative, says Nike, but rather as the first rung on the ladder of economic opportunity that Nike has extended to workers with few options. Otherwise, they'd be out "harvesting coconut meat in the tropical sun," wrote Nike spokesman Dusty Kidd, in a letter to the *Utne Reader.* The all-is-relative response craftily shifts attention away from reality: Nike didn't move to Indonesia to help Indonesians; it moved to ensure that its profit margin continues to grow. And that is pretty much guaranteed in a country where "local standards" for wages rarely take a worker over the poverty line. A 1991 survey by the International Labor Organization (ILO) found that 88 percent of women working at the Jakarta minimum wage at the time—slightly less than a dollar a day—were malnourished.

A woman named Riyanti might have been among the workers surveyed by the ILO. Interviewed by the Boston *Globe* in 1991, she told the reporter who had asked about her long hours and low pay: "I'm happy working here. . . . I can make money and I can make friends." But in fact, the reporter discovered that Riyanti had already joined her coworkers in two strikes, the first to force one of Nike's Korean subcontractors to accept a new women's union and the second to compel managers to pay at least the minimum wage. That Riyanti appeared less than forthcoming about her activities isn't surprising. Many Indonesian factories have military men posted in their front offices who find no fault with managers who tape women's mouths shut to keep them from talking among themselves. They and their superiors have a political reach that extends far beyond the barracks. Indonesia has all the makings for a political explosion, especially since the gap between rich and poor is widening into a chasm. It is in this setting that the government has tried to crack down on any independent labor organizing—a policy that Nike has helped to implement. Referring to a recent strike in a Nike-contracted factory, Tony Nava, Nike representative in Indonesia, told the Chicago *Tribune* in November 1994 that the "troublemakers" had been fired. When asked about Nike policy on the issue, spokesman Keith Peters struck a conciliatory note: "If the government were to allow and encourage independent labor organizing, we would be happy to support it."

Indonesian workers' efforts to create unions independent of governmental control were a surprise to shoe companies. Although their moves from South Korea have been immensely profitable, they do not have the sort of immunity from activism that they had expected. In May 1993, the murder of a female labor activist outside Surabaya set off a storm of local and international protest. Even the U.S. State Department was forced to take note in its 1993 worldwide human rights report, describing a system similar to that which generated South Korea's boom 20 years earlier: severely restricted union organizing, security forces used to break up strikes, low wages for men, lower wages for women—complete with government rhetoric celebrating women's contribution to national development.

Yet when President Clinton visited Indonesia last November, he made only a token effort to address the country's human rights problem. Instead, he touted the benefits of free trade, sounding indeed more enlightened, more in tune with the spirit of the post–Cold War era than do those defenders of protectionist trading policies who coat their rhetoric with "America first" chauvinism. But "free trade" as actually being

practiced today is hardly *free* for any workers—in the U.S. or abroad—who have to accept the Indonesian, Chinese, or Korean workplace model as the price of keeping their jobs.

The not-so-new plot of the international trade story has been "divide and rule." If women workers and their government in one country can see that a sneaker company will pick up and leave if their labor demands prove more costly than those in a neighbor country, then women workers will tend to see their neighbors not as regional sisters, but as competitors who can steal their precarious livelihoods. Playing women off against each other is, of course, old hat. Yet it is as essential to international trade politics as is the fine print in GATT.

But women workers allied through networks like the Hong Kong-based Committee for Asian Women are developing their own post–Cold War foreign policy, which means addressing women's needs: how to convince fathers and husbands that a woman going out to organizing meetings at night is not sexually promiscuous; how to develop workplace agendas that respond to family needs; how to work with male unionists who push women's demands to the bottom of their lists; how to build a global movement.

These women refuse to stand in awe of the corporate power of the Nike or Reebok or Adidas executive. Growing numbers of Asian women today have concluded that trade politics have to be understood by women on their own terms. They will be coming to Beijing this September ready to engage with women from other regions to link the politics of consumerism with the politics of manufacturing. If women in Russia and Eastern Europe can challenge Americanized consumerism, if Asian activists can solidify their alliances, and if U.S. women can join with them by taking on trade politics—the post–Cold War sneaker may be a less comfortable fit in the 1990s.

• • •

19

Widows Banding Together

MARGARET OWEN

On a scorching day in November 1988 more than 500 Indian widows, from many different villages surrounding a town in North Gujarat, gathered in the main square to share common problems, convey a list of their immediate demands to local officials and to work out a plan for future action.

This unique meeting came about through the efforts of Shramjivi Samaj, a local trade union of the poor whose members are mostly women. Everyone was astonished by the daring of the widows, whose militancy defied all traditions. The widows too amazed themselves by their actions, unprecedented in a patriarchal society where widows are seen as inauspicious, and expected to bear their dishonour in silence.

These widows were of all ages, from the very young, some mere children, to the extremely old. They had had enough of the obstructing bureaucracy and were angry about the futility of attempting to obtain the inadequate pensions that some among them—the destitute and the elderly—were on paper entitled to. They wanted legal protection to stay in their homes; to enjoy the inheritance rights given them under the Hindu Succession Law; to retain their dead husband's portion of land to feed themselves and their families. They also asked for proper remuneration for the long hours of work they put in, for example, in the tobacco fields, or as head-loaders, bidi-rollers, or piece-workers. Most of all, they wanted a future for their children, and an end to the dire poverty, degradation and discrimination they experienced due to their widowed state.

Holding informal meetings in each others' houses, at the wells, in the fields, these widows had begun to develop an acute collective awareness of the corruption, the injustice, and the paralysis of the *panchayats* (village councils) which were male-dominated and rarely responded to their complaints. The majority of these widows were illiterate, but by the time the big day arrived they were marvelously articulate, passionately angry, and had shed much of their natural shyness at speaking in mixed company.

The event was a brilliant success. With the help of the local radio and the press, it created public awareness of issues that had barely been brought to light before, and gave the women themselves a new confidence to lobby for change.

One practical consequence of the rally was that many more widows began to apply, with Shramjivi Samaj support, for the pensions which the State had legislated for them. The administration, attacked by the media, and harangued by other political elements, could do nothing but yield to their demands.

Of course the pensions were quite inadequate to support destitute widowed women, and in any case ceased after six months, but the meeting enormously increased public support for the widows, and has spurred other similar movements in other states in India.

SEWA, the Self-Employed Widows' Association, based in Ahmedabad, Gujarat, is the brainchild of the indefatigable Ela Bhatt. It has been a model, for several years, of how poor women can organize themselves for change and economic independence. Recently SEWA has developed new programmes aimed to assist not only their members who have already been widowed, but to prepare all the women for a usually inevitable

From Margaret Owen, "Widows Banding Together," *People and the Planet,* vol. 4, no. 3, 1995, pp. 20–22.
Copyright © 1995 People and the Planet. Reprinted by permission.

widowhood, life on their own or as female heads of households.

Widowhood is inevitable for a majority of women because in all countries widows outnumber widowers, not only because women tend to outlive men even in the poorest countries, but because women usually marry men who are older than themselves.

Since 1994 SEWA members can insure against their husband's deaths for a small premium, and in the event get assistance with funeral expenses, as well as a lump sum or a regular pension. Since it is unlikely that poor developing countries would be able to support any sort of national pension scheme in the foreseeable future, schemes such as SEWA's may influence other grass-roots women's organizations elsewhere to explore how it could be adapted in their localities.

"Ah, if I had only known there was this help when I was widowed," an elderly SEWA group trainer told me. "Now I tell all the younger women to join this scheme. It is only a little extra, but when calamity strikes, one is looked after. And not just materially, we give each other emotional support in times of tragedy." Rudi Ahir has overcome her fears, learnt to deal with government officials and the community now respect her.

SEWA's surveys revealed that many widows are left totally destitute by the time their husbands die; often they have sold all of their assets, such as jewellery, to pay for medicines; are left with heavy debts their men incurred without their knowledge, and are frequently victims of unscrupulous money-lenders who seize their land in satisfaction. Insurance against illness and accident and death of the main breadwinner, for a small premium, can protect women against sudden family disasters.

Lila's husband had cancer. "He was in the hospital and I did not know how to care for him and for the children, cook, clean and earn at the same time. I had to borrow from so many people because I could not earn. In the end I pawned not just my jewellery, but my cooking pots and blankets. I had nothing left for my daughters' dowries. I wish I had become a SEWA member long ago."

Banu is another SEWA member. The custom of her caste required her to stay indoors for six months after her husband died. "Because of this I could not go out to earn. But Sharda, the local SEWA organizer, learnt of my sorrow and visited to offer sympathies. She made me a member, and after that she came often, and went with me to the *Panchayat* to get back my land from my husband's brothers who had wrongfully taken it from me. I took the organizer to other women in the same plight as myself and now we are all working to help each other."

Similar activities are taking place in Bangladesh. There BRAC, the Bangladesh Committee for Rural Advancement, has been training paralegals in many villages where they have projects. The mullahs and traditional leaders disapprove of this new solidarity and independence among women, and show it by threats and harassment. In some villages the BRAC organizers have been ordered to leave, and cease all their activities.

I met about 200 widows in the Maniganj District. Their paralegal is a young widow called Ishrat. It is to her that the women go at the first signs of trouble. Although under Muslim law a woman is entitled to half her brother's portion of her father's estate, and an eighth of the estate of her husband, in practice she rarely receives her due. Disputes with the husband's family over land are commonplace. Often widows find themselves evicted from their homes, and sometimes their children are taken from them.

Ishrat works to reduce confrontation, and achieve a just settlement. Sometimes she will take a small group of women with her and the plaintiff to confront the defendant with his alleged offence, and try to shame him into surrender of the stolen land and deeds. If this tactic doesn't work, she will represent the widow at the *Salish* (village council). If all else fails, she can ask the BRAC legal office to bring the matter to the courts. News of such challenges to the male-dominated establishment spread like wild-fire, so that everyone becomes better informed about widows' legal rights.

In Uganda there are so many widows, victims of the long war, and of AIDS. As elsewhere on the continent, they are frequently victimized as interpretations of ethnic traditions result in their loss of rights that traditionally they should enjoy. There are countless cases all over West, Southern and East Africa of widows "chased off their land," and made destitute. Last year FIDA (the International Federation of Women Lawyers) and the Ministry of Women's Affairs helped groups of widows to organize a demonstration in Kampala before invited judges, police officers, politicians, and traditional leaders. Individual widows stood up and testified about their harrowing experiences and the refusal of officialdom to protect them. Since then any occurrence of "chasing off widows" makes the front page of the national newspaper.

Two co-widows, fired by the current publicity, who had been sexually abused by their brother-in-law and robbed of all their household furniture, collected a group of ten women, pursued him to his house, brandishing farm implements, sticks, and ropes, screaming for revenge and shouting out his transgressions. The terrified brother-in-law jumped out of a back window promising to return everything.

The scourge of AIDS has ruined the lives of millions of women and children all over Africa, but the realization that the vulnerability of women to HIV infection is due to their powerlessness to negotiate the terms of their sexual relations with men, and the necessity to survive as economically independent main breadwinners has done much to strengthen women's groups. In Kampala Philly Lutaya (named after a popular Ugandan singer who "went public" about AIDS before he died) and TASO (The AIDS Support Organization) have empowered AIDS widows, many infected themselves, to defend their rights, and face up to the pandemic by helping each other, providing shelter, food, medicines, work, and loans for school fees and land. TASO also trains AIDS widows in how to counsel PWAs (people with AIDS) and encourages them to "come out." In addition, TASO teaches widows how to write wills so that they can be sure that their children will inherit their property when they die and not lose it under traditional law.

In Northern Ghana the self-styled "Ministry of Widows" is a grass-roots organization that rescues those women who have been banished from their homesteads, and tries to provide them with income-generating activities and shelter. In Kenya, the Widows and Orphans Societies offer a safety-net for AIDS widows who are reduced to begging and prostitution.

Widows' associations in Africa tend still to be welfare and relief orientated. It is in India that the movements have shed the "victim" image, become consciously political and give real meaning to that overused but little understood word "empowerment." In India, discriminated against and impoverished women are seen as people with vast potential for contributing economically and socially to communities and who have the will and the skill to obtain this recognition from government.

At a recent Indian Widows Conference held in Bangalore, 49 widows, some of whom had never left their houses since widowhood, met with academics, planners, and activists, and described their histories and their priorities. In a moving ceremony they adorned each other with kumkum, bangles, and coloured saris, and vowed to take the battle for their human rights back to their villages through their local groups. "From now on, no matter how they revile us, we will work together, help each other, and defy anyone who seeks to oppress us," Goda said. "Now I have so many widows coming to my house that I feel strong. My relatives and neighbours realize that even if I have no husband, I have a large family of sisters to support me."

It is unfortunate that the deprivation of widows has received so little attention from the women's movements around the world. But new attitudes will only come about when widows themselves take collective action. The experiences in India show that widows are often capable of being more independent and militant than most married women. Freedom from conjugal control and the need to earn a living often impels them to be more assertive than their married sisters. They can be real agents of change.

The driving force behind any change has, of course, to be the widows themselves. The means of action must involve proper knowledge of rights, legal battles, public lobbying and criticism of political parties, officials, traditional and religious leaders, and local village councils. Success comes when collective action backs up specific demands.

There will be a widows' panel at this year's NGO Forum in Beijing. Hopefully, it will encourage more banding together of bereaved women who should never be seen only as vulnerable, but as people who have much to contribute and who bring about much needed social change.

• • •

20

First They Came After the Homosexuals

PASTOR MARTIN NIEMULLER

In Germany the Nazis first came for the Homosexuals, and I did not speak up because I was not a Homosexual.

Then they came for the Communists, and I was not a Communist so I did not speak up.

Then they came for the Jews, and I did not speak up because I was not a Jew.

Then they came for the Trade Unionists, and I did not speak up because I was not a Trade Unionist.

Then they came for the Catholics, and I was a Protestant so I did not speak up.

Then they came for me, and by that time there was no one to speak up for anyone.

October. 1945

5

Population Growth

The twentieth century has been extraordinarily successful for the human species—perhaps too successful. As our population has grown from 1 billion to 6 billion and the economy has exploded to more than twenty times its size in 1900, we . . . now face a challenge that rivals any in history: restoring balance with nature while expanding economic growth for the billions of people whose basic needs . . . are still not being met.

LESTER R. BROWN

How large is the world's population? How fast is it growing? As the opening quote indicates, total world population is around 6 billion. World population is continuing to grow rapidly—by 80 million in 1997 alone. However, the annual growth rate has been slowing. It has dropped from a high point of 2.2 percent in 1963 to 1.4 percent in 1997. The United Nations has projected that the world's population will reach 9.4 billion in the next 50 years. Growth varies by region. Most of the projected growth is expected to occur in Asia, with high growth rates also occurring in the Middle East and North Africa (Brown, Renner, and Flavin, 1998, p. 102).

The issue of world population growth provides an excellent example of the social construction of social problems. As discussed in Chapter 1, problems are defined in the course of debates among groups with opposing interests and viewpoints. For much of human history, people did not regard population growth as a social problem. When the world's population was small and resources were plentiful, children were seen as a blessing. More children meant more people who could work and create a prosperous livelihood. Not surprisingly, most of the world's historic religions urged people to reproduce (as the Bible

puts it, "be fruitful and multiply"). It was after the rise of industrial capitalism in Europe that some people began to see population growth as a social issue. In this period, Europe's cities grew enormously as impoverished peasants fled the countryside to the cities in search of work in the rising manufacturing sector. The cities developed vast slums that were nightmares of overcrowding, disease, and pollution. In this context, Thomas Malthus wrote his influential work, *An Essay on the Principle of Population,* which continues to haunt our thinking on the issue. Malthus argued that humanity was in danger because the population was growing much faster than the food supply. Eventually, massive famines would kill people off, correcting the imbalance between population and resources, but civilization would be plunged into chaos.

Population growth has continued to be debated since Malthus's time. Two sociologists, John R. Wilmoth and Patrick Ball, recently studied popular magazine articles about population growth published in the United States between 1946 and 1990. As many as 1,683 articles dealt with this topic. The researchers identified five predominant themes, or "interpretive packages," in these articles, which can be briefly summarized as follows: 1) "Limits to Growth"—population growth threatens the earth's "carrying capacity." There is some maximum amount of people that the earth's resources can support at our present level of consumption. Continued growth will result in impoverishment and environmental catastrophe. 2) "Quality of Life"—population growth contributes to problems such as overcrowding, traffic congestion, social disorganization, and destruction of the environment. 3) "Population Pressure"—overpopulation leads to wars and revolutions. 4) "Race Suicide"—the wealthy, more intelligent peoples of the world, largely European in origin, are being outbred by the poorer, less intelligent peoples, largely Asian, African, or Latin American in origin. 5) "Growth Is Good"—population growth is good because it promotes economic expansion. Interestingly, Wilmoth and Ball did not find a "Resource Imbalance" theme, the idea (to be discussed below) that an imbalance of wealth between the richer and poorer nations, not population growth, is the cause of problems such as poverty and environmental degradation (Wilmoth and Ball, 1995).

Today, population growth continues to be a hotly debated social issue. A number of the articles in this chapter reflect the ongoing controversy over whether population growth is a problem, and if so, how the problem should be defined. Joel Cohen argues in his insightful article, "How Many People Can the Earth Support?" that the question of overpopulation may not be one that can be answered simply by a number or even a range of numbers—it requires a social definition as to the quality of life we wish our world population to have. Cohen's article is followed by an article arguing that it is too simplistic to regard population growth as the cause of global poverty and environmental degradation. In "Too Many Mouths to Feed," Frances Moore Lappé and Joseph Collins look at the problem of poverty and argue that the causal relationship may in fact be the reverse—that poverty leads people to have large families. Lappé and Collins believe that it is not population pressure, but the lack of economic democracy in the poor nations, that results in impoverishment. In "Gray Dawn: The Global Aging Crisis," Peter G. Peterson argues that the "graying" of the world's population, rather than its sheer size, must be considered a serious threat to global well-being.

It was once believed that education on contraceptives ("family planning") was the key solution to curbing population growth. Little attention was paid to the social structures of families, to the gender roles played by men and women, and the consequences of women's inequality for fertility. As a result of the growing strength of the women's movement, awareness has grown that the empowerment of women is critical to slowing population growth. This point is now solidly entrenched in the thinking of United Nations policymakers—it became a leading idea at the 1994 International Conference on Population and Development in Cairo, and also at the 1995 United Nations Fourth World Conference on Women in Beijing. Some of the major points about the relationship between women's rights and population are explored in "Female Empowerment Leads to Fewer Births" by Cameron Barr.

Along with women's empowerment, however, contraception is still seen as an important aid to slowing world population growth. The governments of many poor nations around the world are encouraging family planning. Interestingly, this is occurring in some places even though the traditional religion has been strongly opposed to contraception. In "With Iran Population Boom, Vasectomy Receives Blessing," Neil MacFarquhar writes about the situation in Iran, where 45 percent of the population is under the age of 17.

WORKS CITED

Brown, Lester R., Michael Renner, and Christopher Flavin. 1998. *Vital Signs: The Environmental Trends That Are Shaping Our Future.* New York: W. W. Norton and Company.

Wilmoth, John R., and Patrick Ball. 1995. "Arguments and Actions in the Life of a Social Problem: A Case Study of 'Overpopulation,' 1946–1990." *Social Problems,* 42, 3 (August 1995), 318–343.

QUESTIONS FOR DISCUSSION

1.   Joel Cohen asks, "How many people can the earth support?" What is your own personal answer to this question? State explicitly the criteria regarding quality of life upon which your answer is based.

2.   Does poverty cause population growth, or does population growth cause poverty? After reading "Too Many Mouths to Feed," answer the following: Would redistribution of world resources from the richer to the poorer nations, and from the wealthier classes in the poorer nations to the impoverished masses, appreciably slow the rate of population growth? If so, why? What other social patterns need also to be changed in order to reduce population growth?

3.   Do you agree with Peter G. Peterson that "global aging" is a serious world problem? Why or why not?

4.   Why, according to Cameron Barr, is the empowerment of women key to slowing the world's population growth?

5.   Neil MacFarquhar's article, "With Iran Population Boom, Vasectomy Receives Blessing," reports on a contraception program in one poor nation that is apparently receiving increased popular support. Do you believe that programs such as this are inherently empowering to *women*? Why or why not?

 INFOTRAC COLLEGE EDITION: EXERCISE

As discussed in this chapter, the 1994 United Nations International Conference on Population and Development held in Cairo emphasized that women's rights must be expanded in order to curb population growth. A great deal of controversy exists worldwide over the nature of the relationship between women's rights and government population policies, such as policies encouraging contraception and abortion. Using Infotrac College Edition, look up the following articles:

"Women Endangered, Says World Population Report." United Nations Population Fund Report. *UN Chronicle.*

Spring 1998v35i1. (*Hint:* Enter the search term United Nations Conference on Population and Development, 1994.)

"The Feminist Challenge to Cairo." Melinda Tankard Reist. *The Human Life Review.* Winter 1995v21n1p53. (*Hint:* Enter the search term United Nations Conference on Population and Development, 1994.)

Are government family planning programs inherently empowering for women? Why or why not? Are these the key to slowing population growth? Why or why not?

FOR ADDITIONAL RESEARCH

Books

Brown, Lester R. 1995. *Who Will Feed China?* New York: W. W. Norton & Company.

Brown, Lester R., et al. 1995. *State of the World, 1995.* New York: W. W. Norton & Company.

Cohen, Joel E. 1995. *How Many People Can the Earth Support?* New York: W. W. Norton & Company.

Organizations

Alan Guttmacher Institute
120 Wall Street, Floor 21
New York. NY 10005

(212) 248-1111
www.agi-usa.org

Planned Parenthood Federation of America, Inc.
810 Seventh Avenue
New York, NY 10019
(212) 541-7800
www.plannedparenthood.org

Zero Population Growth
1400 16th Street, Suite 320
Washington, DC 20036
(202) 332-2200
(800) 767-1956
www.asapusa.com/zpg/

ACTION PROJECTS

1.   Visit the United Nations Web site (www.un.org) and go to the section on Population. Find the latest press briefings on population estimates and projections. What are some of the points made? What did you find that was unexpected or surprising? Find the United Nations Population Information Network (POPIN). What are some of the current United Nations programs aimed at slowing world population growth? What was the International Conference on Population and Development? What were some of its key recommendations?

2.   Is the world overpopulated? As Cohen's article pointed out, this is a matter of the quality of life we seek to have. Interview students on your campus to learn their views about the quality of life they seek. Use Cohen's ideas to help you develop a set of ques-tions. Are the students' views realistic, in the light of current world population trends? What needs to be done in order to ensure that everyone on earth has a quality of life resembling that described by the students you interviewed?

3.   Interview a foreign student on your campus who comes from a nation in Africa or Asia that has a high rate of population growth. Learn about the gender roles and gender inequalities in their society. How do these influ-ence the number of children women have? Also find out about cultural beliefs and values regarding contraception and abortion. What effect do these have on the number of children women give birth to? Are these views chang-ing? Why? To this student's knowledge, what efforts have been undertaken by the government to curb rapid popu-lation growth?

21

How Many People
Can the Earth Support?

The answers depend as much on social, cultural, economic and political choices as they do on constraints imposed by nature

JOEL E. COHEN

On April 25,1679, in Delft, Holland, the inventor of the microscope, Antoni van Leeuwenhoek, wrote down what may be the first estimate of the maximum number of people the earth can support. If all the habitable land in the world had the same population density as Holland (at that time about 120 people for every square kilometer), he calculated, the earth could support at most 13.4 billion people—far fewer than the number of spermatozoans his lenses had revealed in the milt of a cod.

In subsequent centuries, van Leeuwenhoek's estimate has been followed by dozens of similar calculations. Around 1695 a Londoner named Gregory King estimated that the earth's "Land If fully Peopled would sustain" at most 12.5 billion people. In 1765 a German regimental pastor, Johann Peter Süssmilch, compared his own figure (13.9 billion) with the estimates of van Leeuwenhoek, the French military engineer Sébastien Le Prestre de Vauban (5.5 billion) and the English writer and cartographer Thomas Templeman (11.5 billion).

In recent decades estimates of maximum population have appeared thicker and faster than ever before. Under the rubric of "carrying capacity" they crop up routinely in environmental debates, in United Nations reports and in papers by scholars or academic politicians trained in ecology, eco-nomics, sociology, geography, soil science or agronomy, among other disciplines. Demographers, however, have been strangely silent. Of the more than 200 symposiums held at the 1992 and 1993 annual meetings of the Population Association of America, not one session dealt with estimating or defining human carrying capacity for any region of the earth. Instead, professional demographers tend to focus on the composition and growth of populations, restricting their predictions to the near term—generally a few decades into the future—and framing them in conditional terms: *If* rates of birth, death and migration (by age, sex, location, marital status and so on) are such-and-such, *then* population size and distribution will be so-and-so.

Such conditional predictions. or forecasts, can be powerful tools. Projections by the U.N. show dramatically that *if* human populations continued to grow at 1990 rates in each major region of the world, *then* the population would increase more than 130-fold in 160 years, from about 5.3 billion in 1990 to about 694 billion in 2150. Those figures are extremely sensitive to the future level of average fertility. If, hypothetically, from 1990 onward the average couple gradually approached a level of fertility just one-tenth of a child more than required to replace themselves, world population would grow from 5.3 billion in 1990 to

12.5 billion in 2050 and 20.8 billion in 2150. In contrast, if (again hypothetically) starting in 1990 and ever after couples bore exactly the number of children needed to replace themselves world population would grow from 5.3 billion in 1990 to 7.7 billion in 2050 and would level off at around 8.4 billion by 2150.

The clear message is that people cannot forever continue to have, on average, more children than are required to replace themselves. That is not an ideological slogan; it is a hard fact. Conventional agriculture cannot grow enough food for 694 billion people; not enough water falls from the skies. The finiteness of the earth guarantees that ceilings on human numbers do exist.

Where are those ceilings? Some people believe that any limit to human numbers is so remote that its existence is irrelevant to present concerns. Others declare that the human population has already exceeded what the earth can support in the long run (how long is usually left unspecified). Still others concede that short-term limits may exist, but they argue that technologies, institutions and values will adapt in unpredictable ways to push ceilings progressively higher so that they recede forever. The differences of opinion are buttressed by vast disparities in calculation. In the past century, experts of various stripes have made estimates of human carrying capacity ranging from less than a billion to more than 1,000 billion. Who, if anybody, is right?

For several years I have been trying to understand the question, "How many people can the earth support?" and the answers to it. In the process I came to question the question. "How many people can the earth support?" is not a question in the same sense as "How old are you?"; it cannot be answered by a number or even by a range of numbers. The earth's capacity to support people is determined partly by processes that the social and natural sciences have yet to understand, partly by choices that we and our descendants have yet to make.

In most of its scientific senses, *carrying capacity* refers to a population of wild animals within a particular ecosystem. One widely used ecology textbook defines it as follows: "Number of individuals in a population that the resources of a habitat can support; the asymptote, or plateau, of the logistic and other sigmoid equations for population growth." Even within ecology, the concept of carrying capacity has important limitations. It applies best under stable conditions and over relatively short spans of time. In the real world, climates and habitats fluctuate and change; animals adapt to their conditions and eventually evolve into new species. With each change, the carrying capacity changes, too.

When applied to human beings, the concept becomes vastly more volatile. I have collected twenty-six definitions of human carrying capacity, all published since 1975. Most of them agree on a few basic points—for instance, that the concept refers to the number of people who can be supported for some period (usually not stated) in some mode of life considered plausible or desirable. Most of the definitions recognize that ecological concepts of carrying capacity must be extended to allow for the role of technology. Most also agree that culturally and individually variable standards of living, including standards of environmental quality, set limits on population size well before the physical requirements for sheer subsistence start to become an issue.

In other respects, however, the definitions vary widely or even contradict one another. How long must a population be sustainable? Does it make sense to speak of local or regional carrying capacity—or do trade and the need for inputs from outside any specified region imply that only a global scale will do? More fundamental, how constraining are constraints? Some definitions deny the existence of any finite carrying capacity altogether, holding that human ingenuity will win out over any natural barriers; others acknowledge that the limits are real but recognize that human choices, now and in the future, will largely decide where those limits fall.

In my opinion, that last point—the interplay of natural constraints and human choices—is the key to making sense of human carrying capacity. The deceptively simple question "How many people can the earth support?" hides a host of thorny issues:

HOW MANY PEOPLE AT WHAT AVERAGE LEVEL OF MATERIAL WELL-BEING?

The human carrying capacity of the earth will obviously depend on the typical material level at which people choose to live. Material well-being includes food (people choose variety and palatability, beyond the constraints imposed by physiological requirements); fiber (people choose cotton, wool or synthetic fibers for clothing, wood pulp or rag for paper); water (tap water or Perrier or the nearest river or mud hole for drinking, washing, cooking and watering your lawn, if you have one); housing (Auschwitz barracks, two men to a plank, or Thomas Jefferson's Monticello); manufactured goods; waste removal (for human, agricultural and industrial wastes); natural-hazard protection (against floods, storms, volcanoes and earthquakes); health (prevention, cure and care); and the entire range of amenities such as education, travel, social groups, solitude, the arts, religion and communion with nature. Not all of those features are captured well by standard economic measures.

HOW MANY PEOPLE WITH WHAT DISTRIBUTION OF MATERIAL WELL-BEING?

An ecologist, an economist and a statistician went bow hunting in the woods and spied a deer. The ecologist shot first, and his arrow landed five meters to the left of the deer. The economist shot next, and her arrow landed five meters to the right of the deer. The statistician looked at both arrows, looked at the deer, and jumped up and down shouting: "We got it! We got it!"

Estimates of human carrying capacity rarely take into account the scatter or distribution of material well-being throughout a population. Yet paying attention to average well-being while ignoring the distribution of well-being is like using an average arrow to kill a deer. People who live in extreme poverty may not know or care that the global average is satisfactory, and the press

of present needs may keep them from taking a long-term view. For example, thanks to genetic engineering, any country with a few Ph.D.'s in molecular plant biology and a modestly equipped laboratory can insert the genes to create stronger, more disease-resistant, higher-yielding plants. If every region has the scientific and technical resources to improve its own crop plants, the earth can support more people than it can if some regions are too poor to help themselves.

HOW MANY PEOPLE WITH WHAT TECHNOLOGY?

The complexities of technological choices often disappear in heated exchanges between environmental pessimists and technological optimists:

Ecologist: When a natural resource is being consumed faster than it is being replenished or recycled, an asset is being depleted, to the potential harm of future generations.

Technologist: If new knowledge and technology can produce an equivalent or superior alternative, then future generations may turn out to be better off.

Taxpayer: Which natural resources can be replaced by technology yet to be invented, and which cannot? Will there be enough time to develop new technology and put it to work on the required scale? Could we avoid future problems, pain and suffering by making other choices now about technology or ways of living? [*No answer from ecologist or technologist.*]

The key to the argument is time. As Richard E. Benedick, an officer of the U.S. Department of State who has also served with the World Wildlife Fund, worried:

While it is true that technology has generally been able to come up with solutions to human dilemmas, there is no guarantee that ingenuity will always rise to the task. Policymakers must contend with a nagging thought: what if it does not, or what if it is too late?

HOW MANY PEOPLE WITH WHAT DOMESTIC AND INTERNATIONAL POLITICAL INSTITUTIONS?

Political organization and effectiveness affect human carrying capacity. For example, the United Nations Development Program estimated that developing countries could mobilize for development as much as $50 billion a year (an amount comparable to all official development assistance) if they reduced military expenditures, privatized public enterprises, eliminated corruption, made development priorities economically more rational and improved national governance. Conversely, population size, distribution and composition affect political organization and effectiveness.

How will political institutions and civic participation evolve with increasing numbers of people? As numbers increase, what will happen to people's ability to participate effectively in the political system?

What standards of personal liberty will people choose?

How will people bring about political change within existing nations? By elections and referendums, or by revolution, insurrection and civil war? How will people choose to settle differences between nations, for instance, over disputed borders, shared water resources or common fisheries? War consumes human and physical resources. Negotiation consumes patience and often requires compromise. The two options impose different constraints on human carrying capacity.

HOW MANY PEOPLE WITH WHAT DOMESTIC AND INTERNATIONAL ECONOMIC ARRANGEMENTS?

What levels of physical and human capital are assumed? Tractors, lathes, computers, better health and better education all make workers in rich countries far more productive than those in poor countries. Wealthier workers make more wealth and can support more people.

What regional and international trade in finished goods and mobility in productive assets are permitted or encouraged? How will work be organized? The invention of the factory organized production to minimize idleness in the use of labor, tools and machines. What new ways of organizing work should be assumed to estimate the future human carrying capacity?

HOW MANY PEOPLE WITH WHAT DOMESTIC AND INTERNATIONAL DEMOGRAPHIC ARRANGEMENTS?

Almost every aspect of demography (birth, death, age structure, migration, marriage, and family structure) is subject to human choices that will influence the earth's human carrying capacity.

A stationary global population will have to choose between a long average length of life and a high birthrate. It must also choose between a single average birthrate for all regions, on the one hand, and a demographic specialization of labor on the other (in which some areas have fertility above their replacement level, whereas other areas have fertility below their replacement level).

Patterns of marriage and household formation will also influence human carrying capacity. For example, the public resources that have to be devoted to the care of the young and the aged depend on the roles played by families. In China national law requires families to care for and support their elderly members; in the United States each elderly person and the state are largely responsible for supporting that elderly person.

HOW MANY PEOPLE IN WHAT PHYSICAL, CHEMICAL AND BIOLOGICAL ENVIRONMENTS?

What physical, chemical and biological environments will people choose for themselves and for their children? Much of the heat in the public argument over current environmental problems arises because the consequences of present and projected choices and changes are uncertain. Will global warming cause great problems, or would a global limitation on fossil-fuel consumption cause greater problems? Will toxic or nuclear wastes or ordinary sewage sludge dumped into the deep ocean come back to haunt future generations when deep currents well up in biologically productive offshore zones, or would the long-term effects of disposing of those wastes on land be worse? The choice of particular alternatives could materially affect human carrying capacity.

HOW MANY PEOPLE WITH WHAT VARIABILITY OR STABILITY?

How many people the earth can support depends on how steadily you want the earth to support that population. If you are willing to let the human population rise and fall, depending on annual crops, decadal weather patterns and long-term shifts in climate, the average population with ups and downs would include the peaks of population size, whereas the guaranteed level would have to be adjusted to the level of the lowest valley. Similar reasoning applies to variability or stability in the level of well-being; the quality of the physical, chemical and biological environments: and many other dimensions of choice.

HOW MANY PEOPLE WITH WHAT RISK OR ROBUSTNESS?

How many people the earth can support depends on how controllable you want the well-being of the population to be. One possible strategy would be to maximize numbers at some given level of well-being, ignoring the risk of natural or human disaster. Another would be to accept a smaller population size in return for increased control over random events. For example, if you settle in a previously uninhabited hazardous zone (such as the flood plain of the Mississippi River or the hurricane-prone coast of the southeastern U.S.),

you demand a higher carrying capacity of the hazardous zone, but you must accept a higher risk of catastrophe. When farmers do not give fields a fallow period, they extract a higher carrying capacity along with a higher risk that the soil will lose its fertility (as agronomists at the International Rice Research Institute in the Philippines discovered to their surprise).

HOW MANY PEOPLE FOR HOW LONG?

Human carrying capacity depends strongly on the time horizon people choose for planning. The population that the earth can support at a given level of well-being for twenty years may differ substantially from the population that can be supported for 100 or 1,000 years.

The time horizon is crucial in energy analysis. How fast oil stocks are being consumed matters little if one cares only about the next five years. In the long term, technology can change the definition of resources, converting what was useless rock to a valuable resource; hence no one can say whether industrial society is sustainable for 500 years.

Some definitions of human carrying capacity refer to the size of a population that can be supported indefinitely. Such definitions are operationally meaningless. There is no way of knowing what human population size can be supported indefinitely (other than zero population, since the sun is expected to burn out in a few billion years, and the human species almost certainly will be extinct long before then). The concept of indefinite sustainability is a phantasm, a diversion from the difficult problems of today and the coming century.

HOW MANY PEOPLE WITH WHAT FASHIONS, TASTES AND VALUES?

How many people the earth can support depends on what people want from life. Many choices that appear to be economic depend heavily on individual and cultural values. Should industrial societies use the available supplies of fossil fuels in households for heating and for personal transportation, or outside of households to produce other goods and services? Do people prefer a high average wage and low employment or a low average wage and high employment (if they must choose)?

Should industrial economies seek now to develop renewable energy sources, or should they keep burning fossil fuels and leave the transition to future generations? Should women work outside their homes? Should economic analyses continue to discount future income and costs, or should they strive to even the balance between the people now living and their unborn descendants?

I am frequently asked whether organized religion, particularly Roman Catholicism, is a serious obstacle to the decline of fertility. Certainly in some countries, church policies have hindered couples' access to contraception and have posed obstacles to family planning programs. In practice, however, factors other than religion seem to be decisive in setting average levels of fertility for Roman Catholics. In 1992 two Catholic countries, Spain and Italy, were tied for the second- and third-lowest fertility rates in the world. In largely Catholic Latin America, fertility has been falling rapidly, with modern contraceptive methods playing a major role. In most of the U.S. the fertility of Catholics has gradually converged with that of Protestants, and polls show that nearly four-fifths of Catholics think that couples should make up their own minds about family planning and abortion.

Even within the church hierarchy, Catholicism shelters a diversity of views. On June 15, 1994, the Italian bishops' conference issued a report stating that falling mortality and improved medical care "have made it unthinkable to sustain indefinitely a birthrate that notably exceeds the level of two children per couple." Moreover, by promoting literacy for adults, education for children and the survival of infants in developing countries, the church has helped bring about some of the social preconditions for fertility decline.

On the whole the evidence seems to me to support the view of the ecologist William W. Murdoch of the University of California, Santa Barbara: "Religious beliefs have only small, although sometimes significant, effects on family

size. Even these effects tend to disappear with rising levels of well-being and education."

In short, the question "How many people can the earth support?" has no single numerical answer, now or ever. Human choices about the earth's human carrying capacity are constrained by facts of nature and may have unpredictable consequences. As a result, estimates of human carrying capacity cannot aspire to be more than conditional and probable: if future choices are thus-and-so, then the human carrying capacity is likely to be so-and-so. They cannot predict the constraints or possibilities that lie in the future; their true worth may lie in their role as a goad to conscience and a guide to action in the here and now.

The following beautiful quotation from *Principles of Political Economy*, by the English philosopher John Stuart Mill, sketches the kind of shift in values such action might entail. When it was written, in 1848, the world's population was less than one-fifth its present size.

> There is room in the world, no doubt, and even in old countries, for a great increase of population, supposing the arts of life to go on improving, and capital to increase. But even if innocuous, I confess I see very little reason for desiring it. The density of population necessary to enable mankind to obtain, in the greatest degree, all the advantages both of cooperation and of social intercourse, has, in all the most populous countries, been obtained. A population may be too crowded, though all be amply supplied with food and raiment. It is not good for man to be kept perforce at all times in the presence of his species. A world from which solitude is extirpated, is a very poor ideal. . . . Nor is there much satisfaction in contemplating the world with nothing left to the spontaneous activity of nature; with every rood of land brought into cultivation, which is capable of growing food for human beings; every flowery waste or natural pasture ploughed up, all quadrupeds or birds which are not domesticated for man's use exterminated as his rivals for food, every hedgerow or

superfluous tree rooted out, and scarcely a place left where a wild shrub or flower could grow without being eradicated as a weed in the name of improved agriculture. If the earth must lose that great portion of its pleasantness which it owes to things that the unlimited increase of wealth and population would extirpate from it, for the mere purpose of enabling it to support a larger but not a better or a happier population, I sincerely hope, for the sake of posterity, that they will content to be stationary, long before necessity compels them to it.

It is scarcely necessary to remark that a stationary condition of capital and population implies no stationary state of human improvement. There would be as much scope as ever for all kinds of mental culture,

and moral and social progress; as much room for improving the Art of Living, and much more likelihood of its being improved, when minds ceased to be engrossed by the art of getting on. Even the industrial arts might be as earnestly and as successfully cultivated, with this sole difference, that instead of serving no purpose but the increase of wealth, industrial improvements would produce their legitimate effect, that of abridging labour. . . . Only when, in addition to just institutions. the increase of mankind shall be under the deliberate guidance of judicious foresight, can the conquests made from the powers of nature by the intellect and energy of scientific discoverers, become the common property of the species, and the means of improving it and elevating the universal lot.

22

Too Many Mouths To Feed

FRANCES MOORE LAPPÉ AND
JOSEPH COLLINS

MYTH: Hunger is caused by too many people pressing against finite resources. We must slow population growth before we can hope to alleviate hunger.

OUR RESPONSE: In all of our educational efforts during the last 15 years, no question has been more challenging than, do too many people cause hunger? We've answered no, but in the eyes of some this is tantamount to irresponsibly dismissing population growth as a problem.

We do not take lightly, however, the prospect of human numbers so dominating the planet that other forms of life are squeezed out, that all wilderness is subdued for human use, and that the mere struggle to feed and warm ourselves keeps us from more satisfying pursuits.

Indeed, to us the question of population is so vital that we can't afford to be the least bit fuzzy in our thinking. So here we will focus on the two most critical questions this myth poses. Are population density and population growth

From Frances Moore Lappé and Joseph Collins, "Myth 3: Too Many Mouths to Feed," *World Hunger: Twelve Myths.* Copyright © 1986 by The Institute for Food and Policy Development. Reprinted by permission of Grove/Atlantic, Inc.

the cause of hunger? And, what is the link between slowing population growth and ending hunger?

If too many people caused hunger, then reducing population density could indeed alleviate it. But for one factor to cause another, the two must consistently occur together. Population density and hunger do not.

China has only half the cropland per person as India, yet Indians suffer widespread and severe hunger while the Chinese do not. Taiwan and South Korea each have only half the farmland per person found in Bangladesh, yet no one speaks of overcrowding causing hunger in Taiwan and South Korea. Costa Rica, with less than half of Honduras' cropped acres per person, boasts a life expectancy—one indicator of nutrition—14 years longer than that of Honduras and close to that of the Western industrial countries.

Surveying the globe, we in fact can find no correlation between population density and hunger. For every Bangladesh, a densely populated and hungry country, we find a Brazil or a Senegal, where significant food resources per capita coexist with intense hunger. Or we find a country like the Netherlands where very little land per person has not prevented it from eliminating hunger and becoming a large net exporter of food.

But what about population growth? Is there not an obvious correlation between rapid population growth and hunger? Without doubt, most hungry people live in Asia, Africa, and Latin America, where populations are growing fastest.

This association of hunger and rapid population growth certainly suggests a relationship between the two. But what we want to probe is the nature of that link. Does rapid population growth cause hunger, or do they occur together because they are both consequences of similar social realities?

Since there doesn't seem to be a correlation between population density and hunger, we propose the following thesis as most probable: hunger, the most dramatic symptom of pervasive poverty, and rapid population growth occur together because they have a common cause.

LESSONS FROM OUR OWN PAST

Let's begin by looking at our own demographic history. As recently as two generations ago, mortality rates in the United States were as high as they are now in most third world countries. Opportunities for our grandmothers to work outside the home were limited. And ours was largely an agrarian society in which every family member was needed to work on the farm. Coauthor Frances Lappé's own grandmother, for example, gave birth to nine children, raised them alone on a small farm, and saw only six survive to adulthood. Her story would not be unusual in a typical fast-growing third world country today.

In the United States, the move to two-child families took place only after a societywide transition that lowered infant death rates, opened opportunities to women outside the home, and transformed ours into an industrial rather than agrarian economy so that families no longer relied on their children's labor. Birth rates fell in response to these changes well before the advent of sophisticated birth control technologies, even while government remained actively hostile to birth control. (As late as 1965, selling contraceptives was still illegal in some states.)

Using our own country's experience to understand rapid population growth in the third world, where poverty is more extreme and widespread, we can now extend our hypothesis concerning the link between hunger and rapid growth: both result where societies deny security and opportunity to the majority of their citizens—where adequate land, jobs, health care, and old-age security are beyond the reach of most people.

Without resources to secure their future, people can rely only on their own families. Thus, when many poor parents have lots of children, they are making a rational calculus for survival. High birth rates reflect people's defensive reaction against enforced poverty. For those living at the margin of survival, children provide labor to augment meager family income. In Bangladesh, one study showed that even by the age of six a boy provides labor and/or income for the family. By the age of twelve, at the latest, he contributes more than he consumes.

Population investigators tell us that the benefit children provide to their parents in most third world countries cannot be measured just by hours of labor or extra income. The intangibles are just as important. Bigger families carry more weight in community affairs. Moreover, the "lottery mentality" is associated with poverty everywhere. In the third world, with no reliable channels for advancement in sight, third world parents can always hope that the next child will be the one clever and bright enough to get an education and land a city job despite the odds. In many countries, income from one such job in the city can support a whole family in the countryside.

And impoverished parents know that without children to care for them in old age, they will have nothing.

They also realize that none of these possible benefits will be theirs unless they have many children, since hunger and lack of health care kill many of their offspring before they reach adulthood. The World Health Organization has shown that both the actual death and the fear of death of a child will increase the fertility of a couple, regardless of income or family size.

Finally, high birth rates reflect not only the survival calculus of the poor, but the disproportionate powerlessness of women. In highly patriarchal societies, a woman's status is largely determined by the number of children she bears—especially the number of sons. And in many societies where women would prefer fewer children, they don't have control over their own fertility. As one doctor in a Mexican clinic explained:

> When a wife wants to . . . limit the number of mouths to feed in the family, the husband will become angry and even beat her. He thinks it is unacceptable that she is making a decision of her own. She is challenging his authority, his power over her—and thus the very nature of his virility.

Even where women are less dependent on men, or where family decisions are shared more equally, restricted access to health services, including birth control, limits women's ability to control reproduction. This is especially true for poor women, for whom access to health services is most limited of all.

TESTING THE THESIS

If our thesis that rapid population growth results largely from the powerlessness of the poor is true, then population growth will only slow when far-reaching economic and political changes convince the majority of people that social arrangements *beyond the family*—jobs, health care, old-age security, and education (especially for women)—offer both security and opportunity.

Substantial empirical evidence lends weight to this thesis. Consider the implications of the following World Bank population statistics covering three-fourths of the world's people who live in 72 countries that the bank designates low and low-middle income.

While average annual population growth rates in all industrial countries have been below 2 percent a year for decades, in these 72 countries only six had reduced their growth below 2 percent by 1982. Of these, only four have experienced a dramatic drop in their growth rate since the 1960s: China, Sri Lanka, Cuba, and Colombia. Although not a country, and therefore not listed in the World Bank statistics, the Indian state of Kerala also reduced its population growth rate below 2 percent. (The remaining two slow-growing poor countries are Haiti and Jamaica, but their growth rates have not fallen since the 1960s.)

Population growth in these four exceptional countries and in Kerala has slowed at twice the rate of the current industrialized countries during their transition from high to low growth. What do these exceptions tell us? What could societies as different as those of China, Sri Lanka, Cuba, Colombia, and Kerala have in common?

Except Colombia, they have all assured their citizens access to a basic diet through more extensive food guarantee systems than exist in other third world societies.

China. Since the 1950s, a basic ration of grain has been one of the "five guarantees" to which needy Chinese citizens are entitled.

Cuba. Rationing and setting price ceilings on staples has kept basic food affordable and available to the Cuban people for nearly 25 years. Under Cuba's rationing system, all citizens are guaranteed enough rice, pulses, oil, sugar, meat, and other food to provide them with 1,900 calories a day.

Sri Lanka. From the postwar period to 1978, the Sri Lankan government supported the consumption of basic foods, notably rice, through a combination of free food, rationed food, and subsidized prices. Since the late 1970s, however, this elaborate food security system has begun to be dismantled. Interestingly, the World Bank does not predict Sri Lanka's growth rate to continue to fall, but rather to rise slightly by the year 2000.

Kerala, India. Eleven thousand government-run fair-price shops keep the cost of rice and other essentials like kerosene within the reach of the poor. This subsidy accounts for as much as one-half of the total income of Kerala's poorer families.

In most of these societies, income distribution is also less skewed than in most other countries. The distribution of household income in Sri Lanka, for example, is more equitable than in Indonesia, India, or even the United States.

This positive link between fertility decline and increased income equity is confirmed by empirical investigations. While one might question the possibility of such neat precision, one World Bank study of 64 different countries indicated that when the poorest group's income goes up by one percentage point, the general fertility rate drops by almost three percentage points. When literacy and life expectancy are added to the income analysis, the three factors explain 80 percent of the variation in fertility among countries.

Several other critical features stand out as we try to understand why, of all low and lower-middle-income societies, these five have experienced such a rapid decline in population growth.

Colombia, not known for its government interventions on behalf of the poor, appears to defy the above preconditions of security and opportunity, but not entirely.

Colombia's infant mortality is well below most lower-middle-income countries. Its health service sends medical interns to the countryside for one year's free service, unlike many third world countries, where medical services barely reach outside the capital city. Colombia has also achieved high literacy rates, and almost all children attend primary school.

Colombia's record also demonstrates that shifting resources toward women, expanding their opportunities, has a much bigger impact on lowering birth rates than an overall rise in income—a general pattern, according to Yale University's T. Paul Schultz. Colombia's women appear to be achieving greater economic independence from men and therefore are becoming better able to determine their own fertility. An unusually high percentage of girls attend secondary school. Colombian women are entering the work force at a rapid pace, and new income from the coffee boom of the 1970s contributed to many rural women's new economic independence.

Perhaps the most intriguing demographic case study—highlighting the several intertwined questions raised in this chapter—is that of Kerala. Its population density is three times the average for all India, yet commonly used indicators of hunger and poverty—infant mortality, life expectancy, and death rate—are all considerably more positive in Kerala than in most low-income countries as well as in India as a whole. Its infant mortality is half the all-India average.

Other measures of welfare also reveal the relatively better position of the poor in Kerala. Besides the grain distribution system mentioned above, social security payments, pensions, and unemployment benefits also transfer resources to the poorest groups. While land reform left significant inequality in land ownership, it did abolish tenancy, providing greater security to many who before were only renters. Thanks to effective farmworker organizing, agricultural wages are relatively high. Literacy and education levels are far superior to other states, particularly for women: the female literacy rate in Kerala is two and a half times the all-India average.

Like Kerala, China fascinates us because it demonstrates that even with scarce resources both hunger and rapid population growth can be addressed by shifting resources to the poorest citizens—especially to poor women.

Valuable lessons are also to be found in the experience of those countries still growing at well over 2 percent yearly, but in which crude birth rates have dropped markedly since the 1960s: Thailand, the Philippines, and Costa Rica. Despite their poverty, critical health and other social indicators distinguish these countries from most third world countries, providing clues to their fertility decline. Infant death rates are relatively low, especially in Costa Rica, and life expectancy is high—for women, ranging between 65 to 76 years. Perhaps most important, in the Philippines and Costa Rica an unusually high proportion of women are educated, and in both the Philippines and Thailand, proportionately more women work outside the home than in most third world countries.

BUT WE DON'T HAVE TIME

In presenting the essence of this thesis—high birth rates result from economic insecurity—to concerned audiences over the years, at least one questioner will invariably respond, "All well and good, but we don't have time! We can't wait for society-wide change benefiting the poor. That takes too long. The population bomb is exploding now."

The implication is that we should do the only thing we can do now: fund and promote family planning programs among fast-growing populations. The rest is pie in the sky.

Our response is twofold. First, demographers will tell you that even if average family size in a fast-growing society were cut by half tomorrow, its population would not stop growing until well into the next century. So every solution, including family planning programs, is a long-term one; there are no quick fixes. The second part of our answer is more

surprising: simply providing birth control technology through family planning programs doesn't affect population growth all that much.

D. J. Hernandez, chief of the Marriage and Family Statistics Branch of the U.S. Bureau of the Census, has reviewed all available research to determine the contribution of family planning programs to fertility decline. Examining the research on demographic change in 83 countries, he concluded that the best studies have found little net effect from family planning programs. Hernandez observed that "perhaps as much as 10 percent but possibly as little as 3 percent of the cross-national variation in fertility change in the third world during the late 1960s and early 1970s was an independent effect of family planning programs."

Naturally, Hernandez has been roundly attacked by family planning proponents. But even the study most cited by Hernandez's critics, a 1978 overview of 94 third world countries, concluded that birth control programs alone accounted for only 15 to 20 percent of overall fertility decline, with largely social and economic factors accounting for the rest.

Our highlighting these findings—which reveal a relatively small impact of family planning programs on population—does not mean, of course, that we belittle their potential value. Making contraceptives widely available and helping to reduce inhibitions against their use are critical to the extension of human freedom, especially the freedom of women to control their reproduction. But these findings do confirm that what is truly pie in the sky is the notion that population growth rates can be brought down to replacement levels through a narrow focus on the delivery of contraceptive technology.

Although the experience of Kerala and Colombia suggests that birth rates can fall drastically while great economic inequalities remain, an overwhelmingly clear pattern emerges from worldwide demographic change. At the very least, critical advances in health, social security, and education must change the lives of the poor—especially the lives of poor women—before they can choose to have fewer children. Once people are motivated to have smaller families, family planning programs can quicken a decline in fertility, but that is all; they cannot initiate the decision to have smaller families.

UPPING THE ANTE

Refusing to admit the implications of these findings, many third world governments and international agendas have responded to the marginal impact of family planning programs by upping the ante: designing ever-tougher programs involving long-term injectable contraceptives, sterilization, and financial incentives and penalties.

One example is the injectable contraceptive Depo-Provera. Although considered too hazardous for general use in the United States, 1 million injections lasting three to six months have been given in Thailand alone. Known short-term side effects include menstrual disorders, headaches, weight gain, depression, abdominal discomfort, and delayed return to fertility. And while long-term side effects will not be known for some time, preliminary studies suggest that Depo-Provera is probably linked to an increased risk of cervical cancer.

Family planning programs that are more aimed at controlling population than enhancing the self-determination and well being of women often fail to offer ongoing, village-level medical care to assist women in choosing appropriate methods and to monitor side effects. So the programs end up being partly self-defeating, because many women just stop using the contraceptives. In one survey in the Philippines, two-thirds of the women who had stopped using the pill and 43 percent who had had the IUD removed blamed medical side effects.

In at least a dozen countries, a variety of material incentives are increasingly used to induce people to undergo sterilization or to use contraception. Defenders of incentive programs stress that they are voluntary. But when you're hungry, how many choices are voluntary? Most telling is that in Bangladesh, where the majority are desperately poor, sterilizations rose dramatically when incentives were increased in 1983.

Moreover, they tend to fluctuate with the availability of food. According to the *Bangladesh Observer,* during the flood months of July to October 1984, sterilizations rose to an unprecedented quarter million, accounting for almost one-fourth of all sterilizations performed between 1972 and 1982.

According to author Betsy Hartmann, who has lived and worked in Bangladesh, the system invites abuse. As in many other countries, doctors and clinic staff are paid for each sterilization they perform and anyone, even outside the health system, can get a special fee for "referring" or "motivating" someone to undergo sterilization.

An extreme example suggests the abuses that can result. Donations of wheat to the poor in the aftermath of severe flooding in 1984 were made conditional on women agreeing to sterilization, Hartmann reports. After the operation, each woman received a certificate signed by a family planning officer vouching for her sterility and entitling her to a sari, money, and wheat.

CHINA'S SOLUTION?

Those who cling to family planning programs as the answer to population growth might do well to heed the current experience of China.

Through a far-reaching redistribution of land and food, assurance of old-age security, and making health care and birth control devices available to all, China achieved an unprecedented birth rate decline. But since 1979, China has taken a different tack. Believing that population growth was still hindering modernization, the Deng Xiaoping government instituted the world's most restrictive family planning program. Material incentives and penalties are now offered to encourage all parents to bear only one child. According to John Ratcliffe of UC Berkeley's School of Public Health:

> Enormous pressure—social and official— is brought to bear on those who become "unofficially" pregnant; few are able to resist such constant, heavy pressure, and most accede to having an abortion. While

coercion is not officially sanctioned, this approach results in essentially the same outcome.

At the same time, Ratcliffe points out, China's post-1979 approach to economic development has undercut both guaranteed employment and old-age security. This has thrown rural families back on their own labor resources, so that large families—especially boys—have once again become a family economic asset.

And what have been the consequences? Despite the world's most stringent population control program, China's birth rates have not continued to fall. If anything they have risen slightly! The message should be unmistakable: people will have children when their security and economic opportunity depend on it, no matter what the state says.

Advocates of more authoritarian measures seem to forget altogether the experience of the other poor societies that along with China have reduced their growth rates to below 2 percent. Recall they are Cuba, Kerala, Colombia, and Sri Lanka. None relied significantly on social coercion or financial incentives. As health care was made available to all, Cuba's birth rates fell, for example, without even so much as a public education campaign, much less financial incentives.

No one should discount the consequences of high population density and rapid population growth, including the difficulties they can add to the already great challenge of development. While in some African countries low population density has been an obstacle to agricultural development, in most countries high population density would make more difficult the tasks of social and economic restructuring necessary to eliminate hunger. Where land resources per person are scarce, social conflict during land reform will be more intense than where resources are plentiful. In Nicaragua's agrarian reform, the luxury of ample farmland—including much unused land—allowed the government to leave many large estates untouched, while still distributing land to the landless. By contrast, El Salvador's much greater population density leaves little land idle;

thus, any significant shift of land to the landless will undoubtedly involve more conflict.

NO TIME TO LOSE

In this brief chapter, we've outlined what we believe are the critical distinctions too often muddied in discussions of population:

- Unchecked population growth is a crisis for our planet not only because it threatens the future well-being of humanity but because of our moral obligation to share the earth responsibly with other forms of life.

- Population density nowhere explains today's widespread hunger.

- Rapid population growth is not the root cause of hunger, but it is—like hunger—a consequence of social inequities that deprive the poor majority—especially poor women—of the security and economic opportunity necessary for them to choose fewer children.

- To bring the human population into balance with the natural environment, societies have no choice but to address the extreme maldistribution of access to survival resources—land, jobs food, education, and health care.

- Family planning cannot by itself reduce population growth, though it can speed a decline. It best contributes to this transition when integrated into village and neighborhood-based health systems that offer birth control to expand human freedom rather than to control behavior.

We believe that precisely because population growth is such a critical problem, we cannot waste time with approaches that do not work. We must unflinchingly face the evidence telling us that the fate of the world hinges on the fate of today's poor majorities. Only as their well-being improves can population growth slow.

To attack high birth rates without attacking the causes of poverty and the disproportionate powerlessness of women is fruitless. It is a tragic diversion our beleaguered planet can ill afford.

23

Gray Dawn

The Global Aging Crisis

PETER G. PETERSON

DAUNTING DEMOGRAPHICS

The list of major global hazards in the next century has grown long and familiar. It includes the proliferation of nuclear, biological, and chemical weapons, other types of high-tech terrorism, deadly super-viruses, extreme climate change, the financial, economic, and political aftershocks of globalization, and the violent ethnic explosions waiting to be detonated in today's unsteady new democracies. Yet there is a less-understood challenge—the graying of the developed world's population—that may actually do more to reshape our collective future than any of the above.

Over the next several decades, countries in the developed world will experience an unprecedented growth in the number of their elderly and an unprecedented decline in the number of their youth. The timing and magnitude of this demographic transformation have already been determined. Next century's elderly have already been born and can be counted—and their cost to retirement benefit systems can be projected.

Unlike with global warming, there can be little debate over whether or when global aging will manifest itself. And unlike with other challenges, even the struggle to preserve and strengthen unsteady new democracies, the costs of global aging will be far beyond the means of even the world's wealthiest nations—unless retirement benefit systems are radically reformed. Failure to do so, to prepare early and boldly enough, will spark economic crises that will dwarf the recent meltdowns in Asia and Russia.

How we confront global aging will have vast economic consequences costing quadrillions of dollars over the next century. Indeed, it will greatly influence how we manage, and can afford to manage, the other major challenges that will face us in the future.

For this and other reasons, global aging will become not just the transcendent economic issue of the 21st century but the transcendent political issue as well. It will dominate and daunt the public-policy agendas of developed countries and force the renegotiation of their social contracts. It will also reshape foreign policy strategies and the geopolitical order.

The United States has a massive challenge ahead of it. The broad outlines can already be seen in the emerging debate over Social Security and Medicare reform. But ominous as the fiscal stakes are in the United States, they loom even larger in Japan and Europe, where populations are aging even faster, birthrates are lower, the influx of young immigrants from developing countries is smaller, public pension benefits are more generous, and private pension systems are weaker.

Aging has become a truly global challenge, and must therefore be given high priority on the global policy agenda. A gray dawn fast approaches. It is time to take an unflinching look at the shape of things to come.

The Floridization of the Developed World

Been to Florida lately? You may not have realized it, but the vast concentration of seniors there—nearly

Just Thirty Years From Now, One in Four People in the Developed World Will Be Aged 65 or Older, Up From One in Seven Today.

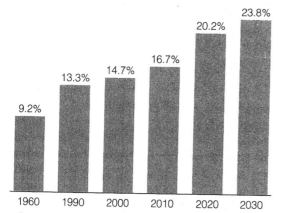

Percentage of the population of the developed world that is elderly (aged 65 & over)

Source: OECD (1996)

19 percent of the population—represents humanity's future. Today's Florida is a demographic benchmark that every developed nation will soon pass. Italy will hit the mark as early as 2003, followed by Japan in 2005 and Germany in 2006. France and Britain will pass present-day Florida around 2016; the United States and Canada in 2021 and 2023.

Societies Much Older Than Any We Have Ever Known

Global life expectancy has grown more in the last fifty years than over the previous five thousand. Until the Industrial Revolution, people aged 65 and over never amounted to more than 2 or 3 percent of the population. In today's developed world, they amount to 14 percent. By the year 2030, they will reach 25 percent and be closing in on 30 in some countries.

An Unprecedented Economic Burden on Working-Age People

Early in the next century, working-age populations in most developed, countries will shrink. Between 2000 and 2010, Japan, for example, will

suffer a 25 percent drop in the number of workers under age 30. Today the ratio of working taxpayers to nonworking pensioners in the developed world is around 3:1. By 2030, absent reform, this ratio will fall to 1.5:1, and in some countries, such as Germany and Italy, it will drop all the way down to 1:1 or even lower. While the longevity revolution represents a miraculous triumph of modern medicine and the extra years of life will surely be treasured by the elderly and their families, pension plans and other retirement benefit programs were not designed to provide these billions of extra years of payouts.

The Aging of the Aged: The Number of "Old Old" Will Grow Much Faster than the Number of "Young Old"

The United Nations projects that by 2050, the number of people aged 65 to 84 worldwide will grow from 400 million to 1.3 billion (a threefold increase), while the number of people aged 85 and over will grow from 26 million to 175 million (a sixfold increase)—and the number aged 100 and over from 135,000 to 2.2 million (a sixteenfold increase). The "old old" consume far more health care than the "young old"—about two to three times as much. For nursing-home care, the ratio is roughly 20:1. Yet little of this cost is figured in the official projections of future public expenditures.

Falling Birthrates Will Intensify the Global Aging Trend

As life spans increase, fewer babies are being born. As recently as the late 1960s, the worldwide total fertility rate (that is, the average number of lifetime births per woman) stood at about 5.0, well within the historical range. Then came a behavioral revolution, driven by growing affluence, urbanization, feminism, rising female participation in the workforce, new birth control technologies, and legalized abortion. The result: an unprecedented and unexpected decline in the global fertility rate to about 2.7—a drop fast approaching the replacement rate of 2.1 (the rate required merely to

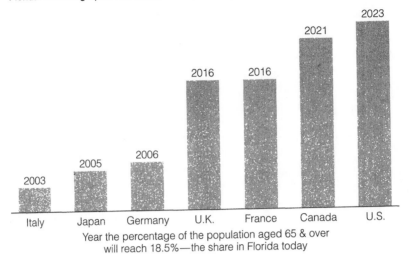

Florida: A Demographic Benchmark Every Developed Nation Will Soon Pass

Year the percentage of the population aged 65 & over
will reach 18.5%—the share in Florida today

Source: OECD (1996)

maintain a constant population). In the developed world alone, the average fertility rate has plummeted to 1.6. Since 1995, Japan has had fewer births annually than in any year since 1899. In Germany, where the rate has fallen to 1.3, fewer babies are born each year than in Nepal, which has a population only one-quarter as large.

A Shrinking Population in an Aging Developed World

Unless their fertility rates rebound, the total populations of western Europe and Japan will shrink to about one-half of their current size before the end of the next century. In 1950, 7 of the 12 most populous nations were in the developed world: the United States, Russia, Japan, Germany, France, Italy, and the United Kingdom. The United Nations projects that by 2050, only the United States will remain on the list. Nigeria, Pakistan, Ethiopia, Congo, Mexico, and the Philippines will replace the others. But since developing countries are also experiencing a drop in fertility, many are now actually aging faster than the typical developed country. In France, for example, it took over a century for the elderly to grow from 7 to 14 percent of the population. South Korea,

Taiwan, Singapore, and China are projected to traverse that distance in only 25 years.

From Worker Shortage to Rising Immigration Pressure

Perhaps the most predictable consequence to the gap in fertility and population growth rates between developed and developing countries will be the rising demand for immigrant workers in older and wealthier societies facing labor shortages. Immigrants are typically young and tend to bring with them the family practices of their native culture—including higher fertility rates. In many European countries, non-European foreigners already make up roughly 10 percent of the population. This includes 10 million to 13 million Muslims, nearly all of whom are working-age or younger. In Germany, foreigners will make up 30 percent of the total population by 2030, and over half the population of major cities like Munich and Frankfurt. Global aging and attendant labor shortages will therefore ensure that immigration remains a major issue in developed countries for decades to come. Culture wars could erupt over the balkanization of language and religion; electorates could divide along ethnic lines, and émigré leaders could sway foreign policy.

GRAYING MEANS PAYING

Official projections suggest that within 30 years, developed countries will have to spend at least an extra 9 to 16 percent of GDP simply to meet their old-age benefit promises. The unfunded liabilities for pensions (that is, benefits already earned by today's workers for which nothing has been saved) are already almost $35 trillion. Add in health care, and the total jumps to at least twice as much. At minimum, the global aging issue thus represents, to paraphrase the old quiz show, a $65 trillion question hanging over the developed world's future.

To pay for promised benefits through increased taxation is unfeasible. Doing so would raise the total tax burden by an unthinkable 25 to 40 percent of every worker's taxable wages—in countries where payroll tax rates sometimes already exceed 40 percent. To finance the costs of these benefits by borrowing would be just as disastrous. Governments would run unprecedented deficits that would quickly consume the savings of the developed world.

And the $64 trillion estimate is probably low. It likely underestimates future growth in longevity and health care costs and ignores the negative effects on the economy of more borrowing, higher interest rates, more taxes, less savings, and lower rates of productivity and wage growth.

There are only a handful of exceptions to these nightmarish forecasts. In Australia, total public retirement costs as a share of GDP are expected to rise only slightly, and they may even decline in Britain and Ireland. This fiscal good fortune is not due to any special demographic trend, but to timely policy reforms—including tight limits on public health spending, modest pension benefit formulas, and new personally owned savings programs that allow future public benefits to shrink as a share of average wages. This approach may yet be emulated elsewhere.

Failure to respond to the aging challenge will destabilize the global economy, straining financial and political institutions around the world. Consider Japan, which today runs a large current account surplus making up well over half the capital exports of all the surplus nations combined. Then imagine a scenario in which Japan leaves its retirement programs and fiscal policies on autopilot. Thirty years from now, under this scenario, Japan will be importing massive amounts of capital to prevent its domestic economy from collapsing under the weight of benefit outlays. This will require a huge reversal in global capital flows. To get some idea of the potential volatility, note that over the next decade, Japan's annual pension deficit is projected to grow to roughly 3 times the size of its recent and massive capital exports to the United States; by 2030, the annual deficit is expected to be 15 times as large. Such reversals will cause wildly fluctuating interest and exchange rates, which may in turn short-circuit financial institutions and trigger a serious market crash.

As they age, some nations will do little to change course, while other may succeed in boosting their national savings rate, at least temporarily, through a combination of fiscal restraint and household thrift. Yet this too could result in a volatile disequilibrium in supply and demand for global capital. Such imbalance could wreak havoc with international institutions such as the European Union.

In recent years, the EU has focused on monetary union, launched a single currency (the euro), promoted cross-border labor mobility, and struggled to harmonize fiscal, monetary, and trade policies. European leaders expect to have their hands full smoothing out differences between members of the Economic and Monetary Union (EMU)—from the timing of their business cycles to the diversity of their credit institutions and political cultures. For this reason, they established official public debt and deficit criteria (three percent of GDP for EMU membership) in order to discourage maverick nations from placing undue economic burdens on fellow members. But the EU has yet to face up to the biggest challenge to its future viability: the likelihood of varying national responses to the fiscal pressures of demographic aging. Indeed, the EU does not even include unfunded pension liabilities in the official EMU debt and

Widening Public Pension Deficits Could Soon Consume
the Economic Savings of the Developed World

2000	2005	2010	2015	2020	2025	2030	2035	2040

0.1%

−0.2%

−0.9%

−1.9%

−3.3%

A deficit swing of
8.6% of GDP would
consume entire G-7
net national savings**

−5.1%

−7.4%

Change from 1995 in the combined G-7 budget balance attributable
to projected public pension deficits, as a percentage of G-7 GDP*

−9.9%

−12.1%

*Assumes no change in taxes and other spending: includes
 interest on prior-year pension deficits
**Assumes all other saving continues at 1985–94 annual rate

Source: OECD (1996); author's calculations

deficit criteria—which is like measuring icebergs without looking beneath the water line.

When these liabilities come due and move from "off the books" to "on the books," the EU will, under current constraints, be required to penalize EMU members that exceed the three percent deficit cap. As a recent IMF report concludes, "over time it will become increasingly difficult for most countries to meet the deficit ceiling without comprehensive social security reform." The EU could, of course, retain members by raising the deficit limit. But once the floodgates are opened, national differences in fiscal policy may mean that EMU members rack up deficits at different rates. The European Central Bank, the euro, and a half-century of progress toward European unity could be lost as a result.

The total projected cost of the age wave is so staggering that we might reasonably conclude it could never be paid. After all, these numbers are projections, not predictions. They tell us what is likely to happen if current policy remains unchanged, not whether it is likely or even possible for this condition to hold. In all probability, economies would implode and governments would collapse before the projections ever materialize. But this is exactly why we must focus on these projections, for they call attention to the paramount question: Will we change course sooner, when we still have time to control our destiny and reach a more sustainable path? Or later, after unsustainable economic damage and political and social trauma cause a wrenching upheaval?

A GRAYING NEW WORLD ORDER

While the fiscal and economic consequences of global aging deserve serious discussion, other important consequences must also be examined. At the top of the list is the impact of the age wave on foreign policy and international security.

Will the Developed World Be Able to Maintain Its Security Commitments?

One need not be a Nobel laureate in economics to understand that a country's GDP growth is the product of workforce and productivity growth. If workforces shrink rapidly, GDP may drop as well, since labor productivity may not rise fast enough to compensate for the loss of workers. At least some developed countries are therefore likely to experience a long-term decline in total production of goods and services—that is, in real GDP.

Economists correctly focus on the developed world's GDP per capita, which can rise even as its workforce and total GDP shrink. But anything with a fixed cost becomes a national challenge when that cost has to be spread over a smaller population and funded out of shrinking revenues. National defense is the classic example. The West already faces grave threats from rogue states armed with biological and chemical arsenals, terrorists capable of hacking into vulnerable computer systems, and proliferating nuclear weapons. None of these external dangers will shrink to accommodate our declining workforce or GDP.

Leading developed countries will no doubt need to spend as much or more on defense and international investments as they do today. But the age wave will put immense pressure on governments to cut back. Falling birthrates, together with a rising demand for young workers, will also inevitably mean smaller armies. And how many parents will allow their only child to go off to war?

With fewer soldiers, total capability can be maintained only by large increases in technology and weaponry. But boosting military productivity creates a Catch-22. For how will governments get the budget resources to pay for high-tech weaponry if the senior-weighted electorate demands more money for high-tech medicine? Even if military capital is successfully substituted for military labor, the deployment options may be dangerously limited. Developed nations facing a threat may feel they have only two extreme (but relatively inexpensive) choices: a low-level response (antiterrorist strikes and cruise-missile response) or a high-level response (an all-out attack with strategic weapons).

Will Young/Old Become the Next North/South Fault Line?

Historically, the richest industrial powers have been growing, capital-exporting, philanthropic giants that project their power and mores around the world. The richest industrial powers of the future may be none of these things. Instead, they may be demographically imploding, capital-importing, fiscally-starving neutrals who twist and turn to avoid expensive international entanglements. A quarter-century from now, will the divide between today's "rich" and "poor" nations be better described as a divide between growth and decline, surplus and deficit, expansion and retreat, future and past? By the mid-2020s, will the contrast between North and South be better described as a contrast between Young and Old?

If today's largest low-income societies, especially China, set up fully funded retirement systems to prepare for their own future aging, they may well produce even larger capital surpluses. As a result, today's great powers could someday depend on these surpluses to keep themselves financially afloat. But how should we expect these new suppliers of capital to use their newly acquired leverage? Will they turn the tables in international diplomacy? Will the Chinese, for example, someday demand that the United States shore up its Medicare system the way Americans once demanded that China reform its human rights policies as a condition for foreign assistance?

As Samuel Huntington recently put it, "the juxtaposition of a rapidly growing people of one culture and a slowly growing or stagnant people of another culture generates pressure for economic and/or political adjustments in both societies." Countries where populations are still exploding rank high on any list of potential trouble spots, whereas the countries most likely to lose population—and to see a weakening of their commitment to expensive defense and global security programs—are the staunchest friends of liberal democracy.

In many parts of the developing world, the total fertility rate remains very high (7.3 in the Gaza Strip versus 2.7 in Israel), most people are very young (49 percent under age 15 in Uganda), and the population is growing very rapidly (doubling every 26 years in Iran). These areas also tend to be the poorest, most rapidly urbanizing, most institutionally unstable—and most likely to fall under the sway of rogue leadership. They are the same societies that spawned most of the military strongmen and terrorists who have bedeviled the United States and Europe in recent decades. The Pentagon's long-term planners predict that outbreaks of regional anarchy will occur more frequently early in the next century. To pinpoint when and where, they track what they call "youth bulges" in the world's poorest urban centers.

Is demography destiny, after all? Is the rapidly aging developed world fated to decline? Must it cede leadership to younger and faster-growing societies? For the answer to be no, the developed world must redefine that role around a new mission. And what better way to do so than to show the younger, yet more tradition-bound, societies—which will soon age in their turn—how a world dominated by the old can still accommodate the young.

WHOSE WATCH IS IT, ANYWAY?

From private discussions with leaders of major economies, I can attest that they are well briefed on the stunning demographic trends that lie ahead. But so far they have responded with paralysis rather than action. Hardly any country is doing what it should to prepare. Margaret Thatcher confesses that she repeatedly tried to raise the aging issue at G-7 summit meetings. Yet her fellow leaders stalled. "Of course aging is a profound challenge," they replied, "but it doesn't hit until early in the next century—after my watch."

Americans often fault their leaders for not acknowledging long-term problems and for not facing up to silent and slow-motion challenges. But denial is not a peculiarly American syndrome. In 1995, Silvio Berlusconi's *Forza Italia* government was buffeted by a number of political storms, all of which it weathered—except for pension reform, which shattered the coalition. That same year, the Dutch parliament was forced to repeal a recent cut in retirement benefits after a strong Pension Party, backed by the elderly, emerged from nowhere to punish the reformers. In 1996, the French government's modest proposal to trim pensions triggered strikes and even riots. A year later the Socialists overturned the ruling government at the polls.

Each country's response, or nonresponse, is colored by its political and cultural institutions. In Europe, where the welfare state is more expansive, voters can hardly imagine that the promises made by previous generations of politicians can no longer be kept. They therefore support leaders, unions, and party coalitions that make generous unfunded pensions the very cornerstone of social democracy. In the United States, the problem has less to do with welfare-state dependence than the uniquely American notion that every citizen has personally earned and is therefore entitled to whatever benefits government happens to have promised.

How governments ultimately prepare for global aging will also depend on how global aging itself reshapes politics. Already some of the largest and most strident interest groups in the United States are those that claim to speak for senior citizens, such as the American Association of Retired Persons, with its 33 million members, 1,700 paid employees, ten times that many trained volunteers, and an annual budget of $5.5 billion.

Senior power is rising in Europe, where it manifests itself less through independent senior organizations than in labor unions and (often union-affiliated) political parties that formally adopt pro-retiree platforms. Could age-based political parties be the wave of the future? In Russia, although the Communist resurgence is usually ascribed to nationalism and nostalgia, a demographic bias is at work as well. The Communists have repositioned themselves as the party of retirees, who are aggrieved by how runaway inflation has slashed the real value of their pensions. In the 1995 Duma elections, over half of those aged 55 and older voted Communist, versus only ten percent of those under age 40.

Commenting on how the old seem to trump the young at every turn, Lee Kuan Yew once proposed that each taxpaying worker be given two votes to balance the lobbying clout of each retired elder. No nation, not even Singapore, is likely to enact Lee's suggestion. But the question must be asked: With ever more electoral power flowing into the hands of elders, what can motivate political leaders to act on behalf of the long-term future of the young?

A handful of basic strategies, all of them difficult, might enable countries to overcome the economic and political challenges of an aging society: extending work lives and postponing retirement; enlarging the workforce through immigration and increased labor force participation; encouraging higher fertility and investing more in the education and productivity of future workers; strengthening intergenerational bonds of responsibility within families; and targeting government-paid benefits to those most in need while encouraging and even requiring workers to save for their own retirements. All of these strategies unfortunately touch raw nerves—by amending existing social contracts, by violating cultural expectations, or by offending entrenched ideologies.

TOWARD A SUMMIT ON GLOBAL AGING

All countries would be well served by collective deliberation over the choices that lie ahead. For that reason I propose a Summit on Global Aging. Few venues are as well covered by the media as a global summit. Leaders have been willing to convene summits to discuss global warning. Why not global aging, which will hit us sooner and with greater certainty? By calling attention to what is at stake, a global aging summit could shift the public discussion into fast forward. That alone would be a major contribution. The summit process would also help provide an international framework for voter education, collective burden-sharing, and global leadership. Once national constituencies begin to grasp the magnitude of the global aging

challenge, they will be more inclined to take reform seriously. Once governments get into the habit of cooperating on what in fact is a global challenge, individual leaders will not need to incur the economic and political risks of acting alone.

This summit should launch a new multilateral initiative to lend the global aging agenda a visible institutional presence: an Agency on Global Aging. Such an agency would examine how developed countries should reform their retirement systems and how developing countries should properly set them up in the first place. Perhaps the most basic question is how to weigh the interests and well-being of one generation against the next. Then there is the issue of defining the safety-net standard of social adequacy. Is there a minimum level of retirement income that should be the right of every citizen? To what extent should retirement security be left to people's own resources? When should government pick up the pieces, and how can it do so without discouraging responsible behavior? Should government compel people in advance to make better life choices, say, by enacting a mandatory savings program?

Another critical task is to integrate research about the age wave's timing, magnitude, and location. Fiscal projections should be based on assumptions that are both globally consistent and—when it comes to longevity, fertility, and health care costs—more realistic than those now in use. Still to be determined: Which countries will be hit earliest and hardest? What might happen to interest rates, exchange rates, and cross-border capital flows under various political and fiscal scenarios?

But this is not all the proposed agency could do. It could continue to build global awareness, publish a high-visibility annual report that would update these calculations, and ensure that the various regular multilateral summits (from the G-7 to ASEAN and APEC) keep global aging high on their discussion agendas. It could give coherent voice to the need for timely policy reform around the world, hold up as models whatever major steps have been taken to reduce unfunded liabilities, help design funded benefit programs, and promote generational equity. On these and many other issues, nations have much to learn from each other, just as those who favor mandatory funded pension plans are already benefiting from the examples of Chile, Britain, Australia, and Singapore.

Global aging could trigger a crisis that engulfs the world economy. This crisis may even threaten democracy itself. By making tough choices now world leaders would demonstrate that they gen- uinely care about the future, that they understand this unique opportunity for young and old nations to work together, and that they compre- hend the price of freedom. The gray dawn approaches. We must establish new ways of think- ing and new institutions to help us prepare for a much older world.

24

Female Empowerment Leads to Fewer Births

CAMERON BARR

DHAKA, BANGLADESH

Batashi, a woman in her 30s wearing gold jewelry and a bright blue sari, stands in front of her tidy cement and packed-mud home in a village called Nabogram. She and millions of women like her in the developing world are help- ing to answer a question that is central to the effort to slow global population growth: Is it enough to hand a poor woman a contraceptive?

Batashi's village, about two hours outside Dhaka, Bangladesh, is perched on earthen ridges that divide rice paddies and keep people above the flood plain during the rainy season. Most people work tiny plots of land they don't own. But Batashi's economic situation is better than average because of small loans and other assistance she has received from a private development agency called BRAC, the Bangladesh Rural Advancement Committee.

She now owns three cows, tethered in a nearby shed, that produce 10 kilograms (22 pounds) of milk every day. She also owns three bicycle rickshaws, one pulled by her husband and the other two rented to men who earn their liv- ing by pulling them.

Do the cows and the rickshaws have any bear- ing on Batashi's decision to use family-planning services? "Yes, family planning is easier for me," she says. Economic opportunity "gives me freedom and . . . gives me power." Batashi has two sons and a daughter. "This is enough," she says, "because I want these three to be educated persons. I don't want my children to be rickshaw pullers."

Batashi's case illustrates why many population analysts now believe that "parallel efforts" and "holistic approaches" to overpopulation—that is, combining the delivery of family-planning ser- vices with programs that improve women's stand- ing in society—are critical to slowing birthrates.

"What's clear," says Sharon Camp, senior vice president of the Population Crisis Committee in Washington, "is that if you can do good family planning and secure improvements in the status of women—legal, economic, social, and political— those things interact so powerfully that you can get declines in birthrates at breathtaking speeds.

That's probably the solution to the world's population problem."

Even so, family-planning bureaucracies in many developing countries have been slow to expand beyond the traditional approach to family planning—providing contraceptives and advice on using them.

"From a program standpoint, the woman is a target to achieve demographic ends," says Saroj Pachauri, a program officer at the Ford Foundation in New Delhi. "It's not her concerns that are paramount."

Population experts have long understood the importance of "female empowerment"—a catchall term that covers women-centered health care, improvements in social and legal status, access to education and jobs, and a fairer distribution of responsibility for children between mothers and fathers.

But during the past 10 years, a growing body of evidence has verified the correlation between empowerment and reducing family size.

One of the best examples in the world is nearby in the southern Indian state of Kerala, where the fertility rate—on average, 2.3 children per woman—is one of the lowest in the developing world. One reason is a female literacy rate of 66 percent, several times the average in India's northern states, according to a United Nations report.

The other reason is the status of women. According to tradition, women inherit land in Kerala, and families pay a "bride price" to the parents of the bride, symbols, the UN Population Fund says, that women are considered an asset and not a liability.

EDUCATION CHANGES OUTLOOK

There is plenty of evidence to show that eight or more years of education for women, because it encourages later marriages and broadens a woman's outlook, results in fewer births. "With education and economic opportunities, women begin to define themselves as citizens, not just family members," says Judith Bruce, a senior associate at the Population Council, a private research group in New York. "Interacting in a wider set of relationships means having access to more objective information [on the costs and benefits of having more children]. It means more equality in the marital relationship."

Research also links increased family planning with economic opportunity. One study looked at a group of 1,600 women members of the Grameen Bank, a Bangladeshi institution that provides small loans to promote small businesses. Fifty-nine percent of the bank members were using some sort of family planning, compared with 43 percent of women in a control group.

Economic insecurity, on the other hand, can have a reverse effect, driving up birthrates since children, especially sons, are considered essential to old-age security.

According to recent research, 60 percent of women in India over age 60 are widows. But some of India's many castes prohibit widows from returning to the village of their birth or joining the lineage of their married daughters.

"Unless she has adult or married sons, she is left virtually on her own," says Marty Chen of the Harvard Institute for International Development, in Cambridge, Mass., and author of the research.

In a village near Batashi's, a woman named Behula says she had three girls before she bore a boy.

"When my three daughters marry, my house will be completely empty," she says. "If I have a boy, he will be present in front of my eyes [and] in the end he will see me in my old age."

"To me it seems so obvious that we really don't have to explain it because women are [the] ones who make the choice," says Amina Islam, a program officer at the UN Population Fund in Khaka. "If you improve their status, they will be in a [better] position to make decisions about their fertility."

Sitting in the government family-planning office for Manikganj District, where Batashi's village is located and where BRAC has been active, two low-level male bureaucrats corroborate the point.

"It is very easy to talk about family planning with a working woman," says one, Motalib Hussain, the district family-planning supervisor. "It is very easy for us because she has already developed herself. [But] it is very difficult to talk about family planning with housewives."

Women's Literacy and Population Growth

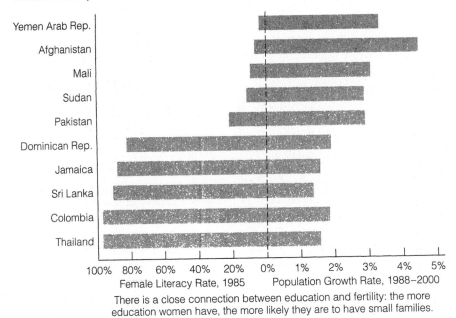

There is a close connection between education and fertility: the more education women have, the more likely they are to have small families.

Source: United Nations Population Fund. Van Pelt/Christian Science Monitor.

At the higher levels of Bangladesh's family-planning bureaucracy, however, there is less regard for the impact that education and economic opportunity can have on fertility. "We are proud of our success," says A. K. M. Rafiquzzaman, director-general of Bangladesh's Family-Planning Directorate, sitting in an office air-conditioned against Dhaka's 100-degree heat.

"Bangladesh has disproved the theory," he asserts, that economic development and literacy are prerequisites to a contraceptive-prevalence rate of 40 percent.

"CAN'T WAIT FOR DEVELOPMENT"

"The main problem," writes Alia Ahmed in her 1991 book, "Women and Fertility in Bangladesh," "is that one cannot wait for socioeconomic development to solve the problem of overpopulation when development itself is being stifled by the high rate of population growth."

But most experts insist that there are limits to relying exclusively on a "supply side" approach to family planning, providing contraceptive services to interested couples.

To produce deep cuts where they count most—in fertility rates—the "demand side" of the equation must be addressed as well by providing the education and economic opportunities that create a desire for smaller families and thus a wider demand for family-planning services.

"It's one thing to reduce fertility rates from five or six children per woman to three or four using traditional family-planning services," says Ronald Ridker, a senior economist at the World Bank in Washington. "But unless there is substantial social and economic progress, including improvements in the status of women, it will be impossible to get from three or four [per mother] to a replacement level of 2.1.

"We can't wait for social and economic progress to occur to accomplish these things," adds Dr. Ridker.

"We have to intervene selectively in the development process to ensure that the conditions that change attitudes toward family size are fostered."

25

With Iran Population Boom, Vasectomy Receives Blessing

NEIL MACFARQUHAR

TEHERAN, IRAN

In the vasectomy department of birth control clinic serving this city's hard-pressed southern quarters, a single letter stamped with official seals lies pressed under the doctor's glass desktop.

Without exception, the wary men shuffling into the office first seek reassurance that the operation will not result in their voices rising or their beards falling out. Then, in this ardent Muslim theocracy, the questions often swing around to whether Islam condones the procedure.

The doctor just taps on the glass. The letter is no simple document. It is a religious edict that the clinic recently solicited from the country's supreme leader, asking him whether men could practice birth control.

In the brief edict carrying the full weight of law, Ayatollah Ali Khamenei listed the withdrawal method, condoms and vasectomies as acceptable contraceptive procedures for men. He scrawled two lines at the end that roughly read, "When wisdom dictates that you do not need more children, a vasectomy is permissible."

The Islamic revolution is trying to curb its population growth. The early vision of producing an Islamic generation that would be a challenge to the West in its very size has been drastically redrawn, overshadowed by the staggering difficulties of managing a population that has vaulted to more than 60 million from 35 mllion since 1979.

"Back then we had just had a revolution that faced threats from both internal and external enemies," said Grand Ayatollah Nasser Makaram-Shirazi, a top spiritual guide. "We wanted to increase the number of people who believed in the revolution in order to preserve it."

The legal marriage age was dropped to 9, and everyone from a popular television preacher to some of the most learned clerics addressing the weekly Friday prayers exhorted Iranians to make babies. As result, at least 45 percent of the population is under 17.

But in the late 1980's, as officials pondered the prospects of educating, housing and employing them all, especially in the face of continued economic difficulties and American trade sanctions, they made a startling policy reversal. And now Iran has one of the world's more aggressive programs to encourage smaller families: to get a marriage license here, you must take a course in family planning.

"Now instead of thinking about the quantity of Muslims we have to think about the quality; we need healthier, more educated and better informed Muslims," Ayatollah Makaram-Shirazi said in an interview, sounding more like an American foreign aid official than a religious leader of a fundamentalist Islamic state. "I would not recommend it for all countries, but we have understood that if we do not control the population we will have problems in economics, education, health and culture that will leave the deprived who made the revolution even more deprived."

Even without the religious edict, soaring inflation and eroding wages would probably have pushed Iranians to have smaller families. But rapid public agitation was made possible by support from influential members of the clergy ready to

Rate of Growth

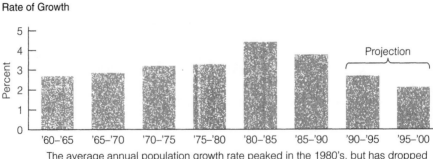

The average annual population growth rate peaked in the 1980's, but has dropped
since the Islamic Government began encouraging family planning.

Total Population

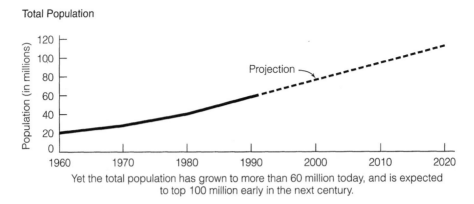

Yet the total population has grown to more than 60 million today, and is expected
to top 100 million early in the next century.

Sources: United Nations; Population Reference Bureau

fight battles over which Koranic verse or which saying of Mohammed carried the necessary weight.

"There was always some question whether family planning was approved by the religious leadership or not," said Dr. Alireza Marandi, Iran's Minister of Health. As a deputy minister in the early 1980's, he recognized that the population growth rate, among the world's highest, was rocketing out of control. But he also knew that any attempt to haul the issue into the public arena might let slip the unstoppable forces of religious intolerance.

So the ministry quietly kept alive a prerevolutionary program of distributing free condoms and I.U.D.'s while maneuvering for an opening. "We could never advertise on television or put anything in the newspapers," Dr. Marandi said. "If we

had offended anyone, the whole thing would have gone down the drain."

He was most sensitive to what Ayatollah Ruhollah Khomeini, the Leader of the Revolution, would think. Knowing that if Ayatollah Khomeini had uttered one word against birth control the whole nation would have tossed away its condoms, he always hesitated to ask. He finally did after the Cabinet approved birth control by one vote in 1988, but the internal opposition was so vehement that the Prime Minister refused to announce the decision.

Ayatollah Khomeini suggested a public discussion. That unleashed a tumult of medical seminars and sent Muslim scholars digging through their texts for religious sanction.

Iranian officials cited two examples of the better-known sayings of Mohammed used to oppose birth control until that time. Once Mohammed said he was proud of those who had a large number of children; another time be mentioned that he hoped that the number of Muslims would outnumber all other faiths by Doomsday. But birth control supporters argued that those were from the early days of the faith when his followers were few, and they found a welter of other sayings by Mohammed and other lesser prophets that small families brought greater ease.

The debate culminated in a 1993 law that enshrined birth control, lifting subsidized health insurance and food coupons for any child after the third. Abortion remained illegal in all cases except when the mother's life was in danger.

But condoms and pills remained free. The state also introduced mandatory prenuptial birth control classes. Before getting a marriage license couples must submit a stamped form indicating that they attended a segregated, hourlong lecture on birth control.

One student in the class at Farman-Farmaian Health Center in south Teheran was Bijan Javar ("Call me Jim"), a 38-year-old Amtrak worker from Newark, who came home to find a bride.

"She's a good one, I got lucky on that," said Mr. Javar, taking the lecture in stride as his classmates chuckled nervously and occasionally asked nontechnical questions like the best way to initiate sex. "This class might seem unusual for someone outside the country, but it is not surprising for Iran."

Indeed, Iranian birth control officials said that even more coercive measures, such as forcing women who have had more than five children to undergo a tubal ligation, are acceptable given the scope of the problem.

"Sometimes we just have to go after people to direct their activities," said Zakiah Rashad, a birth control counselor. The system seems to be working well enough that a recent directive told counselors to be a little gentler, to talk to people about the different methods rather than just ordering them to follow one.

Statistics are something of a lost art in Iran, but experts say the population growth rate has dropped steeply. By some estimates it has declined from about 4 percent in the 1980's—one of the highest rates ever seen anywhere—to about 2.5 percent, though rural families still tend to bear many children. Some in the Ministry of Health assert that the growth rate is now even lower. In any case, there are certain to be some 100 million Iranians early in the next century.

Today new parents worry about their children's prospects, not to mention their own ability to pay for them. About 10 to 15 men demand vasectomies every day at the Farman-Farmaian Health Center. It is both the surest method and free.

Azim Aslawi, 44, glanced nervously at the men walking stiff-legged from the operating rooms, trying to avoid groin pains. In the doctor's office he asked if Islam really permitted such a change. The doctor tapped the glass.

"Now that I have read the letter I am much more willing to do it," said Mr. Aslawi, the father of two. "No one is proud of big families anymore."

6

Environmental Destruction

> *The days of the frontier economy—in which
> abundant resources were available to propel
> economic growth and living standards—are over.
> We have entered an era in which global prosperity
> depends on using resources more efficiently,
> distributing them more equitably, and reducing
> consumption levels overall.*
>
> SANDRA POSTEL

Preserving the earth's environment will be one of humanity's most urgent concerns in the twenty-first century. Although environmental movements have succeeded in raising awareness throughout the world, actions to preserve the global environment still fall short of what may be necessary to continue life as we now know it. An example is the challenge to biodiversity, or the diversity of species of animals and plants. Healthy ecosystems depend on a multiplicity of species that interact with one another in such a way as to sustain the life of the whole. At a recent international summit meeting called by the National Academy of Sciences, environmental scientists warned of a catastrophic loss of biodiversity, if present trends continue. Whereas a species of life once typically could be expected to survive for about a million years, current environmental problems such as habitat destruction, acid rain, or global warming are leading to a survival expectation of only about a thousand years. One scientist has calculated that 75 percent of the earth's bird species will become extinct by the end of the twenty-first century. It appears that a major extinction of life has begun. This is comparable in scope to five previous extinctions but it is the first to be the result of human rather than natural causes (McDonald, 1997, p. 16).

Some environmentalists are warning of a massive environmental collapse. For example, Sandra Postel, the scholar-activist quoted above, argues that the earth's "carrying capacity," or ability to support human beings at our current level of resource use, has already been surpassed. The current rates of depletion of forests, rangeland, fisheries, cropland, and freshwater sources are so rapid as to threaten the continuation of our existing way of life. The causes of this situation—rapid global economic growth, population growth, and rising income inequalities—must be reversed if we want to ensure the well-being of future generations (Postel, 1994, pp. 5–8, 19–21). In "Easter's End," Jared Diamond shows us that this concern is not so far-fetched. He compares the world today with the ancient culture of the Easter Islands. This culture vanished after people with vested interests destroyed the island's trees, plants, and animals in a quest for economic and political gain. The key difference between humanity today and the Easter Islanders, argues Diamond, is that we now have the knowledge to prevent the earth's destruction. David Korten's "The Limits of the Earth" may lead us to question this point. Writing from the conflict perspective, Korten examines the fateful environmental consequences of contemporary philosophies of economic growth and globalization. Powerful corporations, governments, and international agencies such as the World Bank and International Monetary Fund that control the global economy fail to heed the warnings of environmentalists and continue to operate as though the earth's resources are inexhaustible.

Although humanity is so far still surviving, environmental destruction has led to a diminished quality of life around the world. This challenge to the quality of life takes a multitude of forms.

Polluted air and water, for example, are an immediate danger and sicken and kill millions of people, particularly children, in the poor nations each year (Easterbrook, 1994, pp. 60–61). Charles W. Schmidt discusses the impact of toxic environments on children's health in "Childhood Cancer."

Environmental problems today are closely intertwined with human rights issues. Pamela Wellner's "A Pipeline Killing Field" discusses human rights abuses in Burma, where the military dictatorship has displaced villages, conscripted poor and powerless people into forced labor, and cut down forests to build a pipeline, in an effort to assist foreign oil companies contracting to exploit Burma's gas fields. This article illustrates another important point about environmental problems—that they are strongly connected to the cultural survival of indigenous peoples. The habitats being destroyed around the world are often the homelands of cultural minorities whose history in the area pre-dates the rise of the capitalist world economy. Indigenous peoples' resistance to environmental destruction and the destruction of their cultures is a key and growing element in environmental movements throughout the world. This topic is further explored in readings in Chapter 9 on "The Cultural Survival of Indigenous Peoples." Such human rights abuses are not confined to the poorest nations of the world. In the United States, for example, there is a growing movement against "environmental racism."

The chapter concludes with an article examining one of the many environmental movements around the world. "Border Patrol" by Bruce Selcraig discusses the life of Domingo González, a founder of the Coalition for Justice in the Maquiladoras. González seeks to protect workers and communities along the U.S.-Mexican border from severe environmental hazards accompanying the unregulated industrialization of Mexico by global corporations.

WORKS CITED

Easterbrook, Gregg. 1994. "Forget PCB's. Radon. Alar." *The New York Times Magazine,* 9 September: 60–63.

McDonald, Kim A. 1997. "Scientists Refine Estimates of Number of Species and Their Rate of Extinction." *The Chronicle of Higher Education*, 14 November: A17.

Postel, Sandra. 1994. "Carrying Capacity: Earth's Bottom Line." In *State of the World, 1994* ed. Lester R. Bown, 3–21. New York: W. W. Norton and Company.

QUESTIONS FOR DISCUSSION

1. Compare and contrast the situation of the earth today with the situation of the Easter Islanders, as discussed by Jared Diamond. Do you agree with Diamond that we have the ability today to prevent a massive environmental collapse?

2. Why, according to Korten, is the "Bretton Woods system" doomed to failure? What does Korten propose as solution to the problem of the "limits of the earth"? Do you agree with his ideas? Why or why not?

3. Charles W. Schmidt discusses the influence of environmental problems on children's health in "Childhood Cancer." What are some environmental problems in your own community that affect the quality of people's lives?

4. In "A Pipeline Killing Field," Pamela Wellner reports on the connection between human rights abuses and environmental destruction in Burma, and the involvement of powerful Western oil companies. Should the U.S. government issue economic sanctions against Burma, in order to help stop such abuses? Or will isolating this nation contribute to making the situation worse? Various other "top down" and "bottom up" strategies are being proposed to deal with this problem: for example, state and local governments could refuse to do business with the

companies involved, shareholders could put pressure on the companies to change their practices, and consumers could boycott the companies and others owned by the same corporations. Do you believe these are effective strategies? Which, if any, do you prefer? Explain.

5.   Bruce Selcraig discusses a situation closer to home in "Border Patrol." Domingo González has focused on

educating the media and government about environmental problems caused by multinational corporations operating along the U.S.-Mexico border. What are some of these problems? Do you believe these are serious social problems affecting people in the United States, or are these primarily "Mexico's problems"? Explain.

🍷 INFOTRAC COLLEGE EDITION: EXERCISE

The introduction to this chapter refers to the growing movement against "environmental racism" in the United States. Do some additional research to learn more about this movement.

Using the InfoTrac College Edition, search for articles on this topic. (*Hint:* Enter the search term environmental racism, using the Subject Guide.)

Find and read several articles that interest you. What is meant by environmental racism? What are some examples? What are some of the controversies that currently exist over the definition, causes, and solutions to this problem?

FOR ADDITIONAL RESEARCH

Books

Brown, Lester R., Michael Renner, and Christopher Flavin. 1998. *Vital Signs.* New York: W. W. Norton and Company.

Brown, Lester R., et al. 1998. *State of the World, 1998.* New York: W. W. Norton and Company.

Le Breton, Binka. 1993. *Voices from the Amazon.* West Hartford, CT: Kumarian Press.

Organizations

Earth First!
P.O. Box 5176
Missoula, MT 59806
(406) 728-8114

Greenpeace USA
1436 U Street NW
Washington, DC 20009
(202) 462-1177
www.greenpeaceusa.org
www.greenpeace.org

Sierra Club
85 Second Street, 2nd Floor
San Francisco, CA 94105-3441
(415) 977-5500
www.sierraclub.org

World Wildlife Fund
1250 24th Street
Washington, DC 20037
(202) 293-4800
www.panda.org

ACTION PROJECTS

1.   The environmental movement contains many organizations pursuing a variety of strategies. Contact the organizations listed above, and others you may locate through the Internet, to learn more about their goals and strategies. Compare and contrast them. Which of the organizations appear to be the most effective in

relation to solving the world's environmental problems, and why?

2.   Environmentalists around the world often come into conflict with communities seeking economic development. Do these goals conflict in your own community?

Do some background research in the archives of your local newspaper. Identify key organizations and individuals that have been involved in conflicts of this nature—for example, local environmental activists; town, city, or regional planning agencies; developers; and others. Interview people who represent different sides in the conflict in order to learn their viewpoints. What, in your analysis, would be the best solution to the issue?

3.   Find out more about violations of the human rights of environmental activists. Using the Internet, or your library, contact human rights organizations such as Human Rights Watch or Amnesty International (see Organizations listing in Chapter 10, "Racial/Ethnic Conflicts and the Danger of Genocide," for their addresses). Make a list of the violations going on around the world. What is being done about them?

26

Easter's End

JARED DIAMOND

• • •

Among the most riveting mysteries of human history are those posed by vanished civilizations. Everyone who has seen the abandoned buildings of the Khmer, the Maya, or the Anasazi is immediately moved to ask the same question: Why did the societies that erected those structures disappear?

Their vanishing touches us as the disappearance of other animals, even the dinosaurs, never can. No matter how exotic those lost civilizations seem, their framers were humans like us. Who is to say we won't succumb to the same fate? Perhaps someday New York's skyscrapers will stand derelict and overgrown with vegetation, like the temples at Angkor Wat and Tikal.

Among all such vanished civilizations, that of the former Polynesian society on Easter Island remains unsurpassed in mystery and isolation. The mystery stems especially from the island's gigantic stone statues and its impoverished landscape, but it is enhanced by our associations with the specific people involved: Polynesians represent for us the ultimate in exotic romance, the background for many a child's, and an adult's, vision of paradise. My own interest in Easter was kindled over 30 years ago when I read Thor Heyerdahl's fabulous accounts of his *Kon-Tiki* voyage.

But my interest has been revived recently by a much more exciting account, one not of heroic voyages but of painstaking research and analysis. My friend David Steadman, a paleontologist, has been working with a number of other researchers who are carrying out the first systematic excavations on Easter intended to identify the animals and plants that once lived there. Their work is contributing to a new interpretation of the island's history that makes it a tale not only of wonder but of warning as well.

Easter Island, with an area of only 64 square miles, is the world's most isolated scrap of habitable land. It lies in the Pacific Ocean more than 2,000 miles west of the nearest continent (South America), 1,400 miles from even the most habitable island (Pitcairn). Its subtropical location and latitude—at 27 degrees south, it is approximately as far below the equator as Houston is north of it—help give it a rather mild climate, while its volcanic origins make its soil fertile. In theory, this combination of blessings should have made Easter a miniature paradise, remote from problems that beset the rest of the world.

The island derives its name from its "discovery" by the Dutch explorer Jacob Roggeveen, on Easter (April 5) in 1722. Roggeveen's first impression was not of a paradise but of a wasteland: "We originally, from a further distance, have considered the said Easter Island as sandy; the reason for that is this, that we counted as sand the withered grass, hay, or other scorched and burnt vegetation, because its wasted appearance could give no other impression than of a singular poverty and barrenness."

The island Roggeveen saw was a grassland without a single tree or bush over ten feet high. Modern botanists have identified only 47 species of higher plants native to Easter, most of them grasses, sedges, and ferns. The list includes just two species of small trees and two of woody shrubs. With such flora, the islanders Roggeveen encountered had no source of real firewood to warm themselves during Easter's cool, wet, windy winters. Their native animals included nothing larger than insects, not even a single species of native bat, land bird, land snail, or lizard. For domestic animals, they had only chickens.

European visitors throughout the eighteenth and early nineteenth centuries estimated Easter's human population at about 2,000, a modest number considering the island's fertility. As Captain James Cook recognized during his brief visit in 1774, the islanders were Polynesians (a Tahitian man accompanying Cook was able to converse with them). Yet despite the Polynesians' well-deserved fame as a great seafaring people, the Easter Islanders who came out to Roggeveen's and Cook's ships did so by swimming or paddling canoes that Roggeveen described as "bad and frail." Their craft, he wrote, were "put together with manifold small planks and light inner timbers, which they cleverly stitched together with very fine twisted threads. . . . But as they lack the knowledge and particularly the materials for caulking and making tight the great number of seams of the canoes, these are accordingly very leaky, for which reason they are compelled to spend half the time in bailing." The canoes, only ten feet long, held at most two people, and only three or four canoes were observed on the entire island.

With such flimsy craft, Polynesians could never have colonized Easter from even the nearest island, nor could they have traveled far offshore to fish. The islanders Roggeveen met were totally isolated unaware that other people existed. Investigators in all the years since his visit have discovered no trace of the islanders' having any outside contacts: not a single Easter Island rock or product has turned up elsewhere, nor has anything been found on the island that could have been brought by anyone other than the original settlers or the Europeans. Yet the people living on Easter claimed memories of visiting the uninhabited Sala y Gomez reef 260 miles away, far beyond the range of the leaky canoes seen by Roggeveen, How did the islanders' ancestors reach that reef from Easter, or reach Easter from anywhere else?

Easter Island's most famous feature is its huge stone statues, more than 200 of which once stood on massive stone platforms lining the coast. At least 700 more, in all stages of completion, were abandoned in quarries or on ancient roads between the quarries and the coast, as if the carvers and moving crews had thrown down their tools and walked off the job. Most of the erected statues were carved in a single quarry and then somehow transported as far as six miles—despite heights as great as 33 feet and weights up to 82 tons. The abandoned statues, meanwhile, were as much as 65 feet tall and weighed up to 270 tons. The stone platforms were equally gigantic: up to 500 feet long and 10 feet high, with facing slabs weighing up to 10 tons.

Roggeveen himself quickly recognized the problem the statues posed: "The stone images at first caused us to be struck with astonishment," he wrote, "because we could not comprehend how it was possible that these people, who are devoid of heavy thick timber for making any machines, as well as strong ropes, nevertheless had been able to erect such images." Roggeveen might have added that the islanders had no wheels, no draft animals and no source of power except their own muscles. How did they transport the giant statues for miles, even before erecting them? To deepen the mystery, the statues were still standing in 1770, but by 1864 all of them had been pulled down, by the islanders themselves. Why then did they carve them in the first place? And why did they stop?

The statues imply a society very different from the one Roggeveen saw in 1722. Their sheer number and size suggest a population much larger than 2,000 people. What became of everyone? Furthermore, that society must have been highly organized. Easter's resources were scattered across the island: the best stone for the statues was quarried at Rano Raraku near Easter's northeast end; red stone, used for large crowns adorning some of the statues, was quarried at Puna Pau, inland in the southwest; stone carving tools came mostly from Aroi in the northwest. Meanwhile, the best farmland lay in the south and east, and the best fishing grounds on the north and west coasts. Extracting and redistributing all those goods required complex political organization. What happened to that organization, and how could it ever have arisen in such a barren landscape?

Easter Island's mysteries have spawned volumes of speculation for more than two and a half centuries. Many Europeans were incredulous that Polynesians—commonly characterized as "mere savages"—could have created the statues or the beautifully constructed stone platforms. In the 1950s, Heyerdahl argued that Polynesia must have been settled by advanced societies of American Indians, who in turn must have received civilization across the Atlantic from more advanced societies of the Old World. Heyerdahl's raft voyages aimed to prove the feasibility of such prehistoric transoceanic contacts. In the 1960s the Swiss writer Erich von Däniken, an ardent believer in Earth visits by extraterrestrial astronauts, went further, claiming that Easter's statues were the work of intelligent beings who owned ultramodern tools, became stranded on Easter, and were finally rescued.

Heyerdahl and Von Däniken both brushed aside overwhelming evidence that the Easter Islanders were typical Polynesians derived from Asia rather than from the Americas and that their culture (including their statues) grew out of Polynesian culture. Their language was Polynesian, as Cook had already concluded. Specifically, they spoke an eastern Polynesian dialect related to Hawaiian and Marquesan, a dialect isolated since about A.D. 400, as estimated from slight differences in vocabulary. Their fishhooks and stone adzes resembled early Marquesan models. Last year DNA extracted from 12 Easter Island skeletons was also shown to be Polynesian. The islanders grew bananas, taro, sweet potaoes, sugarcane, and paper mulberry—typical Polynesian crops, mostly of Southeast Asian origin. Their sole domestic animal, the chicken, was also typically Polynesian and ultimately Asian, as were the rats that arrived as stowaways in the canoes of the first settlers.

What happened to those settlers? The fanciful theories of the past must give way to evidence gathered by hardworking practitioners in three fields: archeology, pollen analysis, and paleontology.

Modern archeological excavations on Easter have continued since Heyerdahl's 1955 expedition. The earliest radiocarbon dates associated with human activities are around A.D. 400 to 700, in reasonable agreement with the approximate settlement date of 400 estimated by linguists. The period of statue construction peaked around 1200 to 1500, with few if any statues erected thereafter. Densities of archeological sites suggest a large population; an estimate of 7,000 people is widely quoted by archeologists, but other estimates range up to 20,000, which does not seem implausible for an island of Easter's area and fertility.

Archeologists have also enlisted surviving islanders in experiments aimed at figuring out how the statues might have been carved and erected. Twenty people, using only stone chisels, could have carved even the largest completed statue within a year. Given enough timber and fiber for making ropes, teams of at most a few hundred people could have loaded the statues onto wooden sleds, dragged them over lubricated wooden tracks or rollers, and used logs as levers to maneuver them into a standing position. Rope could have been made from the fiber of a small native tree, related to the linden, called the hauhau. However, that tree is now extremely scarce on Easter, and hauling one statue would have required hundreds of yards of rope. Did Easter's now barren landscape once support the necessary trees?

That question can be answered by the technique of pollen analysis, which involves boring out a column of sediment from a swamp or pond, with the most recent deposits at the top and relatively more ancient deposits at the bottom. The absolute age of each layer can be dated by radiocarbon methods. Then begins the hard work: examining tens of thousands of pollen grains under a microscope, counting them, and identifying the plant species that produced each one by comparing the grains with modern pollen from known plant species. For Easter Island, the bleary-eyed scientists who performed that task were John Flenley, now at Massey University in New Zealand, and Sarah King of the University of Hull in England.

Flenley and King's heroic efforts were rewarded by the striking new picture that emerged of Easter's prehistoric landscape. For at least

30,000 years before human arrival and during the early years of Polynesian settlement, Easter was not a wasteland at all. Instead, a subtropical forest of trees and woody bushes towered over a ground layer of shrubs, herbs, ferns, and grasses. In the forest grew tree daisies, the rope-yielding hauhau tree, and the toromiro tree, which furnishes a dense, mesquite-like firewood. The most common tree in the forest was a species of palm now absent on Easter but formerly so abundant that the bottom strata of the sediment column were packed with its pollen. The Easter Island palm was closely related to the still-surviving Chilean wine palm, which grows up to 82 feet tall and 6 feet in diameter. The tall, unbranched trunks of the Easter Island palm would have been ideal for transporting and erecting statues and constructing large canoes. The palm would also have been a valuable food source, since its Chilean relative yields edible nuts as well as sap from which Chileans make sugar, syrup, honey, and wine.

What did the first settlers of Easter Island eat when they were not glutting themselves on the local equivalent of maple syrup? Recent excavations by David Steadman, of the New York State Museum at Albany, have yielded a picture of Easter's original animal world as surprising as Flenley and King's picture of its plant world. Steadman's expectations for Easter were conditioned by his experiences elsewhere in Polynesia, where fish are overwhelmingly the main food at archeological sites, typically accounting for more than 90 percent of the bones in ancient Polynesian garbage heaps. Easter, though, is too cool for the coral reels beloved by fish, and its cliff-girded coastline permits shallow-water fishing in only a few places. Less than a quarter of the bones in its early garbage heaps (from the period 900 to 1300) belonged to fish; instead, nearly one-third of all bones came from porpoises.

Nowhere else in Polynesia do porpoises account for even 1 percent of discarded food bones. But most other Polynesian islands offered animal food in the form of birds and mammals, such as New Zealand's now extinct giant moas and Hawaii's now extinct flightless geese. Most other

islanders also had domestic pigs and dogs. On Easter, porpoises would have been the largest animal available—other than humans. The porpoise species identified at Easter, the common dolphin, weighs up to 165 pounds. It generally lives out at sea, so it could not have been hunted by line fishing or spear-fishing from shore. Instead, it must have been harpooned far offshore, in big seaworthy canoes built from the extinct palm tree.

In addition to porpoise meat, Steadman found, the early Polynesian settlers were feasting on seabirds. For those birds, Easter's remoteness and lack of predators made it an ideal haven as a breeding site, at least until humans arrived. Among the prodigious numbers of seabirds that bred on Easter were albatross, boobies, frigate birds, fulmars, petrels, prions, shearwaters, storm petrels, terns, and tropic birds. With at least 25 nesting species, Easter was the richest seabird breeding site in Polynesia and probably in the whole Pacific.

Land birds as well went into early Easter Island cooking pots. Steadman identified bones of at least six species, including barn owls, herons, parrots, and rail. Bird stew would have been seasoned with meat from large numbers of rats, which the Polynesian colonists inadvertently brought with them; Easier Island is the sole known Polynesian island where rat bones outnumber fish bones at archeological sites. (In case you're squeamish and consider rats inedible, I still recall recipes for creamed laboratory rat that my British biologist friends used to supplement their diet during their years of wartime food rationing.)

Porpoises, seabirds, land birds, and rats did not complete the list of meat sources formerly available on Easter. A few bones hint at the possibility of breeding seal colonies as well. All these delicacies were cooked in ovens fired by wood from the island's forests.

Such evidence lets us imagine the island onto which Easter's first Polynesian colonists stepped ashore some 1,600 years ago, after a long canoe voyage from eastern Polynesia. They found themselves in a pristine paradise. What then happened to it? The pollen grains and the bones yield a grim answer.

Pollen records show that destruction of Easter's forests was well under way by the year 800, just a few centuries after the start of human settlement. Then charcoal from wood fires came to fill the sediment cores, while pollen of palms and other trees and woody shrubs decreased or disappeared, and pollen of the grasses that replaced the forest became more abundant. Not long after 1400 the palm finally became extinct, not only as a result of being chopped down but also because the now ubiquitous rats prevented its regeneration: of the dozens of preserved palm nuts discovered in caves on Easter, all had been chewed by rats and could no longer germinate. While the hauhau tree did not become extinct in Polynesian times, its numbers declined drastically until there weren't enough left to make ropes from. By the time Heyerdahl visited Easter, only a single, nearly dead toromiro tree remained on the island, and even that lone survivor has now disappeared. (Fortunately, the toromiro still grows in botanical gardens elsewhere.)

The fifteenth century marked the end not only for Easter's palm but for the forest itself. Its doom had been approaching as people cleared land to plant gardens; as they felled trees to build canoes, to transport and erect statues, and to burn; as rats devoured seeds; and probably as the native birds died out that had pollinated the trees' flowers and dispersed their fruit. The overall picture is among the most extreme examples of forest destruction anywhere in the world: the whole forest gone, and most of its tree species extinct.

The destruction of the island's animals was as extreme as that of the forest: without exception, every species of native land bird became extinct. Even shellfish were overexploited, until people had to settle for small sea snails instead of larger cowries. Porpoise bones disappeared abruptly from garbage heaps around 1500; no one could harpoon porpoises anymore, since the trees used for constructing the big seagoing canoes no longer existed. The colonies of more than half of the seabird species breeding on Easter or on its offshore islets were wiped out.

In place of these meat supplies, the Easter Islanders intensified their production of chickens, which had been only an occasional food Item. They also turned to the largest remaining meat source available: humans, whose bones became common in late Easter Island garbage heaps. Oral traditions of the islanders are rife with cannibalism; the most inflammatory taunt that could be snarled at an enemy was "The flesh of your mother sticks between my teeth." With no wood available to cook these new goodies, the islanders resorted to sugarcane scraps, grass, and sedges to fuel their fires.

All these strands of evidence can be wound into a coherent narrative of a society's decline and fall. The first Polynesian colonists found themselves on an island with fertile soil, abundant food, bountiful building materials, ample lebensraum, and all the prerequisites for comfortable living. They prospered and multiplied.

After a few centuries, they began erecting stone statues on platforms, like the ones their Polynesian forebears had carved. With passing years, the statues and platforms became larger and larger, and the statues began sporting ten-ton red crowns—probably in an escalating spiral of one-upmanship, as rival clans tried to surpass each other with shows of wealth and power. (In the same way, successive Egyptian pharaohs built ever-larger pyramids. Today Hollywood movie moguls near my home in Los Angeles are displaying their wealth and power by building ever more ostentatious mansions. Tycoon Marvin Davis topped previous moguls with plans for a 50,000-square-foot house, so now Aaron Spelling has topped Davis with a 56,000-square-foot house. All that those buildings lack to make the message explicit are ten-ton red crowns.) On Easter, as in modern America, society was held together by a complex political system to redistribute locally available resources and to integrate the economies of different areas.

Eventually Easter's growing population was cutting the forest more rapidly than the forest was regenerating. The people used the land for gardens and the wood for fuel, canoes, and houses—and of course, for lugging statues. As forest disappeared, the islanders ran out of timber and rope to transport and erect their statues. Life became more uncomfortable—springs and streams dried up, and wood was no longer available for fires.

People also found it harder to fill their stomachs, as land birds, large sea snails, and many seabirds disappeared. Because timber for building seagoing canoes vanished, fish catches declined and porpoises disappeared from the table. Crop yields also declined, since deforestation allowed the soil to be eroded by rain and wind, dried by the sun, and its nutrients to be leeched from it. Intensified chicken production and cannibalism replaced only part of all those lost foods. Preserved statuettes with sunken cheeks and visible ribs suggest that people were starving.

With the disappearance of food surpluses, Easter Island could no longer feed the chiefs, bureaucrats, and priests who had kept a complex society running. Surviving islanders described to early European visitors how local chaos replaced centralized government and a warrior class took over from the hereditary chiefs. The stone points of spears and daggers, made by the warriors during their heyday in the 1600s and 1700s, still litter the ground of Easter today. By around 1700, the population began to crash toward between one-quarter and one-tenth of its former number. People took to living in caves for protection against their enemies. Around 1770 rival clans started to topple each other's statues, breaking the heads off. By 1864 the last statue had been thrown down and desecrated.

As we try to imagine the decline of Easter's civilization, we ask ourselves, "Why didn't they look around, realize what they were doing, and stop before it was too late? What were they thinking when they cut down the last palm tree?"

I suspect, though, that the disaster happened not with a bang but with a whimper. After all, there are those hundreds of abandoned statues to consider. The forest the islanders depended on for rollers and rope didn't simply disappear one day—it vanished slowly, over decades. Perhaps war interrupted the moving teams; perhaps by the time the carvers had finished their work, the last rope snapped. In the meantime, any islander who tried to warn about the dangers of progressive defor-

estation would have been overridden by vested interests of carvers, bureaucrats, and chiefs, whose jobs depended on continued deforestation. Our Pacific Northwest loggers are only the latest in a long line of loggers to cry, "Jobs over trees!" The changes in forest cover from year to year would have been hard to detect: yes, this year we cleared those woods over there, but trees are starting to grow back again on this abandoned garden site here. Only older people, recollecting their childhoods decades earlier, could have recognized a difference. Their children could no more have comprehended their parents' tales than my eight-year-old sons today can comprehend my wife's and my tales of what Los Angeles was like 30 years ago.

Gradually trees became fewer, smaller, and less important. By the time the last fruit-bearing adult palm tree was cut, palms had long since ceased to be of economic significance. That left only smaller and smaller palm saplings to clear each year, along with other bushes and treelets. No one would have noticed the felling of the last small palm.

By now the meaning of Easter Island for us should be chillingly obvious. Easter Island is Earth writ small. Today, again, a rising population confronts shrinking resources. We too have no emigration valve, because all human societies are linked by international transport, and we can no more escape into space than the Easter Islanders could flee into the ocean. If we continue to follow our present course, we shall have exhausted the world's major fisheries, tropical rain forests, fossil fuels, and much of our soil by the time my sons reach my current age.

Every day newspapers report details of famished countries—Afghanistan, Liberia, Rwanda, Sierra Leone, Somalia, the former Yugoslavia, Zaire—where soldiers have appropriated the wealth or where central government is yielding to local gangs of thugs. With the risk of nuclear war receding, the threat of our ending with a bang no longer has a chance of galvanizing us to halt our course. Our risk now is of winding down, slowly, in a whimper. Corrective action is blocked by vested interests, by well-intentioned political and business leaders, and by their electorates, all of whom are perfectly correct in not noticing big changes from year to year. Instead, each year there are just somewhat more people, and somewhat fewer resources, on Earth.

It would be easy to close our eyes or to give up in despair. If mere thousands of Easter Islanders with only stone tools and their own muscle power sufficed to destroy their society, how can billions of people with metal tools and machine power fail to do worse? But there is one crucial difference. The Easter Islanders had no books and no histories of other doomed societies. Unlike the Easter Islanders, we have histories of the past—information that can save us. My main hope for my sons' generation is that we may now choose to learn from the fates of societies like Easter's.

27

The Limits of the Earth

DAVID C. KORTEN

Barely one year ago, President Clinton described the NAFTA and GATT treaties as the cornerstones of his economic and foreign policy and touted them as major accomplishments of his Administration. No more. In his 1996 State of the Union Message he made no mention of either of them.

That tacit acknowledgment of the growing public backlash against economic globalization was echoed only days later at the World Economic Forum's annual meeting in Davos, Switzerland. There, several thousand power brokers of corporate capitalism met under the theme of "Sustaining Globalization." Apparently, the consequences of globalization are proving so devastating that its leading proponents are suddenly deeply worried about public reactions.

The concerns of the Davos participants were amplified in a February 1, 1996, *International Herald Tribune* article written by Klaus Schwab and Claude Smadja, respectively founder/president and managing director of the World Economic Forum. With striking candor and insight, Schwab and Smadja noted that "globalization tends to de-link the fate of the corporation from the fate of its employees" and has created a world in which "those who come out on top win big, and the losers lose even bigger." They warned that a "mounting backlash, especially in the industrial democracies, is threatening a very disruptive impact on economic activity and social stability in many countries."

Maintaining that globalization is irreversible, Schwab and Smadja call on political and economic leaders to find ways of demonstrating to the public "how the new global capitalism can function to the benefit of the majority and not only for the corporate managers and investors." Implicit in their supporting recommendations is a belief that this can be achieved through economic expansion fueled by still more public subsidies to business to increase national competitiveness.

While Schwab and Smadja are certainly right about the negative consequences of economic globalization, they are wrong in claiming that it is inevitable or irreversible. Nor can it be made to benefit the majority of people. Its claims and promises are grounded in a flawed ideology that contradicts basic ecological and social realities. A global economy is inherently unjust, unstable and unsustainable.

The removal of barriers to the international flow of goods and money did not happen as part of a natural evolution, as its advocates claim, nor was it the consequence of inexorable historical forces. The policies that made it happen resulted from conscious choices of a self-interested minority who, over the past half-century, have designed, shaped and now control the institutions that dominate global economic activity.

The defining moment was the infamous Bretton Woods meetings of 1944, at which the dominant ideologies of globalization—accelerated growth through global free trade and deregulation—were institutionalized. At the opening session, U.S. Secretary of the Treasury Henry Morgenthau advocated rapid "material progress on an earth infinitely blessed with natural riches." He asked participants to embrace the "elementary economic axiom . . . that prosperity has no fixed limits. It is not a finite substance to be diminished by division."

Thus Morgenthau set forth one of several underlying assumptions of the economic paradigm that has guided globalization since Bretton Woods. Two of these assumptions are deeply flawed. The first is that growth and enhanced world trade will benefit everyone. The second is that growth will not be constrained by the inherent limits of a finite planet.

By the end of this meeting, the World Bank and the International Monetary Fund were founded, and the groundwork was laid for what later became the General Agreement on Tariffs and Trade. In the intervening years, these Bretton Woods institutions have held faithfully to their half-century-old mandate. Through "structural adjustment programs," the World Bank and the I.M.F. have pressured countries of the South to open their borders and convert their economies from diverse production for local self-sufficiency to export production for the global market. Trade agreements like GATT reinforced these actions, opened the global economy to the increasingly free movement of goods and money, and eliminated controls on corporate behavior.

As we look back fifty years later, we can see that economic growth has expanded fivefold, international trade has expanded by roughly twelve times and foreign direct investment has been expanding at two to three times the rate of trade expansion. Yet, tragically, while the Bretton Woods institutions have met their goals, they have failed in their purpose of bringing prosperity to the people of the world. The earth has more poor people today than ever before. There is an accelerating gap between the rich and the poor. Widespread violence is tearing families and communities apart. And the planet's ecosystems are deteriorating at an alarming rate.

There is a growing consensus outside official circles that the planet's ecological limits, and the economic injustice inherent in the Bretton Woods system, doom that system to ultimate failure and require a radical change of course.

As the founder of ecological economics, Herman Daly, regularly reminds us, the human economy is embedded in the natural ecosystems of our planet. Until the present moment in human history, however, the scale of our economic activity relative to the scale of the ecosystems has been small enough so that, in both theory and practice, we could afford to ignore this fundamental fact. Now, however, we have crossed a historical threshold. Because of the fivefold expansion since 1950, the environmental demands of our economic system have filled up the available environmental space of the planet. In other words, we live in a "full world."

The first environmental limits that we have confronted and possibly exceeded are not the limits to nonrenewable resources (such as oil), as many once anticipated, but rather the limits to *renewable* resources and to the environment's "sink functions"—its ability to absorb our wastes. These limits have to do with the loss of soils, fisheries, forests and water; the absorption of CO_2 emissions; and the destruction of the ozone layer. We could argue whether a particular limit was hit at noon yesterday or will be passed at midnight tomorrow, but the details are far less important than the basic truth that we have no choice but to adapt our economic institutions to the reality of a full world.

The structure and ideology of the existing Bretton Woods institutions are geared to an ever-continuing expansion of economic output and to enfolding national economies into one seamless global economy. The result is to intensify competition for already overstressed environmental space. In a full world, this intensified competition accelerates destruction of the regenerative capacities of the ecosystem on which we and future generations depend; it crowds out all forms of life not needed for immediate human consumption; and it increases competition between rich and poor for control of ecological resources. In a free market, which responds to money, not needs, the rich win this competition every time. We see it happening all over the world: Hundreds of millions of the financially disfranchised are displaced as their lands, waters and fisheries are converted to uses serving the wants of the more affluent.

The market cannot deal with questions relating to the appropriate scale of economic activity. There are no price signals indicating that the poor are going hungry because they have been forced off their lands; nor is there any price signal to tell polluters that too much CO_2 is being released into the air, or that toxins should not be dumped into soils or waters. Steeped in market ideology and acutely sensitive to corporate interests, the Bretton Woods institutions have been unable to give more than lip service either to environmental concerns or the needs of the poor. Rather, their efforts have in practice centered on making sure that people with money have full access to whatever resources remain—with little regard for the broader consequences.

If ecological limits don't doom the Bretton Woods system, growing inequality across the planet will. The United Nations Development Program's *Human Development Report* for 1992 introduces the champagne glass as a graphic metaphor for a world of extreme economic injustice. The bowl of the champagne glass represents the abundance enjoyed by the 20 percent of the world population who live in the richest countries and receive 82.7 percent of the world's income. At the bottom of the stem, where the sediment settles, we find the poorest 20 percent, who barely survive on 1.4 percent of the total income. The combined incomes of the top 20 percent are nearly sixty times larger than those of the bottom 20 percent. Furthermore, this gap has doubled since 1960, when the top 20 percent enjoyed only thirty times the income of the bottom 20 percent. And the gap continues to grow.

These figures actually understate the true inequality in the world, because they are based on national averages rather than actual individual incomes. If we take into account the very rich people who live in poor countries and the very poor people who live in rich countries, the incomes of the richest 20 percent of the world's people are approximately 140 times those of the poorest 20 percent. Perhaps an even more startling expression of inequality is that the world now has more than 350 billionaires whose combined net worth equals the annual income of the poorest 45 percent of the world's population. That gap is growing as well.

In his book *The Work of Nations* (1991) Secretary of Labor Robert Reich explained that the economic globalization the Bretton Woods institutions have advanced so successfully has severed the interests of the wealthy classes from a sense of national interest and thereby from a sense of concern for and obligation to their less fortunate neighbors. A thin segment of the superrich at the very lip of the champagne glass has formed a stateless alliance that defines global interest as synonymous with the personal and corporate financial interests of its members.

This separation has been occurring in nearly every country in the world to such an extent that it is no longer meaningful to speak of a world divided into Northern and Southern nations. The meaningful divide is not geography—it is class.

Behind the ecological degradation and growing inequalities have been several hundred powerful corporations and banks whose ambit is now global. By expanding the boundaries of the market beyond the frontiers of the nation-state, corporations have increasingly moved beyond the reach of government. The structural adjustment programs of the World Bank and the I.M.F. and the global free-trade agreements empowered to override national laws have further weakened governments. As a result, real governance has been transferred from national governments, which at least in theory represent the values and interests of citizens, to transnational corporations, which by their nature serve only the short-term interests of their most powerful shareholders. Consequently, societies everywhere on the planet are less able to address environmental and social needs. Meanwhile, ever greater power is being concentrated in the hands of a very few global corporations. Indeed, where governments once sought to strengthen market competition through antitrust actions, they now often encourage increased economic concentrations through mergers and accusations in the cause of making national corporations "more globally competitive."

The rapid rate at which large corporations are trimming their work forces has created a misleading impression in some quarters that these companies are losing their power. The Fortune 500 firms shed 4.4 million jobs between 1980 and 1993, but during this same period, their sales increased 1.4 times, assets increased 2.3 times and C.E.O. compensation increased 6.1 times. The average C.E.O. of a large corporation now receives a compensation package of more than $3.7 million per year. Those same corporations employ 1/20th of 1 percent of the world's population, but they control 25 percent of the world's output and 70 percent of world trade. Of the world's hundred largest economies, fifty are now corporations.

The Economist recently reported that five companies now control more than 50 percent of the global market in the following industries: consumer durables, automotive, airlines, aerospace, electronic components, electricity and electronics, and steel. Five corporations control more than 40 percent of the global market in oil, personal computers and—especially alarming in its consequences for public debate on these very issues—media. These companies and others like them are the true beneficiaries of the global economy.

The vision and decisions at the Bretton Woods conference have transformed the governance of societies. Nonetheless, sustainability in a growth-dependent globalized economy is what Herman Daly calls an impossibility theorem. What is the alternative? The answer is the *opposite* of globalization. It lies in promoting greater economic localization—breaking activities down into smaller, more manageable pieces that link the people who make decisions to the consequences of those decisions. *It means rooting capital to a place and distributing its control among as many people as possible.*

Powerful interests stand resolutely in the way of achieving such a reversal of current trends. The biggest barrier, however, is the lack of public discussion on the subject. We must begin the process of change by recognizing that our global development models—and their underlying myths—are artifacts of the ideas, values and institutions of the industrial era. Modern corporations have been the cornerstone of that era, concentrating massive economic resources in a small number of centrally controlled institutions. They brought the full power of capital-intensive technologies to bear in exploiting the world's natural and human resources so that a small minority could consume far more than their rightful share of the world's real wealth. Now, as globalization pushes the exploitation of the earth's social and environmental systems beyond their limits of tolerance, we face the reality that the industrial era is exhausting itself—because it is exhausting the human and natural resource base on which our very lives depend.

28

Childhood Cancer

CHARLES W. SCHMIDT

Even as new treatments for childhood cancer continue to bring a cure within tantalizing reach, a mysterious rise in the incidence of the disease is beginning to trouble researchers.

The overall rate of childhood cancer in the United States is rising about 1 percent per year, data from the National Cancer Institute show. Put in perspective, this means that about 690 more cases will be diagnosed in the year 2000 than were diagnosed in 1991. The story is the same in England, Scotland, Germany, and Australia.

With approximately 8,000 diagnoses reported every year, cancer is the most common cause of disease-related death among children in the United States. Although the disease in children has been tentatively linked to many different risk factors—including genetic abnormalities, ultraviolet and ionizing radiation, electromagnetic fields, viral infections, certain medications, food additives, tobacco, alcohol and a number of industrial and agricultural chemicals—conclusive evidence remains elusive.

One important feature of the disease is that it occurs most often in young children—usually those younger than 5. This is notable because it is generally thought that carcinogenic exposures cause cancer only after an extended latent period that can last decades. Many researchers are therefore speculating that certain genetic traits may play a large role in determining which children are at greatest risk.

According to Leslie Robison, professor of pediatrics at the University of Michigan, and holder of the Children's Cancer Research Fund Chair in Pediatric Cancer for the University of Minnesota Cancer Center, these genetically predisposed children may represent a unique population that is particularly sensitive to the cancer-causing properties of carcinogenic substances.

The biggest question on the minds of most researchers is whether the increases are somehow related to toxic chemicals in our air, water and food. This issue is taking on a new urgency in light of emerging evidence that certain environmental chemicals known as endocrine disrupters may be able to exert subtle effects on the ebb and flow of hormonal cycles. Whether or not chemically induced hormonal changes may be related to the development of cancer in children is unknown—and toxicologists are hard at work attempting to devise methods by which these kinds of effects can be evaluated experimentally.

While this area of study represents a sophisticated angle on toxicological research, what is particularly surprising is the lack of even basic knowledge on the toxicity of literally tens of thousands of industrial chemicals, many of which have been discharged, either intentionally or otherwise, into the environment. Could it be that the veritable chemical soup to which children are exposed on a daily basis is somehow related to the increase in cancer incidence?

Many researchers believe that this may be the case.

A NEW IMPERATIVE

Addressing the conspicuous absence of information on the toxicity of so many chemicals, particularly with the respect to childhood exposures, has become a priority at the Environmental Protection Agency. An executive order issued by the Clinton administration earlier this year directed the EPA and other federal agencies to explicitly consider children's health when setting standards for pesticide residues in food, and contaminants in water and air. This change in direction may be ushering in a new era in the standard setting process.

Because children may be especially sensitive to the toxic effects of chemicals, refining our knowledge of their unique exposure patterns will be an important component of the research programs initiated under the executive order. These exposure estimates can be used by regulators at the EPA and others federal agencies to establish safe levels of chemical intake for exposed children.

"Pound for pound, children breathe more air, drink more water, and eat more food than adults," said Philip Landrigan, chief scientist at EPA's newly created Office of Children's Health Protection. "Thus, they are more heavily exposed to toxins present in those media. Children's exposures are further enhanced by their playing close to the ground and their normal hand-to-mouth activity."

Perhaps the most critical period of exposure occurs in the womb. The child's developing organ systems are particularly sensitive to the effects of chemicals, which are passed on via the mother.

Susan Preston-Martin, an epidemiologist and professor of preventative medicine at the University of Southern California, has been studying the role that chemicals known as nitroso compounds may have in the development of brain tumors in infants. Nitroso compounds are found in cured meats such as hot dogs and sausages, as well as in cigarette smoke.

"The fetal animal in particular is exquisitely sensitive to nitroso compounds," she said, referring to laboratory research she has conducted on animals. They can cause brain tumors in a wide variety of species, including primates, and a much smaller dose is required if the exposure is transplacental."

Preston-Martin also has completed a study investigating maternal consumption of nitroso compounds and the risk of pediatric brain tumors in humans, and found that the risk increased with the mother's consumption of cured meat products.

Interestingly, the risk appeared to be reduced among women who took vitamin supplements, particularly vitamins C, A, and E, during pregnancy. Vitamins are known to "scavenge" nitroso compounds from circulation. Researchers are scrutinizing the way in which vitamins may reduce childhood cancer risk.

A number of researchers are also suggesting that the kinds of jobs held by parents may hold important clues to the incidence of cancer in their children.

"The literature is filled with observations of increased or decreased estimates of risk depending upon the occupation of the parents," said Robison, the pediatrics professor.

"We need to follow up on the possibility that certain kinds of childhood cancer may be related to specific parental job categories," added Aaron Blair, an epidemiologist with the National Cancer Institute.

"Leads that need to be followed include all types of cancer and agricultural occupations, pediatric brain cancer with machinists and mechanics, and leukemia with occupations involving exposure to solvents," he said.

The chemicals a parent is exposed to in the workplace may be passed on to children in a number of different ways. In addition to the direct transfer of chemicals from the workplace to the child's environment, children also may be exposed to chemicals that build up in the mother's body and are then passed on to the developing fetus.

Exposures may also cause genetic defects in sperm cells, passing the risk on via the father.

RADIATION AND EMFS

Exposure to high levels of radiation, such as those experienced by atomic bomb survivors and survivors of radiation therapy, is still the best documented cause of childhood cancer. High-level radiation exposures have been linked to a number of diseases including leukemia, a bone cancer known as osteosarcoma and thyroid cancer.

The association with low-level radiation exposures is less conclusive, however.

Among the most compelling lines of evidence linking childhood cancer to low levels of radiation is a study published in the March 1985 issue of The New England Journal of Medicine, which reported that twins X-rayed to determine fetal position had twice the risk of developing leukemia than twins that were not X-rayed.

However, the findings of other studies investigating low-level radiation exposures are not as convincing. Neither exposure to low-level radioactive fallout, living close to a nuclear power plant, nor having parents that work at a nuclear facility has been unequivocally linked to elevated cancer incidence in children.

Researchers are also growing increasingly skeptical that electromagnetic fields, or EMFs, have a role in the development of childhood cancer.

EMFs emanating from high voltage power lines were first linked to leukemia back in the mid-1980s. Dozens of subsequent studies appeared to back up the initial findings, further fanning the public's fear that unseen waves of electricity were invading their homes and making them sick. However, more recent investigations appear to be downplaying the connection.

A recently conducted study headed by the National Cancer Institute and the Children's Cancer Group, a multi-institutional research and treatment organization headquartered in Arcadia, California, failed to identify any association between leukemia and exposure to EMFs.

Nonetheless, Martha Linet, an epidemiologist at the National Cancer Institute and principle investigator of the study, published in the July 1997 issue of the New England Journal of Medicine, cautions that it may be premature to draw any definitive conclusions.

"Even though our study didn't find an association, we're still waiting for the results of two large nationwide studies evaluating EMFs in childhood cancer," she said.

Those studies are one in the United Kingdom headed by Sir Richard Doll, a professor emeritus at Oxford University, and another in Canada being led by Mary McBride and others from the Cancer Control Agency of British Columbia.

Researchers also wonder if some cancers, particularly childhood leukemia, may be related to an as-of-yet unidentified virus. According to this hypothesis, critical exposure to the virus could occur from the mother's infection during pregnancy or after the child is born.

There is some speculation that children in lesser developed countries (in which the rates of leukemia are half of those in the United States) may be less susceptible to a leukemia-inducing virus for the seemingly paradoxical reason that their immune systems are actually stronger than those of children from developed nations.

This is because the generally less hygienic environments in which these children live cause them to be exposed to greater numbers of pathogens. Similar to the way in which a vaccine works, these exposures prime their immune systems at an early age, and render them resistant to infection.

Thus, leukemia could be included in a group of viral illnesses linked to improved hygiene, a group that also includes polio and infectious mononucleosis, among others.

FEW EXPLANATIONS

Discovering what lies behind the increased childhood cancer rates—whether environmental, genetic, or otherwise—has galvanized researchers. An EPA-sponsored conference on childhood cancer held in September in Washington, D.C., drew close to 300 physicians, academic and government researchers, and nonprofit representatives, all assembled for the purpose of designing a multi-million dollar research strategy that will begin early next year.

"The truth is we don't yet know what's behind the increases," the EPA's Landrigan said.

"Even though death rates are down due to improved treatment, the incidence rates have been steadily increasing and have not yet been readily explained."

He added that "it is not likely that the increases are related to improved diagnosis."

The increases are most apparent in the two most common types of childhood cancer: acute lymphoblastic leukemia—a disease in which immature white blood cells run rampant and inhibit the body's ability to mount an effective immune response—and cancers of the central nervous system, particularly a kind of pediatric brain tumor known as a glioma.

National Cancer Institute data show the number of male and female cases of acute lymphoblastic leukemia, or ALL as it is more commonly referred to, increased 25 percent between 1973 and 1990.

This trend now appears to be reversing in boys, although the rates in girls are still on the rise. The increase in brain tumors is even more dramatic: close to 40 percent between 1973 and 1994. In contrast to ALL, which is twice as common in white children as it is in blacks, the increase in brain tumors appears to be split evenly between both races and sexes.

Other cancers, such as a childhood cancer of the kidney known as Wilms tumor and osteosarcoma, may be rising as well. However, the rarity of these diseases makes it difficult for epidemiologist to discern whether or not the increases are statistically significant.

These increases are not readily apparent in Connecticut—which is one of the 14 states whose health department contributes data to the National Cancer Institute for the calculation of national trend estimates.

However, similar to the way in which it is hard to identify trends in the very rare cancers, such as Wilms tumor, trends are also not readily apparent when the numbers are restricted by locations, such as individual states.

Ultimately, uncovering what lies behind the increasing trends will require years of coordinated research. Fortunately, however, childhood cancer is an increasingly curable disease.

Medical research has made great advances in the management and treatment of the illness over the last several decades, and afflicted families can take heart that a diagnosis of cancer in their child may no longer represent the same bleak prospect that it did in the past. The long-term survival rate for ALL, for example, now ranges from 50 percent to 90 percent.

Long-term survival rates for childhood brain tumors range from 73 percent to 85 percent. By contrast, the survival rate for adults with brain tumors is closer to 30 percent.

Dr. Steven Galson, a senior scientist at EPA's Office of Children's Health Protection, believes parents would be best advised to limit their child's exposure to chemicals, and that mothers should take care to reduce their own exposures while pregnant. However, Galson notes that this kind of control is difficult to achieve.

"Multi-vitamin preparations appear to reduce risk," he said. "It seems to be clear that pesticides are carcinogenic, so parents should take extreme care in protecting their kids from exposures to these compounds."

• • •

29

A Pipeline Killing Field

Exploitation of Burma's Natural Gas

PAMELA WELLNER

In Burma, ecological destruction, particularly deforestation, is inextricably linked with human rights abuses. Rich in natural resources and cultural diversity, the country is being ravaged to keep a military dictatorship in power as it wages war against ethnic nationalities and a pro-democracy movement.

The military has ruled Burma since 1962 when it seized power from the democratic government. By the late 1980s, however, a growing democracy movement was causing it to rethink. General Ne Win stepped down as dictator in July 1988 (although behind the scenes he still wields immense political influence) and was replaced by Defense Minister, General Saw Maung who renamed the regime the State Law and Order Restoration Council (SLORC), imposed martial law,[1] changed the country's name to Myanmar, and intensified repression: more than 1,000 pro-democracy demonstrators were reportedly killed in September 1988 and many more incarcerated.

In 1990, SLORC held a national election in which over 80 percent of the seats were won by the National League for Democracy (NLD), led by Aung San Suu Kyi. Refusing to relinquish power, SLORC has since kept Aung San Suu Kyi under house arrest. Thousands of students and pro-democracy leaders have fled to the forest since 1988 and now live as refugees, mainly on Burma's borders with Thailand, Bangladesh, India and China, where many of Burma's indigenous peoples also live. There they have formed a parallel government, the National Coalition Government of the Union of Burma (NCGUB), which is allied with the ethnic minorities—Karen, Mon and others—who have been fighting the Burmese government since independence from Britain in 1948.[2] In the words of 1976 Nobel Peace Laureate Betty Williams, who visited the Thai-Burma border in 1993, SLORC "continues to uproot, rape and murder thousands of indigenous people as it moves to cleanse the border of ethnic insurgents and plunder the country's rich natural resources."

PROSPECTING FOR OIL AND GAS

To fund its military operations, SLORC has not only sold timber and fishing concessions to foreign, mostly Thai, Chinese and Indian, companies, but has also invited multinational companies to prospect freely for oil in Burma and in Burmese waters. Since 1989, at least 10 foreign companies (predominantly from the US, Canada and Europe) have bought exploration licences for mostly onshore concessions in government-controlled areas of central Burma, investing over $400–500 million in the search for oil and gas: indeed, in the last six years, oil companies have provided the largest sectoral block of foreign investment in the country.[3] As the Institute for Asian Democracy stated in its 1992 report, *Towards Democracy in Burma,* "A major find by any will assure the SLORC substantial income for years to come and will further insulate the generals against international pressure."

Initially, however, the oil companies were disappointed, so much so that by 1992, a number of them—including Shell (UK/The Netherlands), Broken Hill Pty (Australia), Croft (UK), Kirkland Oil (UK) and Idemitsu (Japan)—did not renew their contracts to continue exploration because of poor discoveries, high operating costs and other factors. In 1994, Amoco joined the exodus, citing poor financial returns. Although this was never publicly admitted, in at least two cases the withdrawals were prompted by concern over human rights abuses in Burma.

Other companies might well have followed suit had it not been for the discovery of two major natural gas fields in the Andaman Sea, named the Yadana ("jewel") and the Yetagun ("waterfall"). In June 1994, Minister of Energy U Khin Maung Thein put proven reserves in the Gulf of Martaban, site of the Yadana field, at over six trillion cubic feet of natural gas—more than three times original estimates.[4] Encouraged by these finds, some oil companies are continuing their involvement, on and offshore, and a further 18 offshore concessions have been announced by SLORC.

The Yetagun concession, covering 50,000 square kilometres, is now being developed by a consortium of Premier, Nippon Oil and Texaco. In the Yadana field, French oil multinational Total has joined forces with Unocal, a US company, to explore a concession of some 26,140 square kilometres, originally bought by Total for a reported $15 million. When this concession starts producing gas, expected to be in mid-1998, the Myanmar Oil and Gas Enterprise has an option to take up 15 per cent of the project (as it does in the Yetagun field) and the Petroleum Authority of Thailand an option for 26 per cent, with Unocal retaining at least 28 per cent and Total 31 per cent.[5]

RESETTLEMENT AND FORCED LABOUR

To transport the gas from the Yadana field, a pipeline is to be constructed from the Andaman Sea across Tenasserim—a thin strip of Burmese territory which contains some of the least disturbed tropical forest in Burma and is controlled by the Karen National Union and the New Mon State Party—to Thailand.[6]

SLORC has been forcibly displacing villages along the pipeline's route since February 1992, a process which has intensified since October 1993, causing many people to take refuge in Thailand. SLORC is also logging the area with conscript labour to clear a 60 metre swathe for the pipeline and to obtain timber to build supporting roads and SLORC military outposts. In one area, the authorities forced every family to work seven days without payment or food in order to construct or upgrade roads. Villagers have been told that they will receive no compensation for the land (full of coconut and other fruit trees) that has been cleared.

Villages have also been moved to make way for a 160-kilometre extension from Ye to Tavoy of the notorious "Death Railway" built during the Second World War by the Japanese with the forced labour of nationals of Japanese-occupied countries—Thailand, Malaya, Singapore—and Allied prisoners-of-war. Officials of the New Mon State Party believe that, as the railway extension will intersect the pipeline's east-west path, it is being built to transport pipeline equipment and materials and SLORC military units to the area. Since early February 1994, over 60,000 people, mainly Mon and Karen, have been conscripted to work on the railway.[7] SLORC battalions have entered villages and demanded labour, equipment and supplies. The village heads are then responsible for rounding up labourers, usually men and women between the ages of 18 and 60, although it is not uncommon for children as young as eleven and pregnant women to be conscripted as well. If the number of labourers demanded cannot be met, the village is asked to pay a fine of 3,000 kyats per person to compensate for the shortfall. In one town, five bulldozers were stationed to make up for possible labour shortages and the village was charged 10,000 kyats for each hour a bulldozer was in operation.[8]

Conditions in work camps are appalling. Villagers are normally forced to work on shifts of approximately two weeks, and are told to bring their own food, blankets and tools. Workers unable to keep up are beaten. Facilities for human waste disposal and water purification are inadequate, and consequently, many people are dying of intestinal diseases. Sometimes the sick are allowed to go back home, but many die along the way or once they reach home.[9] According to one escaped labourer, each village has to provide at least 20 tons of timber, while the Karen Information Service report that in the Tavoy region each family is required to provide 100 square feet of railway sleeper.[10]

THAI COMPLICITY

SLORC receives moral and financial support for such abuses from two main sources: the Thai government and military and Thai firms; and Western oil companies.

Thailand's rapid economic development has led to severe shortages of natural resources, particularly of timber and fish. In the last few years, Thai companies have therefore turned towards their less-developed neighbour for new supplies, paying about $300 million in licences and fees for various logging, mining and fishing concessions. Many of these concessions are controlled by present or retired military officers who have been given virtually a free hand to exploit Burma's natural resources in return for financing SLORC's arms purchases. These companies have in addition negotiated timber deals with the ethnic nationalities who also need money for weapons.

However, with timber resources near the Thai border diminishing, some sections of the Thai government and the business sector now appear to believe that they will benefit more if they help SLORC "contain" the ethnic groups than if they negotiate with the insurgents for the resources. The prospect of the pipeline has undoubtedly contributed to this shift. Thailand is currently the only contracted importer of gas from the pipeline and will be the main consumer of gas from the Andaman fields. It is thus in the interests of both the Thai government and the Petroleum Authority of Thailand to help SLORC secure the area around the pipeline.

Attitudes towards the ethnic groups living in areas strategic to the pipeline are thus hardening, whilst Thailand's policy towards SLORC has become more accommodating. In September 1993, the Thai foreign minister announced that the Thai government would recommend that Burma be given observer status in the Association of South-East Asian Nations (ASEAN).[11] At the same time, the National Security Council of Thailand has been attempting to secure the border area by pressurizing the various ethnic groups whose lands border Thailand to agree to ceasefires with SLORC. A series of meetings between the New Mon State Party and SLORC to discuss ceasefire conditions have taken place since the beginning of the year, arranged through a Thai businessman with interests in the pipeline, although these discussions have recently stalled.[12] Tentative approaches have also been made to the Karen National Union, which has so far declined to negotiate with SLORC.

Whereas Thailand has, in the past, allowed people fleeing persecution to take refuge on the Thai side of the border, the Thai army is now forcibly relocating refugees back to Burma, including those who have fled from forced labour in the pipeline and railway areas. Once back in Burma, these people face retribution from SLORC and are again taken for forced labour. On 7 April 1993, two refugee camps just south of Nat Ei Taung were burned down by the Thai army. According to an unnamed source quoted in the Thai newspaper, *The Nation,* "this action was probably related to the gas pipeline."[13]

In late February 1994, some 7,000 refugees at Loh Loe camp, one of the largest refugee camps along the border, were forcibly relocated by the Thai army to Halockhani in Burma, just five kilometres from SLORC military base, the field command of the Burmese army's 62nd battalion. On 21 July 1994—the first day of the annual ASEAN meeting in Bangkok—the 62nd battalion attacked Halockhani, but were ambushed by the Mon National Liberation Army. Using Mon refugees as human shields, SLORC soldiers retreated, taking 16 prisoners and burning 50 houses. Some 5,000 refugees again fled to the

Environmental degradation is accelerating in Burma through the various projects SLORC is pursuing to raise foreign exchange—logging, hydroelectricity, mining and fishing. Since 1989 the rate of deforestation has increased dramatically, mainly through logging of the border areas. In 1948, some 74 per cent of Burma—500,000 square kilometres—was covered in forest, whereas today, most sources put the figure at only about 30 per cent. Foreign companies, as well as SLORC and some ethnic groups, profit from the logging: SLORC recently announced, however, that foreign companies would have to shift from logging to developing value-added processes such as plywood mills.

Burma's four main rivers—Irrawaddy, Chindwin, Salween and Sittaung—are viewed as a source of abundant hydropower. Plans have been drawn up with Thai authorities to build dams in the east and southeast of the country, the largest of which would be the Upper Salween Dam, a 166-metre high construction which could produce 4,560 megawatts.

Reports have increased over the last two years of ecological despoilation, disease and killings in the mining regions of north-eastern Shan state, where SLORC is trying to expand the trade in rubies and other precious stones. With over one million displaced people and refugees within Burma, many people have joined the scramble to the ruby mines at Mongshu in Shan state and a disease-ridden jade mine in Kachin state, where a reported half million prospectors and traders from across the country have converged. Vast areas of forests and hillsides have been stripped bare in the process.

Modern trawler fleets from Thailand have fished out large areas of the Andaman Sea since buying concessions from SLORC in 1989. Despite complaints from local fishers, over 280 boats from another eight Thai companies were allowed to buy a new round of contracts in November 1993. The Burmese Fisheries Ministry has recently signed at least nine joint venture agreements to exploit Burma's marine resources, including two shrimp ventures with Mariam Marshall Segal, a New York Investor.

Source: Paradise Lost? The Suppression of Environmental Rights and Freedom of Expression in Burma, Article 19 September 1994

Thai side of the border, refusing to return to Burma until their safety was guaranteed. With the Thai army blocking attempts to provide medical and food assistance to the Mon refugees, they may have little choice but to go back to Halockhani.[14]

MULTINATIONAL COMPLICITY

Without Total's and Unocal's investment, the natural gas pipeline would be unlikely to go ahead. Yet the two companies deny any responsibility for what is happening in Tenasserim and refuse to admit that the construction of the pipeline involves the use of forced labour nor that it is having a severe environmental impact.

Total claims that only the last few kilometres of the pipeline's route are forested. In September 1994, at the Total's Paris headquarters, spokesperson Michel Delaborde said:

> aiding the economic development of the country is a good thing. One can discuss this point. But we are not involved in politics; we are industrialists and everything we do, we will do with complete respect for working conditions and human rights.[15]

Unocal, meanwhile, asserts that it is directly benefiting the people through the provision of employment and better health and training standards for 2,000 local workers, as well as contributing to the growth of democracy.[16] In a *Los Angeles Times* article, however, a member of Unocal's team investigating the pipeline admitted that the team spent one day only in the pipeline area and that their access was restricted by local SLORC military authorities.[17]

An umbrella group of international environmental and human rights groups, the Coalition for Corporate Withdrawal from Burma, asked Unocal shareholders in April 1994 to vote in favour of the company issuing a comprehensive report on its operations in Burma, including the human rights and environmental impact of the Martaban Gulf gas pipeline. Although Unocal's Board of Directors recommended voting against the proposal, 15.3 per cent of the shareholders voted in favor of the resolution—insufficient for the report to be issued, but enough to bring the resolution up at next year's annual shareholders' meeting.

Although US oil companies deny responsibility for SLORC's activities, they may find themselves legally accountable. The Center for Constitutional Rights, a US public interest law firm, claims in a letter to Unocal that the company "could he held legally liable for deaths, injuries, property damage or other harm arising out of your company's operations in Burma." The Center asserts that under US tort law:

> If a corporation enters into a contractual relationship with disreputable parties [such as SLORC], and it is reasonably foreseeable that those parties will hurt someone, the corporation may be held liable for the harm resulting from the business transaction.

The Clinton administration is soon to release a review of US foreign policy with Burma. The review may suggest that economic sanctions be applied against the SLORC regime, or may take a more conciliatory approach. Oil companies are likely to try to convince the administration that, instead of sanctions, a "Code of Conduct" should be drawn up which would allow them to con-

tinue their projects in Burma. At the annual ASEAN meeting in July 1994, however, the US did not support a proposal put forward by Australian Senator Gareth Evans for the adoption of "critical dialogue" with SLORC—a form of "constructive engagement."[18]

The proposal was supported by New Zealand and received the tacit approval of the European Union. British Foreign Secretary Douglas Hurd said in September 1994:

> We follow the European Union line on Burma; we're strongly critical of the abuse of human rights by the SLORC . . . but that should not impede or prevent trade . . . In certain circumstances, trade can actually have a useful effect in opening up the country and showing the need for change inside the country, so there's never been a trade boycott, though there is a European guideline barring political contact.[19]

When Archbishop Desmond Tutu of South Africa was in Thailand in 1993, however, he called for sanctions against SLORC, saying:

> International pressure can change the situation in Burma. Tough sanctions, not 'constructive engagement,' finally brought the release of Nelson Mandela and the dawn of a new era in my country. This is the language that must be spoken with tyrants—for sadly, this is the only language they understand.

One of the few hopes for the safety of the ethnic nationalities living in the pipeline area and for the preservation of the forest around it is for foreign companies to withdraw their investments in Burma. A May 1994 press release of the New Mon State Party, issued just before its third round of peace talks with SLORC, maintained that:

> Ethnic nationalities and ecological diversity are the immediate victims of greed and racism, because the oil and gas corporations as well as multinational investors are assisting the SLORC military junta with technical aid and propaganda to legitimize SLORC rule.[20]

Last year, the Democratic Alliance of Burma, an umbrella organization of political, student and ethnic groups, including the Mon and Karen whose territory the pipeline would run through, issued the following international appeal:

> If they make a pipeline to Thailand, it would go through our last big rain-forest. . . . We have tried to save our forests and made wildlife preserves like Kaser Doo mountain where animals like elephant, rhinoceros, wild giant cattle, tapir and hornbill can live with no disturbance from outside. The pipeline would ruin it completely . . . SLORC can never hope to patrol [all the pipeline]. But if the foreign companies ask it to, the SLORC is greedy enough to make a campaign to kill off every Karen and Mon in the area, so the pipeline might seem safe. We, the Democratic Alliance of Burma, earnestly appeal to the international countries whose oil companies are in this Burmese business to stop it immediately before our people are killed on a pipeline killing field.

NOTES AND REFERENCES

[1] Martial law was officially lifted by 1992, but SLORC has continued to rule under a complex array of security laws which allow military officers sweeping powers of arbitrary arrest and detention. See *Paradise Lost? The Suppression of Environmental Rights and Freedom of Expression in Burma*, Article 19, The International Centre Against Censorship, 33 Islington High Street, London N1 9LH, 1994.

[2] There are at least nine distinct ethnic groups in Burma. Burmans account for two-thirds of those who live in Burma and have dominated the government since independence. The other groups are the Karen, Karenni, Mon, Shan, Kachin, Chin, Rohingya and the Arakanese. For more information, see Smith, M., *Burma: Insurgency and the Politics of Ethnicity*, Zed Books, London and New Jersey, 1993; Lintner, B., *Land of Jade: A Journey Through Insurgent Burma*, Kiscadale Publications, Edinburgh, 1990; Lintner, B., *Outrage: Burma's Struggle for Democracy*, Review

Publishing, Hong Kong, 1989; Aung San Suu Kyi, *Freedom From Fear,* Penguin Books, Harmondsworth (reprinted January 1995).

[3] All Western development aid to Burma was cut off in 1988 in protest at SLORC's seizure of power. Foreign investment, therefore, is coming from private companies, predominantly those based in the US, Thailand, Singapore and Japan. See *Paradise Lost?* op. cit. 1, p. 16.

[4] *Paradise Lost?* op. cit. 1. pp. 16–17.

[5] On 9 September 1994, a memorandum of understanding was signed between Total, Unocal, the Myanmar Oil and Gas Enterprise (MOGE) and the Petroleum Authority of Thailand (PTT) in Rangoon under which the PTT will buy $400 million of gas for 30 years, starting in 1998. The PTT looks set to take up its full option, but there has been no announcement of the percentage of MOGE's participation. See "The Human Rights Pipeline," *Los Angeles Times,* 11 April 1994; Birsel, R., "Burma, Thailand sign gas purchase agreement," Reuters, 9 September 1994; Boonsong Kositchetana, "PTTEP prepared to take up to 30% stake in Burma gas field," *Bangkok Post,* 13 September 1994; "Gas contract with Rangoon overrides everyone's future best interests," (editorial), *Bangkok Post,* 11 September 1994.

[6] The pipeline will pass from the coast near the village of Hpaung-daung Yaw at Heinze bay, through the villages of Onbinkwin, Kanbauk and Zinba village to the Thai-Burmese border, approximately 43 kilometres as the crow flies. The Petroleum Authority of Thailand is responsible for building the 390-kilometre Thai section of the pipeline which will enter Thailand at the village of Pilok in Khanchanburi province, run past Ban I Tong (Nat Ei Taung) down to a 2,800MW power plant operated by the Electricity Authority of Thailand in Ratchaburi province. Gas from the smaller Yetagun field may use this pipeline as well, or may go underwater to Ranong in Thailand. See "Attack on Mon refugees may delay gas pipeline," *The Nation,* 2 August 1994; Boonsong Kositchetana, op. cit. 5; Boonsong Kositchetana, "Consortium complete development plan for Burmese gas field," *Bangkok Post,* 13 August 1994; Fahn, J., "Ranong likely terminal for second Burmese gas pipeline," *The Nation,* 23 August 1994.

[7] Since October 1993, an estimated 120,000 to 150,000 people have been subjected to such slave labour. See "Ye-Tavoy Railway Construction: A Report on Forced Labour in the Mon State and Tenasserim

Division in Burma," New Mon State Party, April 1994.

[8] The Burmese government's official exchange rate of six kyats for one US dollar contrasts with the black market rate of 110 kyats. A villager's annual income is around $25–30 (black market rates).

[9] These conditions have been confirmed in interviews with some SLORC military defectors who were stationed to oversee the construction of the railway.

[10] In February 1994, Mon military intelligence intercepted a message sent by SLORC personnel to their battalion headquarters detailing the amount of land cleared since the middle of the previous month in three labour camps under the supervision of one SLORC battalion. The first camp had cleared 15,132 linear yards, the second 11,000 yards, and the third 9,400 yards.

[11] Because of intense lobbying by Burmese opposition groups and public condemnation within Thailand, however, this initiative was not accepted by other ASEAN members.

[12] The New Mon State Party halted the negotiations because the conditions imposed by SLORC are tantamount to Mon surrender, with SLORC effectively controlling Mon lands. Personal communication,

Faith Doherty, Southeast Asia Information Center, and Fahn, J., "Mons and SLORC," *The Nation,* 7 September 1994.

[13] *The Nation,* Bangkok, 4 May 1993.

[14] *Burma Issues,* August 1994, p. 4; AFP, "Aid agency fears Mons being starved out of Thailand," *The Nation,* 7 September 1994; Amnesty International, *Thailand: Burmese and other asylum-seekers at risk,* London, September 1994.

[15] "Oil Firms Face Heat on Myanmar Investments," *In Asia Today,* 8 September 1994.

[16] *Paradise Lost?* op. cit. 1, pp. 18–19.

[17] *Los Angeles Times,* 26 April 1994.

[18] Personal communication, Michelle Bohana, Institute for Asian Democracy. It is hoped that a UN resolution (A/C 3/48/L70-48 section agenda item 114cc) will be introduced within the next few months, asking the international community to adopt a resolution by consensus that advances previous resolutions, such as that appointing a Special Rapporteur on human rights, and also suggests facilitating the transfer of power to those elected in the 1990 elections.

[19] *The Bangkok Post,* 14 September 1994.

[20] *Paradise Lost?* op. cit. 1, p. 18.

30

Border Patrol

BRUCE SELCRAIG

Buried amidst the bombast and pie charts of last November's "debate" between Al Gore and Ross Perot on the North American Free Trade Agreement was a fleeting moment in which the billionaire Texan held a videocassette up to the camera and told 20 million cable viewers that the documentary showed "a major U.S. chemical plant in Mexico that digs holes in the ground, dumps the chemical waste in those holes,

bulldozes over those holes and contaminates the water supply for the people in that area."

While Perot paused to reload, a squirming Gore pleaded, "Can I respond? Can I respond?" But the moderator, Washington talk-show host Larry King, soon broke for a commercial and never returned to the mysterious chemical plant in the unnamed Mexican town (as we will a bit later).

No one knows how many NAFTAfarians realized what place Perot was talking about, but 2,000 miles away, in the Tex-Mex border town of Brownsville, Texas, the man who brought that chemical plant and the surrounding community to the world's attention was savoring Perot's every word.

For Domingo González, cofounder of the Texas-based Coalition for Justice in the Maquiladoras, which produced the documentary, Perot's plug on national television, albeit vague, was a stunning achievement. "When Perot held up our videotape," González told me later, "it was like hitting a home run. We felt our work was finally paying off."

You may never have heard of Domingo González—that suits him fine—but perhaps more than anyone, this 45-year-old Brownsville native has helped shape world opinion about health conditions along the border, specifically around the foreign-owned *maquiladoras,* or assembly plants, in Brownsville's sister city of Matamoros, Mexico, just across the Rio Grande. In the pre-NAFTA frenzy of the past four years the former migrant worker turned career activist has taken at least a hundred U.S. and foreign journalists, five congressional delegations, and dozens of labor, religious, and environmental groups across the border for a look at the industrial wasteland that is Matamoros.

Journalists from Amsterdam, Tokyo, Madrid, Munich, São Paulo, London, Toronto, Mexico City, and Paris have all come calling on González. When *Rolling Stone* writer William Greider outlined NAFTA's faults two years ago, the first two words of his column were, naturally, Domingo González. When ABC's *Primetime Live* investigated the border's alarming number of anencephalic births—babies born with partial or missing brains, a condition some health experts believe is connected to the *maquilas'* use of mutagenic solvents—González was just off-camera most of the time, leading the way to mothers, doctors, and factory workers.

It was not by accident that many of the world media's bleak portraits of life around the *maquilas* came from Matamoros. González made it easy by offering up articulate victims, corporate bad guys,

cooperative experts, historical context, and, most important, directions through the town's rutted, unmarked alleys—with a minimum of the preachy sales pitch that annoys most mainstream journalists.

"There are few like Domingo," says National Public Radio reporter John Burnett, who has worked extensively along the border. "He not only understands, emotionally and technically, the issues raised by pollution, but in a maze of industrial plants he knows right where to go."

Environmental reporter Dave Harmon, of *The Monitor* in nearby McAllen, Texas, says González serves as a conduit for Matamoros residents too fearful or unsophisticated to contact the U.S. media themselves. "Let's face it," Harmon says. "If you're some gringo down here from Boston and you have to knock out a story in a few days, you can't do it without someone like Domingo."

Indulging reporters' deadlines as well as their frequent misconceptions of the region, the ever-patient González treats each one as if he or she were producing the definitive border exposé. He takes them to see some of the "Mallory children," 70 or so youths with facial deformities and mental retardation whose mothers worked with solvents and PCBs in the 1960s and '70s at the now-closed Mallory capacitor plant. He takes them by an accident-plagued pesticide factory where workers are so close to a neighborhood they can shake hands with residents over their backyard fences. In the shadow of Fortune 500 *maquilas,* González walks reporters past acrid, milky-white ditches laced with xylene, the Rio Grande floating with human excrement, and gritty, oblivious kids playing beneath railroad tank cars carrying ammonia and hydrofluoric acid.

The tours rarely fail to have the desired effect. During one such visit, Ohio Representative Marcie Kaptur and an entourage of journalists were strolling past a rainbow-hued industrial canal when a chicken wobbled by, took a sip from the ditch, and promptly dropped dead at Kaptur's feet. "Gee," she told reporters, "this really tells the story."

Were Kaptur and the journalists being manipulated? The toxic tours tell only *some* of the story, but González doesn't conceal his allegiances or motives, and reporters allow for his bias just as they do for that of the *maquila* managers who profess ignorance about illegal chemical-dumping and child labor. "There's nothing so ugly to reporters," González says, "as to feel they're being set up. Sometimes we do so little 'setting up' that we look totally disorganized. When you're as loose as we are," he laughs. "it's easy to make everything look spontaneous."

Before last year's NAFTA vote González's toxic tour became so popular, he was crossing the old Brownsville bridge over the foul Rio Grande several times a day with camera crews in tow. *Colonia* residents soon became blasé about all the photographers and boom mikes. Not so González's enemies. The resulting negative publicity so upset the Matamoros *maquila* association that, according to Mexican newspaper reporters, the association president asked the city council to have González and five other activists investigated by Mexico's thuggish Interior Ministry. Such "investigations" are not taken lightly in a country where, over the last decade, human-rights groups have documented many government-linked deaths of journalists, labor leaders, and opposition political activists.

"We sort of tease Domingo about the danger," says his friend Rose Farmer, the manager of an Audubon preserve near Brownsville. "But it's a real concern. He is at risk."

"Domingo is really quite a phenomenon because he accomplishes things," says Chris Whalen, editor of the conservative, Washington, D.C.–based newsletter *The Mexico Report*, and one who has investigated human-rights abuses in Mexico. "I'm really surprised the Mexican government hasn't had him killed."

Three years ago, I too was looking for that perfect border metaphor, that community-as-microcosm story that embraces all the elements of the environmental and social disaster that has befallen the region since the *maquila* program brought industrialization 30 years ago. When word-of-mouth eventually led me to González, he flattered me by listening—he never interrupts—as though my journalistic search were daring and novel, which it decidedly wasn't.

He paused in thought to let me know the enormity of this task, then winked and smiled.

"I think I know where to take you," he told me.

Few scenes in the Third World, and nothing in the United States, not even the neighborhoods around the world's largest concentration of petrochemical plants near where I grew up in Houston, prepared me for the sight of a tiny Matamoros *colonia* called Privada Uniones. This is the place that so appalled Ross Perot—and that González makes sure all visiting journalists see.

No more than a patch of land roughly 50 by 200 yards, Privada Uniones contains some 30 homes made mostly of plywood and corrugated tin. The shacks are surrounded by chemical plants, a rail line that supplies them, and a grain warehouse that covers the neighborhood in fumigated-corn dust. The residents of this industrial hell, who all seem to have wracking coughs, don't just live *close* to the chemical plants—their tiny homes virtually adjoin them. On one side is the former site of Quimica Retzloff, whose abandoned pesticide-waste pit is no more than 20 feet from residents' kitchens and yards. Separated from the plant by only a cinder-block wall, the pesticide holding pond would sometimes overflow in heavy rains, seep through the soil and kill the neighbors' gardens. That was, however, the least of their concerns. In 1983, a chemical leak at Retzloff killed most of the *colonia*'s chickens and dogs; in December 1990, two 55-gallon drums of methamidophos pesticide exploded, lofting a chemical cloud over Matamoros that sent 90 people to the hospital. The plant finally closed last year, but the site has never been cleaned.

Opposite Retzloff is the Mexican-owned affiliate of Northfield, Illinois–based Stepan Chemical, one of the United States largest makers of surfactants, which help disperse chemicals in everything from pesticides to toothpaste. Stepan, too, has had problems, experiencing an ammonia leak and an explosion that broke windows and TV screens throughout the *colonia*. (Stepan paid for repairs.) In the summer of 1991, a year after the residents first sought González's help, he and I walked along a ditch that came out of Stepan's property rust-red with chemical wastes. Inside Stepan's fences, maybe a hundred feet from the residents' homes, was an uncovered and unlined toxic dump where workers would empty drums of chemicals.

Privada Uniones was fed up. Community leaders, who had been documenting the tragedies around them since the early 1980s, had appealed unsuccessfully to every level of Mexican authority. They were ignored until González and the U.S. news media showed interest.

The Coalition for Justice in the Maquiladoras had the canal tested by the Boston-based National Toxics Campaign, whose EPA-approved lab helps environmental groups document industrial pollution. The results showed the ditch contained the solvent xylene at 23.2 million parts per billion—roughly 50,000 times the U.S. drinking-water standard. (Xylene can cause brain hemorrhaging as well as lung, liver, and kidney damage.) An organizer for the AFL-CIO, a Coalition partner, began videotaping workers dumping chemical barrels into Stepan's open pit. All of which made compelling footage for the Coalition's video documentary, *Stepan Chemical: The Poisoning of a Mexican Community.*

Meetings followed between Stepan and the community, but little more. Stepan had the canal behind its property filled with dirt, as was the open pit within its gates; however, neither was excavated, allowing whatever soil contamination that was occurring to continue. Mexican officials actually closed Stepan for a few days, but in that nation such measures are widely ridiculed as political shows rather than true law enforcement. What González really wanted was for Stepan to excavate and decontaminate the entire community, relocate the residents, and compensate them for their losses.

For González, the cleanup of the Stepan site, and others far worse along the border, will be the true test of whether Mexico is committed to environmental stewardship or was just putting up a front for the NAFTA campaign. If, as NAFTA's supporters claimed, the pact will produce a bounty that can help fund such cleanup efforts, González reasons that the Stepan site should be a

priority. But so far, the trade pact carries only a pledge of $2 billion to $3 billion, an amount González says would barely make a dent in cleaning up Matamoros alone, where he estimates there may be as many as 23 industrial sites that would qualify for Superfund status in the United States.

Today Stepan officials are adamant that they will never pay damages to the community or help to relocate residents. Charles P. Riley, Jr., chief of manufacturing at Stepan's Illinois headquarters, says the Matamoros plant complies with all Mexican laws, ships its toxic wastes to approved sites, has a new million-dollar wastewater-treatment plant, and is not now, nor ever has been, contaminating the *colonia.*

"It's ironic," Riley told me. "We've actually made the place much safer. I would live in the *colonia* and not have any worries about my health." (No doubt the residents would be happy to make the arrangements.)

Riley says González distorts the truth. "Apparently this is a trait of today's reactionaries and activists," says Riley. "He continues to say Stepan is the largest polluter in Mexico, which is ridiculous." (Later Riley conceded that González has only said Stepan was "one of" Mexico's largest polluters. I asked Riley to provide any documentation of González having made false statements. I've received none.)

Riley believes Stepan was a convenient target for those whose real agenda was the defeat of NAFTA. He has convinced himself that Stepan's problems with González will go away and the community will come to respect the company. Toward that end Stepan has donated furniture, books, and, soccer field to local schools, and has hosted an open house and a dance at the Matamoros plant.

"We had hamburgers, sodas. It was quite a nice thing," Riley told me. "I think we're changing minds."

I last caught up with González on a rainy evening in San Antonio just two weeks before the NAFTA vote in the House of Representatives, the treaty's first showdown on Capitol

Free Trade's Pricetag

For Domingo González, passage of the North American Free Trade Agreement (NAFTA) last year was just a setback in the long war against border pollution. The story begins in 1965, when Mexico established a tariff-free trade zone along its border with the United States. Rock-bottom wages and the Mexican government's disregard of its own environmental laws have been luring foreign manufactures to the region ever since. Today some 1,200 *maquilas* crowd the border from Texas to Tijuana.

The trade agreement extends the border free-for-all to the rest of Mexico. This time, a side agreement tries to mitigate some of the polluters' worst excesses. But it's a weak attempt, according to Larry Williams, director of the Sierra Club's International Program: the ancillary record has inadequate backing and limited bite.

For example, Mexico and the United States have promised $8 billion for border cleanup over the next decade. But that figure depends on bankers ballooning $450 million in promised seed capital into $2 billion in private lending. That may be a long shot, because even the seed money is not guaranteed. If any cash finally does flow, the Sierra Club estimates, the true cost of cleanup will total $20 billion over ten years.

And while the environmental agreement's citizen-input process is the first ever to be included in an international trade agreement, it is made as difficult as possible. Here's how the public would pursue relief from *maquila* pollution under NAFTA:

Individuals or advocacy groups from any NAFTA country can file a complaint with the Commission on Environmental Cooperation (CEC), which includes environmental ministers from Canada, Mexico, and the United States. If the complaint meets certain "threshold determinations," the CEC then informs the Mexican government. Mexico can halt an investigation if it determines that the facility in question is subject to a "pending proceeding," which might simply mean that the government is seeking voluntary compliance from the polluter.

If there's no government objection, the CEC secretariat prepares a "factual record." It must rely solely on public records and is held to no timetable or deadline. Once the draft report is presented to the full CEC, Canada and the United States must both vote to file a complaint against Mexico. That leads to a lengthy dispute-resolution process and the unlikely possibility that sanctions would be levied against Mexico for failure to enforce its environmental laws. Mexico can avoid fines simply by claiming that it doesn't have the money to implement those laws.

The process could drag on for 18 months; the citizens who initiated the complaint with the CEC can do nothing officially to influence its resolution. What they *can* do is become watchdogs, holding the NAFTA governments accountable through the glare of public attention. That, combined with the patience of a Domingo González, might allow trade without environmental trade-offs.

Hill. Without his morning coffee his eyes were still a bit reptilian as he emerged from a hotel lobby and climbed into my waiting rental car. Knowing his reputation for living on the cheap—"I learned mooching from the Farmworkers," he laughs—I had made him an offer he couldn't refuse: a free ride back to Brownsville in return for five hours of highway rumination on the life of an activist.

The previous night, after working all day in Brownsville, he had flown to the Alamo City, spoken at a local college's snoreful NAFTA debate, and doled out sound bites to bored TV reporters. ("They're promising the same things in NAFTA that they did 30 years ago with the *maquila* program.") Then he sat up past midnight over beer and burritos with a fugitive Mexican leftist and Susan Mika, another founder of the Coalition.

Such schedules don't seem to sap González's stamina, though his stout, 5-foot-6 frame now expands a bit at the middle—the result, he confesses, of too many late-night strategy sessions and not enough exercise. "Getting in at two, waking up at six, eggs and coffee—that'll kill ya," he says, between bites of his breakfast taco. His hair is still black as outer space and lies flat to his head; the wide cheekbones and coffee complexion come from his mother's Indio side (His father, who speaks only Spanish, is as fair as a Spaniard.)

González is an increasingly rare individual in self-absorbed America. Free of cynicism and driven by conscience, he has spent the last 25 years, often with great personal sacrifice, helping others fight their battles: first, in his 20s, through Catholic Church charities on the border, then Lyndon Johnson's War on Poverty, and eventually the Quaker-affiliated American Friends Service Committee (AFSC), one of the largest social-service agencies in the country. In a time when many activists are one-issue supernovas or weekend zealots with nice day jobs, González is the real thing: a full-time do-gooder.

González no longer receives a paycheck from the Coalition, because he was uncomfortable with the appearance of being a "paid agitator." The group pays for some of his phone bills and printing supplies, but he lives mainly off a periodic consultancy for the Texas Center for Policy Studies, an Austin-based environmental think tank. He has no savings account, no health insurance, no automobile, TV, furniture, or credit cards. "I am," he says wryly, "recession-proof." Now and then friends send him checks in the mail or buy groceries; reporters on expense accounts are often good for a couple of meals each month. Divorced for some 16 years and the father of three grown sons, González could no longer afford his apartment last year and so shuttled between the homes of his parents, former wife, a brother, and a Brownsville judge. Contacting him by phone was

like trying to dial up Salman Rushdie. (He now shares an older house with several other Brownsville activists.)

One never gets the feeling that González's low-rent lifestyle is being put on display. He does live a simple life, but he's no activist monk. He likes *Star Trek* and the Marx Brothers, even an occasional cigarette, and confesses to watching Dallas Cowboys games on the tube with his sons. "I wish I had had the money," he tells me, "to have given my sons a better education, a better house, all the things families need, but they understand. If I ever come into money, I'd give it all back to them and my friends. They've made everything I do possible."

As we head south, San Antonio's suburban sprawl dissolves into South Texas ranch land of mesquite and prickly pear. With the NAFTA vote imminent, González is worried. "Look at this," he says wearily, unfolding *USA Today* to show a photo of Bill Clinton chatting with Henry Kissinger, below the headline: "Clinton Rolls Out NAFTA's Big Guns."

"I think we may have failed to make NAFTA accessible to everyone," González tells me. "It's not that complicated. Are we going into the future with a corporate democracy, with public relations instead of the truth, with economics as our only consideration? That may sound idealistic, but the discussion should really be on that level." And, he says, correctly anticipating NAFTA's passage, "we didn't have it."

Much of González's credibility about border affairs comes from the fact that when he speaks of the Tex-Mex world, he draws on a lifetime of experience. Born in 1949, in a cluster of about a dozen small-acreage family farms 20 miles east of Brownsville, he is the second-oldest of four children—"the four who made it," he says; six other siblings died at or shortly after birth. His family grew cotton and vegetables until the early 1960s, when dam projects on the Rio Grande reduced the flow of the river below Brownsville to a trickle and increased its salinity to the point that the farmers could no longer irrigate their crops. "The dams killed us off," he says.

His family began following fruit and vegetable harvests in Arkansas, Michigan, and Illinois. It was not until 1964, when the family first went to California's mammoth vineyards, Domingo recalls, that he saw the worst abuses of migrant laborers. "We lived in what we called tin cans. They were made of corrugated aluminum and were terribly hot in the summer. We put carpeting on the roofs and poured water on it, but nothing helped. If Siberia had work camps, they were not worse than these."

One Sunday afternoon, a skinny, 15-year-old González was hanging out in a public park in Lamont, just south of Bakersfield, when he noticed a slight Chicano man striking up conversations throughout the park about workers' rights, a minimum wage, health care—radical things there. "He just started talking to people as though they had all come to hear him," González remembers, still with some awe. "That was my first memory of Cesar Chávez."

The following summer González returned to California's Central Valley with his family and, after work, helped college activists organize the labor camps. González says his family benefited greatly from Chávez's efforts, as panicked growers soon raised the hourly wage from $1.15 to $1.25 to $1.40 to discourage the emerging United Farm Workers. "We made great money in '64, '65, and '66," he recalls. "For the two-month grape harvest we were probably making $700 to $800 a week. We were not downtrodden. We bought a new pickup, came back to Brownsville, and could afford a down payment on a brick home. And we owed it all to Cesar Chávez."

In the next ten years, González would get married (at 19), have three sons, and drop out of college to begin his activist career. But by 1976, when he moved to an AFSC job in Philadelphia, his near-religious commitment to the cause had exacted a large price. His wife, Doris Mae, missed the border so much she returned there the next year with the three children. The marriage ended in 1978. For the next 12 years González stayed in Philadelphia, submerged "in a siege mentality" against Ronald Reagan's dismantling of social programs, all the while agonizing over being separated from his sons.

"I was constantly in debt," he says. "I tried to visit them about every three months or so. I'd plan work in Texas so I could go see them."

When does commitment to a good cause become its own kind of selfishness? I ask him. Why didn't he take a permanent job in Texas so he could be closer to his sons?

He takes so long to respond I'm sure I've offended him, but it is his way to wait until the right words come. "I thought about that, the selfishness, constantly," he finally offers. "I just couldn't leave what I was doing. During the Reagan years we thought we were at war—those times were incredibly hard. When I would go home to Brownsville, eventually it would come time for me to leave, and my youngest son would throw himself in front of the door and scream, 'Daddy, don't go, Daddy, don't go.' It was not easy to live through those moments."

Nearing Brownsville, the rolling ranch land flattens into a humid coastal pool table dominated by citrus, vegetable, and cotton farms. "Ah, the tragic valley," González announces. He's traveled this road, what, 500 times? a thousand? but he savors the unfolding landscape like a modern pioneer.

Amid the garish *casas de cambio* of downtown Brownsville, a town of 60,000 that is 95-percent Latino, we slow down to see a knot of believers placing cards and photos before a tree in someone's front yard—a tree transformed into a shrine, for the image of the Virgen de Guadalupe has been found in its bark. Raised Catholic, González still smiles at this quaint tradition, though it's a painful reminder of the conflicts he has in working among the border's devoutly religious poor. While acknowledging the positive influences of the contemporary Church in Latin America, he despairs over the masses "who are convinced they're worthless, powerless to change their environment."

Undoing this sense of fatalism is one of the largest challenges a border activist faces, but González draws his strength not from thinking he will transform the lives of the poor and the powerless, but from the knowledge that only in trying will he be able to deal with their misery. "There is actually very little you can do for the poor," he says, "but what they can do for you is to re-establish your faith in the world; that no matter how bad things are, there is always happiness and always smiles."

"The other night," he says, trying to explain how he finds motivation, "I was in Matamoros, with a family, and the woman has three kids, beautiful kids; the oldest daughter was born with what looks like a stick for a leg, another has something wrong with her eye, and their little boy was born with a large tumor at the base of his spinal cord. His legs are dried up, he can't walk. Three crippled kids, and yet there we were having the greatest time, laughing, telling stories. I had a British film crew with me and they were dumbfounded.

"Anyone," he says, "who's living a dreary middle-class existence wondering about his or her purpose in life ought to come down here and work with people who have nothing and find out what life is about."

There—he's let the secret out. What sustains many activists in these mythic battles against poverty and corporate neglect and environmental decay is not only altruism or the pursuit of ideology. Sometimes it is merely knowing that a good and just fight cleanses the mind and simplifies one's own internal conflicts. Perhaps, more than anything, it is about enlightment.

· · ·

7

War and Militarism

The world will continue to change dramatically, but fighting a war can destroy us utterly. What we need now are techniques of harmony, not those of contention. The Art of Peace is required, not the Art of War.

MORIHEI UESHIBA

War refers to group-sponsored violence against another group or groups. *Militarism* refers to all the group-sponsored activities involved in preparing for war, such as recruiting and training fighters, producing weapons and other needed resources, and promulgating cultural beliefs and values justifying warfare.[1]

War and militarism are not modern developments—they were present before the current era in less complex, preindustrial societies[2] (Keeley, 1996). Because war is a behavior deeply entrenched in human history, it is tempting to think that we humans are instinctually violent. There is, however, little scientific evidence to support this view. Because *some* human societies are peaceful, we human beings are clearly capable of being nonviolent. It is the *social* arrangements in any group that seem to determine the direction its members will take, whether the path of violence, or the path of peace. Like other social problems in the world, war and militarism have social causes, and thus social solutions (Gil, 1996, p. 78).

Although war and militarism are not modern problems, modern wars are vastly more destructive than the conflicts of earlier times. Modern technology has made weapons ever more deadly. Contemporary conflicts are often especially brutal, with the effects being total destruction of the enemy's social fabric, rather than simply the deaths of the enemy's troops. The brutal nature of today's conflicts is at an extreme in *genocidal* racial and ethnic wars in which the winners seek to physically annihilate the loser's population. Such genocidal wars will be examined in detail in Chapter 10, Racial/Ethnic Conflicts and the Danger of Genocide, as they are such a severe world problem today that they deserve special coverage. Several readings in this chapter emphasize the great human costs of contemporary warfare, with emphasis on the harm done to civilians. Civilians today suffer as much if not more than the combatants. For example, in wars occurring since World War II, 90 percent of the casualties have been among civilians (*U.N. Chronicle*, 1996, p. 10). Civilians continue to suffer, even after the cessation of hostilities. For example, it is estimated that every twenty minutes, land mines left by forces in combat maim or kill someone, somewhere in the world. There are some 110 million land mines buried in the earth in seventy countries (Winslow, 1998, pp. 12–13). Another distressing effect of modern warfare is that it has harmed millions of children. In "Child Soldiers," Mike Wessells reports on the impact of war on children around the world and efforts to help heal children forced into military combat.

What are the social causes of war and militarism? Many social scientists believe that the social roots of violence lie at a deep, cultural level. In many cultures, wars promoting the national interests are seen as just and necessary. Beliefs and values promoting militarism are prevalent in many cultures and reflected in how children are socialized through such means as toys and games, media programs, and school textbooks. There are also powerful vested interests promoting violence, and analysts using the conflict perspective have written extensively about this phenomenon. A pattern existing in the United States and a number of other industrial societies is the *military-industrial complex,* a combination of large organizations pushing for ever-greater weapons production and government military spending. This term was coined by former U.S. President and Army general, Dwight David Eisenhower, who defined it in his farewell address as a "conjunction of an immense military establishment and a large arms industry" and warned of the potential for a "disastrous rise of misplaced power." Vested interests in war and militarism exist in poor as well as wealthy nations. In "Bill Clinton's America: Arms Merchant to the World," Lora Lumpe presents a critical view of the U.S. government's role in encouraging the export of U.S.-manufactured weapons to a variety of governments around the world.

War and militarism are spurred on, not only by vested interests in each state, but also by the very nature of the competitive world system of states. The world system perspective has emphasized that the world is divided into an ever-increasing number of competitive nation-states, each of which defines itself as sovereign, or having the right to control matters within its own borders. This competitive *inter-state system* contrasts strongly with the situation in earlier historical epochs, in which only a few *empires* prevailed. In today's competitive system of states, each state is dedicated to the advancement of what it sees as its own economic, political, and strategic military interests. These interests are usually in line with the economic interests of the nation's upper class (Shannon, 1992, p. 39). Conflicts inevitably arise under such a system, and can escalate to war if not checked; militarism is also fostered under such a system, as each state prepares itself to defend its own interests. This situation is today greatly magnified by the fact that ethnic nations within larger nations around the world are increasingly demanding autonomy. In states that lack democratic processes for conflict resolution, these disputes tend to result in violence (Clay, 1990).

Solutions to the problem of war and militarism are being pursued on a variety of levels. Many people believe that a fundamental solution will require total transformation of cultures around the world. This idea is reflected in the foundation of programs of peace studies and conflict resolution. For example, 250 institutions of higher education in the United States now offer programs in peace studies, and 70 offer a degree in this field (Walsh, 1996, p. 7). Programs teaching children and adults the skills of peaceful conflict resolution are also appearing in schools, churches, and community organizations. Peace education often takes creative forms. In "Bringing the Cost to Light," Blaise Tobia and Virginia Maksymowicz describe a project by artist John Craig Freeman using billboards to teach people about the hidden monetary and environmental costs of Cold War–era militarism. Peace activists have often argued that we must stop encouraging children to play with war toys if we want to bring about peace. Joel Best, a symbolic interactionist, reviews this idea critically in "Too Much Fun: Toys as Social Problems and the Interpretation of Culture." Best suggests we need to do more research into the meanings that children actually assign to their toys. Other people working against war and militarism focus their efforts at a different level—on the social structures with vested interests in promoting war and militarism—and are taking direct actions to protest these interests. Peace movements taking direct action to protest these interests are active around the world.

NOTES

[1] I have developed these definitions by modifying definitions used by Sam Merullo in "War and Militarism," in Calhoun and Ritzer, eds., *Social Problems,* Princeton, NJ: McGraw-Hill, 1993. Merullo defines *war* as "collectively organized violence carried out by political entities (usually nations) through formally organized institutions that is intended to achieve rationally defined goals," and defines *militarism* as "the preparation for conducting war." Merullo's definitions seem to limit warfare to groups organized at the complex level of nation-states. I believe the idea needs to be expanded to encompass collective violence organized by groups with a less complex level of political organization. Keeley (1996) has amassed a good deal of evidence on the warring behavior of premodern

societies, and I believe we need to take this into account in our understanding of the nature of war. (See Note 2 below.)

[2]Keeley (1996) attacks the idea prevalent among social scientists that war did not exist prior to the development of agriculture, settled populations, social class divisions, and state formation. Keeley makes a strong argument, although it is certainly not going to be the last word on this issue, as many theoretical perspectives link the rise of modern social problems, including war, to the world's current state of social evolution. (For example, structural-functionalists see modern wars in the context of social differentiation into more and more complex forms of organization; Marxists link the rise of war to the emergence of a surplus; and the world system perspective contrasts wars within the modern interstate system with those waged by the earlier empires.)

WORKS CITED

Jason Clay. 1990. "What's a Nation? Latest Thinking." *Mother Jones*, 15(7), pp. 28–30.

Dwight D. Eisenhower. 1965. "Farewell Address," in Dwight D. Eisenhower, *Waging Peace, 1956–1961*. Garden City, NY: Doubleday.

David G. Gil. 1996. "Preventing Violence in a Structurally Violent Society: Mission Impossible," *American Journal of Orthopsychiatry*, 66(1) January, pp. 77–84.

Lawrence H. Keeley. 1996. *War Before Civilization*. New York: Oxford University Press.

Thomas R. Shannon. 1992. *An Introduction to the World System Perspective*. Boulder, CO: Westview Press.

U.N. Chronicle. 1996. "Too Soon for Twilight, Too Late for Dawn: The Story of Children Caught in Conflict," 4, Winter, pp. 7–14.

Catherine Walsh. 1996. "More Than 1,200 U.S. Campuses Currently Offer R.O.T.C., But Only 250 Offer a Peace Studies Program," *America*, 174(10), March 23, p. 7.

Philip Winslow. 1998. "Land Mines: The Ordeal of Chisola Pezo." *World*, March/April.

QUESTIONS FOR DISCUSSION

1.   What is the difference between "war" and "militarism"? Are these activities instinctual for human beings? Or socially caused? Explain.

2.   Describe the process through which children become soldiers, according to Mike Wessells. What are some solutions discussed by Wessells? What additional strategies need to be pursued, in your opinion, to prevent or reduce the impact of war and militarism on children?

3.   Lora Lumpe discusses the U.S. military-industrial complex in "Bill Clinton's America: Arms Merchant to the World." How does the military-industrial complex affect your own community or college?

4.   How can we change from a culture that promotes militarism and war to one that promotes peaceful methods of conflict resolution? How important is childhood socialization? Would eliminating war toys make a difference? After reading "Too Much Fun . . ." by Joel Best, what is your own opinion on the impact that war toys have on children?

 INFOTRAC COLLEGE EDITION: EXERCISE

Choose a nation that is currently involved in military conflict (or that was, in the recent past), such as Sudan, Yugoslavia, Congo, or Indonesia. Learn more about the background of the conflict, and efforts under way to resolve it.

Search for information, using InfoTrac College Edition. (*Hint:* Enter as the search terms the name of the nation, plus "and War" (for example, "Sudan and War") using the Subject Guide. Where there are two nations or ethnic groups involved, enter the names of the two—for example, "Indonesia and East Timor," "Israel and Lebanon," "Serbs and Kosovo," and so on.

What are some of the social *structural* causes of the conflict, such as economic and political factors? What are some of the *cultural* factors involved? What are some solutions to the conflict that are being pursued? Do you believe these are adequate? If not, what else do you believe needs to be done?

FOR ADDITIONAL RESEARCH

Books

Barash, David P. 1991. *Introduction to Peace Studies.* Belmont, CA: Wadsworth.

Chomsky, Noam. 1987. *On Power and Ideology, the Managua Lectures.* Boston, MA: South End Press.

Kennedy, Paul. 1987. *The Rise and Fall of the Great Powers.* New York: Random House.

Organizations

American Friends Service Committee
1501 Cherry Street
Philadelphia, PA 19102
(215) 241-7000
www.afsc.org

Sane/Freeze: Campaign for Global Security
1819 H Street NW
Washington, DC 20006
(202) 862-9740
www.webcom.com/peaceact

War Resister's League
339 Lafayette Street
New York, NY 10012
(212) 228-0450
www.wrl@igc.apc.org
www.nonviolence.org/wrl

World Federalist Association
418 7th Street SE
Washington, DC 20003
(202) 546-3950
www.wfa.org

ACTION PROJECTS

1. Interview a student on your campus who took part in a U.S. military action abroad, such as the Panama Invasion or the Persian Gulf War. Why did they decide to join the military? How were they influenced in their decision by parents, counselors, peers, and others? What happened to them during the action abroad, and what were their feelings about it at the time? What, if any, negative consequences have they experienced as a result of their participation? Do they believe the action was justified? Why or why not? What do this student's experiences tell us about the social causes of militarism and war?

2. Make an inventory of conflict resolution programs existing in your community. For example, which schools have programs to teach children how to resolve conflicts peacefully? Which colleges have such programs? Can a college student in your community major in peace stud-ies or mediation? Does your community have a community mediation service? Based on your research, how much progress would you say your community has made toward replacing violent with peaceful methods of conflict resolution?

3. Visit a local toy store and look for toys that promote violence or militaristic values. What age groups are they for? Are they for boys, or girls, or both? What percentage of the toys in the store are toys of this type? Alternatively, view children's programs for a few hours on a Saturday morning. How many incidents of violence are depicted? What characters are involved in the violence? Are they male, or female, or both? Reflect on your observations at the toy store or of the television programs. What does this tell us about American culture, the social construction of gender, and militarism?

31

Child Soldiers

MIKE WESSELLS

While in Sierra Leone a couple of summers ago, I visited Grafton Camp, a facility for recently demobilized child soldiers operated by UNICEF and local partners. Many of the boys, ranging from nine to 16 years of age, had killed people as they fought in a civil war that paused with a fragile cease-fire in 1995. The camp director said that when the youths had been given drugs—most likely, amphetamines—while soldiering, they "would do just about anything that was ordered." Some, he added, were proud of having been effective killers.

These boys, who had shortly before been willing to kill and who had never received an adequate foundation of moral development, danced with enormous energy and played cooperative games under the supervision of the camp's counselors. As I watched, it was sobering to think that under certain conditions, practically any child could be changed into a killer.

But today, it is even more sobering to see once again how easily children who have been denied education and trained for fighting are manipulated by local political leaders. Fighting has resumed in Sierra Leone following a May coup, and many of the combatants are under 18. They have become part of a continuing cycle of violence.

A SOLDIER AT SEVEN

The nature of armed conflict has changed greatly in recent years. The end of the Cold War ushered in an era of ethnopolitical conflicts that are seldom fought on well-defined battlefields. Conflicts are increasingly internal, and they are characterized by butchery, violence against women, and atrocities sometimes committed by former neighbors. More than 80 percent of the victims are noncombatants, mostly women and children.

Increasingly children serve as combatants or as cooks, informants, porters, bodyguards, sentries, and spies. Many child soldiers belong to organized military units, wear uniforms, and receive explicit training, their lethality enhanced by the widespread availability of lightweight assault weapons. Other children participate in relatively unstructured but politically motivated acts of violence, such as throwing stones or planting bombs.

The use of children in armed conflict is global in scope—a far greater problem than suggested by the scant attention it has received: Child soldiers are found from Central America to the Great Lakes region of Central Africa, and from Belfast in the north to Angola in the south.

The problem defies gender boundaries. Girls are often forced into military activity—in Ethiopia, for instance, girls comprised about 25 percent of opposition forces in the civil war that ended in 1991. Typically, sexual victimization is a part of soldiering for girls, many of whom are forced to become "soldiers' wives." After the conflict ends, families and local communities may reject the girls as impure or unsuitable for marriage. Desperate to survive, many former girl soldiers become prostitutes.

The use of child soldiers violates international norms. The U.N. Convention on the Rights of the Child (CRC), signed in 1989 and ratified by more than 160 nations, establishes 15 years as the minimum recruitment age. In fact, most countries have endorsed an optional protocol that boosts the minimum recruitment age to 18 years. But in the face of armed conflict, military units in some nations—whether governmental or rebel—often pay little attention to age.

In Grafton Camp, children were encouraged to draw, and many drew pictures that reflected their war experiences. One showed a house being shelled by artillery. Soldiers fired at the house and at people in the street, who were fleeing.

Inside the house was a man who had been shot. Blood flowed from his mid-section. I asked the artist, a small-for-his-age boy of nine, to tell me about the picture and what it showed. He explained that soldiers (the rebel forces) attacked his village, bombed his house, and came inside and shot his parents. The bleeding man was his father. I did not ask why he had not painted his mother, who had also been murdered.

How old was he when his parents were killed? "Seven," he said. I asked him what happened after the attack. "My parents died—the soldiers told me to go with them so I did."

I asked what he had done in the military. He had "carried things." When I asked if he had killed anyone, he said "No." But when asked if he would have killed someone if told to do so, the strength of his desire to survive showed. "Yes," he said. He would have done "what he had to do."

When asked what he wanted for the future, he said, "I only want to go to school."

CHILD SOLDIERS AND INSECURITY

Child soldiering violates the fundamental rights of children, exploits youth for political purposes, subjects them to slaughter and the ravages of war, and immerses them in a system that sanctions killing. And it also poses formidable security risks for others. A society that mobilizes and trains its young for war weaves violence into the fabric of life, increasing the likelihood that violence and war will be its future. Children who have been robbed of education and taught to kill often contribute to further militarization, lawlessness, and violence.

The use of child soldiers also threatens fragile cease-fires and blocks reconciliation and peace. Not infrequently, conflict continues at the local level even after a cease-fire has been signed. Child soldiers are pawns in local conflicts because they provide a ready group for recruitment by warlords, profiteers, and groups that foment political instability.

The problem is especially severe in developing countries, in which children constitute nearly half the population and in which children are often reared in a system that mixes war, poverty, violence, hunger, environmental degradation, and political instability.

The war in Angola, which ended in 1994 with a cease-fire, began more than 35 years ago as a liberation struggle, became a proxy war in the East-West contest, and left a legacy of about 10 million land mines and several generations who have never known anything other than war.

On a recent visit to Luanda, the capital of Angola, I saw more child amputees in a day than one might see in a lifetime elsewhere, with the exception of Afghanistan and Cambodia. During the most intense fighting, from 1992–94, large numbers of children lost parents and their homes, and they suffered from extreme poverty and hunger. The scars, emotional as well as physical, are deep.

Many Angolan children report nightmares and flashbacks, display heightened aggressiveness, and suffer from hopelessness. Thousands of children—defined as people under 18 years of age—entered the military. For both parents and children, war had become normal.

Hopes for peace in Angola rose in April as a new government of national unity and reconciliation took office, and the cease-fire continued. Violent youths, however, may yet sabotage the cease-fire. Roving gangs of bandits terrorize and rob civilians in rural areas. Many of the bandits are boys who served in the military; they lack education and job skills, but they understand the power of a gun.

Banditry aside, Angola faces the question of how to demobilize and reintegrate into civil society thousands of underage soldiers, many of whom fear rejection by their communities and who lack skills needed to meet their basic needs through nonviolent means.

The problems cannot be addressed through political reforms or peace treaties alone. They require work at the grassroots level to reorient and help former child soldiers adapt to peace.

Unfortunately, tensions in Angola are strong. Underage soldiers provide ready fodder for war. In Angola, as elsewhere, it is the militarization of young people and of society that creates a climate in which protracted armed conflict flourishes.

GLOBAL AND SYSTEMATIC

How widespread is child soldiering? Numbers are hard to come by. The destruction and turmoil of war make it difficult to create and preserve accurate records. Particularly in Africa, many countries have no history of keeping precise birth records.

Beyond that, many military groups, governmental and rebel, make no attempt to document or accurately report the ages of the children they recruit. And former child soldiers are often reluctant to identify themselves because they fear rejection by their communities or retribution from their former commanders—or from those whom they once attacked.

The best estimate—which is admittedly soft—is that in the mid-1990s, there were about a quarter of a million child soldiers, current or recently demobilized.[1] This figure comes from a series of 26 country case studies conducted by Rädda Barnen (Swedish Save the Children) as part of a larger U.N. Study on the "Impact of Armed Conflict on Children."[2]

Because the U.N. study was led by children's rights activist Graça Machel, the former first lady of Mozambique, it is typically referred to as the "Machel Study." One of the main conclusions of the study, released a year ago, is that child soldiering is a global problem that occurs more systematically than most analysts had previously suspected.

The Machel Study showed that in some countries, children constitute a significant percentage of the combatants. In Liberia, for instance, about 10 percent of an estimated 60,000 combatants in the civil war that began in 1989 were children. In El Salvador, children composed 20 percent or more of the FAES (Fuerzas

Armadas de El Salvador). In Afghanistan, 10 percent of the Mujahadeen forces are estimated to have been children under 16 years. In Palestine during the Intifada, nearly 70 percent of Palestinian children are believed to have participated in acts of political violence such as stoning Israeli troops.[3]

Numerical estimates, however, only hint at the damage done to children and to the fabric of the societies in which they live. Children often become part of a system of hatred and killing, even if they do not participate in military activity themselves. In Rwanda, many Hutu children were informants, disclosing the locations of Tutsis and their supporters, who were then slaughtered in the 1994 genocide.

FORCED RECRUITMENT

Children usually become soldiers through coercion, either through mandatory conscription or forced recruitment. When national armies have a manpower shortfall, they may find it convenient not to search too carefully for the accurate birthdate of a conscript. Rebel forces seldom have use for birth records, either. In countries covered by the case studies, government forces as well as rebel forces were often equally likely to use child soldiers.

In Cambodia, says the Machel Study, children who stood as tall as a rifle were often deemed eligible for military service. In Bhutan, local authorities instructed village headmen to bring forward a specified number of people from their respective villages. Children were among the "voluntary recruits."

Manpower-hungry militias often abduct children at gun point. In Afghanistan, Bhutan, Burma/Myanmar, El Salvador, Ethiopia, and Mozambique, soldiers have recruited children forcibly from schools. According to one underage Burmese recruit, government soldiers surrounded his school and arrested 40 to 50 youths between 15 and 17 years of age:

"Our teachers all ran away in fear," says the recruit, quoted in the Burma/Myanmar case study. "We were all terrified. I didn't know what was going on and they didn't explain anything to us."

In Ethiopia, armed militias would surround a public area such as a marketplace, order every male to sit down, and then force into a truck anyone deemed "eligible." At particular risk of abduction were teenagers who worked on the streets selling cigarettes or candy.

Forced abductions, says the Machel Study, were commonly one element in a larger campaign to intimidate communities. Armed groups that abduct children for soldiering are also inclined to go on rape-and-looting rampages while in the villages.

Abductions also can be used as an instrument of war. In Guatemala, for instance, the army singled out young members of the indigenous population for recruitment during its long civil war, thereby pitting them against their cohorts among the rebels. The Mayan community called it "the new genocide."

Militias often use brutish methods to weaken resistance to forcible recruitment. The case study for Uganda reports that people who resisted attacks by the Lord's Resistance Army "would be cut with pangas [machetes]. Quite a number of victims had their lips and ears chopped off in macabre rituals."

To seal off possible avenues of resistance from the children's communities, recruiters may deliberately destroy the bonds of trust between child and community. In Mozambique, for instance, recruiters from RENAMO forced boy recruits to kill someone from their own village.

FEAR AND OBEDIENCE

Abduction is only the first step in a process that uses fear, brutality, and psychological manipulation to achieve high levels of obedience, converting children into killers.

In many countries, child recruits are subjected to beatings, humiliation, and acts of sadism. In Honduras, boys wearing only underwear were exposed by government troops to "the ram," in which they were forced to roll nearly naked on a stony or thorny surface while being beaten or kicked by a squad leader.

In Paraguay, government military trainers beat children with sticks or rifle butts and burned them with cigarettes while verbally mocking them. Those who resisted or who attempted to escape were further brutalized or killed.

A frequently used tactic is to have children learn by doing, which may mean exposing them progressively to violence, numbing them so they might someday commit acts of sadism on fellow humans. Child recruits in Colombia, for example, were forced to cut the throats of domestic animals and drink the blood.

A 14-year-old Mozambican boy, quoted in Mozambique's case study, said of RENAMO forces: "I was told to train. I would run, do head-over-heels, and climb trees. Then they trained me to take guns apart and put them back together again for four months. Every day the same thing. When it was over they did a test. They put someone in front of me for me to kill. I killed."

Few constraints exist on what trainers can do to children, and children themselves may lack the internal constraints against violence that ordinarily develop through exposure to positive role models, a healthy family life, the rewards for socially constructive behaviors, and the encouragement of moral reasoning.

Weakened psychologically and fearful of their commanders, children can become obedient killers, willing to take on the most dangerous and horrifying assignments. In countries such as Uganda, Liberia, and Honduras, child soldiers have served as executioners, and in some countries—notably in Colombia, Peru, and Mozambique—they have been required to perform ritual acts of cannibalism on their victims, acts calculated to instill contempt for human life.

Adolescents are often selected for suicide missions, and some commanders view adolescents as mentally predisposed for such duty. In countries such as Sri Lanka and Burma/Myanmar, child soldiers were given drugs—such as amphetamines and tranquilizers—to blunt fear and pain and then used for "human wave" attacks that resulted in massive casualties. In Guatemala, underage soldiers were used as scouts and land mine "detectors."

Although some commanders complain that child soldiers take excessive risks, slow operations down, and do not seem to understand the dangers they face, many commanders prefer child soldiers because they are highly obedient and willing to follow the most unacceptable orders. As one person said in the Burma/Myanmar case study, "Child soldiers are always very eager to go to the front lines."

UNFORCED RECRUITMENT

Coercion aside, children may join the military for security, a pressing need for unaccompanied children who are vulnerable to nearly every kind of threat. Desperation for food or medical care often drives children into military life. The military may offer children the only path to wages to support themselves or their families. For these reasons, it is meaningless to ever speak of children's involvement in the military as strictly "voluntary."

The quest for national identity, liberation, and a secure homeland animates many armed conflicts. Typically, identity conflicts are saturated with an ideology of liberation struggle that draws a sharp line between Us and Them, glorifies the in-group while denigrating the out-group, and honors high levels of commitment to "the cause." Particularly in conflicts influenced by strong religious ideologies, youth may view the cause as having divine sanction, making it a clear-cut struggle between Good and Evil.

For adolescents still defining their identity, ideology provides direction that is otherwise lacking. In apartheid South Africa, black township youth—the Young Lions—adopted an ideology of liberation, which gave meaning to the harsh realities of their existence and conferred a clear sense of identity and direction.

In Guatemala, many children of landless peasants living in extreme poverty and victimized by repressive regimes embraced an ideology of revolution and joined the liberation struggle. In Rwanda during the early 1990s, the Hutu-dominated government used radio to spread hatred of the Tutsis, who were demonized as murderous outsiders. This helped prepare children for roles as killers in the youth militias in the 1994 genocide.

Many communities glorify war and teach children at an early age to view military activity as prestigious and glamorous. Militaristic values may be transmitted via parades, ceremonies to honor war heroes, and the martyrdom of soldiers.

Media images may also play a part. In Sri Lanka, opposition forces have broadcast Rambo-style TV movies of live combat training.[4] In such contexts, boys learn machismo and come to associate military activity with respect and power—compelling attractions for children who otherwise feel powerless.

While some boys join the military for adventure or to win fame and the respect of other males, others bask in the praise of mothers who express pride in seeing their sons in uniform.

In places such as Northern Ireland, Palestine, or South Africa, now as in the past, peer pressure animates participation in political violence. Youths expect and encourage each other to take part in violent activities, and they attach great value to group loyalty.[5] Having been arrested and tortured are regarded as badges of courage and commitment.

In states such as Chechnya and Ethiopia, families have encouraged sons to join opposition groups as a means of avenging the deaths of family members or of seeking "blood revenge." Families may also encourage sons to join the military for economic reasons, seeing the salary from soldiering as the most likely route to survival.

Children who engage in political violence often have witnessed deaths, torture, or executions. Others have lost parents, had their homes and even their communities destroyed, or have been sexually abused. Even children who have not been physically attacked may feel victimized by assaults on relatives or on their ethnic group.

Psychologically, people who have been victims of violence are at great risk of becoming perpetrators of violence. It's a familiar pattern. I recall a recent visit with three women whose husbands had been shot execution-style while working in the fields in the early 1980s, during a long and brutal civil war. The women now live as a group. One of their sons, now 16, said he did not remember the killings. He was too young. But if the war, which ended in a cease-fire, should resume, he would join a military unit—if it would enable him to avenge his father's death.

In 1992, while I was visiting the Occupied West Bank, a Palestinian father told me how his six-year-old son had gone up the street to the home of an Israeli settler who had recently moved in. The son had no involvement in political violence. But when the son tossed small stones into the settler's garbage can, as if shooting basketballs, the settler stormed out of the house with an automatic weapon and threatened to shoot if the boy returned.

The following month, the father said, his son joined a group of Palestinian youths in throwing stones at Israeli soldiers. "I worry," said the father, "he will be arrested and tortured."

HEALING

In Angola, restoring spiritual harmony through traditional healing is an essential step in helping child soldiers demobilize and reintegrate into their home communities. In many Bantu cultures, people believe that when one kills, one is haunted by the unavenged spirits of those who were killed. Spiritually contaminated, a former child soldier who has killed puts an entire community at risk if he re-enters without having been purified.

In one community, a traditional healer told me a few years ago of a ritual he ordinarily conducts to purify former child soldiers. First, he lives with the child for a month, feeding him a special diet designed to cleanse. During the month, he also advises the child on proper behavior and what the village expects from him.

At the end of the month, the healer convenes the village for a ritual. As part of the ceremony, the healer buries frequently used weapons—a machete, perhaps, or an AK-47—and announces that on this day the boy's life as a soldier has ended and his life as a civilian has begun.

Anecdotal evidence suggests that this kind of purification ceremony helps decrease the stress and fear that gnaws on former child soldiers and helps communities accept young people back. The preliminary evidence also suggests that once young people have been accepted, the community often succeeds in teaching them nonviolent modes of behavior.

Such ceremonies seem to be relatively common in rural areas, not just in sub-Saharan Africa but in indigenous cultures around the world. The healers who practice them are on to something. It is premature and without scientific justification to assume that former child soldiers who have killed or done terrible things are forever "damaged goods" and beyond rehabilitation.

Traditional healing methods may work, in part, because they fit local beliefs. For example, in Guatemala, Mayan people believe that when someone dies, the spirit cannot go to the next life until a burial ritual has been conducted. This is why the exhumations of mass graves now under way in Guatemala are so important to the Mayans.

Many humanitarian assistance and development efforts overlook traditional healing methods, which are dismissed as unscientific. I discovered while working in Sierra Leone that local people were initially reluctant to talk about traditional healing with me, a Western Ph.D. Although traditional methods should not be romanticized or viewed as a panacea, they can be important tools for assisting former child combatants.

Nevertheless, a variety of obstacles impede attempts to address the problem of child soldiering. Warring factions, desperate for more troops, continue to exploit children. In addition, nonstate actors such as armed opposition groups are not signatories to key instruments such as the Convention on the Rights of the Child. Typically, cease-fires and peace treaties include no provisions for the demobilization of child soldiers. Further, cultures vary in their definition of "childhood"; many African societies regard a 14-year-old boy as a man if he has participated in the traditional rite of passage.

Labeling is also a significant problem.[6] Some people, even psychologists and psychiatrists, have written off entire groups of severely traumatized child soldiers as "lost generations." In Nicaragua, according to that country's case study, workers at a center to assist former child soldiers initially feared the children they work with, believing they were "born assassins," "blood-thirsty children," or "human tigers" who "take out people's eyes."

Stigmatizing labels, however, should not obscure the fact that there are tremendous individual differences in children's responses to war experiences, and that many methods—Western and traditional—exist for assisting former child soldiers.

DEMOBILIZATION AND REINTEGRATION

The most immediate healing steps, which generally cannot be taken until after armed conflict ends, involve demobilizing everyone under the age of 18 years, reintegrating them with families and communities, and assisting them in making the transition into civilian life.

Effective demobilization programs provide basic needs, such as food, water, shelter, and security. This is most often accomplished by locating members of the child's immediate or extended family and then reuniting them as soon as possible.

To offer opportunities for healthy development and life in the community, reintegration programs often attempt to place former child soldiers in schools or to provide vocational training that can lead to jobs and financial conditions that mitigate against re-enlistment.

Effective demobilization and reintegration also requires attention to psychological adjustment. Depending on their experiences, former child soldiers may experience flashbacks and nightmares about traumatic events, causing difficulties in concentration that can impair judgment and performance in school. Some former soldiers carry heavy burdens of guilt and worry about what will happen to them.

War-affected children may act out aggressive impulses, creating problems and continuing the spread of violence. Inability to control aggressive behavior is often a problem for children who have been reared in a system of violence, who have few skills for handling conflict nonviolently, and whose moral development may have been limited by early immersion in the military.

It is important to heal the psychological wounds of war, to assist children in coming to terms with their experiences concerning death and violence, to reestablish daily routines that pro-

vide a sense of normalcy and continuity, and to develop values and skills of nonviolent conflict resolution. Nongovernmental organizations and U.N. agencies such as UNICEF have developed several effective programs for achieving these aims.

In Angola, for instance, I work with a multiprovince program that enables adults in local communities to address the emotional needs of war-affected children through a mixture of Western and traditional healing methods.[7]

In the past year, a team organized by Christian Children's Fund and UNICEF has, with funding from the U.S. Agency for International Development, located the families of and successfully demobilized and returned home 83 percent of 2,925 child soldiers in UNITA-controlled areas.

To prepare the communities, the team trained local church people—*Catequistas*—to help parents, teachers, and community leaders understand and deal with the kinds of problems the returning children would face. The *Catequistas* also helped arrange traditional healing ceremonies. About half of the former child soldiers helped by the program are in vocational training, and about a fourth are in school.

STRENGTHENING THE CRC

Although community-based approaches are valuable, the world cannot wait for child soldiering to occur and then try to pick up the pieces afterward. Prevention ought to be the top priority.

An immediate step would be to raise the minimum age of recruitment to 18 years. Although Article 38 of the Convention on the Rights of the Child establishes 15 years as the minimum recruitment age, the U.N. Commission on Human Rights is drafting an optional protocol to the CRC that sets 18 years as the minimum age for compulsory recruitment or for participation in hostilities. Because this protocol enjoys strong support, there is hope for its adoption by the General Assembly.

A key step toward strengthening the CRC is to pressure non-state actors to respect its provisions even though they are not signatories. If only governments adhere to the norms set by the convention, the door is left open to abuses of children's rights by opposition or rebel groups. Pressure to adhere to the standards set by the convention may be applied to both state and non-state actors through careful monitoring by U.N. agencies, nongovernmental organizations, and international media.

Another crucial step is to build commitment to the CRC, the most comprehensive instrument for the protection of children's rights. Although more than 160 nations are parties to the convention, there are several noteworthy exceptions— primarily the United States, which signed it in 1995, but has not ratified it.

The fact that the United States has not ratified the convention, the most widely endorsed human rights instrument in the world, is puzzling to its allies and damaging to its ability to lead on human rights questions. The ratification effort in the United States has been short-circuited by questionable concerns over whether setting the minimum recruitment age at 18 would compromise national security or limit sovereignty.

Another issue in the Senate centers on the fact that the convention outlaws capital punishment for anyone under 18. That raises the concern that ratification of the convention would limit the ability of states to use capital punishment. But perhaps the biggest obstacle to ratification is simply the lack of public awareness. Most people in the United States do not know about child soldiers, which means there has not been much public discussion about the CRC.

Around the world, nations that are parties to the convention must invoke it, not ignore it. They must point out the massive violations of children's rights that occur as a result of armed conflict. There must be enforcement of its basic provisions for safeguarding the physical, social, and psychological integrity of children and for guaranteeing basic rights such as the right to education.

To succeed, prevention efforts need to work toward structural changes that address poverty and oppression, fundamental sources of armed conflict and of much child soldiering.

Connections must also be built between children's rights, arms transfers, and militarization, issues that the peace community and the world at large have tended to address in a fragmentary manner. Only a holistic approach will succeed in ending child soldiering and building healthy social systems that protect children and orient them toward peace.

THE TUNNEL

While visiting the Grafton Camp in Sierra Leone last year, I watched these former soldiers, these boys, these children, at play. They had been robbed of their childhood, exposed to death and suffering at an early age, and some had been made into killers.

And yet, as I observed and as I talked with the boys' counselors, I acquired a new appreciation for human resilience and potential for change. Some of the boys had once cooperated in killing, but now they cooperated in games such as running "the tunnel."

The boys stood in two lines facing each other, with partners in the line joining hands and raising their arms, creating a tunnel through which the first two boys would run.

When they reached the end, they faced one another and locked their hands in the air, becoming part of the tunnel through which the next pair at the front of the line ran.

A child's game, yes. You probably played it, or something like it, yourself. At Grafton Camp, there was much laughter as the boys ran faster and faster through the tunnel and as the tunnel snaked its way around trees and through gardens.

On another level, however, it was a serious game with a psychological dimension. The tunnel existed only through cooperation, the joined hands symbolizing human interconnectedness. The game required trust, because the boys forming the tunnel could have easily collapsed the tunnel, tripping the runners. Or they could have harassed the runners in myriad ways.

However modestly, the game was rebuilding the fabric of trust that the war had ripped apart. While no single game could rehabilitate a former child soldier, the camp itself seemed to offer hope that rehabilitation was possible. After all, some of the counselors were themselves former child soldiers who had been demobilized and who were now working to help younger children make the adjustment back to everyday civilian life.

And yet, what is done in a facility like Grafton is only a drop in an ocean of violence. In fact, Grafton Camp itself closed last summer, a victim of the revived violence in Sierra Leone. Some of the child soldiers at Grafton have been "remobilized." The laughter is long gone.

Children in Sierra Leone are being drawn back into the renewed conflict. Much the same is true in other countries locked in cycles of violence. Although the immediate goal is to protect children in areas of armed conflict, the longer-term goal must be to prevent the wars that lead children to the slaughter.

NOTES

[1] Except where noted, figures are from country studies conducted by a variety of nongovernmental and governmental organizations. These studies are available from Rädda Barmen, and are summarized in Rachel Brett and Margaret McCallin, *Children: The Invisible Soldiers* (Vaxjo, Sweden: Swedish Save the Children, 1996).

[2] The official title of the Machel Study is "Report of the Expert of the Secretary-General Graça Machel, on the 'Impact of Armed Conflict on Children' Document A/51/306 & Add 1." It may be ordered from the Public Inquiries Unit, Department of Information, United Nations, New York, NY 10017. Fax: (212) 963-0071.

[3] Samir Quota, Raija Punamäki, and Eyad el-Sarraj, "The Relations Between Traumatic Experiences, Activity, and Cognitive and Emotional Responses Among Palestinian Children," *International Journal of Psychology*, 1995, vol. 30, p. 291.

[4] Guy Goodwin-Gill and Ilene Cohn, *Child Soldiers: The Role of Children in Armed Conflicts* (Oxford: Clarendon, 1994), p. 31.

[5] Ed Cairns, *Children and Political Violence* (Oxford: Blackwell, 1996), p. 114.

[6] Gillian Straker, *Faces in the Revolution* (Cape Town: David Philip, 1992), p. 13.

[7] Michael Wessells, "Assisting Angolan Children Impacted By War: Blending Western and Traditional Approaches to Healing," in *Coordinators Notebook: An International Resource for Early Childhood Development,* vol. 19 (West Springfield, Mass: Consultative Group on Early Childhood Care and Development, 1996), pp. 33–37.

32

Bill Clinton's America: Arms Merchant to the World

LORA LUMPE

At a Capitol Hill press conference in November 1992, a reporter asked President-Elect Clinton what he would do to "stop the sale of arms from this country around the world." Clinton responded: "I expect to review our arms sales policy and to take it up with the other major sellers of the world as part of a long-term effort to reduce the proliferation of weapons of destruction in the hands of people who might use them in very destructive ways."

Two years, several wars and more than $50 billion of U.S. arms sales later, the White House released the results of its review of conventional weapons export policy. Advocates of both arms control and arms exports had worked to influence the content of the 6-page document, released on February 17, 1995. The arms industry won. "It's the most positive statement on defense trade that has been enunciated by any administration," gushed Joel Johnson, one of the weapons industry's chief lobbyists.

Arms controllers' hopes for U.S. leadership to restrict the trade were based on faith rather than reason. During the two years of the policy review, the Clinton team continued—and in many ways accelerated—the Cold War pro-export practices of the Reagan/Bush administrations. In fiscal years 1993 and 1994, the executive branch (and Congress) signed-off on a staggering $100 billion of government and industry-negotiated arms deals. Moreover, the administration actively assisted industry by subsidizing marketing activities, lobbying foreign officials to "buy American," and financing several billions of dollars of sales.

The "new" guidelines call for business as usual: "the United States continues to view transfers of conventional arms as a legitimate instrument of U.S. foreign policy—deserving U.S. government support when they enable us to help friends and allies deter aggression, promote regional stability, and increase interoperability of U.S. forces and allied forces." Instead of restraint, the policy emphasizes openness in exports. Instead of limiting sales and technology on a regional basis, it promotes "responsible" exports: the U.S. will export only to those countries which it favors and discourage exports by others to those it disfavors. Instead of de-commercializing weapons exports, the government will now explicitly consider the impact on the arms industry in deciding whether to approve a sale. Finally, export decisions will continue to be made on a case-by-case basis, meaning export of anything to anyone is possible.

MARKET TRENDS

There are several annual sources of information on the international arms trade. Each report measures something slightly different. These varying data can be confusing; however, all sources seem to agree on two points. First, they show the arms market is shrinking, due almost entirely to the collapse of the Soviet Union and the end of subsidized arms transfers from the former Soviet republics. However, this claim is based on the accuracy of past U.S. government estimates of Soviet arms transfers during the Cold War. If those estimates were exaggerated for political or other reasons—as were estimates of Soviet military expenditure—then comparisons of today's market with that of, say, 1987 are shaky. Moreover, arms sellers have an interest in suggesting that the market is in decline: it implies that the problem of the international arms trade is taking care of itself.

The second point of agreement—this one indisputable—is that since 1990 the United States has overwhelmingly dominated the market. Proponents of sales often claim that the increase in market share is not due to an increase in U.S. sales but simply to a shrinking "pie." This is not true. U.S. dominance is attributable, in roughly equal parts, to bullish American marketing during and since the Iraq war and to Russia's near withdrawal from the market. Since 1990, U.S. sales activity—through both the government-negotiated Foreign Military Sales program and through industry-negotiated sales licensed by the State Department—has spiked.

In a report issued last July, the Congressional Research Service estimated that Third World countries purchased $20.4 billion of arms in 1993. (The report's definition of "Third World" excludes Turkey, Greece, East European countries and all former Soviet republics.)

According to the report, while U.S. Foreign Military Sales agreements increased only slightly from 1992 to 1993, U.S. market share rose from 56% to 73% of all Third World agreements. The CRS report actually understates the magnitude of U.S. sales, since it excludes arms sales negotiated directly by industry but licensed by the government. In 1993 the U.S. sold weapons to over 140 countries. The Project on Demilitarization and Democracy calculated that 90 percent of the U.S. sales went to countries that were either not democracies or that were human rights abusers. Saudi Arabia and Kuwait were the leading U.S. customers in terms of dollar volume.

Meanwhile, non-U.S. suppliers—often cited in the American press as irresponsible merchants of death—made marginal sales by comparison. Russia's sales fell from $11.8 billion in 1990 to $1.8 billion in 1993. Iran, Syria and the United Arab Emirates were Russia's largest customers.

China sold less than $300 million worth of arms in 1993—less than two percent of the market. After peak sales of $5.8 billion in 1987, it fell from the third-ranked seller in 1990 to sixth place in 1993. China was also the third largest arms importer in 1993, buying $1.3 billion of weapons.

At $2.6 billion in sales, the four largest European suppliers (France, Britain, Germany and Italy) together accounted for 13% of all sales made to the Third World in 1993. This is down from $7.5 billion—29% of the market—in 1992.

UN REGISTER
OF CONVENTIONAL ARMS

- On September 1, 1994, the United Nations released its second annual Register of Conventional Arms, containing data on seven categories of arms imports and exports during 1993. The Register was established in 1991, in response to the Iraq war, to help identify "excessive arms build-ups." Eighty-one UN members submitted information for the 1994 report.

- The report demonstrated the U.S. dominance in the arms market in terms of actual equipment deliveries. In 1993 the U.S. delivered nearly 2,400 tanks, 832 armored combat vehicles, nearly 300 artillery pieces and 100 aircraft, 75 attack helicopters, and 2,900 missiles and missile launchers. The U.S. exported ten times as many tanks as the second largest overall exporter, Germany. Russia delivered 120 tanks, 350 armored vehicles, 14 artillery pieces, 33 combat aircraft, one submarine and no missiles.

- Turkey and Greece—which have had very tense relations of late—were the leading importers, with most of their equipment coming from the U.S. or other NATO nations.

BUYERS CALL THE SHOTS

Surplus arms production here and abroad has created a buyers' market, allowing customers to receive sweeter deals. First and foremost, buyers are extracting better price and financing packages from sellers, dramatically reducing the macroeconomic benefits to selling countries.

A second demand is for the technology to produce weapons. Increasingly, manufacturers are granting licenses to recipient countries to produce subcomponents, components, or entire weapons systems. A prime example is the $5.2 billion Korean Fighter Program deal of 1991. In order to make the sale, U.S. industry was willing not only to send manufacturing jobs overseas but also to risk the creation of new competition in the near term. The security risk of helping to establish new weapons industries abroad takes a back seat to pressures to make the sale now.

Buyers are also demanding higher tech weaponry. In the past few years top-of-the-line systems previously off limits (such as American F-15E "Strike Eagle" and Russian Tu-22M "Backfire" bombers, modern European diesel submarines and supersonic, sea-skimming anti-ship missiles) have been placed on the auction block. This, too, is not without obvious risk to the sellers. Military and intelligence officials repeatedly point to the increasing availability and sophistication of conventional arms as a prime threat to U.S. security. The Director of Naval Intelligence, Rear Admiral Edward Shaefer, testified last summer that "the overall technical threat and lethality of arms . . . being exported have never been higher." CIA Director James Woolsey testified on January 10, 1995, that advanced conventional weapons "have the potential to significantly alter military balances, and disrupt U.S. military operations and cause significant U.S. casualties."

A mix of dangerous security strategies, outmoded diplomatic rationales, and false economic calculations conspires to convince U.S.

policymakers that massive levels of arms exports make sense today. Added to the mix is industry's desire for high profits and organized labor's desire to maintain high-paying jobs.

"RATIONALES"
FOR ARMS SALES

Arms exports continue to be used, as during the Cold War, for both stated and unstated strategic reasons. Recipient nations are said to need U.S. arms in order to take responsibility for their own defense. In reality, the U.S. uses exports and joint military exercises to gain access to overseas bases and to establish the infrastructure and interoperability necessary for U. S. intervention.

Interoperability is a hallmark of the doctrine of "coalition warfare," which the U.S. built up during the Cold War to contain communism. Since the fall of the Berlin Wall, the U.S. has intensified and expanded military ties around the world. According to Pentagon planning documents, instead of arming allies against the Soviet bloc, U.S.-led coalitions are now arming against "regional instability" and "uncertainty."

Further, according to the new arms transfer policy, U.S. arms exports will promote regional stability. The policy statement does not specify exactly how weapons will do this, but presumably it refers to either: a) the creation of a balance of power; or b) the build-up of deterrent capabilities of U.S. allies. However, weapons are more likely to undermine peace and security than to maintain them. Moreover, the geopolitical landscape is so volatile that predicting regime stability and the steadfastness of alliances is impossible. Former U.S. allies—and recipients of U.S. weapons and military training—in Panama, Iraq, Somalia and Haiti became foes.

A third strategic rationale cited in support of arms exports is the need to maintain weapons production lines in case of a future war, The recent spate of mergers and acquisitions in the U.S. arms industry has not reduced output significantly. Production lines for many of America's front-line weapons—e.g., F-15 bombers, F-16

fighters, Apache attack helicopters, and M-1A2 tanks—remain open now only for sales abroad. In other cases, the government is approving new production lines solely for export.

Proponents claim that arms sales allow suppliers to gain and maintain "influence" with recipients. Sellers in the past applied conditions—at least in theory—to weapons purchases. In today's market, however, the buyer is more likely to influence the seller than vice versa. Besides this dubious diplomatic rationale, the U.S. government continues to rely on arms transfers as a one-size-fits-all fix for almost any foreign policy situation. Need to "reward" allies for participating in Desert Storm, peacekeeping in Somalia, or enforcing the no-fly-zone in Iraq? Send weapons. Need to seal a peace agreement? Send weapons, and forgive past military debt as well.

ECONOMIC "RATIONALES"

After the Iraq war, it looked briefly as if the international arms trade was going to be held accountable for enabling, if not fomenting, Iraq's aggression. But the arms export lobby in the United States quickly and effectively headed off the backlash by emphasizing the "jobs" factor. However, while production of most major weapons systems is spread strategically across nearly every state and most Congressional districts, relatively few workers are employed through arms production for export. A 1992 Congressional Budget Office report estimated that sizable reductions in U.S. arms exports to the Middle East, America's largest market, would affect less than one-tenth of one percent of the total work force.

But everyone pays a higher Defense Department (DOD) bill because of these exports. Weapons proliferation, instability and warfare in the developing world are used to justify this year's $250 billion Pentagon request (this excludes $40 billion of other military spending). The development and production of next-generation U.S. weapons are justified now on the basis of weapons being acquired by Third World nations, including those which the United States has sold. Lockheed's lobbying campaign for the F-22 fighter is based on the proliferation of very capable fighters, such as the F-15E, F-16C/D and the F/A-18.

Moreover, arms manufacturers receive vast government subsidies. Taxpayers underwrite the research and development of weapons and employ a Pentagon sales force of several thousand people here and abroad. The DOD spends public money to market U.S. weapons at overseas arms bazaars and nearly $5 billion of public money is given away each year to allow allies to pay for U.S. weapons purchases.

In Belarus, Ukraine, Russia and China the Clinton administration has aggressively promoted and assisted the conversion of arms industries to peaceful pursuits. While visiting Beijing in October, Secretary of Defense William Perry said that it was in U.S. interests to "help these countries resist pressure to make weapons even beyond their needs." However, the administration apparently does not consider this advice valid for the U.S. The Clinton administration's conventional arms transfer policy doesn't refer to conversion and downsizing the U.S. arms industry.

CLINTON'S FAILURE

Over 30 wars are raging around the world today, almost all of them being fought with imported weapons. Given its market dominance, it isn't surprising that U.S. weaponry is finding its way into combat in Afghanistan, Angola, Cambodia, Kashmir, and Somalia to name a few.

Lacking the courage to take on weapons corporations and the Pentagon, and the vision to devise new security paradigms, the Clinton administration has failed to seize the opportunity afforded by the end of the Cold War. Rather than seeking to reduce reliance on force—and building up reliance on the rule of law—the White House has ensured not only much more warfare to come but also killing and destruction at much greater levels.

The long-awaited official policy makes plain that any change in U.S. arms export policy must come from the bottom up. No progress will be made on the issue of limiting the global arms trade without significant grassroots pressure.

WHAT YOU CAN DO

- Educate others in your community about the U.S. role in spreading weapons around the globe. Much information—most of it free—is available from the sources listed below. Obtain and share the information. Speak out and write on these issues.

- Oppose the use of federal taxes to underwrite weapons exports and military training. Let your elected officials know that instead of cutting school lunches they should cut arms export subsidies.

- Ask organizations and professional associations with which you are affiliated to join the national Code of Conduct campaign. This coalition of over 200 national and local organizations works to pass more responsible U.S. arms export policies (see below). To become a co-sponsor, or for more information, contact Scott Nathanson at Peace Action Education Fund, (202) 862-9740 ext. 3041.

- Ask your Congresspeople to co-sponsor "The Code of Conduct on Arms Transfers Act of 1995" (H.R.772/S.326). Call Senators at (202) 224-3121 and Representatives at (202) 225-3121. Or write to: Your Representative, House of Representatives, Washington, DC 20515; or Your Senator, U.S. Senate, Washington, DC 20510.

The Code of Conduct would prohibit arms exports to any government that does not meet the criteria set out in the law unless the President exempts a country and Congress passes a law affirming that exemption. The four criteria a country must meet to be eligible for U.S. weapons are:

- democratic government
- respect for human rights of citizens
- non-aggression (against other states)

- full participation in the U.N. Register of Conventional Arms

The Code's criteria are all primary foreign policy tenets given lip service by past and present U.S. administrations. Nevertheless, 90% of the record $14.8 billion in U.S. arms sales to the Third World in 1993 went to states which didn't meet the Code's criteria. While it wouldn't end all objectionable arms sales, the code would raise the level of scrutiny and force a debate on arms exports to those governments.

FOR MORE INFORMATION

British-American Security Information Council (BASIC)—1900 L Street, NW, Suite 401-2, Washington, DC 20036, (202) 785-1266, e-mail: basicusa@igc.apc.org Focuses on multilateral arms export control initiatives.

Center for Defense Information—1500 Massachusetts Avenue, NW, Washington, DC 20005, (800) CDI-3334 or (202) 862-0700, e-mail: cdi@igc.apc.org CDI has a conventional arms transfer project which produces the Defense Monitor and episodes of "America's Defense Monitor," a public television program on arms production/ export issues.

Council for a Livable World Education Fund—110 Maryland Avenue, NE, Washington, DC 20002, (202) 546-0962 Publishes the monthly Arms Trade News.

The Federation of American Scientists Arms Sales Monitoring Project (307 Massachusetts Avenue, NE, Washington, DC 20002, (202) 546-3300, e-mail: llumpe@igc.apc.org Publishes the Arms Sales Monitor, which reports on U.S. government policies on arms exports and weapons proliferation.

The Arms Project of Human Rights Watch—1522 K Street, NW, Suite 910, Washington, DC, 20005, (202) 371-6592 Focuses on arms transfers to human rights abusing regimes.

Peace Action Education Fund—1819 H Street, NW, Suite 660, Washington, DC 20006, (202) 862-9740 ext. 3041   Peace Action Education Fund coordinates the Grassroots Network Against the Arms Trade and assists citizens in lobbying and bringing local attention to arms production and trade issues.

Project on Demilitarization and Democracy—1601 Connecticut Avenue, NW, Suite 300, Washington, DC 20009, (202) 319-7191 PDD publishes occasional reports on the impact of military spending and arms transfers on countries in the developing world.

William Hartung, World Policy Institute—65 Fifth Avenue, Suite 413, New York, NY 10003, (212) 229-5808 In 1994, Hartung published *And Weapons for All* (HarperCollins), an excellent critique of U.S. arms export policies and practices. He will soon publish a study of the use of U.S. weapons in wars around the world.

• • •

33

Bringing the Cost to Light

BLAISE TOBIA AND
VIRGINIA MAKSYMOWICZ

The trillions of dollars that the United States spent to fight the Cold War are usually condensed into images of high-tech aircraft and missiles, or reports of $600 military hammers. But it is easy to forget that many of these expenditures were, and still are, buried away in secret programs and black budgets out of the public view. Others have been camouflaged within remote or unremarkable landscapes: the midwestern farm fields that harbor missile silos, Washington's Hanford Reserve—or Colorado's Rocky Flats, a scrubby stretch of mesa land between Denver and the Rocky Mountains. Most people drive by the unimpressive industrial buildings with hardly a second glance.

For six months during 1990–91, however, artist John Craig Freeman's five garish red billboards helped call significantly more attention to this now-closed nuclear processing facility. Standing near its entrance gate and proclaiming "Today we made a commitment for 250,000 years," they reminded people that this piece of Colorado landscape was thoroughly saturated with plutonium, the radioactive poison that had been produced in Rocky Flats' factories for over 40 years. In fact, the main production building alone was discovered to have more than 62 pounds of plutonium dust scattered throughout its nooks and ventilation systems. (Scientists have estimated that a few micrograms of plutonium breathed in by a person could be enough to cause cancer.) The 250,000 years to which the billboards refer represents the period it will take for this plutonium to decay into a relatively harmless state—quite a long time indeed, Freeman points

out, since the entire history of human farming spans only about 12,000 years.

What did it cost to produce this plutonium in the frenzy of the Cold War? What will it cost to clean up and contain it during the next decade? What will it cost to secure it during the next 250,000 years? What did it cost many unknowing Americans who were experimented upon without their knowledge or consent, in terms of lost health and income, or pain? Similarly, what did it cost the thousands of soldiers who were positioned near atomic-bomb explosions—many fueled by plutonium from Rocky Flats?

These were the kinds of thoughts that motivated artist Freeman to convert a previously blank set of billboards into a potent communication tool. Freeman undertook the entire project on his own, designing the images on a Macintosh computer, and then printing them out—one letter-sized piece of colored paper at a time—on a laserprinter, and assembling them onto the boards that would cover the five 20-by-30-foot billboards.

With support from Greenpeace, Freeman was able to secure leases on the billboards and get the posters put into place. Ironically, a local environmental group opposed the project, arguing that the billboards were ugly and detrimental to their surroundings, even though the message they carried was noncommercial and ecologically crucial. Ultimately, it was this conflict on the political left that attracted tremendous media attention to Freeman's project, bringing its message to an even wider audience than originally had been envisioned.

From Blaise Tobia and Virginia Maksymowicz, "Bringing the Cost to Light," *The Witness*, vol. 77, no. 5, May, 1994, p. 21. Copyright © 1994 by The Witness. Reprinted by permission.

34

Too Much Fun
Toys as Social Problems and the Interpretation of Culture

JOEL BEST

• • •

Toys are a frequent subject of contemporary claims concerning social problems. Rooted in our culture's longstanding ambivalence regarding leisure and its concerns about children's vulnerability, claims about troublesome toys also reflect anxiety about children's increased susceptibility to non-familial influences, their growing access to toys, and an expanded toy industry, as well as an active social movement sector. Typically, these claims argue that toys represent undesirable values, and that children who play with the toys acquire those values. Parallel arguments may be found in claims regarding other forms of popular and material culture. Interactionists should be wary of making or accepting these claims, because rather than treating children's play as a topic for empirical study, such claims locate meaning in objects, rather than actors.

Among sociologists, symbolic interactionists have been especially interested in children's play. Our founders—Dewey, Cooley, and Mead—as well as such modern scholars as Gregory Stone (1970), Donald Ball (1967), Norman Denzin (1977), and Gary Alan Fine (1987) emphasize play's importance in developing the self and the child's competence as a social actor. At the same time, sociologists worry about play's role in reproducing society's troubling features, such as gender inequality (Thorne 1993).

These days, suspicions about play's dark side often focus on children's playthings—toys. Toys, we are warned, can turn good children into bad actors. These claims—some made by activists, some by sociologists—define toys as social prob-

lems (Spector and Kitsuse 1977). Sociologists of a liberal bent tend to sympathize with claims about the damaging effects of *hazardous toys* (Swartz 1971, 1986), *war toys* (Andreas 1969; Carlson-Paige and Levin 1987, 1990; Regan 1994; Engelhardt 1995), *sexist toys* (Pogrebin 1980; Cunningham 1993; Rand 1995; Varney 1996), or *racist toys* (Wilkinson 1974, 1987). As a result, these claims get a free ride, rarely receiving critical attention from sociologists.

Another recent set of claims offers a way to gain some analytic distance on troubling toys. Conservative Christians worry about *occult toys* "based on some of the very ideas, namely witchcraft, idolatry, emulations, and murders, that God warns against" (Phillips 1986, p. 20; cf. Marrs 1989; Michaelsen 1989; Phillips 1991; Pulling 1989; Robie 1989, 1991). In their view, toys and other popular culture that invoke magic and other (non-Christian) supernatural elements endanger children: "[*Masters of the Universe*] is based on occult practices: magic, sorcery, witchcraft, necromancy, etc. . . . [and] blasphemes God by placing a mortal on the same plane as He" (Phillips 1986, p. 60). This indictment encompasses not only the sword and sorcery imagery of *Masters of the Universe* and *Dungeons and Dragons,* but also the seemingly innocent worlds of *Smurfs* and *Care Bears:*

> On the surface, *Care Bears* teach the children to express their feelings, especially of love, to others. At first, these sound like very good

From Joel Best, "Too Much Fun: Toys as Social Problems and the Interpretation of Culture," *Symbolic Interaction,* vol. 21, no. 2, 1998, pp. 197–212. Copyright © 1998 by Elsevier Science. Reprinted with permission.

ideas, but, they are Humanistic principles, which are in contradiction to God's teachings. Magic and Eastern religious ideals are also prevalent. . . . (Phillips 1986, p. 60)

While these claims are probably unfamiliar to most sociologists, the argument against occult toys parallels more familiar warnings about war toys and sexist toys. Claims about the danger posed by occult toys, then, offer a bit of analytic leverage, a way to help us think more critically about claims that we otherwise tend to accept without question.

In particular, I want to address three topics in this article: first, I ask why toys have become a frequent subject for social problems claims; second, I criticize the content of those claims; and third, I consider the implications of this critique for other sociological interpretations of culture.

WHY WORRY ABOUT TOYS?

Toys stand at the intersection of two themes often featured in American social problems claims. First, our culture's deep-seated ambivalence about play and leisure fosters concerns about *morally controversial leisure;* virtually all sports, popular culture, and other forms of recreation have come under attack at one time or another (Olmsted 1988). Second, toys are used by *vulnerable children* who are the designated beneficiaries of many campaigns to protect the young (Best 1990). These elements reinforce one another: children's exposure to controversial leisure is seen as more troubling than adults' involvement. Thus, while not all toys become targets of moral outrage, a readily available set of cultural resources— our assumptions about questionable leisure and vulnerable children—makes it easy to construct claims about problematic toys.

• • •

[F]our recent trends have inspired increased concern about younger children's leisure, and toys in particular.

First, while our image of the young child nestled within the traditional family may exaggerate how well children were protected from the larger society, children's extra-familial contacts are increasing. In addition to school and church, more young children find themselves spending time outside the home in daycare, preschool, and various organized activities (e.g., sports teams, music lessons, and so on [Adler and Adler 1994]). Moreover, outside influences increasingly intrude into children's homes, particularly via television programming and commercials aimed at children. Thus, young children have more exposure to non-familial organizations and institutions.

These developments reflect forces that both push and pull children toward outside institutions. The principal force pushing children outside the family circle is the growing proportion of mothers in the labor force; fewer families have adults at home throughout the day, available to care for children. At the same time, beliefs that childhood will be enriched by exposure to outside influences pulls many parents to encourage children's extra-familial activities, even as uneasiness about children's increased contact with outsiders fosters fears of abduction, abuse, and other threats to children (Best 1990). And, because outsiders produce and distribute most toys, concerns about children's extra-familial contacts are easily channeled into claims about toys.

Second, today's children own, on average, more toys than children in earlier generations. Ownership of toys, like other consumer goods, spread with growing affluence. At the same time, technological innovations, particularly the development of plastics, fostered the mass production of low-cost toys; and the culture of consumption, by celebrating the purchase and ownership of material goods, justifies buying even more toys (Butsch 1984). These developments are not new; they have continued throughout most of the twentieth century.

Because family members giving gifts account for a large share of toy purchases, shifts in family organization also have affected toy ownership (Owen 1986; Stern and Schoenhaus 1990; Sutton-Smith 1986). Delaying parenthood means that children tend to enter more affluent families, with higher disposable incomes. As the average number of children per family falls, a greater proportion of children are first-born (toy expenditures on first-borns are markedly higher than for their successors). Longer life spans and greater affluence among the elderly mean that there are more grandparents buying gifts for grandchildren. And increased divorce and remarriage increase the pool of adults—e.g., step-parents and step-grandparents—inclined to give toys as gifts. Each of these changes serves to increase toy ownership; together, their effect has been marked. While toy ownership varies by social class, it seems clear that today's children in a given class own more toys than their counterparts in past generations.

Third, transformations in the toy industry, particularly since the Second World War, have altered how children acquire toys (Owen 1986; Stern and Schoenhaus 1990). Toy-buying has become less seasonal; large retail toystore chains that sell toys year-round have largely replaced the department store's toy section, expanded each winter for the holiday buying season. Increasingly, toy manufacturers develop lines of products, so that the purchase of one toy from the line encourages further purchases (e.g., buying a Barbie doll leads to buying additional outfits, dolls, playsets, and so on). Toy packaging routinely includes advertising for other products in the line. Thanks to children's programming on commercial television, toy manufacturers have ample opportunity to reach children via advertising throughout the year. Cable television, with several networks targeting the child audience for much of the day, has increased both children's programming and commercials aimed at children. The deregulation of children's programming by the FCC in 1983 led to the emergence of television cartoon series featuring characters that represent lines of toys (e.g., *Smurfs, G.I. Joe,* and *Teenage Mutant Ninja Turtles* [Kunkel 1988]). Critics view these series as toy commercials disguised as entertainment, especially suited for expandable product lines, so that each televised episode can feature different characters or equipment that can, in turn, be marketed as new toy "action figures," vehicles, and so on.

Exposure to the toy market increases as children grow older. There are virtually distinct toy markets for younger and older children. Toys for infants, toddlers, and even preschoolers are marketed to parents through advertising emphasizing the toys' educational benefits (Nelson-Rowe 1994; Stern and Schoenhaus 1990; Seiter 1993). In contrast, toys are marketed directly to older children, in hopes they will either make their wishes known to adults, or spend their own disposable income: "children under eleven spend an estimated $5 billion annually on themselves, and influence another $50 billion in family purchases" (Stern and Schoenhaus 1990, p. 38). Advertising aimed at children, of course, stresses the satisfactions of owning a particular toy.

This more aggressive marketing reflects long-term social changes. The nineteenth century saw the emergence of specialized toy manufacturers, as well as other developments, such as railroad networks and mail-order catalogs, that fostered toy marketing (Mergen 1992; West 1984). Early reformers worried about these changes, noting that "toy shops [are] in fierce competition to get the latest thing to tempt the unwary . . ." (Pratt 1911, p. 893). These trends continue. Their net effect is that children increasingly consume toys that have been designed, manufactured, distributed, and advertised in a highly commercial marketplace.

Finally, the last third of the twentieth century has featured an active social movement sector, and social movements often make claims about toys. A few social movement organizations (e.g., World Against Toys Causing Harm [WATCH]) focus exclusively on toys; others, such as Action for Children's Television (ACT), make toys a central concern (e.g., ACT's criticisms of toy advertising, the links between television programming and the toy industry, and interactive toys). But even movements that seem only tangentially related to toys often get involved. Social movements critical of particular cultural values warn that impressionable children acquire their values through socialization, and that toys may impart undesirable values (Andreas 1969). Examples of movements offering such critiques include the consumer movement (hazardous toys), the women's movement (sexist toys), the civil rights movement (racist toys), the peace movement (war toys), the vegetarian movement (criticizing "toy replicas of meat foods" [Delahoyde and Despenich 1994]), and conservative Protestantism (occult toys). Movement rhetoric emphasizes children's vulnerability. Thus, the consumer movement attacks the toy industry both for producing toys that endanger children's safety, and for marketing toys through television advertising that children are too young to evaluate. While not all social movements make claims about toys, the rhetorical usefulness of children's innocence and vulnerability, coupled with the link between toys and childhood, make toys attractive targets for several movements' claims.

These developments—children's increased contact with non-familial influences and their growing access to toys, an expanded toy industry, and an active social movement sector—feed into Americans' traditional ambivalence about children's play. In turn, this ambivalence becomes the basis for contemporary claims about troublesome toys.

WHAT'S WRONG WITH TOYS?

• • •

Consider the various critiques of Barbie dolls. An extraordinarily successful commercial toy, Barbie has been the target of social problems claims since her introduction in 1959. Early reactions worried about girls shifting their affections from infant to teenaged dolls: "The child has no innocent companion in its playpen. . . . The effect of predominantly vulgar toys cannot fail to be what Collingwood aptly called the 'corruption of consciousness'" (Langer 1966, p. 48). Barbie dolls might "make a generation of vipers that will cause men to plead for the return of momism" (*Nation* 1964, p. 407) and make girls "less able to achieve the emotional preparation for being a wife and mother that they received from baby dolls" (Winick 1968, p. 208); "[Barbie] may serve to integrate children prematurely into the adolescent subculture and minimize preparation for later adult performances such as those associated with conjugal, parental, and occupational status or position" (Ball 1967, p. 457); or,

. . . will [girls], like Barbie, resist the responsibility of having children, or, following Barbie's lead even more completely, resist the responsibility of marriage and family altogether? . . . Will [they] become . . . frustrated cynics if their private Barbie fantasies do not come true? (Cox 1977, p. 306)

The emerging women's movement shifted the focus of anti-Barbie claims. Formerly indicted for steering girls away from marriage and motherhood, the dolls now came under attack for reinforcing traditional gender stereotypes. Barbie was "empty-headed," with "crushingly limited life choices" that emphasized "passivity" (Maguire 1987, p. 20); she had a "fashion-victim personality" (Jackson 1991, p. 35); she set unattainable standards for beauty (Cunningham 1993; Robie 1989; Urla and Swedlund 1995). Barbie taught girls "the skills by which their future success will be measured: purchase of the proper high-status goods, popularity with their peers, creation of the correct personal appearance, and the visible achievement of 'fun' through appropriate leisure activities" (Motz 1983, p. 122).

Still other interpretations focused on Barbie's sexuality—or denial of sexuality: "What might at first appear to be a sexually suggestive doll . . . is in actuality a model of self-control, using her sexuality to attract men while ensuring that her relationships would remain safely platonic. Far from being a sex goddess, Barbie was the epitome of the unapproachable woman" (Motz 1983, p. 129). Yet,

Barbie and Ken remind us of the nuclear family and the waning age of . . . buying and selling sexual power points like legs and busts. Their continuing popularity indicates that many men and women still see sex as an instrument of power in an industrial and consumer context. But whereas the Barbie Doll used to represent the management of the male audience for marriage, she now personifies the career woman manipulating it for business success. (Berg 1986, pp. 209–10)

Erica Rand, whose *Barbie's Queer Accessories* offers the most subtle analysis of Barbie's meaning, summarizes several contemporary indictments: "Mattel [Barbie's manufacturer] promotes compulsory heterosexuality by making it look like the most natural and attractive choice; it promotes capitalism and the unequal distribution of resources by glamourizing a character with a huge amount of apparently unearned disposable cash and, to understate grossly, a disproportionate amount of luxury items; it promotes ageism and sexism by suggesting that a beautiful young body is a woman's most valuable commodity; it promotes racism by making 'white' Barbie the standard . . ." (Rand 1995, p. 9; cf. Lord 1994).

While most analysts have associated Barbie dolls with undesirable values, the dolls have some defenders. In this view, Barbie is ". . . a truly revolutionary doll, one that inspired, rather than oppressed, young girls' imaginations; . . . the one toy that, since the mid-sixties, has epitomized for little girls the term 'having it all'" (Bracuk 1991, pp. 37–38). Playing with Barbie becomes a form of resistance:

> . . . girls play with their uneducational Barbies as they always have, playing out the "mean babysitter" scenario, madly acting away, with no parent-pleasing values to inhibit their stories. . . . it is her lack of redeeming social value that helps keep her true to the child's sense of play, instead of the parents' worst fears. (Jackson 1991, p. 35)

According to various interpretations, then, Barbie promotes or discourages traditional gender roles, fosters or inhibits sexuality, discourages or encourages girls' independence, establishes or subverts racism and heterosexism, diminishes the importance of religion, celebrates capitalism and consumerism, causes anorexia, and fosters resistance. Barbie, her critics agree, is influential. But how does she exert her influence?

Presumably the children—mostly girls, of course—who play with Barbie glean meanings from the doll. Because Barbie is slender, claimsmakers argue, she therefore teaches girls that being slender is desirable. But how does this teaching occur? There are at least two problems here. The first is that the process by which Barbie influences girls is not spelled out. Do girls define Barbie in positive terms, and therefore decide they want to be thin like Barbie? Or does Barbie somehow constrain their freedom of thought, so that they cannot imagine alternatives to being like Barbie? Or do they come to feel that society demands that they be like Barbie? And how do they come to focus on thinness? Barbie is smiling; why don't girls focus on happiness instead of slenderness as the desired quality? Why not blame Barbie's smile for cosmetic orthodontia, depression, and teen suicide? Barbie's critics rarely bother to explain just how she exerts her influence, how girls glean particular meanings from playing with Barbie. Second, in celebrating slenderness, Barbie hardly stands alone. Many aspects of contemporary culture present slenderness as an ideal for women: the fashion industry, movies and television, advertising, beauty pageants, the fitness and diet industries, pornography, and who knows what else. Even if girls who play with Barbie—and most girls do—grow up to value slenderness, how are we to weigh Barbie's relative influence in this process? And what about boys? Far fewer boys play much with Barbie, yet males, no less than females, grow to value female slenderness. If Barbie play is so important in developing a concern with women being thin, shouldn't men care less?

Most claims about the corrupting influence of toys are assertions. Not only is the process by which toys exert their influence not detailed, but claimsmakers offer virtually no direct evidence that the process even occurs. So far as I know, we have *no* research based on either extended observations of, or intensive interviews with, girls who play with Barbie. Rand (1995), who talked to numerous adult "Barbie survivors" and who is sensitive to the methodological difficulties of talking to children about their toys, relies heavily on her occasional conversations with one girl over a few years. It is, of course, easy to construct claims about the dangerous effects of toys if little or no evidence is required.

INTERPRETING CULTURE

These, then, are the characteristic moves in identifying problematic toys: (1) identify some type of toy; then argue that (2) these toys somehow represent undesirable values, (3) that children who play with these toys acquire those values, (4) which, in turn, leads to undesirable consequences. The toys become "texts" to be "read," their meaning and consequences interpreted by claimsmakers. Note that critics make parallel arguments when they identify rap music, romance novels, pornography, and other cultural objects as social problems.

Often, these arguments are tautological, and therefore incontrovertible. The circularity is barefaced: Occult toys promote occult beliefs. What is an occult toy? One that promotes occult beliefs. Of course, one of the principal comforts of tautology is the secure knowledge that one is right—or at least cannot be proven wrong. Easily made, arresting, and incontrovertible, claims about the dangers of Barbie and other problematic cultural objects have proliferated, even within sociology.

It's not that we don't know better. The sociology of culture has become a productive field, and there are fine ethnographies, often by interactionists, examining the production of culture, particularly popular culture. These studies tend to focus on people in the industry, revealing that creators need to get their works produced, producers must get their products distributed, and distributors have to attract consumers. Studies of popular culture consumers, however, are less common and greatly outnumbered by armchair interpretative analyses that claim to explain what popular culture means to consumers, why they are attracted to it, and how it affects them. Often, these interpretations are critical: they worry that popular culture is crude, debasing, manipulative, or otherwise harmful; and the analysts become claimsmakers, presenting popular culture as a social problem. The literature on the threats posed by Barbie dolls is merely one example of this interpretive turn.

Yet the rare ethnographies of popular culture consumers reveal that analysts' interpretations can be wildly wrong. Janice Radway's *Reading the Romance* (1984) offers an especially nice example.

Her interviews with readers of romance novels show that they enjoy the books for complex reasons that have almost nothing to do with the motives the books' critics imagine readers to have (cf. Halle 1992). The lesson should be clear: instead of trying to discover the harmful meanings hidden in cultural texts, sociologists ought to focus on what we know how to do—watching people and talking to them. When we do this, many of the flaws in claims about troublesome toys become apparent.

Both observations of children's play and surveys asking children about their play preferences discover that children spend—and prefer to spend—most of their play time outside, running around (Mergen 1982, 1992; Sutton-Smith 1986). Favorite toys are things like balls, and favored activities tend to involve lots of running, chasing, and the like. Play with Barbie, GI Joe, and the other commercialized toys that worry claimsmakers tends to rank somewhere down the list.

Moreover, when children do play with toys, they seem to spend much of their time in what analysts may mistakenly consider peripheral activities. For example, my casual observations of my sons' play—and my recollections of my own play as a boy—suggest the importance of what we might call "foreplay"—making preparations to play. One reason that Barbie and G.I. Joe are so popular may be that they come with lots of little accessories—costumes or weapons or whatever. Children seem to spend a very large share of their play time with these toys setting things up, getting organized, getting Barbie or Joe ready to do whatever is going to be done. One might imagine that such foreplay has at least as much to do with learning to be in control of complex situations, planning ahead, and organizing the details, as it does with replicating sexism or militarism.

Even when children actually play with troubling toys, their play may have complicated, contradictory meanings that derive not just from whatever understandings children glean from the toys, but from the actors and the situation:

> The nature of the toy alone cannot tell us whether the player will use it largely to

mimic nature or largely to parody it or both. . . . The toy itself cannot tell us, but our knowledge of play shows us that play's excitements derive from both following the cultural rules for behavior and defying the cultural rules. (Sutton-Smith 1986, p. 251)

But claims about problematic toys ignore this complexity, implying that the nature of toys somehow constrains how children can play with them. Again, analysts interpret what toys mean, often in narrow, even superficial terms. Critics of video games, for example, worry that the stories they depict celebrate violence (e.g., kung fu fighting), sexism (e.g., rescuing the Princess), and so on. But I have often come across four or five boys in our family room, clustered around a video game, one or two playing, while the others kibitz. Their conversation is impressively practical, less along the lines of "Kill, kill!" than "When that happens, you have to press the right arrow and hit the 'B' button twice." Yet critics warn that video games are antisocial and unimaginative:

> The games do little or nothing to help the child develop an inner culture, a sense of self, an awareness that while the world provides challenges and problems, resourcefulness and the use of one's imagination and knowledge of self are an important part of being able to confront those challenges. . . . Compared to the worlds of imagination provided by play with dolls and blocks, [video] games . . . ultimately represent impoverished cultural and sensory environments for the child. (Provenzo 1991, pp. 95–97)

But, when sociologists interview players—or play themselves—they tend to offer more appreciative interpretations: after all, video games may also teach poise, mastery of complex sequences of actions, coping with disappointment and frustration, interacting with computers, and the like (cf. Sudnow 1983; Turkle 1984).

The point is that these are—or at least can be framed as—empirical questions: What do children actually do when they play? What do they think they are doing? Why do they think they are doing it? How do they explain their play to themselves, to their playmates, or to others? What kinds of play do they say they prefer, and why? Are particular forms of play correlated with acquiring particular skills, attitudes, or values, or with performing particular behaviors? Treating these as research questions, devising convincing ways to address them, and actually carrying out the research might well be demanding, boring, and probably frustrating. It is, of course, far easier to "read" the "text" of toys and make pronouncements: This is why Barbie is bad. Such armchair critics may hope to probe the dangers of children's "deep play" with toys, but the result is shallow analysis.

Interactionists like to eschew Grand Theory, but, like other academics, we are not immune to grandiose theorizing. Although our central tenet is that meaning emerges through interaction, we increasingly seem to forget the actors, instead becoming interpreters of culture, deconstructing movies, reading the text of fast food, or discovering dangers in troublesome toys. When we do this, we adopt the hypodermic model of media effects, implying that meaning resides in toys and other artifacts, only to be injected into the user.

• • •

[M]aterial culture's meanings must evolve through social interaction, and, if sociologists hope to understand those meanings, they cannot merely read the object as a text. Such readings, whether by entrepreneurs projecting a market for a new product, intellectuals envisioning future impacts of innovations, or claimsmakers denouncing troublesome toys, ignore the people who construct the object's meanings. It is social actors—ranging from large institutions to individual consumers—who define the uses and effects of airplanes and VCRs and toys. In the case of toys, those actors include manufacturers and distributors, and the people who select and purchase toys, but especially the children who play with them. Toys do not embody violence or sexism or occult meanings. People must assign toys their meanings. Thus, children choose their toys and how they will play with them (a lesson understood by all those parents who refuse to purchase toy guns only to find their children using sticks, fingers, or pieces of toast to shoot their playmates). And, less obviously, claimsmakers discover the dangers lurking in some sorts of play.

By now, it may be clear that my title has a double meaning: claimsmakers worry that some toys encourage too much of the wrong kind of fun, while I have argued that such claims—easily made, yet rarely proven—are a sort of analytic excess—too much academic fun, if you will. Interactionists should know all this perfectly well. We should be fully aware that studying material culture means studying people—not objects. We should be watching children playing with toys, talking to children about their play, or—as I have tried to do here—at least examining how claimsmakers turn toys into social problems, rather than trying to read the meanings embedded in these, or any other, cultural artifacts. Studying people is something we know how to do, something we do well. Culture emerges in actions, not artifacts, and we ought to be where the action is.

Acknowledgment: Kathleen Lowney and Tracy Thibodeau made helpful comments on an early version of this article.

• • •

REFERENCES

Adler, Patricia A., and Peter Adler. 1994. "Social Reproduction and the Corporate Other: The Institutionalization of Afterschool Activities." *Sociological Quarterly* 35:309–28.

Andreas, Carol. 1969. "War Toys and the Peace Movement." *Journal of Social Issues* 25 (1):83–99.

Ball, Donald W. 1967. "Toward a Sociology of Toys: Inanimate Objects, Socialization, and the Demography of the Doll World." *Sociological Quarterly* 8:447–58.

Berg, Dreyer. 1986. "'Transformers,' Barbie Dolls and the Cabbage Patch Kids: Toys, Technology, and Human Identity." *Etc.* 43:207–11.

Best, Joel. 1990. *Threatened Children.* Chicago: University of Chicago Press.

Bracuk, Diane. 1991. "Some Like Her Hot." *Parenting* 5 (August):37–38.

Butsch, Richard. 1984. "The Commodification of Leisure: The Case of the Model Airplane Hobby and Industry." *Qualitative Sociology* 7:217–35.

Carlson-Page, Nancy, and Diane E. Levin. 1987. *The War Play Dilemma: Balancing Needs and Values in the Early Childhood Classroom.* New York: Teachers College Press.

Cox, Don Richard. 1977. "Barbie and Her Playmates." *Journal of Popular Culture* 11 (Fall):303–07.

Cunningham, Kamy. 1993. "Barbie Doll Culture and the American Waistland." *Symbolic Interaction* 16:79–83.

Delahoyde, Michael, and Susan C. Despenich. 1994. "Creating Meat-Eaters: The Child as Advertising Target." *Journal of Popular Culture* 28 (Summer):135–49.

Denzin, Norman K. 1977. *Childhood Socialization.* San Francisco: Jossey-Bass.

Engelhardt, Tom. 1986. "The Shortcake Strategy." Pp. 68–110 in *Watching Television,* edited by Todd Gitlin. New York: Pantheon.

Fine, Gary Alan. 1987. *With the Boys: Little League Baseball and Preadolescent Culture.* Chicago: University of Chicago Press.

Halle, David. 1992. "The Audience for Abstract Art: Class, Culture, and Power." Pp. 131–51 in *Cultivating Differences: Symbolic Boundaries and the Making of Inequality,* edited by Michele Lamont and Marcel Fournier. Chicago: University of Chicago Press.

Jackson, Marni. 1991. "Gals and Dolls: The Moral Value of 'Bad' Toys." *This Magazine* 24:32–33, 35.

Kunkel, Dale. 1988. "From a Raised Eyebrow to a Turned Back: The FCC and Children's Product-Related Programming." *Journal of Communication* 38:90–108.

Langer, Susanne K. 1966. "The Social Influence of Design." Pp. 35–50 in *Who Designs America?,* edited by Laurence B. Holland. Garden City, NY: Anchor.

Lord, M. G. 1994. *Forever Barbie: The Unauthorized Biography of a Real Doll.* New York: William Morrow.

Maguire, Laurie. 1987. "Barbie's Changing Face." *New Society* 18:20–21.

Marrs, Texe. 1989. *Ravaged by the New Age: Satan's Plan to Destroy Our Kids.* Austin, TX: Living Truth Publishers.

Mergen, Bernard. 1982. *Play and Playthings: A Reference Guide.* Westport, CT: Greenwood.

Michaelsen, Johanna. 1989. *Like Lambs to the Slaughter.* Eugene, OR: Harvest House.

Motz, Marilyn Ferris. 1983. "'I Want to Be a Barbie Doll When I Grow Up': The Cultural Significance of the Barbie Doll." Pp. 122–36 in *The Popular Culture Reader,* edited by Christopher D. Geist and Jack Nachbar. Bowling Green, OH: Bowling Green University Popular Press.

Nation. 1964. "The Barbie-Doll Set," 198:407.

Nelson-Rowe, Shan. 1994. "Ritual, Magic, and Educational Toys: Symbolic Aspects of Toy Selection." Pp. 117–31 in *Troubling Children,* edited by Joel Best. Hawthorne, NY: Aldine de Gruyter.

Olmstead, A. D. 1988. "Morally Controversial Leisure: The Social World of Gun Collectors." *Symbolic Interaction* 11:277–87.

Owen, David. 1986. "Where Toys Come From." *Atlantic* 258 (October):65–78.

Phillips, Phil. 1986. *Turmoil in the Toy Box.* Lancaster, PA: Starburst.

___, 1991. *Saturday Morning Mind Control.* Nashville, TN: Oliver-Nelson.

Pogrebin, Letty Cottin. 1980. *Growing Up Free: Raising Your Child in the 80's.* New York: McGraw-Hill.

Pratt, C. L. 1911. "Toys: A Usurped Educational Field." *Survey* 26:893–95.

Provenzo, Eugene F., Jr. 1991. *Video Kids: Making Sense of Nintendo.* Cambridge: Harvard University Press.

Pulling, Pat. 1989. *The Devil's Web: Who Is Stalking Your Children for Satan?* Lafayette, LA: Huntington House.

Radway, Janice A. 1984. *Reading the Romance: Women, Patriarchy, and Popular Literature.* Chapel Hill: University of North Carolina Press.

Rand, Erica. 1995. *Barbie's Queer Accessories.* Durham, NC: Duke University Press.

Regan, Patrick M. 1994. "War Toys, War Movies, and the Militarization of the United States, 1900–85." *Journal of Peace Research* 31: 45–58.

Robie, Joan Hake. 1989. *Turmoil in the Toy Box II.* Lancaster, PA: Starburst.

Seiter, Ellen. 1993. *Sold Separately: Parents and Children in Consumer Culture.* New Brunswick, NJ: Rutgers University Press.

Spector, Malcolm, and John I. Kitsuse. 1977. *Constructing Social Problems.* Menlo Park, CA: Cummings.

Stern, Sydney Ladensohn, and Tod Schoenhaus. 1990. *Toyland: The High-Stakes Game of the Toy Industry.* Chicago: Contemporary Books.

Stone, Gregory P. 1970. "The Play of Little Children." Pp. 545–53 in *Social Psychology Through Symbolic Interaction,* edited by Gregory P. Stone and Harvey A. Farberman. Waltham, MA: Ginn-Blaisdell.

Sudnow, David. 1983. *Pilgrim in the Microworld: Eye, Mind, and the Essense of Video Skill.* New York: Warner.

Sutton-Smith, Brian. 1986. *Toys as Culture.* New York: Gardner Press.

Swartz, Edward M. 1971. *Toys That Don't Care.* Boston: Gambit.

Thorne, Barrie. 1993. *Gender Play: Girls and Boys in School.* New Brunswick, NJ: Rutgers University Press.

Turkle, Sherry. 1984. *The Second Self: Computers and the Human Spirit.* New York: Simon and Schuster.

Urla, Jacqueline, and Alan C. Swedlund. 1995. "The Anthropometry of Barbie: Unsettling Ideals of the Feminine Body in Popular Culture." Pp. 277–313 in *Deviant Bodies: Critical Perspectives on Difference in Science and Popular Culture,* edited by Jennifer Terry and Jacqueline Urla. Bloomington: Indiana University Press.

Varney, Wendy. 1996. "The Briar Around the Strawberry Patch: Toys, Women, and Food." *Women's Studies International Forum* 19: 267–76.

West, Mark Irwin. 1984. "Nineteenth-Century Toys and Their Role in the Socialization of Imagination." *Journal of Popular Culture* 17 (Spring): 107–15.

Wilkinson, Doris Y. 1974. "Racial Socialization through Children's Toys: A Sociohistorical Examination." *Journal of Black Studies* 5:96–109.

___. 1987. "The Doll Exhibit: A Psycho-Cultural Analysis of Black Female Role Stereotypes." *Journal of Popular Culture* 21:19–29.

Winick, Charles. 1968. *The New People: Desexualization in American Life.* New York: Pegasus.

8

The Global Media

*The media represent the same interests that control
the state and private economy, and it is therefore
not very surprising to find that they generally act to
confine public discussion and understanding to the
needs of the powerful and privileged.*

NOAM CHOMSKY

The mass media include a variety of forms of mass communication: television, film, radio, newspapers, magazines, books, and the newest form—the Internet. These media were once nationally (or even regionally or locally) based, but today their ownership, employees, audiences, and funding have become increasingly global. The idea that the mass media are a global social problem is a relatively new one. The media were once regarded in a more positive light. It was only a few decades ago that Marshall McLuhan popularized the optimistic term *global village* to describe one world tied together by new communications technologies. Today, we are just as likely to encounter a negative analysis of the media—witness the titles of just a few of the books published in the 1980s and 1990s: *Manufacturing Consent* (Noam Chomsky), *Inventing Reality* (Michael Parenti), *The Media Monopoly* (Ben Bagdikian), and *The More You Watch the Less You Know* (Danny Schechter).

There are two aspects to the media as a global social problem. The first is a concern about the *content* of the media. For example, for most of us, the media are our primary source of information about the world beyond the boundaries of our direct experience. And as the world becomes increasingly interdependent, more and more of us are seeking to gain accurate knowledge about the world so that we can influence

our fate within it. To the extent that the media present a distorted or biased view of the world, the media are a serious problem. The second is a concern about *control* and *accessibility* of the media. For example, as more and more of the world's peoples become active in efforts to improve global conditions, they seek to have a voice in the media in order to communicate their points of view and contact potential allies. To the extent that the media are inaccessible to those who lack money and power, the media are a serious problem. The Internet, with its extraordinary accessibility—dramatized when the Chinese students at Tiananmen Square were able to communicate with supporters globally despite the Chinese government's tight grip on television and the print media—is a source of excitement and hope for those who have hitherto lacked media access. Many people working for change note with alarm that efforts are now growing to control or shape access to the Internet—for example, by corporations seeking profits, and by governments seeking to curb dissent or censor politically unpopular content.

If we view the media from the order perspective, the media function in a variety of ways to meet positive and important social needs. For example, the media inform us of world events, and they entertain. The media reinforce and reflect cultural values and, in a democratic society, give voice to a variety of competing interest groups and thereby function as a check against the rise of authoritarian government. On a global level, the media can function as an agent of modernization, disseminating modern cultural values and beliefs to traditional societies. From this perspective, the media are problematic if they become dysfunctional—for example, if they distort the news to reflect the views of the political right or the left, or if they reflect traditional rather than modern views. Critics viewing the media from this perspective tend to look optimistically toward the media and democratic governments to take action to correct such dysfunctions. If we view the media from a social conflict perspective, the problems of the media appear more severe and intractable. From this perspective, the media, with the exception of the Internet, are "top down" institutions controlled by large capitalist corporations—in fact, the media *are* corporations. Because the media are capitalist corporations, they seek to shape public opinion in their own interests, which are in direct conflict with those of workers and consumers. The media also tend to support conservative political views and reflect the policies of the governments in power, which in turn usually mirror corporate interests. On a global level, the media are controlled by the corporations of the core, which thus puts them into conflict with the populations in the periphery and semi-periphery. The term *cultural imperialism* has often been applied to the global media, along with other cultural influences from core to periphery such as Western educational systems and scientific philosophies. From the conflict perspective, media problems will probably continue to exist unless the capitalist world economy is fundamentally restructured. If we view the media from a symbolic interactionist perspective, the media are extremely important in the construction of social problems. For example, daily newspapers and television news programs make decisions about what constitutes "news," selecting a limited number of stories from a multitude of possibilities. Thus they set the agenda as to which problems should be seen as important. They also shape public definitions of the problem by giving voice to certain perspectives, while omitting others.

A number of articles in this chapter explore the issues of media content and accessibility from different perspectives. Using the symbolic interactionist perspective, Kathleen S. Lowney and

Joel Best analyze the impact of the Internet on the social construction of social problems in "What Waco Stood For." Lowney and Best researched what ordinary people thought about the U.S. government's actions against the Branch Davidians in Waco in 1993—in comparison to the claims made by government officials, the mainstream press, and survivors of the tragedy. By studying "joke cycles" on the Internet, they provide fresh insights into the uses of this new medium of communication, and the part ordinary people play in constructing social problems. Other articles in this chapter analyze the content and accessibility of the mainstream media from the conflict perspective. In "The Myth of a Liberal Media," Michael Parenti discusses the impact of corporate ownership on media content, debunking the idea that the media contain a liberal bias, and arguing that their bias is in fact conservative. In "Out of Africa!" Ezekiel Makunike examines Western media stereotypes about Africa that narrow our ability to understand this important part of the world. Stewart M. Hoover further analyzes the power structure of the global media in "All Power to the Conglomerate." Carl Wilson, in "Northern Exposure: Canada Fights Cultural Dumping," discusses the situation of Canadian media and artists seeking to preserve Canadian culture in the face of the cultural imperialism of U.S. media giants.

"Bottom up" social movements around the world are struggling to improve access to the media and diversify its content. "Brave New World: Right to Communicate Reshapes Human Connections" is an interview with media activist Howard Frederick, who argues that the "right to communicate" should be seen as a fundamental human right. The chapter concludes with an article that has been widely distributed by the media awareness organization, Center for Media Literacy, to help people empower themselves against media bias. In "Balance Bias with Critical Questions," Patricia Hynds provides a checklist of questions to consider when examining a story in the news.

WORKS CITED

Bagdikian, Ben. 1983. *The Media Monopoly.* Boston, MA: Beacon Press.

Chomsky, Noam. 1988. *Manufacturing Consent.* New York: Pantheon Books.

Parenti, Michael. 1993. *Inventing Reality,* 2nd ed. New York: St. Martin's Press.

Schechter, Danny. 1997. *The More You Watch the Less You Know.* New York: Seven Stories Press.

QUESTIONS FOR DISCUSSION

1.   According to Lowney and Best's research, the Internet gives ordinary people the opportunity to define social problems in ways that may be very different from how they are defined by the media, government, activists in social movements, and so on. Do you believe the Internet is more powerful in this regard than the channels of communication available to ordinary people in the past (for example, telephone, mail, face-to-face communication, and so on)? Explain. Does access to the Internet give people the "right to communicate," as discussed by Howard Frederick in "Brave New World"? Explain.

2.   What are some of the sources of media bias identified by Parenti in "The Myth of a Liberal Media," Makunike in "Out of Africa!" and Hoover in "All Power to the Conglomerate"? Discuss some of the ways that your own views of world events, and those of your family or friends, have possibly been influenced by media bias.

3.   Canadians consume more U.S. than Canadian films, books, television programs, and so on, a phenomenon that exists in other nations as well. Wilson argues in "Northern Exposure" that this is due not to consumers preferring the superior U.S. cultural products, but to the power U.S. corporations have in the world market. What are some reasons Wilson gives in support of this point of view? Do you agree that governments should protect national culture from U.S. cultural intrusion, or are consumers' interests best served by the free market? Explain.

4.   Using the guidelines given by Hynds in "Balance Bias with Critical Questions," read a news story about an international event, or watch the news about the event on television. Did having the guidelines change your perspective on the news? Why or why not?

 INFOTRAC COLLEGE EDITION: EXERCISE

In much of the world, the "four freedoms" discussed by Howard Frederick in this chapter in "Brave New World: Right to Communicate Reshapes Human Connections" (freedom of opinion, freedom of expression, freedom of the press, and freedom of communication) scarcely exist. Learn more about what is happening to *freedom of the press* around the world:

Using InfoTrac College Edition, search for the following article: Ross Herbert and Paul Knox, "Censorship Has Many Faces," *World Press Review,* April 1997 v44n4p16.

(*Hint:* Enter the search terms media and Africa, using the Subject Guide.)

What are some of the vested interests involved in press censorship and violence against journalists around the world? What do you believe can be done to expand freedom of the press in nations where it is restricted? Compare and contrast the situations described in the article with freedom of the press in the United States— that is, in what ways is the situation similar in the United States, and in what ways is it different?

FOR ADDITIONAL RESEARCH

Books

Herman, Edward S., and Robert W. McChesney. 1997. *The Global Media.* Herndon, VA: Cassell.

Lee, Alfred McClung, and Elizabeth Briant Lee. 1972. *The Fine Art of Propaganda.* New York: Octagon Books.

Parenti, Michael. 1993. *Inventing Reality,* 2nd ed. New York: St. Martin's Press.

Organizations

Center for Media Literacy
4727 Wilshire Boulevard, Suite 403
Los Angeles, CA 90010
(213) 931-4177
www.medialit.org

Fairness and Accuracy in Reporting (FAIR/EXTRA!)
P.O. Box 170
Congers, NY 10920-9930
(800) 847-3993
www.fair.org

Institute for Global Communications
Peacenet/Econet/Labornet/Womensnet/Conflictnet
P.O. Box 29904
San Francisco, CA 94129-0904
(415) 561-6100
www.igc.org

Institute for Public Accuracy
65 Ninth Street, Suite 3
San Francisco, CA 94103
(415) 552-5378
www.accuracy.org

ACTION PROJECTS

1. Go to a well-stocked library in your area and look at the current periodicals. Make a list of those you find that deal with global events and issues. Which of them do you find the most informative, and why? Which seem biased, and why?

2. Follow one current event story for a week in a variety of media: a local newspaper, a national newspaper, network news, public television, and public radio news. Also look at a foreign newspaper or newsmagazine. Compare and contrast the way the story is told by the different media. For example, what is presented, and what is left out? Whose viewpoints appear, and whose

do not? What conclusions do you draw about the nature of media bias today?

3. What access do people in your local community have to the media? Make an inventory of the possibilities. Are there public access television channels? Radio stations? Is there an alternative newspaper? Interview leaders of grassroots organizations about the local mainstream media. Are the viewpoints and activities of groups involved in social change accurately represented? Why or why not? What are these organizations doing to gain better access to the mainstream media, or to establish alternative ways to get their views across to the public?

35

What Waco Stood For

Jokes as Popular Constructions of Social Problems

KATHLEEN S. LOWNEY AND
JOEL BEST

In this paper we argue that one way to examine how the public constructs—or does not construct—social problems is by listening to the jokes that they tell. We explore one joke cycle told on the Internet—about the Branch Davidian sect and their deaths—in order to see how the American public constructed this incident. An analysis of the joke cycle shows that—almost from the beginning of the crisis, but especially after the fire and deaths of most sect members—the public labeled the group as deviant. The joke cycle portrayed the Branch Davidians as religious and sexual deviants who were foolish for following David Koresh. Thus the joke cycle imputed responsibility almost entirely to the sect, in sharp contrast to other possible constructions of the events at Waco.

Recent constructionist analyses focus on discourse—the language of claimsmaking (Ibarra and Kitsuse 1993; Maynard 1988; Marlaire and Maynard 1993). But not all forms of social problems discourse attract equal attention from analysts. Constructionists often analyze activists' primary claims, for their words have the power of raw emotion, of the victimized calling out for justice. Or analysts examine secondary claims, when politicians, experts, or the mass media transform the words of primary claimants, sometimes infusing them with new meanings (Best 1990). In either case, such sociological analyses concentrate on the most visible forms of public discourse. Ordinary people's constructions of particular issues—what might be termed tertiary claims—largely go unexamined; we hear the voices of the committed few, but too often ignore the reactions of the many who are less directly involved. Social problems are also constructed in everyday conversations among ordinary citizens, but most analysts neglect this largely invisible form of public discourse.

Tertiary claims have diverse sources. Much research focuses on how primary and secondary claims shape public opinion. And, to the degree that the public is exposed to and influenced by primary and secondary claimsmakers, tertiary claims may borrow information, terminology, and orientations from them. But the public is not passive; its members interpret others' claims within the context of their own interests, experience, and knowledge of other social problems. The resulting tertiary claims do not simply replicate the discourse of primary and secondary claimsmakers; rather, the public constructs social problems on its own terms, although analysts routinely neglect this process.

There are, of course, methods for assessing popular constructions of social problems. Public opinion polls offer some information, but their data usually derive from one or two survey questions, and the wording of those questions can affect the results. Few polls include enough, sufficiently detailed questions to let analysts understand how the public constructs a given issue. Some sociologists have staged focus groups, then analyzed transcripts of their discussions (Gamson 1992; Sasson 1995). Still other analysts treat folklore as a source for popular attitudes toward social problems. Most commonly, analyses of folklore examine contemporary legends (Best 1990; Fine 1992;

From Kathleen S. Lowney and Joel Best, "What Waco Stood For: Jokes as Popular Constructions of Social Problems," *Perspectives on Social Problems,* vol. 8, 1996, pp. 77–97. Copyright © 1996 by JAI Press, Inc. Reprinted by permission.

Turner 1993) or personal-experience narratives (Wachs 1988). This paper uses another form of folklore—the joke cycle—as data for assessing popular constructions of a social problem. It examines jokes about the 1993 events at the Branch Davidian compound outside Waco, Texas.

On February 28, 1993, armed agents of the federal Bureau of Alcohol, Tobacco, and Firearms served a search warrant on the Mt. Carmel headquarters of the Branch Davidians. The warrant alleged that at least some of the sect's members were collecting illegal weapons or stockpiling banned parts of guns. Other accusations, although not detailed in the search warrant, involved physical and sexual child abuse. While stories diverge over just what happened, there was a shootout in which four BATF agents died, and other agents and some Branch Davidians, including their leader, David Koresh, were wounded. Law enforcement officers surrounded the Branch Davidians and ordered them to exit their compound. When they refused, a fifty-one day siege began. Most Branch Davidians remained intractable, despite having their electricity and water cut off, and despite being bombarded with loud noise and music. Koresh continued to debate theological fine points with federal negotiators, while begging for more time to finish his treatise about the end of the world. With patience running out, and the world watching via intense mass media coverage, the government acted. On the morning of April 19, federal agents approached the compound; heavy equipment penetrated the building and tear gas was pumped inside. A fire began within the compound, quickly engulfing it and killing most of the occupants.

Although academic analysts of these events have criticized the government's actions (Taylor and Gallagher 1995; Wright 1995), government policies at Waco had widespread public support. Opinion polls taken on April 20, 1993, the day after the fire, showed that the American public had strong opinions about the incident. When asked "Overall, how much do you blame each of the following for the outcome of yesterday's events in Waco—a great deal, a moderate amount, only a little, or not at all?," ninety-three percent of those surveyed held David Koresh moderately or greatly

responsible. Conversely, only thirty-three percent blamed the FBI to the same degree (McAneny 1993). In fact, the vast majority of those polled thought the FBI had shown restraint in waiting so long to enter the Branch Davidian compound. Fifty-seven percent thought action should have been taken sooner, and another quarter thought the agency had waited long enough (McAneny 1993).

These poll results indicate that most people held critical views of the Branch Davidians. But why? Some analysts argue that the government and the press drew upon public antagonism toward "cults" in constructing the Waco conflict (Richardson 1995). But we can discover more about the public's construction of this issue by examining a very different sort of data. Joke cycles offer another source of tertiary claims. Such cycles are sets of jokes, told about one subject, and clustered together in time; analysts often interpret joke cycles as reflecting popular constructions of social problems, especially forms of social conflict (Dundes 1987). Many recent joke cycles center around disasters, such as the *Challenger* explosion (Ellis 1991; Oring 1987; Simons 1986; Smyth 1986) and the Chernobyl meltdown (Kurti 1988), or personal scandals (e.g., Jeffrey Dahmer, Michael Jackson, and O. J. Simpson [Lamp 1994]). The events at the Branch Davidian compound prompted another such joke cycle.

This paper analyzes nearly four hundred Branch Davidian jokes that circulated on two international electronic humor networks. The first joke was posted to the networks on March 17, 1993, almost three weeks after the BATF raid; the last joke appeared on June 16, some two months after the Waco fire. During these three months, people across the country and around the world shared in the macabre humor of a joke cycle featuring "sick" jokes about fire, death, and sexual deviance. The Branch Davidian joke cycle was one way the public participated in the construction of the Branch Davidian controversy, and these tertiary claims labeled the sect deviant and blamed it for what happened. The jokes helped Americans interpret the tragedy in ways that kept the events at Waco from becoming seen as an instance of some other, very different social problem, such as government abuse of power.

. . .

THE DOMINANT THEME: BRANCH DAVIDIAN JOKES AND THE IMPUTATION OF DEVIANCE

The Branch Davidian siege- and fire-inspired jokes were neither subtle not elaborate. The great majority of the jokes posted to the humor networks were simple two-line riddles which posed a question, followed by a humorous answer. Thus, the cycle's lead joke was:

Q: Do you know what Waco stands for?

A: We Ain't Comin' Out (3/20/93).

Further, most of the riddles' answers involved simple puns, usually references to fire, smoke, or related phenomena:

Q: What is a Branch Davidian hymn?

A: "Come on Baby, Light my Fire" (4/21/93).

This pattern—a predominance of simple riddles and puns—has characterized most other recent joke cycles (Dundes 1987). If such jokes are worth repeating, it is not because they are especially clever, but because they are topical.

In some cases, Branch Davidian jokes simply recycled jokes that appeared in earlier joke cycles. For example:

Q: What were David Koresh's last words?

A: "No, I meant a BUD Light" (4/20/95).

This recycled a joke told following the *Challenger* explosion:

Q: What were Christa McAuliffe's last words?

A: "No, I meant a BUD Light" (Ellis 1991, p. 112).

Similarly, consider the joke:

Q: How many Branch Davidians can you get in a Volkswagen?

A: Two in the front seat. Two were in the back. And the rest were in the ash tray (5/3/93).

This joke recycles a riddle from the Auschwitz joke cycle (Dundes 1987, pp. 19–38). Such recycling indicates that joke cycles can borrow both form (riddles, puns) and even content from the repertoire of familiar, existing jokes. Of course, meaning may be lost in this recycling: the Auschwitz joke specifies a Volkswagen because it is a German car, while the make of automobile adds nothing when the joke is told about Branch Davidians, and variants of the Waco joke featured other makes of cars.

The simple structure and familiar content of these jokes raises the question whether they even merit sociological analysis. Can these jokes reveal anything of sociological significance? Clearly, the cycle's popularity was fleeting, and no one joke can bear the weight of much analysis. However, the number of jokes about the Branch Davidians was very large and, when we examine the joke cycle as a whole, interesting patterns become evident. In particular, the vast majority of jokes in the cycle characterized the Branch Davidians as deviant, and blamed them for the siege and the fire.

The Lead Joke

The cycle's lead joke—"What does Waco stand for?"—was the second joke to appear on the Internet, the one most often told, and the one with the greatest number of variants. The initial pre-fire answer, of course, emphasized the intransigence of the Branch Davidians—"We Ain't Comin' Out." However, following the fire, many variant answers appeared:

A: We All Cooked Ourselves (4/19/93).

A: We're All Cremated Oddballs (4/21/93).

A: Want A Crispy One? (4/21/93).

A: We Are Combustible Occultists (4/21/93).

In all, 38 different answers to the what-does-Waco-stand-for riddle appeared. Where the initial joke depicted the Branch Davidians as resolute or stubborn, the later variants usually made them the butt of the joke. In Klapp's (1962) framework, the post-fire jokes did not characterize the Branch Davidians as heroes (e.g., standing up for their beliefs), but as fools (e.g., causing their own demise) or even villains (e.g., adhering to various deviant beliefs). Later variants on the lead joke labeled the Branch Davidians deviants.

Deviance Identification
Through Joke Telling

Humor bonds joke-tellers and their audience together in a shared moral discourse. But as Francis (1994, p. 148) notes, such bonding is often "at the expense of some excluded person(s)," in this case the dead Branch Davidians. Sharing jokes promotes facile construction of moral membership categories. Bronner (1985, p. 45) argues that sick jokes ". . . ostensibly entertain, but they also significantly explore and manipulate categories of social acceptance." Those who share the joke are "insiders," and those who are joked about are not. There was constant differentiation between insiders and outsiders in the Branch Davidian joke cycle. Most obviously, the joke-tellers and their audience were alive, while the Branch Davidians—the butt of most jokes—were not; the former could "get" the joke, but the latter *were* the joke.

Our content analysis of the joke cycle reveals that most jokes characterized the Branch Davidians as deviant and, in many cases, as responsible for the siege, the fire, and their own deaths. The exact nature of the sect's deviance varied from joke to joke, but the joke cycle as a whole drew a clear distinction between deviant Branch Davidians and other, non-deviant Americans. By posting these jokes, tellers articulated opinions about the normative boundaries of American society, their own location within those boundaries, and the Branch Davidians' status as outsiders. Jokes allowed tellers and listeners to distinguish between normal behavior and deviance, thereby reinforcing shared moral commitments.

Jokes About Religious Deviance   One joke in six (17%) implied that the Branch Davidians were religious deviants. This is not surprising. As a small, marginal sect, the Branch Davidians

defined themselves in terms of their distinctive theological beliefs. They believed David Koresh's interpretations of sacred scriptures to be correct, and all others to be in error. The jokes reversed this claim, depicting the Branch Davidians as in error, breaking religious tenets and holding deviant beliefs. In some jokes, this deviance involved minor rule violations. The most common joke in this category was:

> **Q:** How do you tell the Branch Davidians at a revival meeting?
>
> **A:** They're the ones smoking in the corner (4/21/93).

The pun on "smoking" refers to the fire, but also to behavior some people might find offensive at a religious gathering.

Other jokes about religion alleged more serious infractions. Some implied that the Branch Davidians misunderstood Judeo-Christian tenets, effectively showing their social distance from other Christians who "know their theology." A frequently repeated joke illustrates this religious misinterpretation theme:

> **Q:** What is the new Davidian Branch [sic] holy day?
>
> **A:** Ash Monday. [The fatal fire occurred on a Monday] (4/20/93).

Such jokes utilized familiar religious symbols and showed how the Branch Davidians misunderstood them: one *wears* ashes on Ash Wednesday, instead of turning oneself *into* ashes on a Monday. Another joke involved the degree to which Scripture should be taken literally by believers:

> Yesterday evining [sic] in Indianapolis I saw an old pick-up truck from Texas bearing a bumper sticker reding [sic], "ON FIRE FOR THE LORD!" (6/6/93).

Here the punch line turned on the fact that most conservative Christians interpret that phrase as meaning "spiritually excited" whereas the Branch Davidians apparently took it literally. These jokes showed how the Branch Davidians corrupted familiar religious beliefs by their erroneous interpretations of Biblical Scripture. Joke-tellers and their audiences knew better, and thus held the spiritual high ground.

Other jokes suggested that Branch Davidian beliefs involved more central violations of Christianity's central tenets. In particular, several jokes implied that David Koresh considered himself divine, the equivalent of Christ. Thus, a variant of the what-does-Waco-stand-for riddle had the answer "Well, Another Christ Obliterated" (4/21/93). Other jokes disputed claims of Koresh's divinity:

> After examining the remains of David Koresh, apparently they found a hole in his forehead. No mention of any in his hands or feet, though (5/13/93).

Thus, the Branch Davidians, who viewed themselves as religiously different, were characterized through the joke cycle as religiously deviant.

• • •

A SUBORDINATE THEME: CRITICISM OF THE AUTHORITIES

The bulk of the jokes in the Branch Davidian joke cycle, then, depicted the sect and its members as deviant—holding deviant religious beliefs, associated with prominent deviants, behaving foolishly, involved in sexual deviance, or setting themselves afire. This should not surprise us. Individuals often use humor to label others outsiders: "[T]he fact that there is shared laughter is a social marker that indicates that group members share a common perspective" (Fine 1983, p. 173; see also Arnez and Anthony 1968; Dundes 1987; Freud 1960; Obrdlik 1942). This labeling process reduces sympathy and empathy between insiders and outsiders, thereby increasing the social distance between them. Getting the "punch line" to these jokes acts as a membership identifier, separating in-group members from deviants. Branch Davidian jokes explored and reaffirmed normative boundaries concerning sexuality, religiosity, and the like.

Jokes cycles about the Branch Davidians, Chernobyl, the *Challenger* explosion, and other disasters can be seen as collective attempts to reconstitute a normal, everyday world. Disastrous events, such as children being burned alive in a struggle over religious principles, demand interpretation; disasters must be explained, if not justified. The Branch Davidian joke cycle implied that the explanation for the disaster rested in the sect's deviance. The majority of the jokes portrayed the sect as peculiar, outside the realm of normality. In contrast, the jokes virtually ignored other potential explanations, such as errors on the part of the authorities.

Very few Branch Davidian jokes criticized law enforcement agencies. Rather, the jokes that mentioned law enforcement often reaffirmed the agencies' use of power to settle the standoff:

> **Q:** What did Janet Reno [Attorney General of the U.S.] say to the chief of the FBI?
>
> **A:** "Well done!" (4/21/93).
>
> **Q:** When did the FBI get what they wanted?
>
> **A:** When KorASH finally went out (4/20/93).
>
> New FBI motto: If you can't beat them, broil them (5/3/93).

Only two jokes were directly antagonistic toward law enforcement agencies' policies. Each was posted only once to the networks. The following joke received the most hostile responses by other posters; it was "flamed" in electronic network nomenclature:

> A secretary a [sic] work told me that the Treasury Department is planning to disband the BATF [Bureau of Alcohol, Tobacco, and Firearms] and replace it with the Post Office. The reason given is that postal employees shoot better (5/19/93).

Angry responses ridiculed the incompetency theme at the heart of this joke; they often restated the basic premise of the joke cycle—that it was the Branch Davidians' fault for what had happened—and solely theirs:

> How can you say it was their [BATF's] fault? When parents want to let their own children die, what can cops do to stop it? They were patient, negotiated with him [Koresh] as much as possible. Don't blame them, blame the crazies who couldn't listen instead (4/20/93).

The second joke which implied criticism of the BATF also elicited critical responses:

Q: What was the tragedy when a bus of AFT agents careened off a cliff and exploded on the rocks?

A: There was an empty seat on the bus! (4/3/93).

The news media received even less attention in the joke cycle. In addition to comments about media circuses, three jokes played off the recent news that NBC News had staged scenes for a recent story claiming that GMC trucks tended to burst into flame:

Q: According to NBC, what started the fire at the Branch Davidian site?

A: A GMC truck (5/5/93).

Reluctance to criticize authorities is not a general characteristic of folklore. Analysts often interpret contemporary legends as popular critiques of government or business elites (Fine 1992; Turner 1993). Similarly, some joke cycles directly criticize authority figures (Dundes 1987). Analysts find implicit critiques in other jokes; thus, Oring (1987) interprets the *Challenger* joke cycle as revealing public disenchantment with the news media. In short, the limited criticism of authorities in the Branch Davidian joke cycle stands in sharp contrast to the pointed critiques in other legends and jokes.

Again, no single joke can tell us very much about popular interpretations of the 1993 events involving the Branch Davidians. However, the complete joke cycle is quite revealing. Literally hundreds of jokes suggested that the Branch Davidians were somehow peculiar—flawed or deviant. In sharp contrast, the joke cycle virtually ignored alternative interpretations critical of law enforcement's role in the crisis. In this regard, the joke cycle differs from other public discourse about the Branch Davidians. Throughout the siege, and immediately following the fire, the press contained numerous accounts critical of the authorities' policies (e.g., Beck 1993). These press reports did not absolve the Branch Davidians of all responsibility, but they argued that the federal government might have handled the crisis more effectively. Similarly, scholarly interpretations have criticized the government's actions (Taylor and Gallagher 1995; Wright 1995). And, of course, the 1995 Oklahoma City bombing inspired additional critical assessments of the government's actions.

In contrast, the joke cycle blamed the Branch Davidians almost without questioning the government. These jokes made it easier to accept—or at least not question—what happened at the Waco compound: the authorities had no other recourse but to use force to get the sect to surrender. The lethal result of the government's action was an anomaly—necessary given the confrontation with deviants in Waco, but nothing that might ever be repeated against upstanding citizens. The deaths were unexpected and, however grievous, they were not law enforcement's fault. Many questions about the authorities' actions were raised in the press but ignored by the joke cycle. No one on the electronic networks joked about whether the raid on the Branch Davidian compound was strategically well-planned or even legal, or whether the tactics used, such as playing loud music or cutting off electric power, were the most effective means to get the members out alive. No one joked about the legality of the search warrant. And no one joked about the First Amendment guarantee of religious liberty, or whether law enforcement "bent" if not "broke" that guarantee. Such jokes would not have been as funny, for they would have challenged bedrock assumptions that law enforcement should and does maintain order and protect individual rights.

• • •

CONCLUSION

Although researchers tend to speak of *the* construction of a social problem, any issue is subject to alternative constructions. Analysts have examined coffee drinking and sites for illicit sex as potentially constructable—yet unconstucted—social problems (Troyer and Markle 1984; Ball and Lilly 1984), and noted the possibilities for alternative typifications of constructed issues (Best 1995). However, most constructionist work takes for granted that an issue has a single construction that is revealed in the discourse of elites—activists, politicians, and the press.

This analysis suggests that joke cycles offer a fruitful way of assessing popular constructions of social problems. In contrast to the media coverage—which presented alternative interpretations of the events in Waco, including critiques of government actions—almost all the jokes blamed the Branch Davidians for their own deaths. This reaffirms the findings of post-fire surveys that found public opinion holding the sect responsible for what happened, but the jokes go beyond those survey results, giving a more detailed sense of the ways the tertiary claimsmaking constructed the Branch Davidians as deviant.

The Branch Davidian joke cycle seems to have functioned as a form of preventative social problems work. Holstein and Miller (1993, p. 165) write: "As mundane as it is, and as overlooked in the study of social problems as it might be, the process by which 'nothing' is made of a candidate problem is an important aspect of social problems work." The deaths at Waco could have been ignored, or constructed as evidence that governmental abuse of power was a social problem. Instead, the jokes offered a different interpretation, one that accounted for Waco in terms of bizarre, deviant psychology and religiosity. Joke cycles, such as the Branch Davidian jokes, let sociologists listen to popular social problems talk to learn how social problems are given shape—or left unconstructed.

The events at the Branch Davidian compound led to a variety of forms of public discourse: primary claims (e.g., by surviving Branch Davidians at their trial); secondary claims (e.g., media coverage and statements by various officials); and tertiary claims (including the Branch Davidian joke cycle). At various times, different claims constructed—or failed to construct—the Branch Davidians' behavior and the authorities' actions as social problems. The differences among various constructions produced at different times illustrate that the process of constructing social problems is contingent, flexible, ongoing, and open ended. The Branch Davidian joke cycle reveals that tertiary claimsmaking is an active process, that popular constructions of social prob-

lems are more than simple reflections of elite claims, that tertiary claims deserve more analytic attention than they usually receive.

Acknowledgment: Bill Ellis, Carol Brooks Gardner, James A. Holstein, and Gale Miller made helpful comments on earlier drafts of this paper.

REFERENCES

Arnez, N. L. and C. B. Anthony. 1968. "Contemporary Negro Humor as a Social Satire." *Phylon* 29:339–346.

Ball, R. and J. R. Lilly. 1984. "When is a 'Problem' not a Problem? Deflection Activities in a Clandestine Motel." Pp. 114–139 in *Studies in the Sociology of Social Problems,* edited by J. W. Schneider and J. I. Kitsuse. Norwood, NJ: Ablex.

Beck, M. 1993. "The Questions Live On." *Newsweek* 121(May 3):28–29.

Best, J. (Ed.). 1995. *Images of Issues: Typifying Contemporary Social Problems,* 2nd ed. Hawthorne, NY: Aldine de Gruyter.

___. 1990. *Threatened Children: Rhetoric and Concern about Child-Victims.* Chicago, IL: University of Chicago.

Bronner, S. J. 1985. "'What's Grosser than Gross?': New Sick Joke Cycles." *Midwestern Journal of Language and Folklore* 11:30–49.

Dundes, A. 1987. *Cracking Jokes: Studies of Sick Humor Cycles and Stereotypes.* Berkeley, CA: Ten Speed Press.

Ellis, B. 1991. "The Last Thing Said: The Challenger Disaster Jokes and Closure." *International Folklore Review* 8:110–124.

Fine, G. A. 1992. *Manufacturing Tales: Sex and Money in Contemporary Legends.* Knoxville, TN: University of Tennessee Press.

___. 1983. "Sociological Approaches to the Study of Humor." Pp. 159–181 in *Handbook of Humor Research, Volume 1: Basic Research,* edited by P. E. McGhee and J. H. Goldstein. New York: Springer-Verlag.

Francis, L. E. 1994. "Laughter, the Best Mediation: Humor as Emotion Management in Interaction." *Symbolic Interaction* 17:147–164.

Freud, S. 1960. *Jokes and Their Relation to the Unconscious.* New York: Norton.

Gamson, W. A. 1992. *Talking Politics.* Cambridge, England: Cambridge University Press.

Holstein, J. A. and G. Miller. 1993. "Social Constructionism and Social Problems Work." Pp. 151–172 in *Reconsidering Social Constructionism:*

Debates in Social Problems Theory, edited by J. A. Holstein and G. Miller. Hawthorne, NY: Aldine de Gruyter.

Ibarra, P. R. and J. I. Kitsuse. 1993. "Vernacular Constituents of Moral Discourse: An Interactionist Proposal for the Study of Social Problems." Pp. 25–58 in *Reconsidering Social Constructionism: Debates in Social Problems Theory,* edited by J. A. Holstein and G. Miller. Hawthorne, NY: Aldine de Gruyter.

Klapp, O. E. 1962. *Heroes, Villains, and Fools: The Changing American Character.* Englewood Cliffs, N. J.: Prentice Hall.

Kurti, L. 1988. "The Politics of Joking: Popular Response to Chernobyl." *Journal of American Folklore* 101:324–334.

Lamp, C. 1994. "The Popularity of O. J. Simpson Jokes: The More We Know, the More We Laugh." *Journal of Popular Culture* 28:223–231.

Maynard, D. W. 1988. "Language, Interaction, and Social Problems." *Social Problems* 35:311–334.

Marlaire, C. L. and D. W. Maynard. 1993. "Social Problems and the Organization of Talk and Interaction." Pp. 173–198 in *Reconsidering Social Constructionism: Debates in Social Problems Theory,* edited by J. A. Holstein and G. Miller. Hawthorne, NY: Aldine de Gruyter.

McAneny, L. 1993. "One Day After the Waco Standoff's Tragic Conclusion, Americans Blamed Koresh." *The Gallup Poll Monthly* (April):25–26.

Obrdlik, A. J. 1942 "'Gallows Humor'—A Sociological Phenomenon." *American Journal of Sociology* 47:709–716.

Oring, E. 1987. "Jokes and the Discourse on Disaster." *Journal of American Folklore* 100:276–286.

Richardson, J. T. 1995. "Manufacturing Consent about Koresh." Pp. 153–176 in *Armageddon in Waco: Critical Perspectives on the Branch Davidian Conflict,* edited by S. A. Wright. Chicago, IL: University of Chicago Press.

Sasson, T. 1995. *Crime Talk: How Citizens Construct a Social Problem.* Hawthorne, NY: Aldine de Gruyter.

Simons, E. R. 1986. "The NASA Joke Cycle: The Astronauts and the Teacher." *Western Folklore* 45:261–277.

Smyth, W. 1986. "Challenger Jokes and the Humor of Disaster." *Western Folklore* 45:243–260.

Taylor, J. D. and E. V. Gallagher. 1995. *Why Waco? Cults and the Battle for Religious Freedom in America.* Berkeley, CA: University of California Press.

Troyer, R. J. and G. E. Markle. 1984. "Coffee Drinking." *Social Problems* 31:403–416.

Turner, P. 1993. *I Heard It Through the Grapevine: Rumor in African-American Culture.* Berkeley, CA: University of California Press.

Wachs, E. 1988. *Crime-Victim Stories: New York City's Urban Folklore.* Bloomington, IN: Indiana University Press.

Wright, S.A. (Ed.). 1995. *Armageddon in Waco: Critical Perspectives on the Branch Davidian Conflict.* Chicago, IL: University of Chicago Press.

36

The Myth of a Liberal Media

MICHAEL PARENTI

It is a widely accepted belief in this country that the media suffer from a liberal bias. Television pundits, radio talk-show hosts, and political leaders—including presidents of both parties—help propagate this belief, and their views are widely disseminated in the media. On the other hand, dissident critics—those who maintain that the corporate-owned press exercises a conservative grip on news and commentary—are afforded almost no exposure in this same supposedly liberal media.

Consider the case of David Horowitz. When Horowitz was a radical author and editor of

Ramparts, the mainstream press ignored his existence. But after he and former *Ramparts* colleague Peter Colliers surfaced as new-born conservatives, the *Washington Post Magazine* gave prominent play to their "Lefties for Reagan" pronunciamento. Horowitz and Colliers soon linked up with the National Forum Foundation, which dipped into deep conservative pockets and came up with hundreds of thousands of dollars to enable the two ex-radicals to do ideological battle with the left. Today Horowitz is a rightist media critic with his own radio show, who appears with dismaying frequency on radio and television to whine about how radio and television shut out conservative viewpoints.

Then there are the many talk-show hosts, of whom Rush Limbaugh is only the best known, who rail against the "pinko press" on hundreds of local television stations and thousands of radio stations owned by wealthy conservatives or underwritten by big business firms. To complain about how the media are dominated by liberals, Limbaugh has an hour a day on network television, an hour on cable, and a radio show syndicated by over 600 stations.

Then there are the well-financed right-wing media-watch organizations like Reed Irvine's Accuracy in Media (AIM). In a syndicated column appearing in over 100 newspapers and on a radio show aired on some 200 stations, Irvine and his associates complain that conservative viewpoints are frozen out of the media. Many left critics would like to be frozen out the way AIM, Limbaugh, and Horowitz are.

Not to be overlooked is National Empowerment Television (NET), a new cable network available in all 50 states, offering round-the-clock conservative political commentary. In the words of its founder Paul Weyrich, NET is dedicated to countering media news that "is riddled with a far-left political bias" and "unacceptable" notions about "gender-norming, racial quotas, global warming, and gays in the military."

Political leaders do their share to reinforce the image of a liberal press. During the Iran-contra affair, President Reagan likened the "liberal" media to a pack of sharks. More recently President Clinton complained that he has "not gotten one damn bit of credit from the knee-jerk liberal press." Clinton is confused; almost all the criticism hurled his way by the so-called liberal press is coming from conservatives.

HE WHO PAYS THE PIPER

There is no free and independent press in the United States. The notion of a "free market of ideas" is just as mythical as the notion of a free market of goods. Both conjure up an image of a bazaar in which many small producers sell their wares on a more or less equal footing. In fact—be it commodities or commentary—to reach a mass market you need huge sums of money for production, exposure, and distribution. Those who are without big bucks end up with a decidedly smaller clientele, assuming they survive at all.

Who owns the big media? The press lords who come to mind are William Randolph Hearst, Henry Luce, Rupert Murdock, Arthur Sulzberger, Walter Annenberg, and the like—personages of markedly conservative hue who regularly leave their ideological imprint on both news and editorial content.

The boards of directors of print and broadcast news organizations are populated by representatives from Ford, General Motors, General Electric, Dow Corning, Alcoa, Coca-Cola, Philip Morris, ITT, IBM, AT&T, and other corporations in a system of interlocking directorates that resemble the boards of other corporations. Among the major stockholders of the three largest broadcast networks are Chase Manhattan, J. P. Morgan, and Citibank. NBC is owned outright by GE. The prime stockholder of this country's most far-reaching wire service, Associated Press, is Merrill Lynch.

Not surprisingly, this pattern of ownership affects how news and commentary are manufactured. Virtually all the chief executives of mainstream news organizations are drawn from a narrow, high-income segment of the population and tilt decidedly to the right in their political preferences. Rupert Murdoch was once asked in an interview: "You're considered to be politically conservative. To what extent do you influence the editorial posture of your newspapers?" He responded with refreshing candor: "Considerably . . . my editors have input, but I make the final decisions."

Corporate advertisers exercise an additional conservative influence on the media. They cancel accounts not only when stories reflect poorly on their product but, as is more often the case, when they perceive liberal tendencies creeping into news reports and commentary.

As might be expected, the concerns of labor are regularly downplayed. Jonathan Tasini, head of the National Writers Union, studied all reports dealing with workers' issues carried by ABC, CBS, and NBC evening news during 1989, including child care and minimum wage: it came to only 2.3 percent of total coverage. No wonder one survey found that only 6 percent of business leaders thought the media treatment accorded them was "poor," while 66 percent said it was "good" or "excellent."

Religious media manifest the same gross imbalance of right over left. The fundamentalist media—featuring homophobic, sexist, reactionary televangelists like Pat Robertson—comprise a $2-billion-a-year industry, controlling about 10 percent of all radio outlets and 14 percent of the nation's television stations. In contrast, tens of thousands of liberal and often radically oriented Christians and their organizations lack the financial backing needed to gain media access.

THE PETROLEUM BROADCASTING SYSTEM

A favorite conservative hallucination is that the Public Broadcasting System is a leftist stronghold. In fact, more than 70 percent of PBS's prime-time shows are funded wholly or in major part by four giant oil companies, earing it the sobriquet of "Petroleum Broadcasting System." PBS's public-affairs programs are underwritten by General Electric, General Motors, Metropolitan Life, Pepsico, Mobil, Paine Webber, and the like. One media watchdog group found that corporate representatives constitute 44 percent of the

sources about the economy; activists account for only 3 percent, while labor representatives are virtually shut out. Guests on NPR and PBS generally are as ideologically conservative as any found on commercial networks. Even "Frontline" and Bill Moyers' "Listening to America"—favorite GOP targets—use Republicans far more frequently than Democrats.

Conservatives like Horowitz make much of the occasional muckraking documentary that is aired on public television. But most PBS documentaries are politically nondescript or centrist. Progressive works rarely see the light of day. Documentaries like *Faces of War* (revealing the brutality of the U.S.-backed counterinsurgency in El Salvador), *Building Bombs* (on nuclear proliferation), *Coverup* (on the Iran-contra conspiracy), *Deadly Deception* (an Academy Award–winning critique of General Electric and the nuclear arms industry), and *The Panama Deception* (an Academy Award–winning expose of the United States' invasion of Panama) were, with a few local exceptions, denied broadcast rights on both commercial and public television.

A rightist perspective dominates commentary shows like NBC's "McLaughlin Group," PBS's "One on One" (with John McLaughlin as host), CNBCs "McLaughlin Show" (with guess who), PBS's "Firing Line" (with William F. Buckley, Jr.), CNN's "Evans and Novak" and "Capital Gang," and ABC's "This Week with David Brinkley." The spectrum of opinion on such programs, as on the pages of most newspapers, ranges from far right to moderate center. In a display of false balancing, right-wing ideologues are pitted against moderates and centrists. Facing Pat Buchanan on CNN's "Crossfire" Michael Kinsley correctly summed it up: "Buchanan is much further to the right than I am to the left."

On foreign affairs, the press's role as a cheerleader of the national security state and free-market capitalism seems almost without restraint. Virtually no favorable exposure has ever been given to indigenous Third World revolutionary or reformist struggles or to protests at home and abroad against U.S. overseas interventions. The media's view of the world is much the same as the view from the State Department and the Pentagon. The horrendous devastation wreaked upon

the presumed beneficiaries of U.S. power generally goes unmentioned and unexplained—as do
the massive human-rights violations perpetrated
by U.S.-supported forces in dozens of free-market
client states.

WHY DO CONSERVATIVES COMPLAIN?

If news and commentary are so preponderantly
conservative, why do rightists blast the press for its
supposed leftist bias? For one thing, attacks from
the right help create a climate of opinion favorable to the right. Railing against the press's "liberalism" is a way of putting the press on the
defensive, keeping it leaning rightward for its
respectability, so that liberal opinion in this country is forever striving for credibility within a con-
servatively defined framework.

Ideological control is not formal and overt as
with a state censor but informal and usually
implicit. Hence it works with imperfect effect.
Editors sometimes are unable to see the troublesome implications of particular stories. As far as
right-wingers are concerned, too much gets in that
should be excluded. Their goal is not partial control
but perfect control, not an overbearing advantage
(which they already have) but total dominance of
the communication universe. Anything short of
unanimous support for a rightist agenda is treated as
evidence of liberal bias. Expecting the press corps
to be a press chorus, the conservative ideologue, like
any imperious maestro, reacts sharply to the occasionally discordant note.

The discordant notes can be real. The news
media never challenge freemarket ideology, but
they do occasionally report things that might put
business and the national security state in a bad
light: toxic-waste dumping by industrial firms,
price-gouging by defense contractors, bodies pil-
ing up in Haiti, financial thievery on Wall Street,
and the like. These exposures are more than
rightists care to hear and are perceived by them as
a liberal vendetta.

The conservative problem is that reality itself
is radical. The Third World really is poor and

oppressed; the United States usually does side
with Third World oligarchs; our tax system really
is regressive; millions of Americans do live in
poverty; the corporations do plunder and pollute
the environment; real wages for blue-collar workers definitely have declined; and the rich really are
increasing their share of the pie. Despite their best
efforts, there are limits to how much the media
can finesse these kinds of realities.

The limits of reality sometimes impose limits
on propaganda, as Dr. Goebbels discovered when
trying to explain to the German public how invin-
cible Nazi armies could win victory after victory
while retreating on both fronts in 1944 and 1945.
Although they see the world through much the
same ideological lens as do corporate and government elites, the media must occasionally report
some of the unpleasantness of life—if only to
maintain credibility with a public that is not always
willing to buy the official line. On such occasions,
rightists complain bitterly about a left bias.

Rightist ideologues object not only to what
the press says but to what it omits. They castigate
the media for failing to tell the American people
that federal bureaucrats, "cultural elites," gays, les-
bians, feminists, and abortionists are destroying
the nation, that the U.S. military and corporate
America are our only salvation, that there is no
health-care problem, that eco-terrorists stalk the
land, that the environment is doing just fine—and
other such loony tunes.

SELF-CENSORSHIP

Reporters often operate in a state of self-
censorship and anticipatory response. They frequently wonder aloud how their boss is taking
things. They recall instances when superiors
have warned them not to antagonize big advertisers and other powerful interests. They can
name journalists who were banished for turning in the wrong kind of copy. Still, most
newspeople treat these incidents as aberrant
departures from a basically professional news
system and insist they owe their souls to no
one. They claim they are free to say what they
like, not realizing it is because their superiors

like what they say. Since they seldom cross any forbidden lines, they are not reined in and they remain unaware that they are on an ideological leash.

While incarcerated in Mussolini's dungeons from 1928 to 1937, Antonio Gramsci wrote about politics and culture in his prison notebooks. But he had to be careful not to antagonize the fascist censor. Today most of our journalists and social commentators exercise a similar caution. However, unlike Gramsci, they are not in prison. They don't need a fascist censor breathing down their necks because they have a mainstream one implanted in their heads.

These internalized forms of self-censorship are far more effective in preserving the dominant ideology than any state censor could hope to be. Gramsci knew he was being censored. Many of our newspeople and pundits think they are as free as birds—and they are, as long as they fly around in the right circles.

For conservative critics, however, the right circles are neither right enough nor tight enough. Anything to the left of themselves, including moderate right and establishment centrist, is defined as "liberal." Their campaign against the media helps to shift the center of political gravity in their direction. By giving such generous publicity to conservative preachments and pronouncements while amputating everything on the left, the media limit public debate to a contest between right and center. In doing so, they are active accomplices in maintaining a rightward bent.

On the American political scene, the center is occupied by conservative Democrats like Bill Clinton, who are happy to be considered the only alternative to the ultra-right. This center is then passed off as "liberal." Meanwhile, real liberalism and everything progressive remain out of the picture—which is just what the mainstream pundits, publishers, politicians, and plutocrats want.

• • •

37

Out of Africa!

Western Media Stereotypes Shape World's Portrait

EZEKIEL MAKUNIKE

As a graduate student just arrived from Zimbabwe (then Rhodesia) to study journalism at Syracuse University in 1968–9, I developed the habit of scanning the local papers for news from my home continent.

It was a pretty futile search. I was increasingly dismayed at the near-total lack of news from any part of Africa being presented to Syracuse readers.

I also soon discovered that the little African news that occasionally found the light of day and trickled into the *Syracuse Herald and Journal* was almost always negative. This inspired me to spend some of my free time embarking on a more serious investigation of news selection.

I requested permission from the news departments of those two daily newspapers to glean

From Ezekiel Makunike, "Out of Africa!: Western Media Stereotypes Shape World's Portrait," *Media and Values*, no. 61, winter, 1993, pp. 10–12. Copyright © 1993 by the Center for Media Literacy. Reprinted by permission.

through their wastepaper baskets for telex sheets from wire services containing stories transmitted from Africa. I conducted this search for most of an entire week. While indeed not much was offered by the news services, I was nevertheless surprised to find that much of the little that came in was either "killed" or simply spiked for a more suitable publication date that never came.

When I asked an editor to explain these decisions, he told me that stories on Africa are routinely ignored because of a presumed lack of reader interest.

"You see," he told me, "America does not know Africa well. It never had a colony on that continent. Thus, unless the story has a strong human interest potential, there is no point using it, since no one will read it."

Of course, the editor was both creating a self-fulfilling prophecy and ignoring an obvious fact. The prophecy was simple: White Americans would never become aware of Africa unless they could learn enough about it to be interested, a process the media has a lot to do with.

AMERICAN COLONIALISM

The fact, at once more complicated and highly relevant to contemporary events, was this: American colonialism against Africans was practiced in the American South in the form of hundreds of years of slavery and second-class citizenship. At the time—the late '60s—millions of descendents of those slaves—approximately 11.1 percent of the U.S. population at the time— were beginning to rediscover their African roots. Many of them were (and are still) highly interested in events in Africa. Their history and African background has much to do with current events in the U.S.

The 1960s period of my study also coincided with the height of Cold War politics, with the United States then heavily involved in the Vietnam War. Africa was not of the highest concern in the super-power politics of the time. Yet, news from Africa seemed essential to forming a complete picture of important happenings in the world, and this was lacking.

About the same time, a lengthy newspaper account of an official visit to Africa by Chou En-Lai, then prime minister of the People's Republic of China, appeared in a major paper. The Chinese had promised technical assistance for the construction of a railway line between central Zambia and the Tanzanian seaport of Dar es Salaam, on the Indian Ocean in Southeast Africa. It read something like this:

"Communist China has promised to build a railway line from the black-ruled and land-locked Republic of Zambia to the port of Dar es Salaam, capital of socialist Tanzania. This is meant to reduce dependency on trade routes with and through their southern neighbors, white-ruled Rhodesia (now Zimbabwe) and South Africa. The two black-ruled African countries train and harbor terrorists fighting to overthrow these white-ruled governments."

The implication was that Tanzania and Zambia were being taken for a ride because China didn't possess the technology to execute such a feat of engineering. The promise was seen as simply an example of Communist propaganda.

A few years later, after I had returned to Africa and become the director of the Pan-African journalism institute, the Africa Literature Center in Kitwe, Zambia, I was reminded of this conclusion. My journalism students and I attended the inaugural ceremony of the Tanzania-Zambia Railway Line (Tan Zam Railway). We actually were passengers on the initial trip of this "will never-be-built" line, riding safely for over 1,000 kilometers from Kapiri M'poshi near Kabwe in central Zambia to the seaside in Dar es Salaam. With the help of the "technically backward" Communist Chinese, the feat had actually been completed well ahead of schedule!

Influenced first by colonialism and then by Cold War politics, this contemptuous tone has long shaped and fashioned Western media perceptions of Africa. As I learned very quickly in the U.S., for American readers or viewers to be interested, news out of Africa must be negative. It must conform to the traditional stereotypes in its spot-light on grotesque and sensational events. It must show misery, corruption, mismanagement, starvation, primitive surroundings and, as in the case of Somalia, chaos and outright anarchy.

Foreign correspondents in African capitals and their superiors in the media gate-keeping chain seem to have these perceptions ingrained in them. From newsgathering in Africa to publication and broadcast thousands of miles away, stories about Africa are looked at with these negative lenses. Even more unfortunately, reporters and editors with a broader vision run the risk of having their stories disbelieved and unused. Little wonder they learn to toe the expected line.

This dynamic explains why the life of Africa's varied and diverse countries is missing. We hear about famines and coups, but not the rejuvenation of its cities and the cultural vitality of its village life . . . about oppression and massacres, but not education, economic self-help and political development . . . about poaching and habitat destruction, but not ongoing active efforts at conservation, reforestation and environmental awareness.

Most telling of all, in Somalia and elsewhere, news reports show outside white people feeding and helping black people. They never show black people helping themselves.

As a journalist I understand that "news" is still defined as a usually negative departure from the norm. I also recognize that in the eternal media race for larger circulations and higher ratings, profits and the bottom line dominate concerns about values and ethics.

As in Somalia, the "hit-and-run" mentality of Western media makes it easy to briefly light up trouble spots, while the years of exploitation and deterioration that produced them are left in the dark. The "here today, gone tomorrow" nature of much international reporting, with star newspersons briefly crowding each other at media feeding troughs, then jetting on to the next venue, doesn't help.

By definition such journalists know little of the language and less of the cultures they cover. They certainly never appreciate the subtleties and nuances of local history and interactions that take years to learn. They are neither accustomed nor equipped to observe, understand or explain developmental situations that may change slowly over time.

As a Zimbabwean, my observations are necessarily "out of Africa." But these observations of Western media shortcomings could be applied to many parts of the developing world. Admittedly, the negative patterns of coverage I've described were often conditioned by colonialism and Cold War politics. Unfortunately, they reinforced a pattern of ignorance and distortion that has not changed with the changing political systems. In the case of this news blackout at least, Africa is still a dark continent.

38

All Power to the Conglomerate

If Information Is a Commodity, What Price Is International Understanding?

STEWART M. HOOVER

On October 4, 1957, a rocket took off from an isolated plain near an obscure village in Siberia. Called Sputnik, its unexpected successful penetration of the earth's orbit created a furor in a Western world that feared its advent heralded a Soviet "victory" in the Cold War.

From Stewart M. Hoover, "All Power to the Conglomerate," *Media and Values,* no. 61, winter, 1993, pp. 2–5. Copyright © 1993 by the Center for Media Literacy. Reprinted by permission.

An even more important, yet less heralded, launching took place five years later. Telstar, the pioneer communication satellite, became the first of a swarm of orbiting satellites that power our present worldwide communications network.

Of the two events, our ever-growing ability to forge worldwide connections—and profits—has the greatest impact on the increasingly interdependent world we live in today. As the Cold War fades from view and new relationships emerge, one thing becomes clear: problems and issues from the environment to militarism, ethnic disputes and social unrest now take place in a global, rather than a local or national, context.

More than anything else, this sense of global relationship is a result of international media, a pattern of interconnection whose growth seems to be accelerating faster and faster.

Considered an impossible dream only a few years ago, Cable News Network (CNN) has become a network the whole world watches, a worldwide source of images and sound bites. Whether the subject is a war in the Persian Gulf, starvation in Somalia or an election in the United States, television and film viewers the world over watch the same stories and pictures, newspaper and magazine readers often digest the same basic news wire accounts and, in many cases, consumers even react to the same advertising pitches. In this sense, we do live in a "global village," and, primarily, it is a village defined by the media.

For decades, observers such as Marshall McLuhan predicted that these global boroughs would gradually come to resemble real neighborhoods. The more optimistic among them have maintained that the media will enable us to live and act as though we were actually in contact with people the world over, and that a new, positive, democratic future of global understanding—person to person—will ultimately emerge.

Such an idealistic vision of a media-centered world is attractive. We want very much to believe that the technologies of the media, with their powerful potential to transcend distance and time, can help bind the world together. The concept of Spaceship Earth has achieved important status in our thinking ever since pictures of our home planet, floating in space, were relayed from the

Apollo moon missions in the late 1960s. This view of the earth as a unified physical body has become an icon of the new "global consciousness" of the 1980s and '90s. The possibility that the media might help create the understanding we crave allows us to hope that we can create a cultural unity to match the beauty of this planet.

This is a powerful and persuasive idea. Emerging at the same time that sociologists such as Daniel Bell are describing our society as "post industrial," and others are suggesting that we now live in an "information society," this new role for the media finds the technology of communication at the center of economic and political as well as social and cultural worlds.

In countries from Central Europe to South Asia, media are looked to as potential builders of civic culture—the forum through which diverse ethnic and cultural groups can achieve understanding. At the same time, on continents from North America to East Asia, media are important industries, gradually replacing outdated sectors of the economic sphere as we move into a mediated future. Can media be both things? Can they simultaneously serve the interests of community and human understanding, and promote the interests of commercial and economic development?

These are not new questions. Representatives of the Third World began to question the global role of media decades ago. . . . [F]ormal discussions of the conflict between media freedoms and responsibilities were initiated by Asian, African and Latin American countries—called the Non-Aligned Movement (NAM)—and taken on as a project by United Nations Educational Scientific and Cultural Organization (UNESCO). Years of often acrimonious debate focused on whether the media, as then organized, could serve the interest of global understanding.

MANY VOICES

Implicit in these discussions (called the "New World Order in Communication") was the idea that communicators, in order to authentically contribute to cultural and social understanding, must first serve social and cultural development. That is, before nations and people could enter into global dialogue, they first needed the resources, skills and opportunity to enter that dialogue more or less as equals.

Principal voices in the debate pointed out that severe imbalances in the distribution of global information resources and expertise, often resulted not in a global village of equals, but a global oligarchy of rulers (the nations and institutions of the North) and subjects (the nations and institutions of the South).

A commission of UNESCO impaneled to study this problem under the leadership of Irish statesman Sean MacBride made this assessment of the situation as early as 1978: "We can sum up by saying that in the communication industry there are a relatively small number of predominant corporations which integrate all aspects of production and distribution, which are based in the leading developed countries and which have become transnational in their operations." In the decades since, growth and consolidation of these conglomerates has only accelerated; they are larger and their reach has become greater than ever.

The report of the MacBride Commission, as it was often called, went on to note that, not surprisingly, this corporate dominance favored the objectives of commercial profit over social and cultural development.

In the early 1900s, nearly 15 years after this commission issued its report, the imbalance between the players remains much the same, but developments in technology and political changes have made major changes in the playing field.

EMERGING ROLES

In a post-Cold War world that emphasizes independent development, the need for widespread involvement in communication is greater than ever. Media's role in educating, informing and connecting individuals and peoples can be of crucial importance, often influencing decisions to go to war or make peace. In today's "information age," access to media, and the ability to manipu-

late it, can be a key factor determining poverty or prosperity.

With the old East-West alignments being replaced by more natural regional groupings (and conflicts), the peoples of the South and the newly liberated countries of Eastern Europe and the former Soviet Union must be able to shape cultural and informational resources that are their own. Only then can they fulfill the long-held goal of speaking as equals to the peoples of the developed world. The basic needs have not changed, but they have assumed an added urgency.

What *has* changed dramatically in the past two decades is the structure and management of the media themselves. In successive editions of *The Media Monopoly,* media commentator Ben Bagdikian has documented the accelerating shrinkage of the number of the world's major media companies. As the trend continues, control of the world's media rests in fewer and fewer hands. Five major corporations now dominate:

- U.S.-based Time-Warner, Inc. was created by the much-heralded 1989 merger of Warner Bros., Inc., and Time, Inc. The resulting union joined Warner Brothers film and television entertainment enterprises with Time, Inc.'s magazines (*Time, Sports Illustrated,* and *People,* among others), book publishing, cable television and recording enterprises.

- Bertlesmann of Germany controls holdings in recording (BMG Music and RCA Records), publishing (Bantam Doubleday Dell) and satellite television.

- The Australian-British News Corporation, Ltd., the farflung media network of Rupert Murdoch, controls the single largest newspaper circulation in the world—14 million readers daily in Britain, Australia and North America. Other bastions of Murdoch's empire include the British news service, Reuters, and publications such as the *Financial Times* and the *Economist,* plus European satellite and cable channels.

- Hachette, S.A., a French multimedia conglomerate, publishes magazines such as *Car and Driver, Woman's Day, Home,* and *Elle* in France and books in the U.S. and elsewhere, and also owns television and radio operations in Europe.

- The other major U.S. player, Capital Cities/ABC Inc. is a newspaper and broadcasting conglomerate. (Not incidentally, the 1986 purchase of American Broadcasting Corporation (ABC) by Capital Cities began a major shift in U.S. media ownership that transformed many formerly independent information and entertainment entities into mere cogs in giant corporate wheels.)

Other scholars have shown that such concentration of control is also taking place on less global levels, with regional media conglomerates emerging in Asia, the Pacific and other parts of the Americas. In addition, the end of state media monopolies in Eastern Europe has opened avenues for transnational commercial media there as well.

Economic concentration has coincided with government policies of privatization and deregulation around the world. In the late 1970s and earlier, "the public service model" was the norm for many broadcasting organizations. Typified by the BBC, this philosophy of broadcasting holds that radio and television operate primarily as a public service, that they have a unique role in educating and uplifting the publics which own them.

Until recently, American-style commercial broadcasting (with only a limited responsibility for public service) remained in the minority around the world, as most countries in Europe and former European colonies adopted the public service model when they set up their broadcasting systems. Government ownership of primary media networks (with free speech protections in place at least in democratic countries) was part of this pattern.

This has all begun to change, and the foremost reason for the change is economic.

For decades, the American film industry demonstrated the economic power of marketing entertainment media for export. In the 1960s and '70s, American television programming also began to be sold abroad in large volume.

Filmmakers and television production companies soon learned to appreciate the high profitability of foreign distribution. The resulting deluge of American cultural material was, in fact, one of the prime motivations behind international communication debates. In addition to concerns about news and information available via the media, many people in the developing world were seriously concerned about the impact of these alien images (with their accompanying Western values and social mores) on indigenous cultures.

But the economic logic was not to be denied. Gradually, over the course of the last decade, even British authorities have begun to seriously consider privatizing their media, as it becomes harder and harder to justify public support of an enterprise that could be highly profitable on its own.

In addition, the American philosophy of deregulation has become increasingly popular worldwide. This theory asserts that regulations holding certain media (mainly radio and television broadcasting, cable and television services) to a "public service" standard are no longer necessary because the public's needs for diversity of views and services will be met by the satisfaction of the public wants on a multiplicity of channels in an open marketplace.

The assumption that the public's "needs" and "wants" are the same sounds good, until we consider that the "marketplace model" ignores the likelihood that citizens determining the future of a democracy may "need" to be exposed to cultural products they do not "want" (because they may not consider serious commentary on news and public affairs entertaining and might not seek it out on their own). Ratings and profits should not be the only shapers of social discourse. Of course media programs cannot influence public debate unless they are produced and disseminated.

Another major problem is the likelihood that minority interests will be overridden in the rush to serve majority tastes. On U.S. television, for example, how often have we seen substantive dramas starring nonwhite Americans? How common are political analyses that represent extreme, as well as centrist, shades of opinion? In the international arena, such omissions are even more striking, as groups such as the Kurds and Bosnians fail to register even a blip on the international consciousness until their oppression reaches the level of massacre.

The news media are hardly the only arena for this conflict. The Hollywood film has long dominated worldwide movie-making, but recently local and national film industries have come under increasing pressure from the movie divisions of concentrated, global enterprises. For example, the *New York Times* reported in October 1992 that the Indonesian film industry is losing the competition with American-made films for access to movie houses in Indonesia itself. Similar pressures are felt by indigenous film industries elsewhere in Asia and by television producers throughout the world. As a result, local talents are strangled and any hope of a mutual exchange of ideas is buried in a relentless one-way flood of U.S. information and entertainment goods.

CULTURE AS COMMODITY

Motion picture and music executives have long spoken of records and films as "the product." The concentration and commercialization of today's conglomerates has had an additional, very important consequence. Increasingly, media productions—films, television programs, wire service copy or popular music recordings—are seen as products or commodities to be bought and sold with little or no concern for their cultural content and impact.

In the current round of multilateral trade negotiations called the General Agreements on Tariffs and Trade (GATT), the U.S. and other media-producing countries have made the importance of cultural commodities as export items very clear. The American sentiment seems to be that since we are less and less able to compete in world markets with our manufactured goods, we should continue to dominate in areas where we have always been successful, and media products are one of those areas.

This "commoditization" of media has thus had a very important consequence for the movement toward greater equity in communication resources between the North and the South. Whereas in the 1970s, a country like Indonesia might have moved (and did) to protect its local film industry against devastating foreign competition, in the 1990s such policies can be portrayed by the media powers of the North as restraints on trade, and therefore subject to trade sanctions.

Further, the atmosphere has changed and media and communications enterprises are now viewed very differently. Even countries as small as Barbados have begun to contemplate the notion that television and other media should be operated according to entirely commercial standards, because that is the current global trend.

As demonstrated by such events as the fall of the Marcos regime in the Philippines, the collapse of the Soviet Union and the spur given by publicity to the worldwide environmental movement, the connections of worldwide media can play a positive as well as a negative role in the struggle for self determination and social, as well as economic, progress.

But on balance, the effect of most of these consumerist trends has been a diminution of global dialogue, of access for marginalized voices, of opportunities for the less powerful and more marginal to articulate their own stories via the emerging global media system.

If we are ever to achieve the healthy relationship of the global village envisioned by McLuhan and others, we must turn back the tide moving toward commercial concentration of world media. Otherwise, the commercial voice will not only be the loudest, but the only voice available. Goals of international understanding and cultural integration cannot be met if the cultures involved lack the means to tell their own stories.

• • •

39

Northern Exposure

Canada Fights Cultural Dumping

CARL WILSON

Suppose 97 percent of the movies being shown in Dallas or Cleveland came from Russia or Japan, and radio stations would play American pop music only if required to by law—that the only American TV programming was news and low-budget PBS drama, U.S. magazines got only 20 percent of the space on the newsstands, American films were stocked as "foreign" in video stores and works by Updike, Mailer and Didion were shelved at the back of bookstores as "Americana."

Such a dystopian fantasy isn't pulled from the pages of a cold war thriller about a Sovietized America. It's simply a version of everyday life in Canada, where 60 and 95 percent of the film, television, music and publishing markets are controlled by Americans. It's as if a massive U.S. film festival were going on in every theater, and a tribute to American genius were running perpetually on radio and TV. ("In Canada there is a fear that we are on our way to being functionally annexed," says Dan Johnson, president of the Canadian film distributors' association.)

Canadian artists argue that U.S. domination would be even more dramatic had governments not created subsidies, trade barriers and tax policies to protect some small piece of their own cultural marketplace—and by extension, Canada's identity and voice as a distinct nation. Since the mid-1960s, Canadian culture has been transformed from a nonentity to the country's fourth-largest employer.

This year, however, cultural industries have become the focus of a rash of trade disputes. In March, then-U.S. trade representative Mickey Kantor requested a World Trade Organization reversal of Canadian tax measures against *Sports Illustrated*'s "Canadian" edition. Kantor also put Canada on the hit list of a new "trade enforcement" task force over issues like cable TV licensing, satellite transmissions and book distribution. In late April, acting trade representative Charlene Barshefsky signaled she would take the same tough line against copyright reforms intended to bring extra royalties to the Canadian music industry.

In Canada, the public television network (CBC) has announced it will cease carrying U.S. programming in prime time, and other cultural sectors are lobbying for greater regulation and protection. Such sentiments may seem to run counter to Canada's 1987 bilateral trade agreement with the United States, as well as the North American Free Trade Agreement, but in fact Canadian artists demanded and got an exemption for cultural industries in both deals. In every negotiation since 1987—including current talks to bring Chile into NAFTA—the United States has demanded that the culture issue be reopened, and Canada has refused.

In part this is a problem of perception. "For them it's the entertainment industry. But for us, it is national cultural identity." Keith Kelly of the Canadian Conference of the Arts told the Toronto *Globe and Mail* in January. "They can't accept that. It's like we're speaking a strain of Mandarin." Canadian cultural advocates say that U.S. media corporations' inability to admit any link between cultural industries (as defined in NAFTA) and the cultural life is highly self-interested. "To the U.S. entertainment industry, the U.S. domestic market includes Puerto Rico, Guam and Canada," says Johnson. "This is a sixty-year standing insult."

At least NAFTA is bringing this clash to the surface. Sandy Crawley, president of Canada's Association of Cinema, Television and Radio Artists, says his American counterparts are to get the picture. "People have had to start acknowledging that cultural industries exist, because they're defined in the agreements. The Screen Actors Guild is starting to understand what we mean when we talk about cultural development. They used to say 'Culture? What culture? This is showbiz!'"

In the halls of cultural power, however, that recognition has yet to dawn. In 1991, U.S. trade representative Carla Hills infuriated Canadians when she tried to soothe fears about NAFTA by assuring that Canada could still have "fairs . . . and that sort of thing." One of the standard formulas to decry cultural trade barriers has been, "This isn't about culture, it's about money." The U.S. recording industry has accused Canada of cloaking a "culture of greed" behind the rhetoric of national identity. Time Warner complains of "discrimination" in publishing regulation. And Jack Valenti, Hollywood's head lobbyist and the man that Canadian culture advocates most love to hate, says Canada's "draconian" protections are spreading "a contagion effect around the world."

Dan Johnson argues that "the U.S. protects what it considers fragile industries, like lumber, or national-interest industries like government procurement. In Canada, film is a fragile industry. Culture is the national interest. Why? Because we are another country. There is a border. We are not just more Ohio."

Canada has been wary of creeping colonialism for more than a century, and culture has always been the sorest point. In the days before government subsidies and trade protections created a domestic book industry, Canadian authors were often pressured by their U.S. publishers to transpose their plots from, say, Saskatchewan to North Dakota. And it is a long-standing truism that Canadian actors and musicians have had to move to the United States to get recognition at home, because of U.S. control of the publicity and marketing

machines. This year's Canadian-born Grammy winners Shania Twain and Alanis Morrisette, for example, were simply following in the footsteps of Neil Young and Joni Mitchell when they moved south and signed to American labels.

Nevertheless, since the 1960s, Canada has become more competitive in home-grown films, theater, literature and music. Many of these businesses struggle to stay afloat, but have helped produce figures like novelists Michael Ondaatje and Margaret Atwood, singer Jane Siberry, the Kids in the Hall comedy troupe, director Robert Lepage and filmmakers Atom Egoyan (*Exotica*) and Denys Arcand (*Jesus of Montreal*). All these artists remain based in Canada, producing distinctly non-American works to international acclaim. Quebec artists benefit from a ready-made francophone audience and have developed a loyal public at home and in France.

Still, an overall imbalance remains. The NAFTA exemption protected Canadian artists' achievements but did nothing to expand their influence or give them greater access to American audiences. Free-trade critics also point to several other flaws in the agreement. The exemption clause is immediately followed by the provision that the United States may react with measures "of equivalent commercial effect" to any Canadian cultural policy it deems "inconsistent" with the thrust of NAFTA. Under NAFTA sanctions could have been imposed in the *Sports Illustrated* case, for example, without any arbitration process. The United States took it to the W.T.O. instead as a diplomatic concession.

In addition, several important pieces of cultural legislation in Canada were shelved or watered down while the trade agreements were being negotiated. One long-promised measure would have guaranteed 15 percent of film distribution revenues to Canadians by preventing Hollywood studios from buying up Canadian rights to foreign and independent films. Its final form was so gutted that the distributors themselves lobbied for it to be killed. Tax breaks for domestic moviemakers and a preferential postal rate for Canadian magazines have also been drastically reduced. (The United States had threatened to scrap free trade if such types of policies were not abandoned.)

Vincent Mosco, professor of mass communications at Ottawa's Carleton University, says NAFTA also mortgaged the future by tying culture to particular technologies. If the Internet becomes a major carrier of arts programming and information, for example, government efforts to insure Canadian content—or public systems like Freenets in any of the three countries—might well be actionable as distortions of trade. Under NAFTA, the commercialization of information technology is effectively mandatory.

However, the present battles are not about new technology, or even new policies. Take the *Sports Illustrated* dispute. Canada moved almost thirty years ago to ban imports of what are known as "split-run" magazines. These are special editions of U.S. magazines to which one or two Canadian stories have been added. The publisher calls it a "Canadian" edition and can sell advertising in Canada at bargain rates, since most costs have already been covered by the main edition of the magazine. The Canadian magazine industry argues that it already competes with U.S. magazines, but that split-runs are unfair, the cultural equivalent of dumping. If *Newsweek* did it, it could kill *Maclean's* weekly. An *Utne Reader* split-run could devastate small liberal publications like *This Magazine* and *Canadian Forum*.

In the early nineties, Time Warner realized that if it beamed content electronically to a Canadian printer, it could circumvent the ban. Thus the "Canadian" edition of *Sports Illustrated* was born. Because there is no Canadian general-interest sports magazine, Time Warner argued that it was not competing with anyone. Canada argued that the precedent was disastrous and slapped an 80 percent excise tax on the magazine's advertising profits. The matter will be heard at the W.T.O. over the next year.

The United States is not only the largest cultural exporter in the world—entertainment is the second-largest U.S. export industry, after aerospace—but one of the least enthusiastic importers. Less than 2 percent of time on U.S. television screens is given to foreign programming, for example. Jim Millier, editor of a U.S. publishing trade journal, noted in 1991 that though Americans were constantly "yapping" about Canadian regulations, "when foreign firms come in here, [Americans] too start talking about limiting foreign investment."

Just as there are Americans who willfully misunderstand Canada's interests, there are Canadian nationalists apt to blame every negative development since 1987 on free trade. The past decade in Canada has been marked by cuts to cultural programs and public broadcasting, met with shouts of American conspiracies. Overall, however, few Canadians wish to bar U.S. culture. It is so integral to Canadian life that the country's great satirists and critics (from the Second City TV troupe to Marshall McLuhan and cyberpunk writer William Gibson) have made a specialty of mocking and dissecting American icons. Yet a deep wariness of American power remains, and suspicion about the motivations of Canadian elites in consorting with it.

As Vincent Mosco sees it, what's at stake is not American incursion but the leveling effect of corporatism. "There is an ideological predisposition against the public sphere, an ideological and economic pressure from private interests in each country. On culture, that strategy is more complex because there is a substantial constituency that supports a strong Canadian presence in the arts," he says. This analysis gives some inkling of why Prime Minister Jean Chrétien and Trade Minister Art Eggleton—neither of them seen in Canada as a friend of the arts—have made so much of their defense of Canadian culture. Federal and provincial debt-reduction fervor has decimated the budgets of Canadian arts councils, subsidy programs and the Canadian Broadcasting Corporation. International posturing is meant to compensate for the fact that the Liberal government in Ottawa has broken its election promise to protect these institutions.

With Canada's cultural policy so loaded with contradictions, some of its tough stances end up backfiring. A case in point was the controversy this winter over the U.S.-based Borders book chain.

Borders gave up early this year after months of negotiations on bringing its big-box superstores to Canada. Canadian publishers welcomed the concept of a new retailer, but wanted Borders to respect Canadian-held distribution rights over U.S. and foreign titles. The publishers depend heavily on their distribution business to subsidize homegrown product, and feared Borders would bring in books from its warehouses in the States. The publishers asked only that Ottawa make a clear statement that copyright rules would be strictly enforced. Instead, the issue was confused by Canada's sole bookstore chain, Chapters, which lobbied against Borders being permitted in the country at all. By wrapping itself in the flag, Chapters effectively camouflaged its interest in market control. Ottawa vacillated, Borders withdrew and Canada was left with a monopoly. The result: If Chapters doesn't stock a particular book, it might not be sold in Canada at all.

In such cases, American complaints that Canada is protecting business, not culture, do have some validity. Media concentration in Canada is even more severe than in the United States. Canadian conglomerates like the Rogers cable company and the Southern newspaper chain appeal for relief from competition rules by claiming there is a need for Canadian behemoths to square off more equally against American ones like Paramount or MCA. In Canadian cable, three companies have 68 percent of the audience, up from 36 percent in 1983. Two corporations control more than half the daily newspaper circulation. Government intervention has also created a monopoly in direct-to-home satellite services, reducing potential buyers of independent TV productions and prompting some consumers to contract illegally with satellite companies over the border.

"Capital may be rapacious," says Mosco, "but it exists on both sides of the border. The free trade agreement is a convenient excuse to permit increased integration and concentration among Canadian companies. Rogers provides Canadian content only because the market wants it and the government requires it. A more effective nationalist policy would be to permit foreign companies but under tougher Canadian content rules. That would be genuine free trade."

Although its market of 26 million consumers is attractive, Canada is most important to U.S. media as the front line of planetary expansion. As "globalization" Americanizes the world with McDonald's franchises and *Baywatch,* other countries are beginning to follow Canada in protecting their own. France made film rights a central issue in the last round of GATT negotiations. The late French director Bertrand Tavernier, although a lover of American film, defended the French position: "Do we want to subject people to one language, to one culture, to one way of seeing things?" The French had a name for their fear of being erased by American images and ideas: They called it Canadianization. They also (a bit naïvely) said that what they were seeking was something like Canada's cultural exemption in NAFTA.

Mexico did not get an explicit cultural exemption in the deal, but was allowed to retain nationalist cultural policies that go much further than Canada's. Thirty percent of movies on the Mexican screen must be Mexican; on TV, news announcers and program hosts must be Mexican nationals. Still, Mexican artists and writers are beginning to fear NAFTA's effects. At a conference in March, Quebec International Affairs minister Sylvain Simard discussed the possibility of a joint Mexico–Quebec strategy for minority-language cultural protection. This is the kind of domino effect that gets Jack Valenti speaking of an infection in the world.

Americans who feel affronted by the world's seeming hostility might pause and consider how little access they have to their own major media. U.S. citizens' perspectives are not well reflected in the stories that corporate magazines, films and TV tell either. One thing lacking in the debate on culture and free trade is a non-nationalist strategy to bolster grass-roots cultural efforts throughout the free trade area and beyond.

If there were such a movement, it might take its lead from Guillermo Verdecchia, a South American–born, Vancouver-based playwright who won Canada's Governor General's Award in 1993 for *Fronteras Americanas,* a meditation on borders, culture and polyglot identity. In his acceptance speech, Verdecchia took note of the competing philosophies:

"Thanks to all who have done something about the fact that your local video store stocks Canadian films in the foreign film section; thanks to all the administrators who subsidize art in Canada with skill and intelligence; and to anyone who ever did anything for free. Thank you to everyone who wrote something beautiful; to everyone who goes to the theater; to everyone who ever lent me a book; and thank you to everyone who recognizes that there is more to a city, a people, a nation, more to human endeavor than that which can be measured in the terms dictated by the goddamn, so-called free market."

40

Brave New World

Right to Communicate Reshapes Human Connections

AN INTERVIEW WITH HOWARD FREDERICK

The author of Global Communication and International Relations . . . *and a lecturer in politics at the University of California/Irvine, Howard Frederick was formerly PeaceNet Director at the Institute for Global Communication in San Francisco. He also heads the International Communication Section of the International Association for Mass Communication Research and is a long-time communication activist.*

M&V: *Why should we be concerned about communication and human rights?*

Frederick: As educators and communicators, practitioners and producers, scholars and activists, we are concerned about how new media technologies are affecting all aspects of life and culture. This is especially true when we look at how the human right to communication and information has evolved.

M&V: *Why do you say evolved? Isn't it absolute and unconditional?*

Frederick: Not really. If you look at history, there have been three stages of communication rights. The earliest was *freedom of opinion*. Guaranteed first in ancient Greece, finally won in the Age of Enlightenment, freedom of opinion is a prerequisite for true democracy. But the advent of the world's first mass medium, printing, showed that it was not enough simply to hold an opinion. The "Gutenberg revolutionaries" found themselves continually subject to censorship and repression when they tried to express their opinion on royal and religious doctrines. This led to *freedom of expression,* guaranteed for the first time in the English Bill of Rights. The "Fourth Estate," the press as an institution, continued to grow in influence. By 1789 the U.S. Constitution guaranteed a third communication and information right, *freedom of the press.*

From Howard Frederick, "Brave New World: Right to Communicate Reshapes Human Connections," *Media and Values,* no. 61, winter, 1993, pp. 19–20. Copyright © 1993 by the Center for Media Literacy. Reprinted by permission.

> Everyone has the right to freedom of opinion and expression; this right includes freedom to hold opinions without interference and to seek, receive and impart information and ideas through any media and regardless of frontiers.
>
> *Article 19, Universal Declaration of Human Rights*

M&V: *It didn't stop there?*

Frederick: It's important to remember that these freedoms were guaranteed in only a few countries during the 19th Century. Then came the world wars and all these rights were crushed by totalitarianism and war propaganda. But after World War II, the entire community of nations reaffirmed and enshrined these rights in the famous Article 19 of the Universal Declaration of Human Rights.

M&V: *Where do we stand now?*

Frederick: Here we are, 45 years after the Universal Declaration. Vast changes have transformed the world from an agricultural and manufacturing society into a global information society. Communication is now so instantaneous that we speak of instant history and world public opinion. The struggle for true democracy has shown us that we need new channels of communication for *civil society,* that part of our lives that is neither market nor government but is so often overwhelmed by them. New technologies can decentralize and democratize communication flow, were it not for the media monopolies.

The problem is that the very powers that obstruct civil society at the local level—markets and governments—also control most of the world's communication flow. There is an increasing gap between the world's information rich and information poor populations. A handful of immense corporations dominate the world's media and exert a homogenizing influence over ideas, culture and commerce. Government monopolies still control a huge share of the world's airwaves and telecommunications flows. Markets and governments simply do not provide reliable and diversified information so essential to democracy.

M&V: *Doesn't the Universal Declaration protect people against monopolies?*

Frederick: Only partially. The right to communicate requires four components: (1) the opportunity to read and watch whatever is available (freedom of opinion), (2) the ability to express oneself freely (freedom of expression), (3) a diversity of sources to choose from (freedom of the press) and (4) media distribution channels that give the widest and most diverse segments of the population the ability to reach others. Article 19 safeguarded only the first three conditions. One crucial component is missing: *access to media channels.* But access is controlled everywhere by market and governmental monopolies. People must have resources to satisfy their compulsion to communicate in the practice of democracy. Those who wish to use the media channels should have fair and equitable access to them without discrimination so that they can take part in public affairs or exercise any other human right and fundamental freedom.

M&V: *Where did this idea come from and what kind of actions are being taken to make these rights more widespread?*

Frederick: Actually, the idea of the right to communicate originated here in the United States at the University of Hawaii in the 1960s and now has a considerable institutional history. Some of its supporters convene yearly in the "MacBride Roundtable," named after Sean MacBride, whose UNESCO commission studying global communication problems during

1976–1980 supported the right to communicate. Today's "MacBride Movement" is led by London-based World Association for Christian Communication (WACC). WACC's efforts can be seen in the National Council of Church's statement. . . . Here we see the premier organization representing 42 million American Protestant and Orthodox churches declaring that the right to communicate is a basic human right.

M&V: *Tell us more about the MacBride Roundtable. Why are its actions important?*

Frederick: WACC helped organize the latest MacBride Roundtable in Brazil in August 1992. Former "cold war adversaries" Leonard Sussman (Freedom House) and Kaarle Nordenstreng (International Organization of Journalists) found themselves in agreement on the reasonable and balanced nature of the MacBride Report and its continued validity. UNESCO remains paralyzed because of U.S. withdrawal, Nordenstreng said, so the movement is shifting to citizens' organizations. Sussman stated his conviction that new communications technologies will ultimately produce a new information order with access for everyone.

Armando Rollemberg, current president of the International Organization of Journalists, acknowledged that many organizations in the former socialist bloc, including his own, were used by the USSR to defend state control of the media. One crucial mistake, Rollemberg noted, was not defending the free flow of communication through a plurality of communication systems worldwide.

M&V: *What is happening now on communication rights? What are you hoping for?*

Frederick: In 1993 we have the World Conference on Human Rights (Vienna, June 14-25, 1993), the first U.N. human rights conference in 25 years. The World Conference has been called to review the 1948 Universal Declaration of Human Rights and to seek ways to improve it. I'm hoping that the "right to communicate" will be added to the Universal Declaration on Human Rights. Properly supported, the "right to communicate" can release us from a greater tyranny than any monarch's—the barriers that prevent human beings from communication with one another.

41

Balance Bias with Critical Questions

PATRICIA HYNDS

When I read the daily newspaper or watch the evening news, what is missing often upsets me as much as what is there. What is missing frequently seems to be the truth.

Of course, as the Chinese proverb frequently quoted by the late Latin American journalist, Penny Lernoux, says, "There is your truth, there is my truth and there is the truth."

From Patricia Hynds, "Balance Bias with Critical Questions," *Media and Values,* vol. 50, spring, 1990, pp. 5–7. Copyright © 1990 by the Center for Media Literacy. Reprinted by permission.

Most of us are convinced that "my truth" is "the truth." But as we move into the '90s, more and more sophisticated news management techniques will be used to convince us that someone else's "truth" should be ours. After living in Latin America most of the past nine years, I am convinced that there is a great deal of untruth, some of it deliberate, in what is presented in U.S. media about the rest of the world and even about domestic issues. However, with a little practice, you can learn to recognize the subjective underpinnings of a story.

One important point to remember is that objective reporting is a myth. Every reporter brings to the story his/her own biases and world view. Each reporter has to make choices in writing the story: what to include, what to leave out, what sources to use. A few well-placed adjectives, a few uses of "alleged" or "so-called" can cast a definite ideological twist.

Two reporters can see the same event very differently. I experienced this in a dramatic way when the pope visited Nicaragua in 1983. While many U.S. reporters, especially those arriving and departing with the pope, saw crowds "jeering and heckling" the pontiff, others saw a very different reality—poor Nicaraguans concerned at the continuing loss of their loved ones in the contra war and frustrated at what they felt was the Holy Father's refusal to respond to their pain.

The struggle to appear balanced can obscure "the truth," and it often rests on shaky assumptions. One is the principle that if two perspectives are totally opposed, the truth must lie somewhere in the middle. Another principle stresses that the media must never appear one-sided. Thus, much violence in Third World countries and elsewhere is presented as innocent civilians caught in the crossfire between two equally repugnant forces—even in the face of clear evidence of greater levels of abuse by one side.

Another version of the distorted idea of balance requires that every quote that contradicts previous norms, assumptions—or U.S. policies—must be countered by a quote from the administration or a "Western diplomat" or "high official source." This appearance of balance usually leaves the reader hopelessly lost.

The effort to appear objective frequently results in just the opposite, a weighted coverage favoring the current political "party line," or at least not challenging the conventional perspective. Even in domestic coverage, reliance on official sources and the distorting effect of prejudices and fears can lead to the kind of injustice that occurred in late 1989 when William Bennett, a black, was arrested for a murder apparently committed by the white Charles Stuart. Stuart used fears about minority crime to avoid suspicion for the murder of his wife. The Boston media's dependence on police sources and automatic assumptions about racial tensions helped create a false picture.

VARIED VIEWS

Many factors impede the transmission of accurate information, including changes taking place in the media itself, from more and more outlets owned by fewer and fewer corporations/conglomerates to TV coverage that focuses on the 30-second sound bite instead of description or analysis. In newspaper coverage, superficial, but very popular treatments such as the headline format developed by *USA Today*, work against critical analysis that could challenge official propaganda.

Also, reporters increase their access to sources when they write material that meets source approval, and lose it when they challenge the assumptions of those sources.

Less and less often do major networks or newspapers, let alone local media, station correspondents overseas for any length of time. So when an international story breaks, reporters fly in with no background on the issue, often without speaking the language or understanding anything of the area's history or culture. The result is a too-easy reliance on "official" sources.

The missing voices of activists and grass roots sources are a cause for great concern even when establishment bias is unintentional. But reporters' dependence on authorities makes them—and by extension media consumers—particularly vulnerable to deliberate attempts to mislead by governments and agencies. A case in point was the Office of Public Diplomacy, set up in the State Department during the Reagan administration to drum up support for the contras. Supervised by Oliver North, press releases were created that deliberately put out false information. In point of fact, reporters should consider the axes being ground by any government office of information, but all too often their accounts are taken at face value.

READING BETWEEN THE LINES

Judging the accuracy of what we read, see or hear in the media is not easy. However, it is possible to ask critical questions, to help determine whether or not we are getting the whole story.

1. *Who are the sources and what are their perspectives? How many are U.S. officials? How many are "unnamed"? How many are critics of U.S. policy?* Be suspicious of accounts that come only from high officials and of any reports released during a military emergency. (Journalist were banned from Grenada until the U.S. invasion was over, and press pool reports were reviewed by the military during the invasion of Panama.) But more subtle influences occur all the time—for example, are foreign "person on the street" interviews only with those who are well-educated or who speak English well?

2. *Are significant questions left unasked or unanswered? Is the political, social, economic or historical context missing?* This is especially important in the early coverage of a breaking event. In the aftermath, many questions will be raised and criticisms made, but what will be remembered is what we first saw or heard. To the U.S. public, the event is an issue only as long as it is a headline or lead story. The fleeting appearances and disappearances of African famine from the news are one example. When coverage occurs, usually around a crisis situation, it seldom examines underlying social and political causes. If they are mentioned, the U.S. role in them is rarely questioned.

3. *Do quotes seem abridged or out of context?* The *New York Times* quoted U.S. church worker Jennifer Casolo, upon her release by the Salvadoran military, saying, "I don't think I suffered." According to *Newsday,* she said, "I don't think I have suffered as terribly as the thousands of Salvadorans have suffered here." Quite a difference.

4. *Does coverage seem to offer a partial or selective history of events?* References to returning Panama to "democracy" and to Eric Del Valle as the "last elected president of Panama" are a case in point. In actuality, Del Valle was both instituted and removed from office by General Manuel Noriega. The last elected president of Panama was Nicolas Ardito Barletta. He was elected in May 1985, in elections—defended by the U.S.—that were so fraudulent he was referred to as "El Fraudito."

5. *Are exaggerated or rhetorical claims reported uncritically without journalistic scrutiny?* Loaded adjectives or phrases like "terrorist," "left-wing," "freedom fighters" (or President Ronald Reagan's "evil empire" remarks) should alert us to short-hand rhetoric that conveys an ideological point of view. One of the most overused and misused words in political coverage is "democracy." Democratic-style government can be organized in a variety of ways; the U.S. model is only one. We can be suspicious when the press repeats phrases over and over as if they held the same meaning for everyone.

6. *What stories or events are not covered?* For example, during the media's recent focus on Eastern Europe, Brazil's presidential elections passed virtually unnoticed, even though there are more Brazilians than Poles, Hungarians and Czechs put together. Many parts of the world are covered by U.S. media only if American troops are there, if the U.S. has major economic interests or if the administration is waging a propaganda offensive. It is a sad journalistic truism that many Third World countries only make the news if there are "coups or earthquakes."

No one can form an opinion from stories that are never covered. That's why alternative sources of news are so important. But the critical reader can learn to read between the lines of brief and biased accounts. Asking the right questions about what we see and hear is the first step toward making the informed political choices on which the freedom we value so highly depends.

• • •

9

The Cultural Survival
of Indigenous Peoples

*I insist . . . that symbolic acts are not enough for
indigenous peoples. Our civil, political, economic,
social and cultural rights cannot wait any longer.*

RIGOBERTA MENCHU

On every continent, *indigenous peoples* are
struggling against forces that threaten to
destroy them. Indigenous peoples are the
original inhabitants of areas of the world that have
been incorporated economically and politically by
more powerful peoples. Included are the peoples
of areas that were conquered and colonized by the
capitalist states during expansion of the capitalist
world economy over the last 500 years. Also
included are peoples who lost their independence
to socialist states such as China or the Soviet
Union (or the precapitalist empires that preceded
them) and have yet to regain it, even as these
socialist nations now take a capitalist path.
According to one estimate, there are 6,000 indigenous groups comprising some 600 million people
in the world today, about 10 to 15 percent of the
world's population. This includes such diverse
groups as the native peoples of the United States,
Canada, Central and South America; the aborigi-
nals of Australia; the Tibetans of China; the Penan
people of Sarawak in Malaysia; and the Eskimos of
the Russian North (Clay, 1993, pp. 65–68).

Amazingly, despite 500 or more years of invasion, displacement, genocidal colonial policies,
destruction of natural habitats, and forced cultural
assimilation, indigenous peoples have survived,
and their movements for human rights and self-
determination are enjoying growing support. As a

reflection of this, the United Nations declared
1993 to be the International Year of the World's
Indigenous Peoples and drafted the Declaration
on the Rights of Indigenous Peoples. Yet, as indicated in the introductory quote from Rigoberta
Menchu, Guatemalan Indian leader and winner
of the Nobel Peace Prize, such support must
become more than symbolic if indigenous peoples are to truly survive and thrive. Indigenous
cultures remain under attack on many fronts. The
articles in this chapter have been selected to illustrate some of the major forces confronting
indigenous cultures today, their destructive impact
on indigenous communities, and the growth of
indigenous peoples' resistance movements, despite
the powerful forces arrayed against them. It is
perhaps not surprising, given the history of conflict between indigenous peoples and the colonial
nations, that most of this material reflects the conflict perspective in sociology.

A prime engine of destruction of indigenous
cultures is clearly the relentless growth of the capitalist world economy. Indigenous peoples everywhere are pitted against corporations that seek to
exploit their homelands for profit. The ancestral
lands of indigenous peoples are rich sources of
minerals; timber; land that can be cleared to graze
cattle, grow commercial crops, or construct golf
courses; and rivers that can be dammed for giant
hydroelectric projects to provide power to urban
consumers. In their quest for profits, corporations
and collaborating governments are displacing
indigenous peoples, wrecking the environments
that have sustained their way of life, and even

killing those people who attempt to organize resistance to the onslaught. In "Our Homes Are Not Dumps: Creating Nuclear-Free Zones," Grace Thorpe discusses the impact of the nuclear power industry on native peoples in North America. In "Death of a People: Logging in the Penan Homeland," Wade Davis describes the struggle of the Penan people on the island of Borneo in the Malaysian state of Sarawak to preserve their homelands—part of Malaysia's extensive rain forests—against the vested interests of the logging industry.

These two articles also illustrate another important point about the issue of cultural survival. The economic and political forces threatening the survival of indigenous cultures are very often the same forces that threaten the survival of the earth's natural environment. Many indigenous peoples have historically seen themselves as caretakers of the earth, and their religions and philosophies envision human beings living in harmony with nature. This contrasts sharply with the worldview of capitalist corporations, which places domination and exploitation of the natural world for short-range profit higher in the hierarchy of values than preservation for future generations. It seems likely that indigenous peoples will conserve the environment if they are able to survive to do so. Their resistance movements, if successful, thus have fateful and hopeful implications for the survival of the earth, and thus humanity as a whole.

In the wake of many years of economic disruption, indigenous communities in many parts of the world have undergone profound social disintegration. Social problems that plague many indigenous communities include high rates of alcoholism, suicide, domestic violence, poverty,

and unemployment. For example, in South America, a rapidly increasing suicide rate among indigenous peoples has attracted global public attention. Diana Jean Schemo reports on one situation in "Indians in Brazil Wither in an Epidemic of Suicide." In various parts of the world, the social disintegration of indigenous communities is closely connected to the problem of forced cultural assimilation. For example, in the United States, government policies coerced indigenous people to conform to the dominant culture of the conquering state. The poem "Looking for Indians," by Cheryl Savageau, is a poignant expression of the loss of indigenous culture and the contemporary quest for identity by U.S. Native Americans.

The attacks on indigenous communities have led to many forms of resistance. Winona LaDuke's "Like Tributaries to a River" discusses Native American movements in the United States, with emphasis on native environmental organizing. The chapter continues with an example of successful economic development. Some indigenous peoples are managing to successfully compete in the capitalist economy. In "How to Succeed in Business: Follow the Choctaws' Lead," Fergus M. Bordewich outlines the steps taken by the once-impoverished Choctaw nation to become one of Mississippi's largest employers. Although such cases are an exception on the world scene, Bordewich concludes that native history "is, after all, not only a story of wars, removals, and death, but also one of compromises and creative reinventions of Indian communities continually remaking themselves in order to survive." With growing self-determination, the cultural survival of indigenous peoples will undoubtedly take many forms.

WORKS CITED

Clay, Jason W. 1993. "Looking Back to Go Forward: Predicting and Preventing Human Rights Violations," in

Marc S. Miller, ed. *State of the Peoples.* Boston, MA: Beacon Press.

QUESTIONS FOR DISCUSSION

1.   What are some of the social forces threatening to destroy indigenous cultures, as discussed in "Our Homes Are Not Dumps," "Death of a People," and "Indians in Brazil Wither in an Epidemic of Suicide"? Are there any indigenous peoples in your own community who face similar conditions? Compare and contrast what you know about their situation with the examples discussed in the above-mentioned articles and other readings in this chapter.

2.   What does Cheryl Savageau mean by "looking for Indians"? Could a similar poem be written about people in the United States from other cultural backgrounds—for example, European-Americans who came here from Ireland, Italy, or Poland; African Americans, Latinos, Asian Americans, and others? Why or why not?

3.   What does Winona LaDuke mean by "like tributaries to a river"? Why have Native Americans emphasized environmental organizing?

4.   Compare and contrast the efforts made by the organizations discussed by LaDuke with those made by the Choctaw people, as discussed by Bordewich in "How to Succeed in Business." Why are success stories such as the Choctaws' exceptional in the United States and worldwide?

 INFOTRAC COLLEGE EDITION: EXERCISE

In 1995, the United Nations held a World Conference of Indigenous Peoples. Learn more about what happened at this conference:

Using InfoTrac College Edition, search for the following article: "Last Chance for First Peoples." Stephen Mills. *Omni.* March 1995 v17n6p62. (*Hint:* Enter the search term indigenous peoples, using the Subject Guide. Go to Subdivisions: Conferences, Meetings, Seminars.)

What are some of the problems affecting indigenous peoples discussed at the conference? What are some of the solutions under discussion? What role has the United Nations been playing in the search for solutions? What role has the United States been playing?

FOR ADDITIONAL RESEARCH

Books

Jaimes, M. Annette, ed. 1992. *The State of Native America.* Boston, MA: South End Press.

Miller, Marc S., ed. 1993. *State of the Peoples.* Boston, MA: Beacon Press.

Wolf, Eric R. 1982. *Europe and the People Without History.* Berkeley, CA: University of California Press.

Organizations

Akwe:kon Journal
Cornell University
400 Caldwell Hall
Ithaca, NY 14853
(607) 255-4308
http://nativeamericas.aip.cornell.edu

Center for World Indigenous Studies/Fourth World Institute
1001 Cooper Point Road SW, Suite 140-214
Olympia, WA 98502
Toll free for U.S. calls: (888) 286-2947
Washington State or international calls: (360) 754-1990
www.cwis.org

Cultural Survival
215 First Street
Cambridge, MA 02142
(617) 621-3818
www.cs.org

Human Rights Watch
485 Fifth Avenue
New York, NY 10017
(212) 972-8400
www.hrw.org

ACTION PROJECTS

1.   Go to the Web site for Cultural Survival (www.cs.org) and find "Active Voices." This is an information center that contains bulletins alerting the public about immediate situations affecting particular peoples, and notifying us as to what we can do to be of assistance. Read the current bulletin(s). What are the forces threatening the survival of these peoples? What kind of assistance is asked for? Get together with some other students and take part in the activities asked for. Reflect on what you learned from your experience in relation to the issue of cultural survival of indigenous peoples.

2.   From the books or Web sites of organizations listed above, identify one particular people struggling for cultural survival. This could be a group in the United States, or a people elsewhere in the world. Do research on the Internet or in the library to learn more about their history, culture, and current situation. What are some of the main features of the culture of this people? How do they make their living? What are some of the global economic, political, and cultural factors threaten-

ing their survival? What are they doing to fight back against these forces?

3.   If you are located in a region where there is a Native American nation, contact the nation, and invite someone to come teach the class about the nation's history, culture, and current situation. Before the speaker comes to your class, prepare a list of questions to ask when the speaker opens the talk up for questions and discussion. Reflect on what you learned from this experience about the issue of cultural survival of indigenous peoples. *And/or:* If you are located in a region where there is a Native American museum, make arrangements for a class field trip to the museum. Reflect on what you learned from this experience about the issue of cultural survival of indigenous peoples. (Note: If you are on the East Coast, there are two substantial museums—the Smithsonian's National Museum of the American Indian, in New York City, and the Mashantucket Pequot Museum and Research Center, in Mashantucket, Connecticut. However, many smaller museums with excellent information exist throughout the United States, as well.)

42

Our Homes Are Not Dumps

Creating Nuclear-Free Zones

GRACE THORPE

The Great Spirit instructed us that, as Native people, we have a consecrated bond with our Mother Earth. We have a sacred obligation to our fellow creatures that live upon it. For this reason it is both painful and disturbing that the United States government and the nuclear power industry seem intent on forever ruining some of the little land we have remaining. The nuclear waste issue is causing American Indians to make serious, possibly even genocidal, decisions concerning the environment and the future of our peoples.

I was a corporal, stationed in New Guinea, at the end of World War II when the first atomic bomb was dropped on Hiroshima. The so-called "nuclear age" has passed in the beat of a heart. As impossible as it seems, this year will mark the fiftieth anniversary of that first blast. The question of what to do with the waste produced from commercial and military reactors, involved in weapons manufacture and the generation of nuclear energy, has stumped the minds of the most brilliant physicists and scientists since "Little Boy" was detonated above Japan on August 6, 1945.[1] No safe method has yet been found for the disposal of such waste, the most lethal poison known in the history of humanity. It remains an orphan of the nuclear age.

In rich areas, people have the leisure time to organize and easy access to media and elected representatives. For this reason, the nuclear industry is talking about locating disposal sites in poor regions. Indians are being deluged by requests. Devastation due to nuclear energy, however, is nothing new to Indian peoples.

Between 1950 and 1980, approximately 15,000 persons worked in uranium mines. One-fourth of these were Indian. Many of these mines were located on lands belonging to the Navajos and the Pueblos. In 1993, Dr. Louise Abel of the Indian Health Service disclosed that, of the 600 miners tested who had worked underground for more than a year, only five qualified for payments under the Radiation Exposure Act of 1990. By 1994, only 155 uranium miners and millers or their families had been awarded compensation, less than half the claims filed at that time.[2] Radiation from tailings piles, the debris left after the uranium is extracted, has leached into groundwater that feeds Indian homes, farms, and ranches. High concentrations of radon gas continually seep out of the piles and are breathed by Natives in the area. Background levels of radiation are at dangerous levels. Thus Indians living near the mines face the same health risks as those working underground.[3]

In 1973 and 1974, two nuclear power reactors commenced operation at Prairie Island, Minnesota, only a few hundred yards from the homes, businesses, and child care center of the Prairie Island Mdewankanton Sioux. The facility

---

[1] "Little Boy" was the name given to the first device, dropped on Hiroshima by the bomber Enola Gay. Three days later, on August 9, 1945, "Fat Man" was released by Bockscar and detonated above Nagasaki. See *Bockscar 'ended' WW II, Denver Post*, Mar. 19, 1995.

[2] PETER H. EICHSTAEDT, YOU POISON US: URANIUM AND NATIVE AMERICANS 151, 170 (1994).

[3] *Id.* at 142–46.

From Grace Thorpe, "Our Homes Are Not Dumps: Creating Nuclear Free Zones," *Natural Resources Journal*, vol. 36, fall, 1996, pp. 715–723. Copyright © 1996 by the Natural Resources Journal. Reprinted by permission.

was on the site of the ancient Indian village and burial mound, dating back at least 2,000 years. On October 2, 1979, a 27-minute release of radiation from the plants forced evacuation of the facility, but the Tribe was not notified until several days later. By 1989, radioactive tritium was detected in the drinking water, forcing the Mdewankanton to dig an 800-foot deep well and water tower, completed in 1993. Prairie Island residents are exposed to six times the cancer risk deemed acceptable by the Minnesota Department of Health.[4]

By 1986, the problem of nuclear waste disposal had become acute. The U.S. Department of Energy began to explore the possibility of locating a permanent nuclear repository in Minnesota's basalt and granite hardrock deposits. Among the sites considered was the White Earth Reservation in the northwestern part of the state. The Anishiaabe who live there took the government's interest seriously enough to commission a study of the potential impact. The Minnesota legislature responded by passing the Radioactive Waste Management Act, stating that no such facility could be located within the state without the express authorization of the legislature.[5]

The following year, however, Congress voted to locate the permanent repository at Yucca Mountain, about 100 miles northwest of Las Vegas, Nevada, on land belonging to the Western Shoshone. Plans called for the opening of the facility in 2010. The Nuclear Waste Policy Act set in motion a nationwide search for a community that would accept a temporary storage site, until Yucca Mountain came online.[6] Indian tribes again were specifically targeted.

One by one, tribes who considered accepting the so-called Monitored Retrievable Storage (MRS) facility on tribal land decided against it. Today, of the 17 tribes who began discussions and study, only three remain: the Mescalero Apache of New Mexico, the Goshutes in Utah, and the Fort McDermitt Reservation in Nevada (which houses both Paiutes and Western Shoshones). In addition, Pojoaque Pueblo in New Mexico announced in March 1995 that it was considering locating the MRS on tribal lands. This move, however, was an overt power-play to persuade the New Mexico legislature to halt a bill that would expand gambling in the state to the detriment of the Pojoaque's own gaming interests. According to Pojoaque Governor Jacob Viarrial, if the public does not want his Tribe to store the waste, they should put pressure on the lawmakers to put a halt to the expansion of gaming off reservations.[7]

The National Environmental Coalition of Native Americans (NECONA) was formed in 1993 in Las Vegas to lobby against the MRS or any nuclear waste disposal on Indian lands and to encourage Native Nations to declare themselves Nuclear Free Zones instead. As the number of tribes considering the MRS dwindled, pressure on Washington mounted. NECONA persuaded U.S. Senator Jeff Bingaman of New Mexico, who had been one of the moving forces behind the Radiation Exposure Compensation Act for uranium miners, to oppose the MRS on the energy and appropriations committees. As a result, Congress withheld funding for the program.[8]

With the federal government out of the MRS-construction business, but with the problem of waste disposal still unresolved, utilities began to get desperate. Dozens of plants would be forced to shut-down or find alternative sources of fuel unless a temporary storage site were boated in the near future. Thirty-three utilities, accounting for 94 reactors, began seeking a location. Led by Northern States Power (NSP), the consortium approached Minnesota about locating a facility adjacent to the NSP plant at Prairie Island. Although the plant supplies 15 percent of the state's electricity, "not a single kilowatt reaches the Mdewankanton community it borders."[9]

[4]Jeff Amstrong, *Prairie Island confronts Nuclear Threat,* THE CIRCLE, Apr. 1994, at 16–17.

[5]*Id.* at 16.

[6]George Johnson, *Nuclear Waste Dump Gets Tribe's Approval in Revote,* N.Y. TIMES, Mar. 11, 1995, at 6.

[7]*MRS Plans Back on Burner: Pueblo Pursues Own Nuclear Waste Plan,* INDIAN COUNTRY TODAY, Mar. 16, 1995, at 2.

[8]Johnson, *supra* note 6, at 6.

[9]Armstrong, *supra* note 4, at 16.

The Prairie Island Sioux had applied for a Phase I MRS grant, which provided DOE funds for initial feasibility studies. According to tribal officials, however, the application was tactical. The intent was to use the government's own money to prove that neither an MRS nor a nuclear power plant should be located at Prairie Island.[10] One study showed that the cancer risk would be 23 times greater than the state standard. At the time of the NSP initiative, a survey showed that 91.6 percent of the tribe opposed construction of the MRS.[11] The tribe fought the NSP proposal before the legislature and won. They subsequently declared the Prairie Island Reservation a Nuclear Free Zone.

Meanwhile, NSP has signed an agreement with the Mescalero Apache to move ahead with development of an MRS, in New Mexico. Under the terms of the agreement, the tribe was to seek two 20 year licenses to store up to 40,000 metric tons of spent nuclear fuel. Total revenues over the 40-year life of the facility, estimated at $2.3 billion, would bring as much as $250 million in benefits to the Tribe.[12] The Tribal Council believed that it could proceed with the program by its own authority. It was confident enough of victory to put the issue to tribal members in the form of a public referendum. According, however, to a Native newspaper, *The Circle,* opponents of the storage facility considered the Mescalero tribal government, headed by Chairman Wendell Chino, "dictatorial," and likely to conduct a campaign of intimidation and vote fraud if a referendum takes place.[13]

I used to work for the National Congress of American Indians when Wendell Chino was its chair. He's been Mescalero chairman since 1962, and he has done great things for the people there; I respected him. He is tough, however, and can be a very imposing figure. The sad thing is that the Mescalero don't need this nuclear waste. They have a five-star resort, a casino, two ski lifts, and a sawmill. They have wonderful resources for forestry. Everybody thinks, Ah, the poor Apache, they need this development, but they don't.[14]

The referendum took place on January 31, 1995. The Mescaleros voted down the MRS by a vote of 490 to 362. Shortly after the vote, however, a petition began circulating, calling for a new election. According to Fred Peso, the vice chairman, "A group of grass-roots people presented the petition to the tribal council." Peso blamed "outside interference from environmentalists and other anti-tribal groups" for the defeat of the proposal.[15] In reality, Wendell Chino's powerful political machine was behind the petition. The Tribal Government controls jobs, housing, schools, and the court system. One of the organizers of the petition drive, Fred Kaydahzinne, is director of the federally subsidized tribal housing program. As Rufina Marie Laws, one of the referendum's opponents, stated, "it was real hard for people to turn him down."[16] Petition organizers gathered more than 700 signatures calling for a new vote. When a second ballot was held on March 9, 1995, the measure passed 593 to 372.

There is a great deal of uncertainty as to what will happen now at Mescalero. Opponents of the MRS could seek yet another referendum. They have stated that they will appeal the second vote to the Tribal Court, but they are not optimistic. The State of New Mexico has prohibited transport of spent nuclear fuel on state highways, in an attempt to derail the proposal. Vice Chairman Peso has announced that the Tribe will proceed with licensing applications and technological studies. Officials of NSP have announced that they will move ahead with plans for the project. Contracts are being finalized, and licensing is

---

[10]*Id.* at 17.

[11]*Id.* at 16.

[12]Harlan McKosato, *Mescalero Nuclear Site Back on Track,* INDIAN COUNTRY TODAY, Mar. 16, 1995, at 1.

[13]Armstrong, *supra* note 4, at 17.

[14]According to a recent article by D. C. Coles, a Chiricahua Apache, despite these successful business ventures, unemployment remains a problem. Unemployment was estimated at 30 percent, with much of the rest of the work force underemployed. Health and education levels remain below the national averages. D. C. Cole, *Apache,* in NATIVE AMERICA IN THE TWENTIETH CENTURY: AN ENCYCLOPEDIA, 46 (Mary David, ed. 1994). The nuclear project is anticipated to produce between 200–300 jobs. *See* McKosato, *supra* note 12, at 1.

[15]Johnson, *supra* note 6, at 6.

[16]*Id.*

anticipated to be concluded by December 1996.[17]

If the Mescaleros withdraw, there are the Skull Valley Goshutes in Utah and the tribes at Fort McDermitt standing right behind them. Both reservations are isolated, and unemployment is a problem on both. At the moment, Fort McDermitt seems to be out of the running because it straddles the Nevada state line. The law says that the MRS and the permanent site cannot be in the same state, but that could change. The Goshutes already have waste incinerators, nerve gas plants, and a bombing range bordering their lands. There is a feeling of indifference about the MRS among the few people who live on the reservation. They have signed an accord with Richard Stallings, a federal negotiator charged with locating a temporary storage site, to provide a framework for further talks, and the University of Utah has agreed to undertake a feasibility study with the utilities.[18]

We should also not believe that the problem is limited to the United States. First Nations in Canada are facing the issue. An article in the free trade agreement between Canada and the United States prohibits Canada from preventing nuclear waste coming into the country. The Meadow Lake Cree in Saskatchewan are in discussion with the Atomic Energy of Canada Ltd. (AECL), a corporation of the Canadian government, concerning becoming a permanent repository.

According to recent reports, they have also held negotiations with the Mescalero to become the storage site for wastes temporarily housed at the proposed Arizona facility. Meanwhile AECL continues to market nuclear technology throughout the Americas.[19] The situation in Mexico is terrible. They have very little environmental regulation. At NECONA, we hear reports of "jelly babies," babies born without any bones, due to environmental contamination.

Tribal officials at Mescalero and other reservations that have considered the MRS contend that the issue is one of sovereignty. They use the issue of sovereignty against the environment. It is a very tough tightrope to walk. How can you say to a tribe, "Hey, you shouldn't be doing this. You should be protecting the earth." Then they would turn around and reply, "Hey, we can do as we please. This is Indian sovereignty." In one sense, they would be right. Allowing utilities to build MRS facilities on our lands, however, is not truly an expression of sovereignty. Those supporting such sites are selling our sovereignty. The utilities are using our names and our trust lands to bypass environmental regulations. The issue is not sovereignty. The issue is Mother Earth's preservation and survival. The issue is environmental racism. The purpose of NECONA is to invite tribes to express their sovereign national rights in a more creative way in favor of our Mother, by joining the growing number of tribal governments that are choosing to declare their lands Nuclear Free Zones.[20] Fred Peso at Mescalero has declared, "It is ironic that the state continues to fight our tribe [over the MRS] when New Mexico has enjoyed the benefits of nuclear projects since 1945."[21] The real irony is that after years of trying to destroy it, the United States is promoting Indian national sovereignty—just so they can dump their waste on Native land.[22]

The DOE and the utilities have said that it is natural that we, as Native peoples, should accept radioactive waste on our lands. They have convinced some of our traditionalists that as keepers of the land they must accept it. As Russell Means has said, however, "We have always had our false

---

[17]McKosato, *supra* note 12, at 1–2; *see also Apache Continue with Nuclear Dump Plan,* INDIAN COUNTRY TODAY, June 8, 1995, at A2.

[18]*See Goshutes Sign Nuclear Waste Agreement,* NEWS FROM INDIAN COUNTRY, Dec. 1994, at 5.

[19]Joyce Nelson, *Candu Diplomacy and NAFTA's Nuclear Agenda,* Z MAGAZINE, June 1995, at 30–32.

[20]*See,* Grace Thorpe, *Statement to the National Congress of American Indians,* Sparks, Nov., Sept. 1, 1993 [hereinafter *Statement*]. Currently 15 tribes have passed resolutions declaring their lands to the Nuclear Free Zones. The first was the Sac and Fox Nation of Oklahoma, of which Ms. Thorpe is a member. In addition, through the efforts of Norma Kassi, the Yukon Territory in Canada has declared itself a Nuclear Free Zone as well.

[21]Johnson, *supra* note 6, at 6.

[22]Russell Means, Comment at the North American Native Workshop on Environmental Justice, Iliff School of Theology, Denver, Colo. (Mar. 17, 1995).

prophets."[23] The government and the nuclear power industry attempt to flatter us about our abilities as "earth stewards." Yet as I declared to the National Congress of American Indians in 1993, "It is a perversion of our beliefs and an insult to our intelligence to say that we are 'natural stewards' of these wastes."[24] The real intent of the government and the utilities is to rid themselves of this extremely hazardous garbage on Indian lands so they are free to generate more of it.

Our traditional spiritual leaders have warned us for hundreds of years about taking resources from the earth. They have warned that the earth will become unbalanced and be destroyed. In one of the stories the Navajos have about their origins, they were warned about the dangers of uranium. The People emerged from the third world into the fourth and present world and were given a choice. They were told to choose between two yellow powders. One was yellow dust from the rocks, and the other was corn pollen. The [People] chose corn pollen, and the gods nodded in assent. They also issued a warning. Having chosen the corn pollen, the Navajos were to leave the yellow dust in the ground. If it was ever removed, it would bring evil.[25]

Wherever there are uranium mines, wherever there are nuclear power plants, and wherever our people have been downwind on nuclear tests, the cancer rate goes up. Among the Western Shoshone in Nevada as a result of nuclear testing, many of the people now have thyroid cancer. They are dying a younger death. They have leukemia, which was unheard of in earlier times. In Minnesota, archaeologists excavating Prairie Island thousands of years in the future could be exposed to levels of radiation high enough to cause cancer.[26] Pollution and toxic waste from the Hanford nuclear weapons facility threatens all Native peoples who depend on the Columbia River salmon for their existence.[27] A few years ago, a vial of nuclear material

the size of a human little finger was lost on the road from Los Angeles to Sacramento. An SOS went out to all the newspapers and radio and television stations about this little silver vial: "If you find it, don't pick it up. Alert us immediately. If you pick it up and put it in your pocket for two days, you'll get sick. If you keep it a week, it can kill you. If you breathe the equivalent of 100th of a grain of salt, it can cause lung cancer."

Now those who visited all these horrors upon us want us to accept their nuclear waste, too. Darelynn Lehto, the vice president of the Prairie Island Mdewankanton, testified before the Minnesota State Senate during the fight against MRS there, "It is the worst kind of environmental racism to force our tribe to live with the dangers of nuclear waste simply because no one else is willing to do so."[28] Why do we tolerate it? How long can we tolerate it? What kind of society permits the manufacture of products that cannot be safely disposed of? NECONA is currently lobbying Congress for a bill that will say simply, "Nothing is to be manufactured, used, or reproduced in the United States that cannot be safely disposed of." Is that too simple a thing for a legislator to understand? Probably it is, but it makes sense, doesn't it?

Spent nuclear fuel is permeated with plutonium, the principal ingredient in atomic weapons. Plutonium has a half-life of 24,360 years. Significant amounts would therefore remain active for more than 50,000 years. The so-called permanent repository proposed for Yucca Mountain is designed to hold canisters containing nuclear waste for only 10,000 years. The steel containers holding the material would disintegrate long before the radioactivity had decayed.[29]

Yucca Mountain, however, is nowhere near on its way to becoming the permanent repository. It was originally to have begun receiving waste in 1998, but near unanimous opposition in Nevada slowed the process. In 1992, an earthquake measuring 5.6 on the Richter scale struck the area, raising additional questions as to the

---

[23]Grace Thorpe, *Radioactive Racism? Native Americans and the Nuclear Waste Legacy*, INDIAN COUNTRY TODAY, Mar. 16, 1995, at A5.

[24]Thorpe, *Statement, supra* note 20.

[25]EICHSTAEDT, *supra* note 2 at 47 (quoting Anna Rondon, Nov. 1992).

[26]Armstrong, *supra* note 4, at 16.

[27]David Rich Lewis *Environmental Issues. in* Davis. at 189.

[28]Darelynn Lehto, testimony before Minn. State Senate, Mar. 29, 1994, *quoted in* THE CIRCLE, Apr. 1994, at 17.

[29]William J. Broad, *Scientist Fear Atomic Explosion of Buried Waste*, N.Y. TIMES. Mar. 5. 1995. at 17.

site's viability.[30] Most recently, scientists at the Los Alamos National Laboratory in New Mexico raised the possibility that wastes buried at the Nevada location could explode after the steel container canisters dissolve, setting off a nuclear chain reaction.[31]

These factors make the targeted date of 2010—when Yucca Mountain currently is estimated to be accepting shipments of waste—look improbable. Mescalero tribal officials, in obtaining their tribe's permission, emphasized that their proposed facility was strictly temporary and that at no time would the tribe take possession of the fuel.[32] What will happen, however, if Yucca Mountain does not come online as projected? What if no permanent storage site is available at the end of the MRS' 40 years of a "temporary" storage? New Mexico Attorney General Tom Udall has raised similar questions. He fears that the state "may ultimately have to pick up the pieces."[33] Indians suspect we know who will be left holding the bag.

The debate over nuclear waste has already done serious damage to harmonious relationships among our people. Why must we go through this divisive agony again.[34]

As a mother and a grandmother, I am concerned about the survival of our people just as Mother Earth is concerned about the survival of her children. There is currently a moratorium on construction of nuclear power plants in the United States. There is also current legislation, however, that would allow new building if arrangements are made for the waste. Is this the legacy that we want to leave for our children and for our Mother Earth? The Iroquois say that in making any decision one should consider the impact for seven generations to come. As Thom Fasset, who is Iroquois, reminds us, taking such a view of these issues often makes us feel we are alone, rolling a stone up a hill. It keeps rolling back down on us.[35] That may be the only way, however for us to live up to our sacred duty to the land and to all of creation.

---

[30]Armstrong, *supra* note 4, at 16–17.

[31]Broad, *supra* note 29, at 1, 18.

[32]McKosato, *supra* note 12, at A2.

[33]Johnson, *supra* note 6, at 6.

[34]Thorpe, *Radioactive Racism?, supra* note 23, at A5.

[35]Thom White Wolf Fassett, North American Native Workshop on Environmental Justice, Iliff School of Theology, Denver, Colo., Mar. 17, 1995.

43

Death of a People

Logging in the Penan Homeland

WADE DAVIS

It is just after dawn and the sound of gibbons runs through the forest canopy. The smoke of cooking fires mingles with the mist. A hunting party returns, and the movement of the men reveals that they have killed a wild pig. One dart and the people eat for a week.

From Wade Davis, "Death of a People: Logging in the Penan Homeland," in *State of the Peoples* by Cultural Survival, pp. 23–31. Copyright © 1993 by Cultural Survival. Reprinted by permission of Beacon Press. Boston.

This mountaintop, where generations of Penan have come to pray, looks out over a pristine rain forest, past the clear headwaters of one of Sarawak's ancient rivers to distant mountains that rise toward the heart of Borneo. There on the horizon, coming over the mountains from seven directions and descending into the valley, are the scars of advancing logging roads. The nearest is six miles from this encampment of nomadic Penan. When the wind is right, you can hear the sound of chain saws, even at night.

Virtually no place in Penan territory is free of the sound of machinery. If the Sarawak government continues to have its way, this valley will be laid waste within a year, and the people will be forced from the land.

Straddling the equator and stretching 800 miles east to west and 600 miles north to south, Borneo is the third-largest island on earth. Six major rivers and hundreds of smaller streams drain the isolated center of the island, where the mountains rise to over 13,000 feet. Eighty percent of Borneo is blanketed by extraordinary rich tropical rain forest. Three countries claim parts of the island, with Indonesian Kalimantan encompassing the southern two-thirds; to the north the oil-rich sultanate of Brunei is flanked by the Malaysian states of Sarawak and Sabah.

Malaysia is the world's leading exporter of tin, palm oil, rubber, and pepper—and tropical timber. Sarawak encompasses roughly 38 percent of Malaysian territory. Among some 27 distinct ethnic groups in the state, the Melanau and Malay comprise one-fifth of the population of 1.2 million; another 30 percent of the people are Chinese or recent immigrants from Southeast Asia.

Close to half the population is Dayak, a term that refers to more than a dozen indigenous peoples, including the Iban, Bidayuh, Kenyah, Kayan, Kedayan, Murut, Punan, Bisayah, Kelabit, and Penan. The Penan, in northeastern Sarawak, number about 7,600, of whom 25 percent are settled. The remainder are semi-settled or nomadic and depend on the rain forest for most of their needs. Of an estimated 100,000 indigenous people who roamed the forests of Sarawak at the turn of this century, only the nomadic Penan remain.

"FROM THE FOREST, WE GET OUR LIFE"

Related in spirit to the Mbuti pygmies of Zaire and the wandering Maku of the Amazon, the Penan long depended on the forest for food. As hunters and gatherers, they moved through the immense wooded uplands that give rise to the myriad affluents of the Baram River in Sarawak's Fourth Division. Isolated groups of Penan ranged east into Indonesian Kalimantan and north into Brunei.

Due in part to a remarkable variety of soil types, complex geology, dramatic topography, and a broad range of climates, the forests of the Penan are among the richest, most diverse on earth. They may, in fact, represent one of the oldest living terrestrial ecosystems. Moreover, Borneo has remained remarkably stable: its forests have been essentially undisturbed for millennia. Until this century, human impact was slight.

Within the traditional Penan homeland are all the major forest types to be found inland from the coast in Borneo. These forests harbor a great many endemic species. No fewer than 59 genera and 34 percent of all plant species in the world are found only on Borneo. The fauna includes 30 unique birds and 39 endemic mammals, including such rare and endangered animals as the Sumatran rhino and the orangutan. One survey of 22 acres identified over 700 species of trees, more than have been reported for all North America.

For the Penan, this forest is alive, responsive in a thousand ways to their physical and spiritual needs. Its products include roots that cleanse, leaves that cure, edible fruits and seeds, and magical plants that empower hunting dogs and dispel the forces of darkness. There are plants that yield glue to trap birds, toxic latex for poison darts, rare resins and gums for trade, twine for baskets, leaves for shelter and sandpaper, and wood to make blowpipes, boats, tools, and musical instruments. All these plants are sacred, possessed of souls and born of the same earth that gave birth to the people. "From the forest," they say, "we get our life."

The Penan view the forest as an intricate, living network. Imposed from their imagination and experience is a geography of the spirit that delineates time-honored territories and ancient routes that resonate with the place names of rivers and mountains, caves, boulders, and trees. As much as myth or memory, the landscape links past, present, and future generations.

Stewardship permeates Penan culture, dictating the manner in which Penan use and share the environment. This notion is encapsulated in *molong*, a concept that defines both a conservation ethic and a notion of ownership. To *molong* a sago palm is to harvest the trunk with care. *Molong* is climbing a tree to gather fruit rather than cutting it down, harvesting only the largest fronds of the rattan, leaving the smaller shoots so they may reach proper size in another year. Whenever the Penan *molong* a fruit tree, they place an identifying sign on it, a wooden marker or a cut of a machete. The mark signifies effective ownership and publicly states that the product is to be preserved for harvesting later. In this way, the Penan acknowledge specific resources—a clump of sago, fruit trees, dart-poison trees, rattan stands, fishing sites, medicinal plants—as familial rights that pass down through the generations.

Identifying psychologically and cosmologically with the rain forest and depending on it for their diet and technology, the Penan are skilled naturalists, with sophisticated interpretations of biological relationships. A recent and cursory examination of their plant lore suggests that the Penan recognize over 100 fruit trees, 50 medicinal plants, 8 dart poisons, and 10 plant toxins used to kill fish. These numbers probably represent but a fraction of their botanical knowledge.

LOGGING IS

"GOOD FOR THE FOREST"

Today the Penan and their Dayak neighbors occupy the front lines of perhaps the most significant environmental struggle of our era—the effort to protect the integrity of the world's forests. The rate of deforestation in Malaysia is the highest in the world. In 1983, Malaysia accounted for 58 percent of the total global export of tropical logs, with over 90 per-

cent of the wood going to Japan, Taiwan, and South Korea. By 1985, three acres of forest were being cut every minute of every working day.

With primary forests in peninsular Malaysia becoming rapidly depleted, the industry increasingly has turned to Sarawak. Between 1963 and 1985, 30 percent of the forested land of Sarawak was logged. In 1985, 670,000 acres were logged, providing 39 percent of Malaysian production and generating $1.7 billion in foreign exchange. Another 14.3 million acres—60 percent of Sarawak's forested land—are held in logging concessions.

The banks of the Baram River in Sarawak's Fourth Division are lined for miles with stacked logs awaiting export. Although petroleum accounts for a far larger percentage of Sarawak's export earnings than timber, revenues from the oil fields flow almost entirely to the federal government. By contrast, the Sarawak government controls the forestry sector in the state. As of 1985, licensed logging concessions in Sarawak totaled 14.2 million acres; 23 percent had been exploited and the rest was scheduled for logging.

On paper, Sarawak has one of the world's most experienced and well-funded forestry departments, and its forest policy is impressive. In practice, forest management serves the interests of the ruling elite, which uses its control of the licensing of logging concessions as a source of personal wealth and a means of retaining economic and political power. The authority to grant or deny logging concessions lies strictly with the minister of forestry. Between 1970 and 1981, and since 1985, the highly coveted forestry portfolio has been retained by the chief minister.

That the Ministry of Forestry is used for political and personal financial gain became evident in April 1987. During an election campaign, Chief Minister Datuk Patinggi Hagi Abdul Taib Mahmud announced he was freezing 25 timber concessions totaling 3 million acres, all belonging to relatives and friends of former Chief Minister Tun Abdul Rahman Yakub. Estimates of the value of these holders ranged from $9 billion to $22 billion. In retaliation, Tun Abdul Rahman Yakub revealed the names of politicians, friends, relatives, and associates connected to Datuk Patinggi Hagi Abdul Taib Mahmud, who together controlled 4 million acres. Between them, these two factions controlled a third of Sarawak's forested land. Ironically, the two antagonists are themselves related.

The granting of logging concessions has been, in effect, a means of creating a class of instant millionaires, and nearly every member of the state assembly has become one. With a resource worth billions of dollars, the stakes are high. In a recent election, political parties spent over $24 million competing for a mere 625,000 votes. The only car factory in Sarawak produces BMWs.

Ultimate responsibility for exploiting the rain forest lies with powerful Japanese trading houses. Japan depends on Malaysia for 85 to 90 percent of its tropical wood imports; half of Sarawak's production goes north to Tokyo. In a 1984 speech, quoted to Evelyn Hong's *Natives of Sarawak,* then Malaysian federal minister Leo Moggie acknowledged that "Sarawak timber companies are dependent on these [Japanese] trading houses for their intricate line of credit." Japanese banks provide start-up loans to local logging companies. Japanese companies and Japanese aid finance the purchase of bulldozers and heavy equipment to extract the logs. Japan provides the insurance and financing for the Japanese ships that clog the South China Sea. Once sawn in Japan, most of the wood produced by the oldest and perhaps richest tropical rain forest on earth goes into throwaway packing crates and disposable construction forms for pouring concrete.

Studies by the World Wildlife Fund suggest that selective logging as practiced in the hill forests of Sarawak removes about 34 percent of the natural cover. Yet industry advocates maintain that selective cutting doesn't hurt the forest in the long term. Minister of the Environment and Tourism James Wong has even stated that logging is "good for the forest." When presented with scientific information suggesting otherwise, Wong replied, "I will not bow to experts. I am the expert. I was here before the experts were born."

In theory, selective logging has far less environmental impact than the clear-cutting typical of temperate rain forests. In contrast to the desolation throughout the Pacific Northwest of North America, logged areas of Sarawak remain green, rapidly flushing out with secondary vegetation that creates an illusion of paradise. But this masks the difficulty of extracting, in an environmentally sound manner, a few select trees from a given area of tropical rain forest. In practice, most logging in Sarawak occurs with little planning and no technical supervision. Decisions on how the trees will be cut and how they will reach the specified landing areas lie strictly with the "faller" and the operator of the bulldozer or skidder. Working on a contract basis with their income dependent on their production, these men, often poor, uneducated, and far from home, fighting off hunger with a chain saw, place little importance on the environmental implications of their actions.

Arriving at a setting, the bulldozer operator establishes a landing and then follows the faller from log to log, skidding them one at a time, expanding his trail as the faller works deeper into the cutting block. The faller drops the trees in the direction most convenient to him. To reach them, the bulldozer must carve long, winding, and even circular tracks into the forest floor. With time at a premium, the bulldozer is constantly on the move. Every activity—turning or lifting the logs to attach the cables, pushing smaller logs together, maneuvering the bulldozer into place to begin the haul—further damages the forest.

In many parts of Sabah, skid trails and landings have laid bare over 40 percent of the forest floor. As logging removes the forest canopy, exposing the soil to rain, the compaction of the ground by the extractive process reduces the soil's capacity to retain water. This dramatically increases erosion, which is further exacerbated by road-building techniques that pay little attention to drainage or grade.

In just a few years, the indigenous peoples of Sarawak have seen their clear streams choked with sediment and debris. The federal government's own five-year plan states that "soil erosion and siltation have become Sarawak's main water-pollution problem." In many parts of the state, rivers are permanently turbid; the impact on fish is disastrous.

"WE WILL FIGHT BACK"

In 1987, Dayak anger over logging exploded. After seven years of appealing in vain to government to end the destruction of their homelands, the Penan, on February 13, 1987, issued a firm and eloquent declaration:

> We, the Penan people of the Tutoh, Limbang, and Patah rivers regions, declare: Stop destroying the forest or we will be forced to protect it. The forest is our livelihood. We have lived here before any of you outsiders came. We fished in clean rivers and hunted in the jungle. We made our sago meat and ate the fruit of trees. Our life was not easy but we lived it contentedly. Now the logging companies turn rivers to muddy streams and the jungle into devastation. Fish cannot survive in dirty rivers and wild animals will not live in devastated forest. You took advantage of our trusting nature and cheated us into unfair deals. By your doings you take away our livelihood and threaten our very lives. You make our people discontent. We want our ancestral land, the land we live off, back. We can use it in a wiser way. When you come to us, come as guests with respect. . . .
>
> If you decide not to heed our request, we will protect our livelihood. We are a peace-loving people, but when our very lives are in danger, we will fight back. This is our message.

Like the scores of letters, appeals, and petitions sent by Dayak peoples to state and regional authorities, this proclamation was ignored, so the Penan took direct action. On March 31, 1987, armed with blowpipes, they blocked a logging road in the Tutoh River basin. In April, Kayan at Uma Bawang blockaded a road that pierced their territory.

In every instance, the barriers were flimsy, a few forest saplings bound with rattan. Their strength lay in the people behind them. These human barricades—made up of men, women, and children, the old and the young—which began as a quixotic gesture, an embarrassment to the government, grew into a potent symbol of resolve. Within eight weeks, operations in 16 log-ging camps had been brought to a halt. By October, Penan, Kayan, and Kelabit communities had shut down roads at 23 sites. In all, some 2,500 Penan from 26 settlements took part. For eight months, despite hunger, heat exhaustion, and harassment, the indigenous peoples maintained defiant, yet peaceful, blockades, disrupting the logging industry and frustrating authorities.

The dramatic action electrified environmentalists in Malaysia and abroad. Press coverage in Australia, Europe, and the United States stimulated concern that grew steadily into a sustained international protest. The Malaysian and Sarawak governments responded defensively, imposing severe restrictions on the media. Military and security forces were brought into play, and police joined the logging companies in dismantling the blockades.

In October 1987, Prime Minister Mahlathir Mohamad invoked the Internal Security Act to incarcerate 91 critics of his regime. Among those detailed was Harrison Ngau of Sahabat Alam (Friends of the Earth) Malaysia, a Kayan environmentalist and the most vocal supporter of Dayak resistance. At the same time, 42 Kayan of Uma Bawang were arrested. They were accused of obstructing the police, wrongful restraint, and unlawfully occupying state lands. The last charge, in particular, was received bitterly. After all, the people of Uma Bawang had established a blockade on their own land to protect their legally recognized rights.

Although the dramatic police action temporarily ended the blockades, it also precipitated a battle that exposed the essential illegality of the logging. According to the Sarawak land code, native customary rights are inviolable. Since the state had granted logging concessions without a clear demarcation of customary lands, the rights of thousands of Dayak had, by definition, been compromised. On July 26, 1987, a Kayan, charged with obstructing a public thoroughfare, was acquitted: the magistrate concluded that he had blocked a road on customary land and had acted in a legitimate defense of his rights.

To protect the logging industry, the state government took legislative action. In November 1987, it added Amendment S90B to the Forest Ordinance, making it an offense for any person to obstruct the flow of traffic along any logging road. Penalties for violating Amendment S90B include two years imprisonment and a fine of over $2,000.

The injustice of Amendment S90B was obvious. Lolee Mirai, Penan headman of Long Leng, spoke of the purpose of the amendment: "We, who have rights to the land, were, instead, arrested, and not the timber companies who have caused damages to our land and properties. The law protects only the companies and causes us to suffer more. The law is not good. It unjustly allows outsiders and the logging companies to come and damage our land."

The government believed that blockades would never again disrupt the flow of timber. There were wrong. In May 1988 blockages went up near Long Napir, halting the logging operations of Minister Wong. Two more blockades sprang up in the Upper Baram in September, and four more in October. Between November 1988 and January 1989, blockades occurred at seven sites, and the Sarawak Forestry Department arrested 128 Dayaks, mostly Penan.

By mid-1989, the legislation, repeated arrests, and long and expensive trials appeared to have broken the Dayak resistance. Sporadic blockades were quickly dismantled. Then, in September, indigenous peoples in 19 communities in the Upper Limbang and the Baram erected 12 new barricades. By the end of the fall, 4,000 Dayaks had shut down logging in nearly half of Sarawak.

By early 1990, however, the logging industry, supported by all the power of the Malaysian government, had recovered. To this day the blockades in Sarawak continue.

IN A SINGLE GENERATION

After years of futile lobbying and peaceful protest, the Penan look to the outside world for support. Just before the September 1989 blockades, 80 indigenous leaders signed a joint statement: "We ask for help from people all over the world. We are people with a proud culture and way of life that is build on our forest and land. Don't take our forest and culture and dignity away. We thank everyone

who thinks of us and helps us, even though you are so far away. It is knowing this that keeps us alive."

Without international pressure, Sarawak is unlikely to recognize the rights of the Penan or protect their forest homeland. Authorities have indicated that they intend to maintain the forests of Sarawak as the exclusive preserve of the state and the domain of the political elite. These same authorities have made it clear that they will tolerate no opposition. Encounters with the police grow more brutal with every new blockade.

It is imperative that the global community respond to this situation. If the Penan are to have the opportunity to choose their own destiny, their forest homeland must be protected. Moreover, the interests of the Penan as well as those of neighboring Dayak peoples must be balanced with the need to protect the biological integrity of the land, now delineated by Gulung Mulu National Park, which lies at the heart of Penan territory.

The creation of a biosphere reserve is an obvious and appropriate solution. A biosphere reserve combines forest preservation with the subsistence needs of surrounding communities. Typically, a reserve consists of a series of concentric zones, with a core of permanently protected forest at the center. Moving outward is a series of increasingly intensive utilization zones. In the first zone, indigenous people can hunt, collect medicinal plants, and harvest natural products. In the next zone, people may farm and gather wood. Settlement occurs in a third zone, which acts as a buffer from encroaching development.

Local initiative and direct involvement of national and regional authorities are critical to establishing and maintaining a biosphere reserve. Fortunately, Sarawak can meet both conditions. In 1987, an intra-governmental report of the State Task Force on Penan Affairs called for establishing two biosphere reserves for the nomadic Penan. In 1990, the Penan Association endorsed the concept, substantially increasing the proposed boundaries to surround Gunung Mulu National Park and to include a large portion of the northeastern section of the Fourth Division.

To date, however, the Sarawak government has neither implemented the task force's recommendations nor endorsed the biosphere-reserve concept. Instead, government representatives often suggest that environmentalists and anthropologists want to sequester indigenous peoples in living zoos. "No one," Wong stated, "has the ethical right to deprive the Penan of their right to socioeconomic development and assimilation into the mainstream of Malaysian society."

Wong's statement is true, but so is its corollary. Penan surely have the right to determine the degree to which they enter Malaysian society, even as they respect their obligation to protect the integrity of Penan civilization. That many Penan still desire to pursue their traditional subsistence activities is evident in the Penan Association's numerous public statements.

The issue, then, hangs in the balance. The creation of a biosphere reserve, the meaningful recognition of Dayak rights, and the adoption of sustainable forestry practices all will depend on the actions of each and every one of us who cares about the fate of the Penan, other Dayak peoples, and their forest homelands. Ultimately, we are all responsible.

Today, throughout Sarawak, the sago and rattan, the palms, lianas, and fruit trees lie crushed on the forest floor. The hornbill has fled with the pheasants, and as the trees fall in the forest, a unique way of life, morally inspired, inherently right, and effortlessly pursued for centuries, is collapsing in a single generation.

• • •

44

Indians in Brazil Wither in an Epidemic of Suicide

DIANA JEAN SCHEMO

DOURADOS, BRAZIL

Haunted by memories of the elderly aunt, the nephew and the cousins who have killed themselves, Valério Vera Gonçalves dances hard over the earth and chants to the heavens. He wears a crown of pink feathers, and his entreaties mix with the roar of a nearby tractor on the edge of this boom town: "Go find a good road; chase away the evil spirits," he calls out.

For Mr. Gonçalves, who runs the Kaiowá Indians' house of worship here, prayer is the only defense against an epidemic of suicides that have plagued his people over the last few years.

"We have to dance very hard and pray," said Mr. Gonçalves, 48. "I feel as if I'm losing pieces of myself."

Over the last 10 years, some 200 Kaiowá are thought to have killed themselves in this southern region near the border with Paraguay, and anthropologists say they are now seeing suicides among other tribes. While the general rate of suicide among Brazilians is roughly 1 in 28,000, last year as many as 56 of the 28,000 Kaiowá in this region died in what are presumed to be suicides.

For the Kaiowá, a subgroup of the Guaraní Indians, the suicides seem to be tied to their estrangement from the land, on which their traditional life of fishing, farming and worship depended. A peaceful people who tend to withdraw from confrontation, the Kaiowá have, in the span of 75 years, suffered the kind of losses of land, people and culture that have characterized the Indian encounter with Europeans for 500 years.

When Europeans arrived here, Brazil was home to an estimated five million Indians. At the start of this century, their number had fallen to one million. Now, there are 230,000.

Since 1945, the Kaiowá have watched their land shrink from 25,000 square miles—roughly the size of West Virginia—to 172, and their language and their rituals disappear with the arrival of white colonists and the increase in the number of religious missionaries. In Bororó, which is near Dourados and the center of the surge in suicides, some 6,000 Kaiowá live on 10 acres.

The wave of suicides has prompted other tribes to gather here, offering prayers and shares of their crops. Government agencies have promised food and seeds, but so far, the Indians here say, Government help has not reached them.

Late last year, Justice Minister Nelson Jobim pledged to return nearly 4,000 acres to the Kaiowá. But signaling the explosiveness of the contest for land here, federal and military police protected Mr. Jobim as he announced the areas to be turned over. The 40 colonists whose land was earmarked have vowed not to leave; no land has been returned.

Although the Government Indian protection agency is conducting a census that aims to return the Kaiowá to some of their historic lands, that goal has become more unlikely under a decree that President Fernando Henrique Cardoso signed in January. The decree allows non-Indians to challenge pending and future allocations of land to Indians. Since it was issued, there have been six challenges to Kaiowá lands here.

The Kaiowá live on the margins of Dourados, an agribusiness center started in the 1940's with a land redistribution that gave each colonist family 75 acres to farm. Now, grain silos and neatly planted fields stretch for miles, while the Kaiowá go hungry, unable to eke out a living on less than an acre per person.

As a result, the Kaiowá men leave for work in distant sugar cane fields and distilleries. The women, left behind, clean houses, sell trinkets, beg and sometimes turn to prostitution. The most common suicide tales begin with husbands who became alcoholics while away returning home and fighting with their wives.

"Land is much more than simply a means of subsistence," wrote Alcida Ramos, in his book "Indigenous Societies." "It is the support for a social life that is directly linked to their system of belief and knowledge."

With missionary churches springing up to replace the Kaiowá's religion, and schools teaching Portuguese instead of their native Tupí-Guaraní, the suicides are a stark illustration of the Kaiowá's decline.

"It would disintegrate any people," said Orlando Silvestri Zimmer, who runs the local Indigenist Missionary Council, a Roman Catholic organization that works in behalf of Indian rights.

Mr. Zimmer complained that 28 religious groups, most of them evangelical Protestant, have flocked to the area in recent years to convert the Kaiowá. His own group, he said, does not try to convert Indians.

Anastásio Moura, a Kaiowá who struggles to grow rice and a few crops to feed his family on a half acre, said his people felt spiritually and physically drained. "There's nothing left to take away," he said.

In an effort to build up commercial farming, the Brazilian authorities also imported the more assimilated Terena Indians from the north, who have traditionally dominated the Kaiowá, and have taken over some of the most fertile land.

With a branch, Mrs. Moura traced a rectangle in the dirt to show the land where her parents once grew rice, manioc and potatoes. It was more than seven acres, she said. She then drew a smaller box in one corner denoting her mother's current property. After her father died, she said, the local leader and another Terena demanded three-quarters of the land.

"They said if she didn't give it over, they would beat her up and burn her house," Mrs. Moura said.

Many here complained that corruption and the power of the captains were responsible for the loss of lands and the failure of Government aid to reach them. The captains, Indians who were historically appointed by the military as go-betweens with white society, are now elected, but their authority has subverted the influence of religious leaders and the traditions they represent.

There are also signs that some suicides may mask murders. According to researchers at the State University of Mato Grosso do Sul, as many as six presumed suicides may have been murders, implicating local authorities in a drive to push Indians off the land. Authorities in Dourados deny the accusations.

The Kaiowá do not themselves ascribe the suicides to specific actions by the Brazilian Government, although they do link it to living near white people.

"They have bad thoughts," said Paulito Aquino, a religious chief who gave his age as 110, and who wore the traditional tambeta, a reedlike stick, through his chin.

"They are in their houses, but they're thinking, 'Those Indians, they might as well die,'" said Mr. Aquino.

45

Looking for Indians

CHERYL SAVAGEAU

My head filled with tv images
of cowboys, warbonnets and renegades,
I ask my father
what kind of Indian are we, anyway.
I want to hear Cheyenne, Apache, Sioux,
words I know from television
but he says instead
Abenaki. I think he says Abernathy
like the man in the comic strip
and I know that's not Indian.

I follow behind him
in the garden
trying to step in his exact footprints,
stretching my stride to his.
His back is brown in the sun
and sweaty. My skin is brown
too, today, deep in midsummer,
but never as brown as his.

I follow behind him like this
from May to September
dropping seeds in the ground,
watering the tender shoots
tasting the first tomatoes,
plunging my arm, as he does,
deep into the mounded earth
beneath the purple-flowered plants
to feel for potatoes
big enough to eat.

I sit inside the bean teepee
and pick the smallest ones
to munch on. He tests
the corn for ripeness
with a fingernail, its dried silk
the color of my mother's hair.
We watch the winter squash grow hips.
This is what we do together

in summer, besides the fishing
that fills our plates unfailingly
when money is short.

One night
my father brings in a book.
See, he says, Abenaki,
and shows me the map
here and here and here
he says, all this
is Abenaki country.
I remember asking him
what did they do
these grandparents
and my disappointment

when he said no buffalo
roamed the thick new england forest
they hunted deer in winter
sometimes moose, but mostly
they were farmers
and fishermen.

I didn't want to talk about it.
Each night my father
came home from the factory
to plant and gather,
to cast the line out
over the dark evening pond,
with me, walking behind him,
looking for Indians.

46

Like Tributaries to a River

WINONA LADUKE

WINONA LADUKE

Bimaadizwin o'ow nibi
Jiigibiig nindana kiimin
Jiigibiig ningaganoonaanaan Gizhe Minitou
Jiigibiig Ninbabaamadizimin
Nin Miki go-imin wisiniwin minikwewin
    nibi kaang
Nin dinawamaaganag ayaawag nibi kaang
Gi-bizhigwaadenimoa Gizhe Manitou maji-
    mashkii
atooyegnibikaang Ji-ganawendamang nibi
    gigi minigoomin
omaa gidakiiminaan

Water is life
We are the people who live by the water.
Pray by these waters.
Travel by these waters.
Eat and drink from these waters.

We are related to those who live in the
    water.
To poison the waters is to show disrespect
    for creation.
To honor and protect the waters is our
    responsibility
as people of the land.

SOURCE: Translated from the Anishinabe by Marlene Stately

We live off the beaten track, out of the mainstream, in small villages on a vast expanse of prairie, in dry desert lands, or in thick woods. We drive old cars, live in old houses and mobile homes. Our families are important to us, and there are usually some small children and relatives around, no matter what else is going on. We seldom carry briefcases, and we rarely wear suits. You're more likely to find us meeting in the local community center or in someone's home than in a convention center or at $1,000-a-plate fund-raisers. We are the common face of Native environmentalism.

We organize in small groups, with names like Native Americans for a Clean Environment, Diné CARE (Citizens Against Ruining our Environment), Anishinabe Niijii, and the Gwich'in Steering Committee. We have faced down huge waste dumps, multinational mining and lumber companies, and the U.S. Office of the Nuclear Waste Negotiator. We are underfunded at best, and more often without funding at all; yet individuals throughout the Native nations keep fighting to protect our lands for future generations.

We have close to 200 grassroots Native organizations in North America resisting the environmental destruction of our homelands. Most of these groups are small, perhaps five to ten volunteers working out of their homes. Many operate in remote areas without phones or cars. There are another 500 or so environmental organization in the environmental justice movement and collective networks coalescing around regional and international agendas, groups such as California Indians for Cultural and Environmental Protection, Southwest Network for Environment Economic Justice, Indigenous Environmental Network.

Despite our meager resources, we are winning many hard-fought victories on the local level:

- In 1991, Diné CARE defeated Waste Tech's plans to build a "recycling center"—actually a toxic-waste incinerator—near Dilkon, Arizona, in the western portion of the Navajo Reservation. A year later, it was working on tribal forestry issues. The Navajo Nation's own logging company proposed clearcutting the Chuksa Mountains. In a battle waged over several years, and costing at least one life (activist Leroy Jackson, called the Chico Mendez of the Navajo, died under suspicious circumstances during the struggle). Diné CARE finally succeeded in halting the clearcutting.

- The Gwich'in in Alaska have fought— successfully, so far—to keep oil-drilling from invading their homeland and dispersing the Porcupine caribou herd.

- The Good Road and Natural Resource Coalitions, both Lakota groups, mobilized Pine Ridge and Rosebud reservation residents to stop plans for toxic-waste dumps on their reservations in South Dakota.

- Native Action, a grassroots organization in Montana, protected the sacred Sweet Grass Hills from desecration. The Canadian company Pegasus Gold operates the Zortman-Landusky mine complex near the Fort Belknap Reservation; one mine has lopped off the top of one of the mountains most sacred to tribes in the region. Native Action won an extension on a moratorium on gold mining in the sacred hills that was to have expired in August 1995.

- In a ten-year battle against mega-dams on James Bay, the Cree people built a sophisticated international coalition that succeeded in securing a moratorium from Hydro Quebec on a project called James Bay II.

- The White Earth Land Recovery Project on the White Earth Reservation in Minnesota has restored over 1,000 acres of land within the reservation, defeated several ecologically and culturally destructive development proposals, and begun to restore the traditional land-based economy that depended on local products like maple syrup and wild rice.

After centuries of attempts to remove us from our land, we are still here. We are not about to go away.

Native environmental groups have a commitment and tenacity that springs from place. "This is where my grandmother's and children's umbilical cords are buried. . . . That is where the great giant lay down to sleep. . . . That is the last place our people stopped in our migration here to this village." Our relationship to land and water is continuously reaffirmed through prayer, deed, and our way of life. Our identity as human beings is founded on creation stories tying us to the earth, and to a way of being, *mino-bimaatisii-win,* "the good life." Our intergenerational residence in place reaffirms that relationship and knowledge. The earth is our Mother; it is from her we gain our life.

Native peoples have courageously resisted the destruction of the natural world at the hands of colonial, and later, industrial society, since this destruction attacks their very identity. This resistance has continued from generation to generation, and provides the strong core of today's Native environmentalism. This is why 500 or more federally recognized reservations and Indian communities still exist, why one-half of our lands are still forested, much in old growth, and why we continue the work of generations past by opposing clearcutting, nuclear-waste dumping, dams like the Kinzua in Pennsylvania and those on the Columbia River, and other threats to our lives and land.

To understand Native environmentalism, it's a good idea to redraw your mental geography of Native America. Over 700 Native communities dot the continent. In the United States, Native America covers 4 percent of the land; Native people are the second-largest collective landholders after the federal government.

To the north, in Canada, the situation is different. Reservations, or "reserves," are generally smaller in size (Canada's Indian Act was incredibly stingy). But the northern communities do retain vast areas by treaty or agreements for hunting and trapping. North of the 50th Parallel, 85 percent of the population is Native.

With this new map in mind, consider that:

- Most environmental struggles over the boreal forest in Canada involve Native peoples. The Brazilian rainforest is being cut down at the rate of one acre every 9 seconds; one acre of Canada's forests disappears every 11 seconds. Approximately the same number of Native people live in each.

- According to Worldwatch Institute, 317 reservations in the United States are threatened by environmental hazards, from toxic-waste dumping to clearcutting to radioactive waste. Two-thirds of all "domestic" uranium resources are on Indian lands, as is one-third of all western low-sulfur coal. Sixteen proposals to dump nuclear waste have targeted reservations, and over 100 proposals have been

made in recent years to dump toxic wastes in Indian communities. Few reservations have escaped environmental degradation.

- The Association on American Indian Affairs lists 77 sacred sites nationally that have been disturbed or desecrated through resource extraction and development activities.

- Eighty million acres of Alaskan offshore oil lease sales lie in water surrounding Native coastal villages.

- Over 1,000 uranium mines sit abandoned on Diné land, leaking radioactive contaminants into the air and water. On the same reservation is the largest coal strip mine in the world, operated by Peabody Coal Company. Diné teenagers have a cancer rate 17 times the national average.

- The Western Shoshone in Nevada are the most bombed nation in the world, with more than 700 atomic explosions over the past 45 years. Now the federal government is studying Yucca Mountain, sacred to the Shoshone, as a dumpsite for high-level nuclear wastes.

- Faced with encroaching development and state building codes that ruled their traditional "chickees" substandard housing, the 300-member Seminole Nation in Florida is fighting for its survival, through litigation and a possible land acquisition with support from a private foundation.

"I have fished here forever, through my ancestors," says Margaret Flint-Knife Saluskin, a Yakama fisherwoman. She stands by an ancestral fishing scaffold at Lyle Point on the Columbia River in Oregon—where the Yakama are fighting plans for a huge housing development. "This is where the fish come to give up their spirits. It is a sacred place. I will not come to fish for my family next to tennis courts and the swimming pools of luxury home developments."

In the Northwest, virtually every river is home to a people, each as distinct as a species of salmon. The Clatsop are from north of the Columbia; the Tillamook Siletz, Yaquina, Alsea, Siuslaw, Umpqua, Hanis, Miluk, Colville, Tututni,

Shasta, Costa, and Chetco: all people living at the mouths of salmon rivers. One hundred and seven stocks of salmon have already become extinct in the Pacific region, and 89 others are at high risk. The stories of the people and the fish are not so different; environmental destruction threatens the existence of both.

Susana Santos, an artist and fisherwoman turned Greenpeace campaigner, is from the Tygh band of the Lower Deschutes River. The Tygh today include a scant five families, trying to survive as a people as they struggle to keep up their traditional way of life and their relationship to the salmon.

"That's why I came back to fish," says Santos. "I wanted to dance the Salmon, know the Salmon, say good-bye to the Salmon. Now I am looking at the completion of destruction, from the *Exxon Valdez,* the Trojan Nuclear Power Plant, Hanford, logging, and those dams."

"Seventeen fish came down the river last year. None this. The people are the salmon and the salmon are the people. How do you quantify that?"

In 1992, the Environmental Protection Agency set allowable levels of dioxin discharge from paper mills in the Northwest; this standard was based on a human consumption level of 6.5 grams of fish a day, the national average. Agency officials knew that Native American, Asian Americans, and other low-income people in the area consumed up to 150 grams of fish a day— creating a cancer risk of 8,600 per million. The risk allowable by statute is one per million.

In January 1990, the Yakama Indian Nation enacted a resolution calling for an end to the use of the chlorine bleaching process by the pulp-and-paper industry. In testimony before the Oregon Environmental Quality Commission, Harry Smiskin of the Tribal Council of the Confederated Tribes of the Yakama Indian Nation said, "The Yakama Indian Nation is not interested in discussing with you or anyone how much TCDD or any of the other hundreds of toxic chemicals can be dumped into the Columbia River by the pulp-and-paper industry. The Yakama Indian Nation does not want to debate mixing zones for toxic pollution, nor do we want to debate whether a mile, two miles, or the entire Columbia River is water-quality-limited, and

what that means for beneficial uses. Those topics wrongly assume that it is okay to dump pollution into the river that can impact the health of our fish and the health of our people."

I live on a reservation in northern Minnesota called White Earth, where I work on land, culture, and environmental issues through an organization called the White Earth Land Recovery Project. We, the Anishinabeg, are a forest people, meaning that our creation stories, instructions, and culture, our way of life are entirely based in the forest, from our medicine plants to our food sources, from forest animals to our birchbark baskets.

Yet virtually my entire reservation was clearcut at the turn of the century, providing the foundations for major lumber companies, including Weyerhaeuser, and setting in motion a process of destruction that has continued for nine decades in our community.

"I cried and prayed our trees would not be taken from us, for they are as much ours as is this reservation," said Wabunoquod, an Anishinabe on the White Earth Reservation in 1874.

In 1889 and 1890, Minnesota led the United States in lumber production, with the state's northwest as the leading timber source. Two decades later, 90 percent of our land was in non-Indian hands, and our people were riddled with diseases, with many leaving to become refugees in the cities. Today, our forests are just beginning to recover. But the process of recovery is far from complete: three-fourths of all tribal members remain off-reservation, and 90 percent of reservation lands still remain in non-Indian hands.

When Potlatch, a major lumber corporation, announced that it would double and triple the size of its pulp-and-paper mills in our region, we resisted. Potlatch—and Blandin, and International Paper, and other companies—not only destroy our trees, they destroy the foundation of our culture.

All this is to say that native communities are not in a position to compromise, because who we are is our land, our trees, and our lakes. This is central to our local and collective work. This is also why the conflicts remain in Native America between corporate interests and our traditional ways, and between segments of our community who embrace the values of industrial society and those who continue to embrace traditional values. There is no comfortable compromise in these situations, and corporations that look at a project and a bottom line often find themselves with a snapping turtle as an adversary—a creature reluctant to let go. Native communities can bring up a broad set of issues when opposing a project: tribal jurisdiction, federal trust responsibility, treaty rights, reservation of cultural and religious freedoms, the economic benefit or harm to a village or community, and grave-site protection.

While there is great potential for these strategies, there has been a very awkward courtship between Native environmental groups and larger environmental organizations. For instance, the Nature Conservancy came to my reservation, bought 400 acres of land, and gave it to the state of Minnesota. The Sierra Club has openly opposed transfers of public land to tribes on occasion. In addition, animal-rights groups have opposed traditional harvesters of seal, beaver, fish and whales. These conflicts make it cumbersome to try to build alliances, as most Native groups view the larger, more politically powerful environmental groups with suspicion after meeting them on opposing sides in lawsuits of proposed legislation. The truth is that an "alliance" does not mean that Native communities must embrace all facets of an environmental group's agenda, or vice versa. The important thing is to build trust. Mainstream environmentalists would do well to build relationships and alliances with Native groups—but first they must be willing to listen to our point of view.

Over the past 20 years, Native organizations and communities have crafted a diverse and comprehensive strategy for defending their ways of living and charting a course for the future. This blueprint has taken some organizations and nations to various United Nations forums and into a decade-long process to secure an international declaration on the rights of indigenous peoples, as well as to codify traditional laws and practices, join in litigation in the white man's courts, and restore traditional economic packages.

This past summer the Indigenous Environmental Network held its sixth annual Protecting Mother Earth Gathering at the Cherokee Nation in North Carolina. Founded in 1990 at the Diné village of Dilkon, Arizona, the Network has helped often isolated and widely scattered Native groups find common ground on issues and philosophies, such as the reaffirmation of traditional and natural laws, recognition of environmentally sound lifestyles and livelihoods, and the promotion of indigenous voices in the environmental movement.

In 1995, grassroots activists, traditional elders, and organizers from the Albuquerque-based Native Lands Institute developed an "Indigenous Environment Statement of Principles," written for working in the Native community, to help integrate the alienated terminology of technological society (risk assessment, resource management, mitigation) with traditional values. This document calls for using Native language and culture in assessing the environment, rather than the terms of the dominant culture.

Native environmental activists in northern Wisconsin who have worked for decades on fishing, mining, logging, and other issues in northern Anishinabeg territory are proposing a Seventh Generation Amendment to the United States (and tribal) Constitution that would embody the principle of the Haudenosaunee (Iroquois) Six Nations Confederation (and our own Anishinabe) that "every deliberation we make must be in consideration of the impact on the seventh generation from now."

Across the continent, on the shores of small tributaries, in the shadows of sacred mountains, on the vast expanse of the prairies, or in the safety of the woods, prayers are being repeated, as they have been for thousands of years, and common people with uncommon courage and the whispers of their ancestors in their ears continue their struggles to protect the land and water and trees on which their very existence is based. And like small tributaries joining together to form a mighty river, their force and power grows. This river will not be damned.

• • •

47

How to Succeed in Business

Follow the Choctaws' Lead

FERGUS M. BORDEWICH

Philadelphia, Mississippi, is the kind of place that seemed to survive more from habit than reason after the timber economy that was its mainstay petered out in the 1950s. There is a scruffy, frayed-at-the-edges look to the empty shop fronts and the discount stores where more vibrant businesses used to be, but by the standards of rural Mississippi, Philadelphia counts itself lucky. "Kosciusko and Louisville, they have to wait to buy a tractor or, sometimes, even to meet their payrolls," boasts the mayor, an amiable former postman by the name of Harlan Majors. "And they don't have a fire department worth a hoot. I have 16 full-time firemen."

Philadelphia's trump, the thing that those other towns will never have, is Indians. "Our best industry by far is the Choctaw Nation," Majors says. "They're our expansion and upkeep. They employ not only their own people, but ours too. It has never been as good as it is now. Our economy depends on them. If the tribe went bankrupt, we'd go into a depression."

For generations the Choctaws were a virtual textbook example of the futility of reservation life. Over the last quarter-century, however, the 8,000-member tribe has defied even its own modest expectations by transforming itself from a stagnant welfare culture into an economic dynamo, and one of the largest employers in Mississippi. Choctaw factories assemble wire harnesses for Ford and Navistar, telephones for AT&T, and audio speakers for Chrysler, Harley-Davidson and Boeing. The tribe's greeting card plant hand-finishes 83 million cards each year.

Since 1991, the tribe has operated one of the largest printing plants for direct-mail advertising in the South. Sales from the tribe's industries have increased to more than $100 million annually from less than $1 million in 1979. As recently as 15 years ago, 80 percent of the tribe was unemployed; now, having achieved full employment for its members, nearly half the tribe's employees are white or black Mississippians. Says William Richardson, the tribe's economic development director, "We're running out of Indians."

The quality of life for the great majority of Choctaws has measurably improved. The average income of a family of four is about $22,000 per year, a sevenfold increase since 1980. Brick ranch houses have largely supplanted the sagging government-built bungalows amid the jungle of kudzu-shrouded oaks and pines that forms the heart of the Choctaws' 22,000-acre reservation. The Choctaw Health Center is among the best clinics in Mississippi, while teachers' salaries at the tribal elementary schools are 25 percent higher than at public schools in neighboring, non-Indian towns. "They're willing to buy the best," says a non-Indian teacher who formerly taught in Philadelphia. The tribal television station, the primary local channel for the region, broadcasts an eclectic daily menu that includes thrice-daily newscasts and Choctaw-language public service shows on such diverse topics as home-financing and microwave cooking.

The Choctaws are also a national leader in transferring the administration of federal programs from the Bureau of Indian Affairs (BIA) to the tribes. Virtually everything once carried out by the bureau—law enforcement, schooling, health care, social services, forestry, credit and finance—is now performed by Choctaw tribal bureaucrats. "We're pretty well gone," says Robert Benn, a courtly Choctaw who was the BIA's local superintendent until his recent retirement.

His sepulchral office was one of the last still occupied in the bureau's red-brick headquarters in Philadelphia. "We've seen our heyday. The tribe is doing an exemplary job. They're a more professional outfit than we ever were."

Throughout the sprawling archipelago of reservations that makes up modern Indian country, tribes like the Choctaws are demolishing the worn-out stereotype of Indians as permanent losers and victims, and effectively killing, perhaps with finality, what historian Robert J. Berkhofer Jr. aptly termed the "white man's Indian," the mythologized figure whose image, whether confected by racism or romance, has obscured the complex realities of real Native Americans, from *The Last of the Mohicans* to *Dances With Wolves.* For the first time in generations, Indian tribes are beginning to shape their own destinies largely beyond the control of whites: revitalizing tribal governments, creating modern economies, reinventing Indian education, resuscitating traditional religions and collectively remaking the relationship between the United States and the more than 300 federally recognized tribes.

To be sure, in terms of overall statistics, Indian country continues to present a formidable landscape of poverty and social pathologies. On some reservations, unemployment surpasses 80 percent. Rates of alcoholism commonly range higher than 50 percent. Indians are twice as likely as other Americans to be murdered or to commit suicide, and five times more likely to die from cirrhosis of the liver. In spite of increased access to education, 50 percent of Indian young people drop out of high school. There is no cure-all for these problems, but for the first time since the closing of the frontier, responsibility for finding solutions rests increasingly in Indian hands.

Without viable tribal economies, however, self-determination is likely to remain little more than a pipe dream. A few Indian communities have reaped astonishing profits from legalized tribal gambling, which has grown into a $6 billion industry, accounting for about 2 percent of the $330 billion that Americans legally bet each year. By 1994, more than 160 tribes were operating some form of gambling activity, including 40 full-fledged casinos, in 20 states. The tiny Mashantucket Pequot Tribe, whose Connecticut casino grosses about $800 million annually, half again as much as Donald Trump's Taj Mahal, has repurchased tribal land and provided scholarships and medical coverage for members. The tribe has also contributed $10 million to the Smithsonian's National Museum of the American Indian.

Tribal "gaming," as it is rather delicately known, is not a panacea, however. Although rumors of mob involvement have been largely disproved, some tribes have squandered their earnings. Moreover, it is likely that gambling will taper off as an important source of tribal revenue by the end of the decade, as states grant gambling licenses to other groups.

Other tribes have been blessed with abundant natural resources, which they are now able to exploit in their own interest for the first time. Between 50 percent and 80 percent of all the uranium, between 5 percent and 10 percent of all the oil and gas reserves, and 30 percent of all the coal in the United States lie on Indian lands. Many tribes own rights to water whose value is dramatically increasing. More than 90 tribes have land that is densely forested, while millions of acres of leased tribal grassland provide pasturage for ranchers, and millions of acres more are leased to farmers.

Today, the Navajos of Arizona and the Jicarilla Apaches of New Mexico, among others, operate their own tribal oil and gas commissions to regulate production on their lands. The Southern Ute tribe of Colorado has set up its own oil production firm. One of the most innovative tribes, the Confederated Tribes of Warm Springs, in Oregon, operates three commercial hydroelectric dams and an extensive forestry industry, as well as a textile plant that has produced sportswear for Nike and Jantzen and beadwork for export to Europe, a luxury vacation lodge, and a factory that recently began manufacturing fireproof doors from diatomaceous earth—or fossilized sea creatures.

However, the experience of the Mississippi Choctaws has made clear that even the most poorly endowed tribes, with able and determined tribal leadership, a pragmatic willingness to cooperate with non-Indians, some federal support and the ability to raise capital, can hope to remake themselves into viable communities able to compete in the modern American economy.

The origin of the Choctaws is mysterious. Some say that they arose pristine from the earth at Nanih Waiya, the Mother Mound of the Choctaws, a man-made hill north of the modern reservation, in Winston County. "After coming forth from the mound, the freshly made Choctaws were very wet and moist, and the Great Spirit stacked them along the rampart, as on a clothesline, so that the sun could dry them," as one story has it.

Throughout documented times, the Choctaws were mainly an agricultural people, raising corn, beans, pumpkins and melons in small plots. However, exhibiting an instinct for business that was probably far more prevalent among Native Americans than history records, they raised more corn and beans than they needed for their own use and sold the surplus to their neighbors. Like their neighbors and sometime enemies, the Cherokees, Chickasaws, Crees and Seminoles, the Choctaws gradually adopted European consumer goods, styles of agriculture and schooling, as well as less-savory practices, such as the exploitation of African slaves. By the early 19th century the Choctaws and these neighboring tribes became known collectively as the "Five Civilized Tribes" of the Southeast.

However, the relentless pressure of settlement steadily whittled away at the Choctaws' lands until, in 1830, in the poignantly named Treaty of Dancing Rabbit Creek, the tribe reluctantly relinquished what remained of its land in the East, most members agreeing to remove themselves to the Indian Territories, where their descendants still inhabit the Choctaw Nation of Oklahoma.

Originally, about one-third of Mississippi's Neshoba County was allotted to those Choctaws who chose to remain in the East. By mid-century, however, virtually all of it had passed out of Choctaw hands, sometimes legally, but often through fraud and extortion. Virtually without exception, the Choctaws were reduced to an impoverished life of sharecropping, living scattered among the forests of oak and pine. In time, their numbers were swelled by others who drifted back from the Indian Territories, disillusioned by the anarchy of tribal politics there and the difficulties of life on the frontier.

Ironically, the tripartite racial segregation that deepened as the 19th century progressed only strengthened the Choctaws in their traditions, language and determination to be Indian in a part of America where, for all intents and purposes, Indians had simply ceased to exist. Rather than send their children to schools with blacks, the Choctaws refused to send them to school at all. In 1918, when the federal government winkled out enough land from private owners to establish the present-day Choctaw reservation, nearly 90 percent of the tribe were full-bloods. Most spoke no English at all.

The story of the Choctaw revival is inseparable from that of Phillip Martin, the remarkable chief who has guided the tribe's development for most of the past 30 years. Martin is a physically unimposing man, short and thick-bodied, with small, opaque eyes and thinning hair that he likes to wear slicked back over his forehead. Beneath a grits-and-eggs plainness of manner, he combines acute political instincts with unflagging tenacity and a devotion to the destiny of his people. "He's like a bulldog at the postman—he just won't go away," says Lester Dalme, a former General Motors executive who has managed the tribe's flagship plant, Chahta Enterprise, since 1980. "At the same time, he'll give you the shirt off his back whether you appreciate it or not. He truly loves his people. He can't stand even one of his enemies to be without a job."

By all rights, Martin's fate should have been as gloomy as that of any Choctaw born in the Mississippi of 1926. "Everybody was poor in those days. The Choctaws were a bit worse," he recalls.

As a boy, he cut pulpwood, herded cows and picked cotton for 50 cents per 100 pounds. In those days, Choctaw homes had no windows, electricity or running water. Alcoholism and tuberculosis were endemic. Few Choctaws had traveled outside Neshoba County, and many had never even been to Philadelphia, only seven miles away. The etiquette of racial segregation was finely modulated. Although Choctaws were not expected to address whites as "sir" and "ma'am" or to step off the sidewalk when whites passed, they were required to sit with blacks in movie houses and restaurants. "But we never had enough money to eat in a restaurant anyway," Martin says, with irony, in his porridge-thick drawl.

Martin earned a high school diploma, rare among Choctaws of that time, at the BIA boarding school in Cherokee, North Carolina. His first experience of the larger world came in the Air Force at the end of World War II. Arriving in Europe in 1946, he was stunned by the sight of starving French and Germans foraging in garbage cans for food. White people, he realized for the first time, could be as helpless as Indians.

At the same time, he was profoundly impressed by their refusal to behave like defeated people and by their determination to rebuild their lives and nations from the wreckage of war. He wondered, if Europeans could lift themselves back up out of poverty, why couldn't the Choctaws? When he returned to Mississippi, he quickly learned that no one was willing to hire an Indian. Even on the reservation, the only jobs open to Indians were as maintenance workers for the BIA, and they were already filled. Martin recalls, "I saw that whoever had the jobs had the control, and I thought, if we want jobs here we're going to have to create them ourselves."

He eventually found work as a clerk at the Naval Air Station in Meridian. He began to take an interest in tribal affairs, and in 1962 he became chairman of the Tribal Council at a salary of $2.50 per hour. In keeping with the paternalistic style of the era, the BIA superintendent presided over the council's meetings. He also decided when tribal officials would travel to Washington and chaperoned their visits there, as Indian agents had since the early 19th century. Says Martin, "I finally said to myself, 'I've been all over the world. I guess I know how to go to Washington and back. From now on, we don't need the superintendent.' So after that we just up and went." Martin became a fixture in the Interior Department and the halls of Congress, buttonholing agency heads and begging for money to replace obsolescent schools and decrepit homes, and to pave the reservation's corrugated red-dirt roads.

The tribe's first experience managing money came during the War on Poverty in the late 1960s, when the Office of Economic Opportunity allowed the Choctaws to supervise a unit of the Neighborhood Youth Corps that was assigned to build new homes on the reservation; soon afterward, the tribe obtained one of the first Community Action grants in Mississippi, for $15,000. "That $15,000 was the key to all the changes that came afterward," says Martin. "We used it to plan a management structure so that we could go after other federal agency programs. I felt that if we were going to handle money, we had to have a system of accountability and control, so we developed a finance office. Then we won another grant that enabled us to hire accountants, bookkeepers, personnel managers and planners."

The Choctaws remained calculatedly aloof from both the civil rights movement of the 1960s and the Indian radicalism of the 1970s. Martin says, "We didn't want to shake things up. Where does it get you to attack the system? It don't get the dollars rolling—it gets you on welfare. Instead, I thought, we've got to find out how this system works." Eighty percent of the tribe's members were on public assistance and receiving their food from government commodity lines. "It was pathetic. We had all these federal programs, but that wasn't going to hold us together forever. I knew that we had better start looking for a more permanent source of income." It would have to be conjured from thin air; the reservation was devoid of valuable natural resources, and casino gambling was an option that lay far in the future.

In key respects, Martin's plan resembled the approach of East Asian states like Singapore and Taiwan, which recognized, at a time when most developing countries were embracing socialism as the wave of the future, that corporate investment could serve as the driving force of economic development. Martin understood that corporations wanted cheap and reliable labor, low taxes and honest and cooperative government. He was convinced that if the tribe built a modern industrial park, the Choctaws could join the international competition for low-skill manufacturing work. In 1973, the tribe obtained $150,000 from the federal Economic Development Administration to install water, electricity and sewer lines in a 20-acre plot cut from the scrub just off Route 7. "It will attract somebody," Martin promised. For once, he was dead wrong. The site sat vacant for five years.

With his characteristic tenacity, Martin began writing to manufacturers from one end of the United States to the other. He kept on writing, to 150 companies in all, until one, Packard Electric, a division of general Motors, offered to train Choctaws to assemble wired parts for its cars and trucks; Packard would sell materials to Chahta Enterprise, as the tribe called its new company, and buy them back once they had been assembled. On the basis of Packard's commitment, the tribe obtained a $346,000 grant from the Economic Development Administration and then used a Bureau of Indian Affairs loan guarantee to obtain $1 million from the Bank of Philadelphia.

It seemed, briefly, as if the Choctaws' problems had been solved. Within a year, however, Chahta Enterprise had a debt of $1 million and was near bankruptcy. Production was plagued by the kinds of problems that undermine tribal enterprises almost everywhere. Many of them were rooted in the fact that, for most of the tribe, employment was an alien concept. Workers would abruptly take a day off for a family function and not show up for a week. Some spoke no English. Others drank on the job. Many were unmarried women with small children and had

no reliable way to get to work. The tribe's accountants had already recommended selling everything off for 10 cents on the dollar.

The man to whom Martin turned was Lester Dalme, who was then a general supervisor for GM and who had been raised in rural Louisiana with a virtually evangelical attitude toward work. "My mom taught us that God gave you life and that what you're supposed to do is give Him back your success," says Dalme, a trim man now in his 50s whose office at Chahta Enterprise is as plain as his ethics. Dalme remembers facing the plant's demoralized workers. "They had no idea how a business was run, that loans had to be paid. None of them, none of their fathers, and none of their grandfathers had ever worked in a factory before. They had no idea what quality control or on-time delivery meant. They thought there was a big funnel up there somewhere that money came down. They thought profit meant some kind of plunder, something someone was stealing." Dalme told them, "Profit isn't a dirty word. The only way you stay in business and create jobs is to make a profit. Profit is what will finance your future."

Dalme cut back on waste, abolished some managerial perks and put supervisors to work on the assembly line. Day care was set up for workers with small children; old diesel buses were organized to pick up those without cars. Dalme told employees that he would tolerate no alcohol or hangovers in the plant. He kept an average of three of every ten people he hired, but those who survived were dependable workers. He saw people who had been totally destitute begin to show up in new shoes and clothes without holes, and eventually in cars. After six or seven months, he saw them begin to become hopeful, and then self-confident.

Workers speak with an almost redemptive thrill of meeting deadlines for the first time. Wayne Gibson, a Choctaw in his mid-30s who became a management trainee after several years on the assembly line, recalls, "Factory work taught us the meaning of dependability and punctuality. You clock in, you clock out. It also instilled a consciousness of quality in people. You're proud of what you do. When I was on the production line and I had rejects, it really bothered me. I had to explain it the next day. We're proud of coming in here and getting that '100 percent zero defects' rating."

Chahta Enterprise has grown steadily from 57 employees in 1979 to more than 900 today. Once the tribe had established a track record with lenders, financing for several more assembly plants and for a modern shopping center followed. In 1994 the Choctaws inaugurated Mississippi's first inland casino as part of a resort complex that will include a golf course and a 520-room hotel. "Now we're more into profit centers," says William Richardson, a former venture capitalist from Jackson who was hired by Martin to function as a sort of resident dealmaker for the tribe. "We're as aggressive as hell and we take risks."

And so, today, the Choctaws have achieved virtually full employment. Increasingly, the jobs that the tribe has to offer its members are technical and intellectual, as engineers, business managers, teachers and statisticians; the tribe is, in short, creating for the first time in history a Choctaw middle class.

The scene at the Choctaw Manufacturing Enterprise, just outside Carthage, Mississippi, is typical enough at first glance. Although the building is architecturally undistinguished—just a low, white-painted rectangle hard by cow pastures and pinewoods—it is modern and spacious, and well-ventilated against the withering summer heat. Inside, workers perch at long tables, weaving wires onto color-coded boards that will become part of Xerox photocopiers. It is slow work; as many as 300 wires must go into some of the harnesses and be attached to up to 57 different terminals. Painstakingly, in deft and efficient hands, the brown and green wires are made to join and bifurcate, recombine and intertwine again in runic combinations. As they work, the long rows of mostly women listen, as do factory hands in similar plants anywhere in America, to the thumping beat of piped-in radio, and swap gossip and news of children, and menus for dinner. Across the floor, at similar tables, others are assembling telephones and putting together cir-

cuit boards for computers and audio speakers, and motors for windshield wipers.

But in another sense, the factory floor is remarkable and profound. The faces bent over the wires and phones and speakers record a transformation that no one in Mississippi could have envisioned 40 years ago when Phillip Martin came home from the military looking for any kind of job. The faces are mostly Choctaw, but among them are white and black faces, too, scores of them, all side by side in what was once one of the poorest backwaters of a state that to many seemed second to none in its determination to keep races and classes apart.

"I don't like what this country did to the Indians: it was all ignorance based on more ignorance based on greed," Martin says, in his meditative drawl. "But I don't believe that you have to do what others did to you. Ignorance is what kept us apart. We'd never have accomplished what we did if we'd taken the same attitude. I don't condemn anyone by race. What kept us down was our own lack of education, economy, health care—we had no way of making a living. I believe that if we're going to fit in this country, we'd better try our best to do it on our own terms. But we also have to live with our neighbors and with our community. We all have a common cause here: the lack of jobs and opportunities has kept everyone poor and ignorant. If we can help local non-Indian communities in the process, we do it. We all depend on one another, whether we realize it or not."

For the Choctaws especially, the mere fact of work is a revolutionary thing in a place where there was no work before. In 1989, there were four Choctaws in the Carthage plant's management; now there are 13. "The next generation will be able to manage their own businesses," says Sam Schisler, the plant's CEO, a freckled Ohioan in mauve trousers and a navy blue polo shirt who, like Lester Dalme, joined the Choctaws after running plants for Packard Electric. "I'm happy to manage myself out of a job."

There is something more. The audio speakers whose parts have been imported from Thailand and the circuit boards that have come from

Shreveport are not glamourous, but they are symbolic: the children of the share-croppers for whom a visit to Philadelphia, Mississippi, was a major undertaking have begun to become part of the larger world. "We'll be building the circuit boards ourselves at some point," Schisler says.

The plant, the humid pastures and the pinewoods lack the topographical drama of the rolling prairie and sagebrush desert that are the most familiar landscapes of Indian country. But the red clay of Neshoba County has endured a history no different in its essentials from that of the homelands of the Iroquois, the Sioux, the Paiutes or the Apaches. It too was fought over and mostly lost and, until a few years ago, was equally, even ineradicably one might have said, stained with hopelessness. It is today a land of redemption; not the exotic redemption of evangelical traditionalists who would lead Indians in search of an ephemeral golden age that never was, but a more prosaic and sustainable redemption of a particularly American kind that comes with the opportunity to work a decent job, and with knowing that one's children will be decently educated and that the future will, all things being equal, probably be better than the past.

Indian history is, after all, not only a story of wars, removals, and death, but also one of compromises and creative reinventions of Indian communities continually remaking themselves in order to survive. In the course of the past five centuries, Indian life has been utterly transformed by the impact of European horses and firearms, by imported diseases and modern medicine, by missionary zeal and Christian morality, by iron cookware, sheepherding, pickup trucks, rodeos and schools, by rum and by welfare offices, and by elections, alphabets and Jeffersonian idealism, by MTV and *The Simpsons,* not to mention the rich mingling of Indian bloodlines with those of Europe, Africa and the Hispanic Southwest. In many ways, the Choctaw revolution, like the larger transformation of Indian country in the 1990s, is yet another process of adaptation, as Native Americans, freed from the lockstep stewardship of Washington, search for new ways to live in the modern world.

Racial/Ethnic Conflicts and the Danger of Genocide

Nothing can prepare one for the experience of genocide . . . there are no words that can do justice to the anguish of the survivors and the cruelty of the killers.

RAKIYA OMAAR

Genocide is probably the most baffling social problem of the modern world. It seems unfathomable, incomprehensible that in these times large numbers of people can be murdered because of their race, ethnicity, or political affiliations. The words above are from a human rights activist who investigated the 1994 genocide in Rwanda—a situation in which hundreds of thousands of ethnic Tutsis were brutally slaughtered. From the 1970s to the 1990s, major genocidal conflicts occurred all over the world—in Africa, not only in Rwanda, but also in Burundi; in the Americas, in Guatemala; and in Europe, in the former Yugoslavia.

Genocide has existed throughout much of recorded history (Chalk and Jonassohn, 1990), but was only first defined and treated as a crime after World War II, when people in nations around the world were horrified to learn of the Nazi genocide against Europe's Jews, Gypsies, and others. In 1948, the United Nations established the Convention on the Prevention and Punishment of the Crime of Genocide. This Convention's definition of genocide is widely accepted. According to the Convention, genocide refers to:

> any of the following acts committed with intent to destroy, in whole, or in part, a national, ethnical, racial, or religious group, as such:

(a) Killing members of the group;

(b) Causing serious bodily or mental harm to members of the group;

(c) Deliberately inflicting upon the group conditions of life calculated to bring about its physical destruction in whole or in part;

(d) Imposing measures intended to prevent births within the group;

(e) Forcibly transferring children of the group to another group.

It was once believed that the modern industrial and scientific era would be a time of increasing harmony among ethnic groups. Order and conflict perspectives once agreed on this point. From the order perspective, it appeared that as societies became more functionally interdependent with industrialization, this would lead to a growing homogeneity of culture. As people became more and more like one another, interethnic conflicts would diminish. From the Marxist perspective, it appeared that as workers around the world united into anticapitalist revolutionary movements, they would abandon old ethnic hatreds. While there is no doubt that the global economy has produced a growing homogeneity in consumer culture—people everywhere drink Coke, drive Toyotas, and bank with Citibank—these optimistic views of world cultural harmony have been shattered over the last few decades by case after case of conflict, including genocidal conflict, among ethnic groups. And new technologies, rather than inspiring unity, have been harnessed to spreading racial hatred.

The first reading in this chapter, "163 and Counting . . ." by the Southern Poverty Law Center, is a report on the dramatic rise of Internet sites devoted to racial hatred. This is followed by "Rwanda: Death, Despair and Defiance," by the African Rights Organization, which summarizes this organization's monumental report on the 1994 genocide in Rwanda. An important point made in this article is that ethnic ties can be politically manipulated. The conflict between ethnic Hutu and Tutsi cannot be seen as the result of ancient tribal hatreds; rather, these groups have been profoundly transformed by centuries of contact with powerful colonial states and corporations. "Political manipulation of ethnicity is the main culprit for today's ethnic problem."

Genocide in recent times has also been motivated by political hatreds completely unrelated to concerns of race, ethnicity, religion, or nationality. Since the formulation of the U.N. Convention, there has been considerable controversy over whether the Convention's definition is inclusive enough to deal with all the kinds of large-scale murder that need to be prevented and punished. Should it not also be defined as genocide when governments kill large numbers of their citizens based on their political views? R. J. Rummel has suggested, for example, that we broaden the concept and change it to *demicide,* to include "the intentional killing of people by government." *Demicide* would include "cold blooded government killing . . . (beyond the U.N. definition) starving civilians to death by a blockade; assassination of supposed sympathizers of antigovernment guerrillas; purposely creating a famine; executing prisoners of war; shooting political opponents; or murdering by quota" (Rummel, 1995, pp. 3–4). Chalk and Jonassohn in their 1990 work, *The History and Sociology of Genocide,* broaden the definition of genocide to encompass politically motivated mass murders: "a form of one-sided mass killing in which a state or other authority intends to destroy a group, as that group and membership in it are defined by the perpetrator" (Chalk and Jonassohn, 1990, p. 23). Thus, the 1975–1978 killing of 1 to 2 million Cambodian citizens by the governing Khmer Rouge party—

perpetrators and victims both belonging largely to the same ethnic group—would be seen as genocide under these broader definitions.

Another controversy over the definition of genocide is the failure of governments or the United Nations to deal specifically with the situation of the many female victims of mass rapes and sexual torture that have occurred during genocidal conflicts. The United Nations has only recently begun to recognize rape as a war crime, and to bring perpetrators to justice (*The Day,* 1996).

Scientists have increasingly debunked the popular notion that there are superior and inferior "races," an idea that has accompanied many of the genocides in modern times. The popular idea of race is that people can be divided into distinct biological groups, based on observing their external physical features such as skin color, type of hair, or shape of the eyes. According to this belief, there are superior and inferior races, with the superior races entitled to possession of greater wealth and power. Modern genetic science has demonstrated that there is in reality only "one human race." The world's populations have become so intermixed that it is generally not possible to identify peoples that are totally distinct. The notion that there are different races is a *social construct.* Genetically, individuals within one so-called "race" are often more different from one another than they are from individuals of another "race." That is, "races" are real only in the sense that people in society *believe* that they are and treat one another differently based on this belief. In "The Outsider Looks In: Constructing Knowledge About American Collegiate Racism," Sylvia R. Tamale discusses her research into how "everyday racism" is constructed in interaction among college students.

Many social scientists today prefer to use the term *ethnic group* or *ethnicity* to replace the term *race,* because what people mean when they use the term *race* is often *ethnicity.* An ethnic group is a group of people who share a common culture, history, and, perhaps, territory. (The "national" and "religious" groups mentioned in the U.N. Convention may be seen as ethnic groups.)

Clearly, the notion of *ethnicity* can lead to unequal treatment, even genocide, just as well as the notion of race. However, this concept is often more useful in understanding people's lives. For example, if we use the racial classification used by the 1990 U.S. Census, people in the United States who have an ancestor who came from Africa will be defined as "Black." Yet the social distinctions most salient in understanding their lives might be an ethnic identification—such as Haitian, Dominican, Nigerian, or Puerto Rican—all of which are different from that of the descendants of Africans brought to the U.S. mainland during the centuries of slavery.

What are the prospects for creating a world free of genocide? At first glance, the situation seems gloomy. The number of ethnic groups, for example, involved in serious conflict has increased dramatically in the last few decades, leading one scholar to suggest that the twentieth century will go down in history as the "Age of Genocide" (Stoett, 1995). International and regional intervention to prevent and stop genocides have been less than effective (De Waal and Omaar, 1995; Harff, 1995; Ignatieff, 1995; Kuperman, 1996; Stoett, 1995).

There are some good reasons to be hopeful, however. One reason is that there is a growing understanding that we are in fact all "one human race." In the United States, for example, there is a movement among multiracial people to stop government racial classifications, which is leading to changes in how the U.S. Census is conducted. In "Census and the Complex Issue of Race," Ellis Cose discusses the current controversy over racial classification in the U.S. Census. Another reason is that the existence of multiple ethnic groups in a society does not necessarily mean that genocide will inevitably result. In some multiethnic societies (for example, Switzerland), people have managed to coexist peacefully and to share the society's resources equitably. In others, there may

be historic and pervasive ethnic conflicts, even violent ones, yet these conflicts have not escalated into mass killings (for example, Northern Ireland, South Africa). One scholar of genocide has argued that the presence of peacekeeping forces and the pressure of international public opinion have prevented the sharp conflicts in Northern Ireland and South Africa from escalating to genocide (Kuper, 1981, p. 209). Human rights organizations around the world have criticized the United Nations and regional bodies for failure to intervene swiftly and decisively to prevent mass murders, rapes, or "ethnic cleansing" in Rwanda and the former Yugoslavia. If the world learns from these cases, future situations may have more positive outcomes.

Another reason to be hopeful is that violent conflict is less likely to erupt in states with democratic and peaceful methods of conflict resolution (Rummel, 1995). Supporting democratization movements around the world is one way to help bring about a genocide-free twenty-first century. A final point is that the current upsurge in ethnic conflicts may be associated with particular sociohistorical circumstances that will not stay the same in the future. Ted Robert Gurr studied 233 significant ethnic groups around the world and examined the context of all the conflicts they were involved in over the past several decades. Many of these were conflicts over state power as states (for example, the former Yugoslavia) were reconstructed when the Cold War ended. Gurr argues that this era of transition is now coming to an end, and we may expect the situation in these states to stabilize. Many of the conflicts also erupted in new, weaker states in Africa that were undergoing political transition. Gurr argues that regional and international intervention may be able to mediate and manage such conflicts in the future, as we learn more about designing successful strategies for conflict resolution (Gurr, 1994).

WORKS CITED

Chalk, Frank, and Kurt Jonassohn. 1990. *The History and Sociology of Genocide.* New Haven, CT: Yale University Press.

The Day. June 28, 1996. *New York Times* News Service.

De Waal, Alex, and Rakiya Omaar. 1995. "The Genocide in Rwanda and the International Response," *Current History,* April 1995, 156–161.

Gurr, Ted Robert. 1994. "People Against States: Ethnopolitical Conflict and the Changing World System." *International Studies Quarterly,* 38, 347–368.

Harff, Barbara. 1995. "Rescuing Endangered Peoples: Missed Opportunities," *Social Research,* 62(1) Spring 1995, 23–40.

Ignatieff, Michael. 1995. "The Seductiveness of Moral Disgust," *Social Research,* 62(1), Spring 1995, 77–97.

Kuper, Leo. 1981. *Genocide.* New Haven, CT: Yale University Press.

Kuperman, Alan J. 1996. "The Other Lesson of Rwanda: Mediators Sometimes Do More Damage Than Good," *SAIS Review,* 16(1) Winter–Spring 1996, 221–239.

Rummel, R. J. 1995. "Democracy, Power, Genocide and Mass Murder," *Journal of Conflict Resolution,* 39(1), March 1995, 3–26.

Rummel, R. J. 1997. "Is Collective Violence Associated with Social Pluralism?" *Journal of Peace Research,* 34(2), 163–175.

Stoett, Peter J. 1995. "This Age of Genocide: Conceptual and Institutional Implications," *International Journal,* Summer 1995, 594–618.

QUESTIONS FOR DISCUSSION

1.   As reported in "163 and Counting . . .," hate groups have developed a large and growing number of sites on the Internet. Speculate on the importance of hate groups' use of the Internet to spread their ideas. What do you believe can or should be done about this? For example, should governments or Internet access providers ban such sites? Why or why not?

2.   What is genocide? How can this best be defined? Is the United Nations definition too limited?

3.   Explain what is meant by the statement in "Rwanda: Death, Despair and Defiance" that "political manipulation of ethnicity is the main culprit for today's ethnic problem." What is the significance of this? For example, in the search for solutions to the problem, what difference does it make whether we regard genocide as the result of ancient tribal hatreds, or as the result of political manipulation?

4.   "Race/ethnicity are *social constructions*." Explain this, using examples from the articles by Tamale and Cose. How is "everyday racism" constructed on your own campus or in your community?

 INFOTRAC COLLEGE EDITION: EXERCISE

As discussed in this chapter, the United Nations did not include *politically* motivated murders in its definition of genocide in the 1948 Convention on the Prevention and Punishment of the Crime of Genocide. *Why* did this happen? Using InfoTrac College Edition, look up the following article:

"The Crime of Political Genocide: Repairing the Genocide Convention's Blind Spot." Beth Van Schaack. *Yale Law Journal.* May 1997 106n7p2259–2291. (*Hint:* Enter the search term Beth Van Schaack.)

Read the article, especially Section I, "The Origins of the Geneva Convention and Its Blind Spot." (You may also want to read Section II, "The Cambodian Experience," if you would like to know more of the background of this case of politically motivated genocide.) If the United Nations were to debate this issue again today, what do you believe would happen? Would the world's nations want to change the definition of genocide to include politically motivated murders? If so, why? If not, why not?

FOR ADDITIONAL RESEARCH

Books

Chalk, Frank, and Kurt Jonassohn. 1990. *The History and Sociology of Genocide.* New Haven, CT: Yale University Press.

Fein, Helen. 1979. *Accounting for Genocide.* Chicago, IL: University of Chicago Press.

Kuper, Leo. 1981. *Genocide.* New Haven, CT: Yale University Press.

Organizations

Amnesty International USA
322 8th Avenue
New York, NY 10001
(212) 807-8400
www.amnesty.org

Human Rights Watch
485 Fifth Avenue
New York, NY 10017
(212) 972-8400
www.hrw.org

Klanwatch Project of the Southern Poverty Law Center
400 Washington Street
Montgomery, AL 36101
(205) 264-0286
www.splcenter.org/klanwatch.html

Race Unity and the Institutes for the Healing of Racism
Whitcomb Publishing
32 Hampden Street
Springfield, MA 01103
http://bounty.bcca.org/rel/race_unity

ACTION PROJECTS

1.   From the books listed above, identify one historical case of genocide of interest to you. Do some additional research using the Internet or library to learn more about the case. What were some of the main features of the culture of the people involved? What was the historical context in which the genocide occurred? What were the economic, political, and social forces that led up to the genocide? How did the people resist the genocide? Did they ultimately survive? What influenced their survival or nonsurvival? What lessons can we learn from this case to help prevent future genocides?

2.   Identify one current/recent case of genocide in the world—for example, Rwanda, Kosovo, Guatemala. Do some additional research using the Internet or library to learn more about the case. What are some of the main features of the culture of the people involved? What are the economic, political, and social forces that

led up to the genocide? How did the people resist it? What, if anything, was done by international organizations—governments, the United Nations, or other world bodies—to prevent the genocide? What has been done since the genocide happened to bring the perpetrators to justice? How effective has this been? What lessons can we learn from this case to help prevent future genocides?

3.   What is it like to live through a genocidal conflict? Find out if there are any survivors of a twentieth-century genocide living in your region. Interview someone, or invite someone to come speak to the class, in order to learn firsthand about the impact of genocidal conflict on the lives of individual human beings. Starting points to locate a possible speaker are local synagogues and churches, peace and justice organizations, or refugee assistance groups.

48

163 and Counting . . .

Hate Groups Find Home on the Net

MARK POTOK

In less than three years, hate on the Internet has exploded. Thirty-four months after the March 1995 day when former Klansman Don Black put up the first neo-Nazi site on the World Wide Web, there were 163 active sites spewing racial hatred.

The sites include 29 espousing Klan beliefs; 39 by neo-Nazis; 27 by racist Skinheads; 25 by proponents of Christian Identity, an anti-Semitic and racist religion; and 43 others pushing a hodgepodge of ideologies based on hate.

Almost half those Web sites represent actual groups—organized associations that can be contacted or joined, or from whom racist materials can be ordered. These are not merely lone malcontents serving up hate from a bedroom computer.

This count by the Intelligence Project is conservative. It includes only sites that were active in January 1998, and only those that contain explicitly racist or anti-Semitic material. The count does not include Holocaust denial sites—pages that are implicitly anti-Semitic, as they assume a Jewish conspiracy to cover up the true facts of the Holocaust, but pass themselves off as scholarly revisionism. And it is limited to sites based in America, although many others worldwide are available here.

The count also excludes sites put up by Patriot groups—militias, common-law courts and others—even though some of these groups endorse racist beliefs. A list of Patriot sites will be published in the spring issue of the *Intelligence Report*.

Frighteningly, many of the new sites are aimed directly at children.

For example, the World Church of the Creator, a virulently neo-Nazi group with 33 chapters, recently put up a page specifically aimed at kids. Its title page ("Creativity for children!") looks for all the world like some kind of Sesame Street for haters. Its aim: "To help the younger members of the White Race understand our fight."

"RAHOWA!" it ends. RAcial HOly WAr.

That is in line with a general trend on the hate pages. The racist movement has realized its future lies in the next generation . . . , children enamored of the Web and its colorful presentations. It has targeted these kids, and it is increasingly successful at drawing even those from well-to-do backgrounds.

Net hate sites run the gamut. They feature easy-to-reproduce Third Reich posters. They offer hundreds of violently racist and anti-Semitic jokes and cartoons. Many include "chat rooms" where racists trade news and views. One offers a real-life video of the harassment of an apparently retarded black man. There are dozens of pages featuring pin-up Skinhead women with names like "Katrina." Other sites explain the Byzantine Biblical interpretations of Christian Identity. . . .

These sites may seem arcane curiosities. But the fact is, they are slickly packaged propaganda that have given racists an audience of millions.

Five years ago, a racist group had to struggle financially, find a sympathetic printer and work long hours writing and editing to produce a pamphlet that might reach 100 people. Today, a lone racist can quickly pull down copy from other sites, package it using high-quality photos and graphics that are already available on the Net, and create a page that's accessible worldwide—often for no money at all.

The Net has given racists other advantages as well:

- Encrypted messages, chat room talk, e-mail communications and the propaganda put up on Web sites all give racists an empowering sense of community. Even lone racists, with no co-religionists nearby, feel they are part of a movement.

- Free encryption technology makes inter-group communication easy. Where such codes were once easily breakable, new technology is far more secure.

- E-mail messages are increasingly being used to send hate-mail to unsuspecting victims. While only one case has gone to court so far, incidents are rising fast.

- Net sites give groups the ability to market their wares—anything from Klan robes and Hitler mugs to paramilitary manuals and other publications—and raise revenue as never before. Racist white power bands, formerly limited to insiders and subscribers to certain magazines, use Net audio tracks to attract new customers.

- For those inclined to violence, the Net offers a wealth of information—from instructions on building an ammonium nitrate bomb to methods for converting semi-automatics to fully automatic weapons—that can be accessed in minutes.

The interest of the far right in computers is not new. In the late 1980s, former Texas Klansman Louis Beam was already building computer bulletin boards that racist groups used to communicate with one another. But the Net added another dimension, a virtual world in which hate groups easily could appeal to the uninitiated.

Don Black, a former Klan leader who served three years in prison for plotting to overthrow a Caribbean island government, was the first hate propagandist to recognize the potential of the Internet. After learning to operate and program computers in prison, he emerged to set up a Web site that is still active. Now, most major hate groups have Net sites, many of them containing pages and pages of propaganda.

This upsurge would come as no surprise to Don Black.

"There's a potential here to reach millions," he said years ago. "I think it's a major breakthrough. I don't know if it's the ultimate solution to developing a white rights movement in this country, but it is certainly a significant advance."

49

Rwanda: Death, Despair and Defiance

AFRICAN RIGHTS ORGANIZATION

There can be no introduction to an account of the crime of genocide and the murder of political opponents in Rwanda. Nothing can prepare one for the depth of inhumanity manifest by the killers. Likewise, there can be no simple conclusion, so vast is the crime against humanity.

• • •

From African Rights Organization, "Summary" from *Rwanda: Death, Despair and Defiance*, pp. xviii–xxxiii.

BACKGROUND TO MASS POLITICAL MURDER AND GENOCIDE

The genocide of the Rwandese Tutsi and the murder of many moderate Hutus was a political strategy adopted by a clique of powerful people at the centre of the government of Rwanda. Their plan was to hold on to power at all cost.

The war in Rwanda has often been presented as a "tribal conflict." This is highly misleading. Hutus and Tutsis existed a century ago, but the two categories were defined in very different terms in those days. They were far less mutually hostile. Colonial rule and its attendant racial ideology, followed by independent governments committed to Hutu supremacy and intermittent intercommunal violence, have dramatically altered the nature of the Hutu-Tutsi problem, and made the divide between the two far sharper and more violent. In short, political manipulation of ethnicity is the main culprit for today's ethnic problem.

In the 1970s and '80s, Rwanda was an authoritarian state ruled by President Juvénal Habyarimana with a single party, the National Republican Movement for Democracy (MRND). At that time, the "Tutsi factor" was largely absent in Rwandese politics. The principal power struggle was between the northern-based MRND and a Hutu-led political opposition based in the centre and south, sharpened following Habyarimana's reluctant opening-up to political parties. After the October 1990 invasion of the Rwandese Patriotic Front (RPF), a group largely formed of Tutsi refugees in neighbouring countries, President Habyarimana and his coterie faced the prospect of electoral loss at the hands of the civilian parties, as well as the threat of military defeat by the RPF.

The political and military opposition, with the support of neighbouring countries, aid donors and ultimately the U.N. Assistance Mission to Rwanda (UNAMIR), set in train a peace and democratisation process that seemed destined to remove the MRND from office. Those in power were determined to block this transition. They tried every trick, including political assassination. It is in the context of the political tensions surrounding this momentum towards peace and democracy that the 6 April coup d'état and the subsequent assassination of political opponents must be seen.

The killing was meticulously organized in advance. The principal instruments used included the Presidential Guard, the army, gendarmes, the civil administration and the extremist MRND militia known as interahamwe. These forces were developed from late-1990. They were expanded, trained and armed. During 1990–93 they had many opportunities to practise their methods. There are several well-documented massacres during this period, and also very many political assassinations.

Rwandese Hutu extremism was born at the centre. The extremists were men and women in control of all the organs of state power, which, over the months and years, they directed into their genocidal project. "Hutu extremism" is a bland name for a genocidal ideology, the provenance of which is examined.

The killing was an attempt to unite the Hutu behind an extremist platform dedicated to the eradication of the Tutsi and all moderate political opposition. The extremists played upon deep fears and frustrations in the Hutu populace, at all levels. Economic crisis was threatening prospects of unemployment for many salaried people. There was a serious problem of access to land for young men. Rwandese popular political culture has, since independence, played upon myths of the Tutsi as historic "invaders" of a different race, and perpetrators of injustice. Playing on these themes, the extremists' propaganda proved highly effective. The journal *Kangura* and the Radio-Télévision Libre des Mille Collines (RTLM) were particularly efficient at pushing the Hutu extremist message.

• • •

THE KILLERS

Rwanda's main killers are known. They include the members of the self-appointed interim government, senior army officers, senior civil administrators, and the men who organized and commanded the interahamwe throughout the country. The ideologues who justified the genocide and incited the population are also well-known. African Rights' report lists the names of those primarily responsible for the killing.

The genocidal orders were passed down through the administrative and military hierarchies. The killings were no spontaneous outburst, but followed instructions from the highest level.

At an intermediate level, very many local administrators, members of the professions, businessmen, soldiers and gendarmes were instrumental in organizing the killings. The involvement—often indirect, sometimes brutally direct—of many schoolteachers, doctors and medical staff, judges, employees of international organizations and priests in the slaughter raises profound questions for the integrity of these professions in the future.

Ordinary people joined the killing through various motives—greed, fear of pressure from above, and outright coercion. The interahamwe were sent to rural areas not just to kill, but to force the local people to kill. Often, people were compelled to kill their neighbours or members of their own families. The extremists' aim was for the entire Hutu populace to participate in the killing. That way, the blood of the genocide would stain everybody. There could be no going back for the Hutu population; Rwanda would become a community of killers.

• • •

POLITICAL KILLINGS

When President Habyarimana was killed on 6 April, most politicians immediately feared the worst. There could be no serious efforts to organize resistance because communications were cut, a curfew was imposed and roadblocks mounted. These strategies also made escape very difficult.

Within hours, the first wave of killings was unleashed. The lists of opposition figures to be assassinated had been prepared in advance. The Presidential Guard hunted down opposition politicians in their homes. Many were killed;

those who escaped largely did so through sheer luck; for example, they were not at home when the assassins came. The report contains detailed testimonies from politicians who managed to evade the killers.

The second target of the extremists was any form of political dissent. A chief target was independent journalists. During the preceding years, several independent journals had played a vital role in exposing corruption and abuses by the government. Numerous journalists were murdered. Human rights activists were also targeted. Senior civil servants, especially those involved in the judicial system, were killed.

• • •

A POLICY OF MASSACRES

The largest number of killings were carried out in huge massacres across the country. As the killings began, frightened people—mainly Tutsi but also Hutu—fled to hospitals, schools, churches, stadiums and other places of hoped-for sanctuary. The interahamwe also encouraged people to congregate in these places.

The scars of the massacres that followed are still evident. In particular, churches are still stained with blood, and scarred with damage inflicted by bullets and fragmentation grenades. In some, holes have been torn in the walls so that the killers could throw their grenades in. Many are also the sites of mass graves. Often, bodies remained unburied weeks afterwards.

The testimonies that African Rights obtained from the survivors are detailed, compelling and consistent. They fit precisely with the concrete evidence available and they concur with one another.

The massacre at the Parish of Ntarama, Bugesera, is typical of many in which thousands died. People—mainly Tutsi but also some moderate Hutu—gathered in the church compound. They deterred their attackers for some days, by throwing stones, until the militia (who were armed mainly with machetes) were reinforced by soldiers and armed policemen. The parish was then assaulted, and when resistance had been broken, the interahamwe carried out the mass killing with their machetes. The survivors who tried to run away were hunted down. Another massacre at Nyamata nearby was also carried through as a military operation, involving close military cooperation between the different killing institutions.

The first large-scale massacres were committed within days of the death of the President. Killing began in Gikongoro with a large number of isolated attacks on Tutsis, forcing people to congregate in parishes and schools. Massacres in Gikongoro included Kibeho, where people were killed in the parish, school and hospital. Another major massacre was the Parish of Kaduha; later, refugees and patients were murdered at the health centre. Another large-scale massacre was the Parish of Muganza.

In Butare, there were no large-scale killings for the first twelve days. This was due to the fact that it was a stronghold for the opposition, and especially because of the efforts of the préfet, Jean-Baptiste Habyarimana, who was the only Tutsi préfet. Many refugees fleeing the killings of Gikongoro and Greater Kigali sought refuge there. Unfortunately, on 19 April Habyarimana was removed and replaced by men ready to carry out the policy of massacres. Immediately, the mass killings began. Because of the concentration of refugees, the numbers killed were among the largest in all of Rwanda. In Cyahinda, perhaps twenty thousand people died. In Karama, between thirty-five and forty-three thousand are estimated to have been killed. Another large massacre was at a ranch and agricultural research centre, ISAR, in Songa. It is important to stress that these massacres were military operations spearheaded by the army.

Gitarama was another préfecture that was initially calm. But there too the killers succeeded in undermining local opposition, and introducing interahamwe from elsewhere. Several massacres are documented.

In Kibungo, the killings began almost instantaneously. In Gahini, Rukara, Zaza and elsewhere there were mass killings. In Kigali city there were a number of large massacres, several of which are documented.

Kibuye, which had the largest Tutsi population, was the region most devastated by the genocide. In a few weeks, nearly a quarter of a million people were killed. They perished in Gitesi—at the Parish of Kibuye and Gatwaro Stadium. Two thousand people were bulldozed to death at the Parish of Nyange in Kivumu. Thousands were murdered at the commune office of Rwamatamu, the Parish of Mubuga in Gishyita and at several Adventist Churches—in Ngoma, Muhomboli and Murangara.

Next to Kibuye, Cyangugu suffered the most from the policy of extermination. Huge massacres were carried out in parishes throughout the préfecture—at Nyamasheke, Shangi, Mibilizi, Nkanka and amongst others, Hanika. People were also murdered in health centres and schools attached to these parishes.

• • •

HUNTING THE TUTSI

Apart from the huge massacres, hundreds of thousands of Tutsis were killed individually in a genocidal frenzy. They were hunted down in their houses, chased from hill to hill, or stopped and slaughtered at roadblocks. This began immediately that the news of the President's death was known. In préfectures as far apart as Kibungo and Cyangugu, the killing of Tutsis began within hours. People survived only by extraordinary measures, such as hiding in pits or in ceilings for weeks or months.

The first target of the killers was Tutsi men and boys. Many families saw all their menfolk killed. Even the smallest boys were not spared— the extremists would say that they were killing the future generation of Tutsi soldiers. African Rights' report contains the testimony of a woman whose baby, aged only ten hours, was killed because he was a boy.

Educated Tutsi men and women were particular at risk. Staff members of international organizations were targeted. The university was "cleansed."

The killings were carried out with extraordinary cruelty. The killers burned people alive, often in the ceilings of their houses where they were hiding. People were thrown, dead or alive, into pit latrines. People were compelled to kill their relatives. Those who wanted to escape a slow and painful death by machetes had to pay the killers for the privilege of being shot with a gun.

The survivors ran for swamps and hills. Here too they were hunted, and suffered the extremes of human degradation.

Others took refuge in camps, such as the Hotel Mille Collines, the churches of St. Paul's and St. Famille in Kigali, the bishopric of Kabgayi in Gitarama, Kamarampaka Stadium and the camp of Nyarushishi in Cyangugu. Here too they were subject to the depredations of the killers, who entered to select the people they wanted to kill. In the death camps in Kigali, UNAMIR offered at best uncertain protection. The Amahoro stadium was shelled, and attempts to evacuate people from the churches were highly dangerous.

One of the consequences of the hunt for Tutsis has been that families became separated. Many people still do not know what has happened to their relatives.

• • •

RAPE AND ABDUCTION OF WOMEN AND GIRLS

Rape is a weapon of war and genocide; nowhere more so than in Rwanda. Many of the Tutsi women who survived did so at the price of being abducted and raped by their captors. Rape was very, very common. African Rights' report contains detailed testimonies from women who have been forcibly abducted and raped, the first detailed evidence for this crime.

The interahamwe regarded Tutsi women as the spoils of the killing. Captors regarded raping the women as their right. There are even cases of more humane militiamen hesitating to rape women they had abducted, but then coming under intense pressure from their families to rape the women. Some captors took abducted women as "wives," apparently with the intention of keeping them after the genocide had been completed.

Often, as the focus of the killers increasingly switched to murdering women, these women were sought out, or were handed over by their captors, and then killed. Some women were tortured and raped, other women who had been horribly wounded were also raped.

Like women the world over, the victims of rape in Rwanda are ashamed and fear being ostracised. Many are reluctant to talk about their experiences. Rape was used to shatter the social bonds that hold the Tutsi community together. Its effects will be felt for years, if not decades.

There is an acute need for counselling services. Many raped women were also made pregnant. Unable to terminate the pregnancy for a variety of reasons, many are now forced to live the additional trauma of having children borne out of these brutal rapes.

• • •

VIOLENCE AGAINST CHILDREN

The extremes of the cruelty of the killers is attested to by the extent of the violence against children. Many Tutsi children have been killed. Others suffered appalling wounds received in the massacres or at roadblocks.

Children have not just been killed in massacres along with their parents. They have been deliberately sought out and murdered. There are cases in which the killers have massacred adults and children separately—attacking the assembled children with their machetes and clubs. Children have been selected and shot. They have been hunted down. They have been buried alive or thrown wounded into latrines.

Schools, children's homes and orphanages were attacked. Eighty-two youngsters were killed in an attack on a school in Kibeho, Gikongoro. Other massacres are also detailed.

Rwanda is full of orphans and physically handicapped children. But the psychological traumas are greater still. Many children could not talk for a long time.

It is particularly disturbing to note that schoolteachers and others employed in education were among the killers. In the massacre at the Parish of Rukara, twenty-five schoolteachers have been accused of complicity. The implications for the future integrity of the teaching profession are profound.

• • •

THE ATTACK ON THE CHURCH

The church was a particular target of attack for the killers. Churches were attacked not only because people congregated in churches but in order to desecrate them and to destroy people's confidence that the church could protect them.

Many priests and nuns were killed. This was for many reasons: many priests were known as critics of human rights abuses, and for their connections to churches abroad. The very first massacre of the genocide occurred on the morning of 7 April in the Centre Christus in Kigali, where seventeen priests, lay persons and staff of the centre, both Tutsi and Hutu, were massacred. Thereafter, the ranks of the priesthood were decimated. Tutsi priests and nuns in particular were targets. All denominations were affected.

The RPF also killed churchmen: four soldiers were responsible for the murder of thirteen Roman Catholic priests including four bishops on 2 June. The evidence however points to the conclusion that they were acting against orders, and that the killing was not officially sanctioned.

Some priests were conspicuous in their efforts to resist the killers. For example, Fr. Célestin Hakizimana of St. Paul's church in Kigali continuously confronted the killers in a brave attempt to protect the refugees who had taken shelter in his church.

The institutional church, however, was conspicuous by its failure to take a firm or timely stand against the mass killing. This must be seen in the context of a long history of political compromise. The Roman Catholic and Anglican hierarchies in particular were very close to the Habyarimana government. The Roman Catholic Archbishop served on the central committee of the MRND for fourteen years. While some other churchmen strove for a dialogue between the government and opposition, the two archbishops stuck firmly to a pro-government line.

This line continued essentially unchanged after 6 April. Rwandese church leaders were very reluctant to condemn the genocide, and did so only belatedly and in a very half-hearted manner. Meanwhile, a disturbingly large number of priests, pastors and some nuns assisted the killers. They betrayed the hideout of Tutsi colleagues and refugees to the killers. They refused to provide a sanctuary to the hunted. They spent time in "meetings" with the principal killers of their region. For some churchmen, participation was brutally direct. Abbé Thaddée Rusingizandekwe, a teacher at the Grand Seminary of Nyakibanda in Butare, was one of the leaders in the massacre against the Parish of Kibeho in Gikongoro.

The silence of church leaders and the complicity of some priests, pastors and nuns, presents the churches with a huge challenge. If the churches are to retain credibility in Rwanda— and indeed more widely—it is essential that this complicity be dealt with frankly.

• • •

ATTACKS ON HOSPITALS

In their determination to destroy whatever remained of a moral order in Rwanda, the extremists also turned their fury on hospitals. Many people were killed in hospitals; there were massacres in Kibuye, Kigali, Butare, Nyamata, Cyangugu, Rwamagana, Kibeho, and elsewhere. Patients were murdered, or forced out of the hospital to be killed at the gate. Red Cross ambulances were attacked. A school of nursing was attacked. Even people who took refuge in the maternity clinics were attacked and massacred. Psychiatric patients were murdered.

Many doctors and medical staff were killed. Tutsi and moderate Hutu health professionals were hunted down and murdered.

While many health workers struggled to maintain their neutrality, and to carry out their work under extreme difficulty, others were sadly compromised by the killing. A large number of doctors, nurses and medical assistants were extremely active among the killers. As in the case of teachers and the church, rebuilding institutional integrity for the medical profession is an imperative for Rwandese doctors.

· · ·

RESISTING THE KILLING

One of the most persistent myths about the killings in Rwanda is that they swept over the entire country with no resistance. This is far from the truth. Under the most difficult and dangerous circumstances imaginable, many ordinary Rwandese did their utmost to resist the genocidal slaughter. They did this in the teeth of relentless propaganda that denigrated the basic human values of compassion and solidarity, and exalted killing as a civic duty. The propaganda also insisted that Hutu extremism was militarily victorious, so that those who resisted it had no certainty that their values would ever triumph.

Many Tutsis who survived owed their lives to Hutu friends and neighbours, and even strangers, who took great risks to protect them. Throughout the country, ordinary Hutu people concealed Tutsis in their houses and farms, often with great ingenuity. They knew that the price of being discovered was probably death, and some did indeed pay with their lives for their humanity.

Some administrators and local policemen did their best to protect people against the killers, but they were too few and too exposed to last for long. In less conspicuous ways, the resistance of ordinary people was crucial to the survival of those Tutsis who are still alive in much of Rwanda.

The first response of many communities to the disaster was to set up communal patrols, including members of all ethnic groups. The extremists tried pressure, bribery and force to break the bonds of communal solidarity, and in most places they succeeded. Resistance did not usually last long.

As they gathered together, Tutsi refugees often formed self-defence groups. Armed only with sticks and stones, and an occasional firearm, these groups held out for days or weeks at most. The exception was more than fifty thousand refugees who gathered at the hilltop of Bisesero in Kibuye. They fought back until the arrival of the French forces; by then, only about two thousand were still alive. The inter-ahamwe usually called in the army or Presidential Guard to overcome any organized resistance.

It is essential for the self-esteem of the Rwandese people that the role of those who resisted the killing be emphasized.

· · ·

THE RWANDESE PATRIOTIC FRONT OFFENSIVE

The RPF was a party to the conflict with the government. It was not a party to the genocide. There is no evidence for any policies by the RPF aimed at systematically killing civilians. There have certainly been human rights abuses, including reprisal killings by RPF soldiers, but the RPF has stated its commitment to investigating these and bringing those responsible to book.

The military advance of the RPF was the single most important reason for the end of the killings. The RPF victory opens the real prospect of punishing those guilty of genocide. While there is no doubt that the 1990 RPF invasion was critical to raising the political temperature in Rwanda, this does not justify abuses in response. The argument that the final RPF offensive, launched in the afternoon of 7 April, somehow "provoked" the genocide is not consistent with the evidence. The RPF military victory is what brought the genocide to an end.

Throughout the war, the United Nations and various other international organizations were consistently calling for a ceasefire. There is no evidence that a ceasefire would in any way have contributed to the cessation of the killing of civilians. On the contrary, a ceasefire would have suited the government in that it would have prolonged its life, and hence its genocide.

The advance of the RPF has contributed to some serious humanitarian problems, such as the huge flow of the refugees into Tanzania and Zaire.

But these could only have been avoided at the cost of halting the genocide, a task that no other force could have accomplished. This raises important legal and moral questions.

• • •

REFUGEES

Some of the largest and most dramatic refugee crises of recent history have occurred on Rwanda's borders. They have caused immense humanitarian emergencies. However, the humanitarian aspect is not the focus of this report; rather it is concerned with the way in which many of the killers have become refugees, and have continued to exert a powerful influence in the camps. Prominent killers have taken control of the refugee camps, and are actively intimidating the refugees in an effort to discourage their return to Rwanda. A number of refugees have been killed. Almost the entire Rwandese army is in exile in Zaire. Extremist politicians are continuing to agitate the population and call for an armed invasion of Rwanda.

• • •

THE INTERNATIONAL RESPONSE

The international community created high expectations of its ability to act in Rwanda, when it despatched the two thousand seven hundred strong UNAMIR force to the country in December 1993 to guarantee the transition to peace and democracy. In April 1994 it betrayed all the hopes that it had raised. The international response to the crisis displayed a near-total lack of moral leadership.

The crisis had been brewing for a long time. The U.N. was well aware of the capacity for mass violence in Rwanda, as there had been repeated warnings. But when the storm broke, it appeared to be caught unawares.

At first, there was total confusion. This confusion was partly orchestrated by the interim government, which sought to portray the issues as,

one, "uncontrollable tribal killing" and, two, the war with the RPF, thereby concealing its own political murders and policy of massacres. In substantial part, the international media followed the "tribal anarchy" line. This smokescreen of confusion was slow to clear, allowing the killers to proceed largely undisturbed. The international media, with its automatic tendency to blame violence in Africa on what it calls "age-old ethnic enmities," must examine its record closely.

Belgium has alternately played an honourable and dishonourable role in Rwanda. Its colonial legacy notwithstanding, the Belgian government was at the forefront of pushing for peace and democratisation in the early 1990s. It had the largest contingent of troops in the country. But, after ten soldiers were killed on 7 April, it led the international stampede out of Rwanda. Confused and cynical, the members of the U.N. Security Council followed Belgian prompting and voted to withdraw all but a token U.N. force from Rwanda.

Thereafter, France took the leading activist role on Rwanda at the U.N. France is a long-time ally of the former government and remained close to the Hutu extremists even during the height of the genocide. France was reluctant to push for any action that might embarrass the interim government, while the French secret service continued to supply arms illicitly to the government.

The United States was in a position to take a moral lead on Rwanda. After some encouraging signs in the first days after 6 April, it conspicuously failed to do this. The U.S. government was instrumental in blocking and slowing down U.N. attempts to become re-involved in Rwanda during May and June. The U.S. objected to attempts to brand the crime as genocide, and placed a whole succession of petty conditions on the despatch of U.N. forces. The U.S. made no attempt to expel the Rwandese ambassador until after the fall of the interim government—and then tried to dress its cynical gesture as an act of moral outrage. The rapid reaction to the humanitarian crisis in Goma, Zaire, merely shows the moral bankruptcy of the former policy over the genocide.

African countries. particular Tanzania, and the OAU played an important role in the 1991–94 peace process. The OAU was quick to condemn the killings in very strong terms, but subsequent attempts to resolve the crisis came to nothing, in part because of the withdrawal of U.N. troops. Zaire however has proved a loyal ally to the extremists, giving them refuge and support. Zairian government policy is so erratic and cynical that it is difficult to predict what the next Zairian move will be.

On the ground, the U.S. Special Representative, Jacques Roger Booh-Booh, compounded the errors in New York by insisting that the priority was a ceasefire, and trying to maintain diplomatic neutrality between the two sides—a policy that in effect conceded unwarranted legitimacy to the genocidal interim government.

The French intervention appears to have been launched for a range of motives, including playing to the domestic humanitarian constituency and reassuring francophone African leaders that France would remain loyal. The worst potential result of the invention was the strengthening of the extremists in power. That was fortunately averted as a result of the outright military victory of the RPF shortly after French troops arrived. Operation Turquoise brought some modest benefits, but also considerable solace to the killers. The manner in which the operation was decided upon and launched casts the gravest doubts on the integrity of the U.N.

• • •

THE DEMAND FOR JUSTICE

Most Rwandese are adamant that justice must be done. The truth must be told and the killers brought to book. Only when this is done can reconciliation and national reconstruction begin. An end to the culture of impunity is an absolute prerequisite for the respect for human rights in the future Rwanda.

None of these tasks will be easy. They will be slow and difficult, and will not obliterate emotions—including the desire for vengeance—that may lead to incidents of violence. But the cost of doing nothing, or trying to forget the past, will be far higher. The extremists in exile are already trying to misrepresent the past and conceal and justify what they have done, as the basis for mobilizing a new extremist community. They are busy cultivating contempt for the rule of law, based on cynicism about the prospect of trials for crimes against humanity. If Rwanda remains bitter and divided, they will have fertile ground to advance their cause. Hutu extremism will be defeated only when all Rwandese have fully and openly faced the past, and there is widespread confidence that justice prevails.

There is an international legal obligation to bring prosecutions for the crime of genocide. Genocide has been committed and the Genocide Convention obliges all states to punish those responsible. Enough evidence has been amassed to extradite those primarily responsible from Zaire, Tanzania, Kenya, Belgium, Cameroon, France and other countries and bring them to court.

In the aftermath of the genocide, African Rights advocated that the trials be conducted in a Rwandese court, rather than an international tribunal sitting outside. International assistance is essential to ensure that the trials take place, with full respect for due process of law, and in guaranteeing the impartiality of the court. But we believe that it would have been better to hold the trials in Rwanda, in front of the Rwandese people.

Holding trials in this manner will be the first exercise in truth and justice for most of the Rwandese population. It should be an exemplary process that will help build confidence in the rule of law. It will also direct the anger and bitterness of the Rwandese population towards those responsible for the genocide, thus helping to contain the impulse towards indiscriminate revenge, and helping to create a new image of national identity based on precepts of truth and justice, rather than ethnic chauvinism.

It is unlikely that more than a small minority of the killers can be brought to court. But that should not preclude other forms of justice. The most basic form of justice is exposure. Truthtelling throughout the country is an absolute prerequisite for the return of any form of normality.

The credibility, not just of the Rwandese judiciary, but also of all international organizations committed to principles of human rights and humanitarianism rides on bringing the genocidal criminals to trial. If the killers escape unpunished, it will be an advertisement for the hollowness of human rights rhetoric and the importance of human rights organizations. It will also be an advertisement for the impunity of political criminals, no matter how immense their crime.

. . .

50

The Outsider Looks In

Constructing Knowledge About American Collegiate Racism

SYLVIA R. TAMALE

This study tackles one of the most complex and intriguing issues in contemporary society, namely, the phenomenon of racism. Instead of examining the structural dimensions of racism, it focuses on the interpersonal "everyday racism" that occurs among students. Using the University of Minnesota as a case study, the study employs qualitative research methods to offer new perspectives on everyday racism as perceived through the eyes of a Black foreign female student. Popular portrayals of the midwestern United States present a relatively liberal milieu where racism only subtly affects social relations, and where there is "zero tolerance" for the politics of exclusion. However, the findings of this study illustrate that everyday racism is alive and well in the collegiate environment. Epistemological issues are elaborated, arguing for the position of an interpretive and reflexive rather than a positivist approach to social research.

INTRODUCTION

This paper examines the daily, seemingly innocuous interactions between students at the University of Minnesota and illustrates how "everyday racism" manifests itself in such an environment. The study offers a different perspective on the subject matter at hand. In classical anthropology and ethnographic sociology Western scholars study non-Western societies and subordinate groups within Western societies. For an African researcher to conduct an ethnographic study on racism in the U.S. is to take the lens and turn it on the photographer.

The issue of everyday racism in eastern Africa, where I come from, is little more than an academic subject with the vast majority of the population rarely confronting its realities. My experience in the United States was, therefore, a rude awakening to the fact that my skin color was some kind of caveat emitting negative vibrations and occasioning spontaneous reactions from people around me. Suddenly, I became extremely conscious of the chocolate-brown skin I had always taken for granted. At the same time I became keenly aware of the color of everybody around me. In this racially-stratified society I soon discovered that theorizing about racism was one thing; "living it" quite another.

My limited experience in the U.S. and an initial fascination with the issue of racism raised a number of sociological and ethnographic questions: How does racism manifest itself among

From Sylvia R. Tamale, "The Outsider Looks In: Constructing Knowledge About American Collegiate Racism," *Qualitative Sociology,* vol. 19, no. 4, 1996, pp. 471–495. Copyright © 1996 by Plenum Publishing Corporation. Reprinted by permission.

college students? How does it affect their behavioral interaction? What "coping mechanisms" do students of color develop to handle the problem? Subconsciously, I hope that this study would help unburden some of the heavy load that was beginning to weigh my heart down.

The primary purpose of this paper is to focus on the following methodological question: what role did I play as a researcher in influencing the construction of knowledge about the research question? Although it will not provide a full exposition of my findings on collegiate racism in the United States, some descriptions from my extensive fieldwork and interviews will be given in illustration of the phenomenon.

• • •

METHODS

The study lasted a total of 20 weeks. I spent much of the winter and spring of 1994 conducting field research at a study/eating area situated in a busy location on campus. The field site was ideal for my study for a number of reasons: First, it was a place where students of all races and ethnic backgrounds congregated on a daily basis. These included: White, Blacks, Asians, Hispanics, American-Indians and Arabs. This heterogeneity provided an ideal setting to observe inter-racial interactions among students. Secondly, the size of the field site (approximately 40 by 80 feet) was attractive in that it was neither too large to inhibit close observation nor too small to limit generalizations. The arrangement of the tables, i.e., clusters and singles, facilitated the easy observation of patterns as it provided a natural discriminant sitting arrangement and spatial manipulation. Five rows of four-table clusters were fixed to the floor and occupied a rectangular portion in the middle of the room. More tables run along the four walls in a single row of singles and doubles to make a total of forty-four. On either side of the 2.5 square-feet tables was a chair attached to the table by a hollow black iron bar. A 45-degree swivel of the chairs allowed the users to negotiate their way around the crammed tables before they could

settle down comfortably. Thirdly, the site was ideal for its multi-faceted roles. Because students went there to eat, study, chat or simply wile the time away between classes, it offered a unique opportunity to observe students in a variety of social interactions. Finally, as a student myself, I blended in easily with the "subjects" thus minimizing the problem of obtrusiveness.

The first half of the project was devoted to participant observation. For at least two hours every day, five days each week, I "hung out" at the field site at different times of the day. As a covert observer, I spent approximately 150 hours observing students' behavioral interaction. During this period I was very alert to nonverbal communication between students. In her classic book, *Body Politics,* Nancy Henley (1977) conducts a psychoanalytic study of how nonverbal behavior bound to people's power relationships. She notes: "The 'trivia' of everyday life—touching others, moving closer or farther away, dropping the eyes, smiling, interrupting—are commonly interpreted as facilitating social intercourse, but not recognized in their position as micropolitical gestures, defenders of the status quo—of the state, of the wealthy, of authority, of all those whose power may be challenged" (1977: 3). Subsequently, in order to gain a better understanding of "what's going on" I conducted intensive nonstructured interviews with fifteen students who were regulars at the field site representing the whole spectrum of the racial cosmos at the University of Minnesota. The interviewees constituted a "theoretical sample" of students consciously selected to represent all the racial/ethnic groups within the student body. All interviews were recorded and transcribed. The names of all informants have been altered to provide anonymity.

• • •

ANALYSIS OF MY PLURAL ROLES

Understanding who I am is an essential precursor to the issue of how the multifaceted roles I occupy affected the course of the research, my interaction with informants, and the study's findings: My formative years were largely an African middle-class experience. The oldest of six children, I went to school in Uganda and became a lawyer. I obtained a master of laws from Harvard in 1988, and then returned to Uganda where I worked as a law lecturer for five years. In 1993, I returned to the United States to pursue Ph.D. studies in Sociology and Feminist Studies. Now thirty-two years old and married, many people have commented that I could easily pass for a regular American undergraduate. During my previous stay in the United States, I socialized mostly with international students from the "third world" and had only a minimal interaction with White people.

As a Black/Foreigner

In 1959, John Howard Griffin (an Anglo-American) used medication, ultraviolet rays, and dyes to acquire the "appropriate" skin color when he wanted to study what it *really* felt like to be a Negro in the south (Griffin 1961). Although my skin pigmentation is naturally Black, in many ways, I shared Griffin's anxieties, naivete and discoveries in the course of my field research. Like him, I was consciously "living" racism first hand for the first time. Prior to the study, my outsider status had "protected" me from a lot of the pain and degradation that comes with a heightened sensitivity to racism. The study taught me what other people of color in this country already knew: that racism is as much a part of American life as apple pie and baseball, and that it is multifaceted. Racism can be vertical or present itself in the form of cross-racial hostility and runs on a continuum from greatly blatant to extremely subtle.

The color of my skin was an important asset to this inquiry as it facilitated my process of knowing. At the field site, I deliberately sat on the side of the room which had joint double tables simply to observe which people were more likely to take the free table adjacent to mine. At lunch time, the room would often fill to capacity with the only free table being the one next to mine. I observed many White and Asian students enter with trays of food, scan the room for a place to sit, their eyes resting on the free table next to mine for the briefest of moments and then decide to leave. The

few who were bold enough to approach the free table often asked in the "Minnesota Nice" style, "Is this table free?" before sitting down. It was at times like these that I was acutely aware of the lens-turned-on-the-photographer metaphor referred to in the introduction.

As a foreigner, it took me some time to figure out the "Minnesota Nice" facade. In my initial analysis I misinterpreted it thus:

> White students seem to be overly sensitive to any action that may be interpreted as racist. That is why they were bending over backwards to ask the irrational and superfluous question, "Is this table taken?" when they find a vacant table next to one occupied by a minority student. It is their way of proceeding cautiously to ensure that minority students do not translate any of their actions as rude or racist!

Had I been a native of this country I probably would have quickly picked up this behavior as exhibiting the well-known "Minnesota Nice" behavior. In other words, had I been an insider, I would have been able to "decipher the unwritten grammar" of the Minnesota Nice conduct. The truth subsequently became more apparent when I noticed that the question was not only directed to minority students but even to fellow White students. As the study progressed, I was also to discover that White students were not as self-reflexive about their racist behavior as I had given them credit in the analysis quoted above.

Not only did Whites and Asians avoid sitting next to me, but when I sat next to them, many of them vacated their tables within a few "polite" minutes of my arrival. This behavior was not restricted to me personally but I observed a pattern with most Black students who took tables next to White students. For those others who were not too racist to flee, I normally noticed their slight but visible shifting and rearrangement when a Black person sat next to them. I recorded one such incident in my field notes, thus:

> On returning from re-heating my lunch, I found two White women sitting at the table next to mine. As I sat down I noticed their slight but visible shifting and

rearrangement on realizing that I was the person occupying the table next to theirs. One of them, who had placed her bag and jacket on the chair across from me, nervously asked, "Is this stuff in your way?" This was a superfluous question since my bag was already resting on the bench next to me and I was not carrying anything that would call for such concern on her part. The question was probably part of the discomfort they felt by my presence.

During my observations I also noticed that White people have a distinctive way of staring at people of color; like one would stare at something "exotic." This was especially true for minorities whose physical appearance differs most from that of Whites such as American-Indians and Blacks. The following field note illustrates this point:

> There is a White man sitting in the northeast corner of the room who has been staring at me for some time. When I stare back he quickly shifts his stare but as soon as I look away, he resumes staring. He is about 35 years old, has brown hair and a Hitler-type mustache. . . . As he stares, his elbows are resting on the table and he is covering his mouth with both fists. He is very cool and collected.

The words I used to describe the man's mustache reflected the frustration and anger that I felt at the time. My response to his persistent stare had immediately invoked feeling of nazism, racism, and xenophobia. His stare was not flirtatious; it was not my femininity that he was ogling. It was a curious kind of stare—as a child would stare at animals in a zoo. Shifting his stare every time I stared back at him denied me subjectivity and interaction while giving him power over me. Here, my anxieties as a Black foreign researcher in a predominantly White setting were accentuated and this, no doubt, colored the interpretation that I attached to the incident.

As the same time, my "Blackness" sometimes rendered me invisible to "paler" students. On Valentine's day, for example, three Asian students entered the field site with bundles of Valentine roses for sale. They approached virtually every table aggressively soliciting buyers. Below, I quote my melancholic field note:

> I was the only Black person in the room at the time and when they got to my table, they looked right through me and walked past me! It was as if I was not there. . . . All I could do was follow their receding backs with fiery eyes.

Rollins (1985) argued that one of the ways that racism manifests itself is by the dominant group ignoring the very existence of minorities. In this case it was manifested through cross-racial hostility.

In the course of the interviews, it was quite evident how my different attributes influenced the ways informants responded to my questions, which in turn affected the knowledge they passed on to me. Race relations in the U.S. are such that whenever two people of different races are interacting, the issue of race occupies a prominent place in their consciousness for the duration of the interaction. Race was always salient and part of the dynamic in my interviews, because of and in spite of the subject matter of the study. The fact that Black is at the base of the racial hierarchy means that it receives the biggest dose of racism in this country. This made every informant from the lighter races quite sensitive to what they said concerning issues of racism. For example, Kelly, a White female whom I interviewed was quick to assert that people's skin-color was not a big deal as far as she was concerned despite the fact that I had observed her several times avoiding minorities and meticulously selecting whom she sat next to at the field site. She went on to say:

> I really have a problem with the media lately, especially you know . . . local news and the stuff in there. The portrayal of minorities . . . I think it's just completely disgusting. Especially criminal stuff. African-American men are plastered on the T.V. . . . they portray them as the only criminals around.

Kelly could have said this out of genuine concern. However, what suggested to me that she might have expressed such sentiments simply because they had a "politically correct" flavor to them (especially to a Black interviewer), was the way she spoke these words directly to the tape recorder. Her words seemed to convey the message, "Let it go on record that I think there's racism out there; but I have nothing to do with it. I am on your side."

In her study on White women and racism, Ruth Frankenberg (a White woman) also found that it was not always easy to get her subjects to talk about issues of race and racism. She attributed this primarily to the sensitivity to the topic arguing that, ". . . in a racially hierarchical society, White women have to repress, avoid, and conceal a great deal in order to maintain a stance of 'not noticing' color" (Frankenberg 1993: 33). This suggests that the sensitivity of White interviewees to the issue of racism is not limited to situations where the interviewer is Black. However, in my own study, there were various cues which suggested that subjects were reacting to me, a Black female foreigner, thus confounding the complex and sensitive subject matter. More examples from my interview with Kelly will help clarify this point.

At another stage, I asked Kelly, "When you were growing up, do you remember ever getting minority visitors at home?" Kelly uncrossed her legs and crossed them again, stared up into space for a long while before responding, "Um, it's hard to remember. It makes me think . . . Gee, there had to be at least some . . . you know, but I can't remember any." Perhaps Kelly would have responded in a similar round-about fashion had the same question been posed by say, a White interviewer. But the tell-tale signs were there to show that I, personally was a big source of her uneasiness. While she did not want to tell an outright lie, Kelly felt uncomfortable giving a direct answer because she thought that by so doing, her family would be portrayed as racist to a Black interviewer. Averting her gaze from me and uncrossing her legs were two vital pieces of evidence of Kelly's discomfit in "exposing" herself to a Black foreigner.

Similar efforts to gain my approval and to demonstrate how "liberal" Kelly was can be gleaned from the following example: Question: "What do you think of students from other racial backgrounds?" Response: "I really like to have people around from diverse backgrounds. It's really boring to have people . . . you know everybody is like you. . . ." My rather amused analysis note to this response was: "So for Kelly, racial diversity is nothing more than having a Neapolitan ice-cream!" This time Kelly spoke while directly looking at me. Her beseeching eyes seemed to call for my approval. At that point, I felt kind of sorry for Kelly and offered a nonverbal acknowledgment by way of nodding.

Instances demonstrating how my skin-color influenced the information given to me were numerous. The following excerpt from an interview with Maria, a native of Mexico provides another example.

S: Would you date someone from a different race than yours?

M: I dated a White man once.

S: Would you date a Black guy?

M: (pause) um . . . yes, I think so. I think so, (my emphasis) especially after taking this class on Race, Class and Gender . . . Before, my mind used to be very closed.

S: And would your folks have any problem if you introduced a Black guy to them as your fiancé?

M: Mmm. . . . I don't know.

On the one hand Maria could have been rethinking her personal life in the context of an important class; on the other hand the question made her uneasy. In the margin of my field notes next to Maria's response to the question: "Would you date a Black guy?," I pencil-scribbled, "gaze dropped" and "higher voice pitch." These actions may indicate either her cross-racial hostility with dating or her uncertainty and how the class challenged her assumptions. It could have been both, but clearly my personal attributes had a lot to do with these reactions.

The above illustrations aptly demonstrate how my skin-color placed constraints on the amount and quality of information elicited from White informants. The subject matter of the study made Whites least likely to be my confidantes.

The prospect of talking to a Black person about race relations among students of the University of Minnesota seemed too much for some students. Whereas White students were very eager to be interviewed (not wanting to appear racist, perhaps?), Asians were particularly suspicious of the motive behind such an exercise. Many of those I approached requesting an interview flatly refused to take part. Their refusals were revealing. Culture could be part of the reason why many Asians were reluctant to talk to me. But it could also have been due to the location that Asians occupy on the racial hierarchy. Although Asian students fall within the category of racial minorities, their position in American society is significantly more privileged than say, that of Blacks, Hispanics, and American-Indians. Indeed, as several informants revealed to me, they are regarded as the "model-minority" in the United States. Of course this stereotypical descriptor is a myth for many Asians in the United States and cannot be generalized across the different south-east Asian nationalities. However, it is arguable that many of the Asians in college are in many ways more privileged than other minorities. Such elevated status, therefore, would put them in a tricky situation, making it difficult for them to talk about racism to a minority student who ranks at the very bottom of the racial hierarchy.

My shared status with Black students had the opposite effect. They immediately regarded me as an insider and forged a kind of collaboration that struck an instant rapport in our conversations. Had I been a White researcher, for example, I would have missed the meanings of some nonverbal communication that Black students share amongst themselves. For example, when I made eye contact with Black students many would acknowledge me by a nod, waving or saying hello regardless of the fact that [we] were total strangers. There would have been no way for a non-Black observer watching us in this interaction to know that we did not even know each other. Neither would such a researcher have known that these gestures of solidarity or comradeship are simply one survival mechanism that Black students employ to deal with racism around them. Indeed, those small gestures of acknowledgment and camaraderie often made the dreary and dispiriting exercise at the site a lot more bearable.

Black students spoke to me without inhibition. While they often volunteered anecdotal experiences of racism, it was taboo for White students to even discuss third-party incidents of racism that they may have witnessed. Aisha, an African-American woman, responded this way to my compliment on how smart she looked: "Well thank you . . .," then she leaned toward me and lowered her voice, "you know, today is my T.A. day and I have to look good in order to feel good and to act good. There's a way that these *White students* (my emphasis) look at you . . . perceive you . . . it's a psychological thing and I take advantage of it." In my analysis notes following this data entry I interpreted Aisha's words thus: "This is a good example to illustrate the fact that students of color at the University of Minnesota have to make extra effort to legitimize themselves in order to combat the extant stereotypes about their lot." Although Aisha did not mention it, I suspect that her gender also had something to do with her worries; the gender-hierarchy which pervades our societies leads most women in authoritative positions to be conscious of their gender. However, in Aisha's case the racial factor entered the equation to compound her worries.

Bayo, an African male student, said:

> My experience here . . . you know . . . you read about all of this stuff and until you come over here and experience it you'll never know how bad it is. You know . . . we think . . . I used to think that the Blacks here whine a lot, complaining about racism all the time. But when you come over here you experience it.

When Bayo used the phrase "you know," it was not merely a "stumbling inarticulateness" but intended as a verb (cf. Devault 1990). As African-to-African, he expected me to know what he was talking about. Further, by including me in the phrase "we think," Bayo assumed or implied our common "outsider-within status." He took it for granted that I (as a Black African) held the same preconceptions about American society as he did. He was not only talking to me as a researcher, but as a person who shared similar experiences and someone who might empathize with the experience of African-Americans. And in fact, because Bayo's words were reminiscent of my own experience, I knew exactly what he was talking about.

Even non-Black foreigners drew upon my outsider-status in understanding their experiences of everyday racism in this society. A good illustration of this is my interview with Li, an Asian informant who related an emotional episode he experienced about a month prior to our interview. About half-way through the interview, Li looked at me with moistened eyes and a flushed face and said:

> Usually the way they express racism is more subtle. This White woman didn't say something expressly but she just winced . . . actually I asked her some questions and after she answered my question she, you know . . . (screws up his face) made some facial expression. I need to talk to my advisor or the special committee. There is some committee that you can complain to, I think I need to . . . you know, yes I need to. I'll always remember this. *If they don't express it (racism) then you can feel and that's another story, you know . . .* (my emphasis)

Although Li was ten years my senior, I was a stranger he could safely confide in. As a fellow foreigner and a person of color, he assumed that I would understand precisely what he was talking about. And I did. "Feeling racism" as Li put it, was another concept that I could relate to very well as a minority person. It is part of the reason why it is sometimes impossible to offer cogent proof of its manifestation. Like a thick cloud, it envelopes space even where no blatant act of racism has occurred. Many students of color described this phenomenon to me.

Standpoint theorists would argue that as a Black person I am more likely to understand everyday racism than say, a White-American would. This is because I am equipped with knowledge of both my own context and that of the dominant group—so as to survive in their society. Uma Narayan (1990) analyzes and critiques standpoint theory by arguing that this so-called "epistemic advantage" is a double-edged sword. She points out that while people living under various forms of oppression may be more likely to have a critical perspective on their situation than the dominant group, there is a down side to the "double vision." In my case, the negative side of the epistemic advantage presented itself at two levels. First, the experience of "turning the lens on the photographer" in this research placed a heavy toll on my personal life. As the study progressed, I developed a heightened sense of racism which was conflated with a deep sense of resentment, degradation and marginality.

At another level, the down side of my epistemic advantage meant that I could not always carry out a perfect "dialectical synthesis" of the subject matter at hand. While my status brought insight and opened some doors, it may have closed others. There are various ways that my subjective perspective could have influenced the collection of data. For example, my interpretation of the behavior of three students of color—Musa (African-American male), Yoko (Asian female), and Lisa (African-American female)—who regularly broke the norm by exclusively hanging out with White students, may have been due to my peculiar biases. I analyzed their behavior thus:

> The obvious explanation for this is that these minority students desperately need to identify with the dominant, privileged group. That is their way of dealing with the discrimination that they face in their daily lives. By constantly keeping the company of Whites (regardless of how patronizing the relationship may be), they create a gulf between themselves and their oppressed lot, thus creating a false sense of being different from them.

Despite the fact that I based my conclusions on my observations of who the dominant party was in the relationships of all three cases, the explanation that I purport to be "obvious" may indeed be inaccurate. Subjectivity of this nature is present in all research. Another example of the limitations of my epistemic advantage arises from the exclusiveness of my focus. By focusing on one social space, I could not see other spaces. There exist more social and political locations within the University (e.g., the public interest research group) where students of color and White students work together on particular issues.

• • •

REFERENCES

Devault, M. (1990). Talking and Listening From Women's Standpoint: Feminist Strategies for Interviewing and Analysis. *Social Problems* 37(1), 96–116.

Griffin, H. J. (1961). *Black Like Me.* Houghton Mifflin Company.

Henley, N. (1977). *Body Politics: Power, Sex and Nonverbal Communication.* Englewood Cliffs, NJ: Prentice-Hall.

Narayan, U. (1990). The Project of Feminist Epistemology: Perspectives from a Nonwestern Feminist. In A. Jaggar & S. Bordo (Eds.), *Gender/Body/Knowledge: Feminist Reconstructions of Being and Knowing.* New Brunswick: Rutgers University Press.

Rollins, J. (1985). *Between Women, Domestics and Their Employers.* Philadelphia: Temple University Press.

51

Census and the Complex Issue of Race

ELLIS COSE

Racial classification has always been a serious—if maddeningly imprecise—business, determining who gets access to which wing of the American dream. Slave or freeman, citizen or alien. "Tell me your color," America proclaimed, "and I'll tell you your place."

The nation has long outgrown the days when slavery was a color-coded calling and naturalization was restricted to "aliens being free white persons," but the legacy of color consciousness remains. Too often our culture embraces the notion—even as our laws reject it—that all races are not equal. Nowhere is that assumption more apparent than in our accepted definition of race. For whereas one drop of "black blood" is generally considered sufficient to render one black, whiteness is not so easily bestowed.

Many people are now suggesting that the time has come to reject racial definitions rooted in slavery (and in the drive to re-create a sharp black-white divide that miscegenation had, in some respects, erased). Isn't it time, some critics ask, to rethink, and perhaps dismantle, a racial categorization system whose primary function is to separate and divide? What's the point, they ask, in forcing people into black, white, yellow, or red boxes that cannot possibly accommodate America's growing racial diversity, particularly when the black box is fundamentally different from the others, carries the full baggage of slavery, and defies all common sense? Why, they ask in effect, must a person with any degree of black African ancestry be forced to pretend that no other racial heritage counts?

The questions come most insistently and most urgently from those who make up what has been dubbed the "multiracial movement" and who resent being asked to deny a major part of who they are. Though the issue of racial classification is important to multiracial persons of all conceivable backgrounds, it has assumed a particular importance for those whose heritage is, to some degree, black. For unlike Americans of other races, blacks have largely been defined by the so-called one-drop rule: the presumption that a small percentage of black ancestry effectively cancels out any other racial claim.

Sorting out the matter of racial identification is not the only, or even the most important, task facing the U.S. Bureau of the Census as the decennial census approaches. Correcting the chronic undercounting of certain groups is, in some sense, a larger problem. But the issue of racial categorization may well be the most explosive issue on the table. People have strong feelings about how they are grouped, particularly when it comes to race; and often people's sense of where they belong is very different from the place where others tend to put them.

Race is such a subjective and squishy concept that there is no objective way of determining who is right. Moreover, as philosopher and artist Adrian Piper has observed: "The racial categories that purport to designate any of us are too rigid and oversimplified to fit anyone accurately." At a time when the very idea of racial categories is under assault, the Census Bureau is charged with carrying out what may well be an impossible task.

Piper's own sense of the absurdity of America's concepts of distinct racial groups arises from being a "black" person whom many people assume to be "white." She seemed so white to her third-grade private school teacher that the woman wondered whether Piper knew that she was black. Piper, of course, is only one of a long ling of people who have found their appearance to be at odds with what America insisted that they were.

In 1983, for instance, an appeals court ruled that a Louisiana woman must accept a legal designation of black, although by all outward appear-ances she was white. The woman, Susie Guillory Phipps, who was then forty-nine, had lived her entire life as a white person. Upon hearing the court's decision, she told a *Washington Post* reporter: "My children are white. My grandchildren are white. Mother and Daddy were buried white. My Social Security card says I'm white. My driver's license says I'm white. There are no blacks out where I live, except the hired hands." Phipps had discovered that the state considered her black on obtaining a copy of her birth certificate in order to get a passport. Her attempt to change the designation eventually led her to court. A genealogist who testified for the state uncovered ancestors Phipps knew nothing about and calculated that she was 3/32 black. That was sufficient to make her black under a Louisiana law decreeing that a person who was as little as 1/32 black could not be considered white.

That Louisiana law was unique in this modern era in writing racial classifications into law, but the acceptance of the notion that "black" encompasses virtually everyone with black African ancestry is widespread. It is embodied not only in census data but in civil rights law. Consequently, many people who care about such laws find the matter of re-examining racial categories (especially of who belongs in the "black" category) to be unsettling.

Not that anyone believes that the current categories—spelled out in Office of Management and Budget Directive No. 15—reflect the true diversity of who Americans are. The population of the United States, as even defenders of the present system will acknowledge, consists of much more than four racial clusters (American Indian or Alaskan native, Asian or Pacific islander, black, white) and one relevant ethnic group (Hispanic). An array of spokespersons for an assortment of ethnically or racially interested organizations have proposed that the current categories be changed, or at least expanded. They have made arguments for Middle Easterners to be seen as something other than white, for Hawaiians to be grouped with Native Americans, and for Hispanics to be made into a separate racial (as opposed to ethnic) group. The most intriguing argument, however, comes from those who insist that the Census Bureau should sanction a multiracial category that would, at a minimum, encompass the children of those who come from different racial groups.

Susan Graham, a white woman married to a black man in Roswell, Georgia, told the U.S. House Subcommittee on Census, Statistics, and Postal Personnel that she was not at all happy with census bureaucrats who had told her that biracial children should be assigned the mother's race because "in cases like these, we always know who the mother is and not always the father." She apparently was not so much angered at the suggestion of uncertain paternity as at the fact that no present category reflected her biracial children's full heritage. Instead of making her choose between black and white boxes, she said, the Census Bureau ought to have a category called "multiracial."

As Lawrence Wright reported in the July 25, 1994, *New Yorker* magazine, the proposal "alarmed representatives of the other racial groups for a number of reasons, not the least of which was that multiracialism threatened to undermine the concept of racial classification altogether."

Without question, the current categories are, in many respects, arbitrary. They reflect the conventions of a slave-holding past and serve the needs of various political agendas, but they have nothing to do with the science of genealogy, or for that matter, with science at all. Nor is their meaning always clear. Directive No. 15 instructs that "the category which most closely reflects the individual's recognition in his community should be used for purposes of reporting on persons who are of mixed racial and/or ethic origin." But what about those people who, like Susan Graham, believe multiracial is the only label that fits? As the incidence of interracial marriage increases, more and more people will no doubt find themselves in Susan Graham's shoes.

In 1992, in Bethesda, Maryland, several hundred such people came together for the "first national gathering of the multiracial community," as described by Bijan Gilanshah, in the December 1993 *Law and Inequality* journal. Gilanshah saw

the meeting as an important development in the evolution of a growing social movement. Multiracial individuals, he wrote, existed in a "state of flux." The gathering was only one sign of many that they were "demanding clarification of their nebulous social and legal status and seeking official recognition as a distinct, powerful social unit with idiosyncratic cultural, social and legal interests."

Gilanshah pointed to several bits of evidence in support of his conclusion. Nearly 10 million people had elected to place themselves in the "other" category in the 1990 census, he noted, and interracial unions had sharply increased, seemingly tripling between 1970 and 1990, even as a host of self-described multiracial organizations were springing up around the country. Many of the multiracial activists, he noted, had intense feelings on the subject. "For the multiracial movement, failure of the government to include a multiracial category would result in cultural genocide," argued Gilanshah.

Julie C. Lythcott-Haims, writing in the *Harvard Civil Rights Civil Liberties Law Review* for the summer 1994, made a similar argument. Her primary concern was with adoption policies aimed at ensuring that children are placed with adoptive parents of their own race. She objected to such policies, pointing out that race matching cannot always work "because millions of children are born not merely of one race." Lythcott-Haims went on to make a broader argument whose implications stretched beyond her immediate concern of adopted children. The Census Bureau, she said, should accept a multiracial category "based on the logic that if people must be categorized according to race, these categories should be more accurate."

Lythcott-Haims was especially emphatic in her rejection of the so-called one-drop rule. The assumptions underlying it, she wrote, are "blatantly racist because the central premise is that 'Black blood' is a contaminant while 'White blood' is pure." Moreover, she noted, some multiracial people are uncomfortable declaring one of their racial lines to be better or more worthy of acknowledgment than the other: "The Multiracial person can hardly advocate the superiority or inferiority of one race without touching off a potentially damaging identity struggle within herself."

For Lythcott-Haims, the quest of multiracial individuals for their own racial designation was anything but a trivial pursuit. "If you identify as Black or Asian and our society officially classified you as White, how would you feel? If you identify as White and society required you to call yourself Latino, how would you feel? . . . If we send in our forms but the Census chooses not to recognize us for what we are, it is as if we do not officially exist," she wrote.

Without question, Lythcott-Haims and Gilanshah are correct in noting the absurdity of the current classification scheme. It is folly to force people to try to fit into narrow boxes that do not reflect their real complexity or their true sense of identity. But that does not necessarily mean that a multiracial box would make racial categories any less absurd or render them any more accurate.

For one thing, a multiracial designation does not really categorize a person racially. It simply indicates that a person fits into more than one category. Consequently, it could end up being even less precise than the groupings we have now. If the rubric applies to anyone of racially diverse ancestry, it could conceivably apply to most Americans now considered "blacks" as well as to a substantial portion of those who belong to other recognized racial groups.

Even if it is taken (as many proponents would like) to apply only to those with two parents of recognizably different racial stock, it is not a very accurate descriptor. A multiracial box (unless it spelled out what racial heritages were subsumed by the designation) would put the offspring of a white person and a Native American in the same pigeonhole as the offspring of a black person and a Chinese American. While the offspring of both unions would certainly be multiracial, not many Americans would consider them to be of the same race. In all likelihood, society would still consider the black-Asian child to be black (or perhaps mixed) and the white–Native American child to be white. They would probably, in any number of circumstances, be treated quite differently.

A *New York Times* poll of 1991, for instance, found that 66 percent of whites were opposed to a relative marrying a black person, whereas 45 percent were opposed to a relative marrying a Hispanic or Asian person. Clearly, in the eyes of many of those respondents, all multiracial families are not created equal. If part of the purpose of census classifications is to permit the government to determine how various groups are treated, aggregating groups whose only common denominator is that their parents are racially different would not do much to advance that purpose.

But even assuming one could agree that all people with parents of different racial stock should be considered members of a new race called "multiracial," what about those people who do not really care for that designation? What about the numerous offspring of black and white unions, for instance, who insist on calling themselves black? Providing them with a multiracial box would not guarantee that they would check it. And what about the children of "multiracial" parents? If the designation only applied to the first generation, would children of multiracial people become (like many light-skinned "blacks" who are clearly of mixed ancestry) monoracial by the second generation? Or would those children twenty years from now be fighting for yet another redefinition of race?

And what about Hispanics? Obviously many Latinos find the current categories lacking. They are uncomfortable with a system that insists that they define themselves either as black or white. In the 1990 census, roughly half of the Latinos in California described themselves as neither white, black, American Indian, nor Asian. They were "other." But that does not mean that a "multiracial" grouping is the solution.

Certain Latino leadership organizations, after all, prefer a designation that would set Hispanics apart from other groups racially, not one that would throw them into one huge category that combines all people of mixed racial heritage. Moreover, many Latin American cultures recognize an array of racial delineations much more complex than those acknowledged in the United States. Brazil, for instance, once had more than forty different categories. And Latin societies have come up with an array of terms to note the differences in those who are light-skinned and "European looking" and those who are not.

In an essay "Empowering Hispanic Families: A Critical Issue for the '90s," Frank F. Montalvo observed: "At the heart of the Hispanic experience in the United States is a form of racism that both binds light and dark Latinos to each other and divides them into separate groups. Race may prove to be a more pernicious element in their lives than are linguistic, cultural and socioeconomic differences." It is not at all clear that allowing Latinos the option of describing themselves as multiracial would resolve the problems many have with the current categories or would allow Latinos to express the true racial complexity of their cultures.

In an article in *Transition* entitled "Passing for White, Passing for Black," Adrian Piper tried to explain the phenomenon of people whom the United States had designated as "black" deciding to live their lives as "white": "Once you realize what is denied you as an African-American simply because of your race," wrote Piper, "your sense of the unfairness of it may be so overwhelming that you may simply be incapable of accepting it. And if you are not inclined toward any form of overt political advocacy, passing in order to get the benefits you know you deserve may seem the only way to defy the system."

The suspicion in some quarters is that the new emphasis on a multiracial category may be motivated by a desire to escape into a more socially congenial category, at least for those who face the alternative of being forced into the black box. The multiracial category, after all, is not really new. The rise of the mulatto category during Colonial times was an acknowledgment that the offspring of black–white unions were not necessarily either black or white. But after emancipation the in-between status of mulattoes "threatened the color line," as Gilanshah, among others, has noted. So the mulattoes were pushed into the black category.

Gilanshah obviously does not equate the new multiracial group with the privileged mulattoes of yesteryear. But in many respects, the language of

the multiracial lobby invites such a comparison. Many advocates of the new designation see multiracial individuals as ambassadors between groups. Gilanshah, for instance, argues that society would benefit from having multiracial people who are uniquely positioned to be "sensitive, objective negotiations of inter-group racial conflict." But to assume that only designated multiracial people can be a bridge between races is to assume that others cannot be. It is also to reawaken recollections about the middleman role of American mulattoes and Latin American mestizos, groups who were assigned a status lower than that of whites but higher than that of the groups with whom the whites had mixed. Their purpose, at least in part, was to reduce ethnic tensions and to keep people in their assigned places.

The Reverend Jesse Jackson has looked not to American history but to South Africa and its "colored" class to find an analogue to the multiracial category. Sociologist and anthropologist Pierre L. Van Den Berghe makes the same comparison: "It boggles my mind that the United States, in the late 20th Century, is [considering] reinventing the nonsense that South Africa invented 300 years ago," Van Den Berghe told a reporter for the *Los Angeles Times*. A separate multiracial category, he argued, would further "the inanity of race classification." Moreover, he saw the category as redundant. African Americans, he asserted, are in fact already "mixed-race."

None of this is to say that there is anything wrong with people defining themselves as a multiracial. In many respects, it is certainly a better descriptor than black, white, American Indian, or Asian. And indeed, during the next year, as the Census Bureau tests its various racial indicators, the multiracial category should be given its due. The bureau should evaluate whether a new multiracial category would increase the census's clarity or simply heighten confusion. Certainly, if multiracial persons are allowed to describe themselves as multiracial and then forced to specify how, the quality of the information gathered should not diminish and would perhaps be enhanced.

Still, adoption by the census of a multiracial box is not likely to accomplish much of what its proponents seem to seek. It would not prevent Americans from assuming that people who "look black" are black, whatever their other heritage. It would not provide a reliable anchor in racial identity, since multiracial is not so much an identity as an acknowledgment of multiple heritages. And it would not change current thinking that divides people into often opposing racial and ethnic groups.

It would not, for instance, persuade the members of the National Association of Black Social Workers, who have been outspoken for more than twenty years in advocating that "black" children go only to "black" families, to shift their position suddenly. Nor is it likely to mean a change in the practice of discrimination in employment or elsewhere. It certainly would not help, and might well hurt, enforcement of laws dealing with housing discrimination, employment discrimination, and voting rights.

In the *New Yorker* magazine, Lawrence Wright observed: "Those who are charged with enforcing civil-rights laws see the multiracial box as a wrecking ball aimed at affirmative action, and they hold those in the mixed-race movement responsible." Wright wonders about the practical effect: "Suppose a court orders a city to hire additional black police officers to make up for past discrimination. Will mixed-race officers count? Will they count wholly or partly?"

Proponent of the multiracial classification obviously do not tend to see themselves in such a light. They are not so much making a political protest as a personal statement about identity. And the question they raise is clearly an important one and, in some sense, an inevitable one.

Certainly it is possible to envision a "multiracial" box, perhaps in addition to an "other" box, that would not undermine civil rights laws or launch demands for special status for a multiracial "race." The option would simply allow people to describe themselves as what they perceive themselves to be. And if that box also forced people to designate in what ways they were mixed, the information could eventually be aggregated in whatever way would be useful. In short, the purposes of the data collection process could be served, while people were allowed to make a statement about their personal identity. No census box, however, will solve the larger problems of race in this country. Nor will a census box resolve anyone's sense of racial alienation or provide a secure racial identity.

Ideally, one day we will get beyond the need to categorize. Certainly, geneticists and other scientists are concluding that racial categories make little sense. But the tendency to categorize is strong and will clearly be with us for a while. For the time being those who are struggling with such issues might be well advised to heed the words of Adrian Piper, who declared, "No matter what I do or do not do about my racial identity, someone is bound to feel uncomfortable. But I have resolved that it is no longer going to be me."

• • •

11

Global Crime

It is better enjoying a short life
than being poor forever.

A COLOMBIAN DRUG DEALER

Sociologists define *crime* as the violation of criminal laws enacted by governments. Crimes include not only violations of the laws of nations, but also violations of international laws, laws that have been agreed to by more than one nation. The types of crimes are as varied as the law itself. The study of crime as a social problem encompasses *street crimes* (crimes involving violence or property, such as murder, rape, assault, robbery), the so-called *crimes without victims* (crimes in which participation is voluntary, such as the illegal drug trade, some forms of prostitution, gambling, pornography), *white-collar crimes* (crimes carried out by middle- and upper-class people in the course of their profession, such as embezzlement, stock manipulation, Medicaid fraud), and *hate crimes* (crimes based on prejudices of gender, race, ethnicity, sexual orientation). In studying global crime, it is useful to classify crime as to whether it is *individual* or *organizational*. *Individual crimes* are those carried out by individuals acting alone. *Organizational crimes* include *organized crime* (crimes committed by organizations established for the specific purpose of profiting through illegal activities, such as the Mafia), *corporate crime* (crimes committed by corporations for the purpose of enhancing profits), and *political crime* (crimes committed by groups for political purposes, such as Watergate, vote-purchasing, terrorism, torture or disappearing of political prisoners). Crime, like other social problems, can be analyzed from various perspectives. In the first reading in this chapter, "Nothing Bad Happens to Good Girls," Esther Madriz uses the social constructionist perspective to analyze how the fear of crime in the United States is a form of social control that limits women's lives.

This chapter focuses on organizational crime, which has taken on important global dimensions in the modern world. It begins with a consideration of the global trade in illegal drugs. This highly profitable industry involves all three types of organizational crime—organized crime (syndicates that manufacture and distribute the drugs), corporate crime (money laundering through financial institutions), and political crime (protection by corrupt governments and militaries). The global trade in illegal drugs is an enormous industry employing and selling to millions of people worldwide. Although illegal, it resembles in many respects the legal industries of the capitalist world economy. The industry lends itself well to analysis in the world system perspective. Rural peasants in periphery nations produce the raw materials such as coca and poppy, which are manufactured into consumption drugs such as cocaine and heroin in urban areas, and distributed to customers in the core and semi-periphery. Processes of unequal exchange operate throughout the production and distribution process, with the peasant producer at the bottom receiving only a tiny portion of the product's value, and the leaders of the organized crime networks ("cartels" or "syndicates") at the top receiving an immense portion. In "Who Benefits (and Who Doesn't)," Clarence Lusane succinctly outlines the social structure of the global drug trade.

During much of the late twentieth century, public concern with crime was focused on individual violent crime and organized crime. In recent years, social movements of consumers, workers, environmentalists, and health advocates have drawn increasing attention to the problem of corporate crime. As the economy has globalized, so has corporate crime. The cost to people everywhere is enormous—social scientists studying crime generally believe its cost is far larger than the losses due to street crime. Robert Sherrill's "A Year in Corporate Crime" examines the negative impact of crimes committed by multinational corporations. His account is based on reports of corporate crimes published in *The Wall Street Journal* during 1996.

Political crimes abound in the modern world. One area of particular concern to human rights activists are those that occur within the criminal justice systems of many nations. Such institutions are often involved in committing political crimes against inmates. Such crimes may violate not only national laws, but also international laws, such as the United Nations Universal Declaration of Human Rights or the provisions of the Geneva Convention on the treatment of prisoners of war. In "Pay Now, Pay Later," Christian Parenti discusses state policies in the United States of forcing prisoners to pay for needed medical treatments. These are currently under challenge as a violation of the Constitution's prohibition of "cruel and unusual punishment."

What can be done to stop the illegal drug trade, corporate crime, and political crime? Research indicates that culture and social institutions may be shaped to prevent crime. At least some societies in the world seem to be successfully insulating themselves from the global crime wave associated with the drug industry. In "Japanese Say No to Crime," Nicholas D. Kristof describes the research by sociologist David H. Bayley into why Japan is still the "safest of the industrial nations."

QUESTIONS FOR DISCUSSION

1.   According to Madriz, the fear of crime in the United States is a form of social control over women that perpetuates gender inequality. What are some examples of this process, from Madriz's research? Can you think of examples from your own experiences or observations?

2.   There is a growing consensus that the U.S. government's "War on Drugs" has not been very successful. Explain why, drawing on Lusane's analysis in "Who Benefits (and Who Doesn't)," as well as ideas from the world system perspective and other theoretical perspectives in sociology.

3.   Reflecting the public's perception that corporate crime is less harmful than street crime, corporate crime is less frequently detected, and when detected, less harshly punished. Why do you believe corporate crime has not been regarded as a serious social problem? Analyze this, using ideas from the symbolic interactionist, functionalist, and conflict perspectives.

4.   What is "peonage"? Do you agree with Christian Parenti that forcing prisoners to pay for costs of their incarceration, such as medical care, is peonage? Is this a violation of the U.S. Constitution's prohibition against "cruel and unusual punishment"?

5.   Could Japan's approach to crime and criminals, as discussed in "Japanese Say No to Crime," be transferred to the United States? Why or why not?

 INFOTRAC COLLEGE EDITION: EXERCISE

The United States incarcerates a higher percentage of its citizens than other nations, and considerable controversy exists over why this is the case, and whether this is a positive or negative phenomenon. Using InfoTrac College Edition, look up the following articles in order to learn more about the issue:

"Singapore West: The Incarceration of 200,000 Californians." Mark G. Koetting; Vincent Schiraldi. *Social Justice,* Spring 1997, v24n1p40. (*Hint:* For this, and the two articles below, enter the search term prisons, using the Subject Guide.)

"More Money Spent on Prisons Than Colleges." *Jet,* March 17, 1997, v91n17p23.

"America's War on Drugs: Unprecedented Success or Casualty of Failed Policy?" John R. Weekes. *Corrections Today,* October 1998, v60n6p8.

(You can look for additional articles on the issue by using the search term Narcotics, Control of, using the Subject Guide.)

Why does the United States have such a high incarceration rate? What are some of the social problems resulting from this policy? Has the boom in America's prisons helped solve the nation's crime problem?

FOR ADDITIONAL RESEARCH

Books

Coleman, James William. 1994. *The Criminal Elite.* 3rd edition. New York: St. Martin's Press.

Costanzo, Mark. 1997. *Just Revenge: Costs and Consequences of the Death Penalty.* New York: St. Martin's Press.

Lusane, Clarence. 1991. *Pipe Dream Blues: Racism and the War on Drugs.* Boston, MA: South End Press.

Organizations

Alternatives to Violence Project
P.O. Box 300431
Houston, TX 77230
(713) 747-9999
www.avpusa.org

Drug Policy Foundation
4455 Connecticut Avenue, Suite B-500
Washington, DC 20008
(202) 537-5005
www.dpf.org

INFACT
256 Hanover Street
Boston, MA 02113
(617) 742-4583
www.infact.org

National Coalition to Abolish the Death Penalty
1436 U Street NW, Suite 104
Washington, DC 20009
(202) 387-3890
www.ncadp.org

ACTION PROJECTS

1.   In a variety of other nations, the death penalty has been abolished, because it is considered a violation of human rights. The United States is the only industrialized nation to have the death penalty. Why? Make an analysis of this question after doing some library research and contacting organizations on the list in this chapter. (Organizations listed in the chapter on Democracy and Human Rights may also be helpful.)

2.   Contact INFACT and learn more about its international campaign against the tobacco industry. If you feel this is a campaign you can support, organize a group of

students on campus to publicize it, and help in other ways requested by the organization. Reflect on what you learned from this experience about corporate crime.

3.   How has the global drug trade affected your community? What are some of the human costs? What are some solutions being pursued? Do some research in the library and by interviewing knowledgeable people. Given what you are learning in this course, how effective are the solutions being pursued in your community? What else needs to be done to solve the problems in your community created by the global drug trade?

52

Nothing Bad Happens to Good Girls

ESTHER MADRIZ

The possibility of being the victim of a crime is ever present on my mind; thinking about it is as natural as . . . breathing. . . . The most unsettling crime is that of being cornered and attacked by a group of youths while riding in an empty train car and being violently hurt. . . . There is no place to run, no help available, and being at the mercy of this pack of animals.

MICHAELA, A FORTY-YEAR-OLD MIDDLE-CLASS
WHITE WOMAN WHO LIVES IN UPPER
MANHATTAN, NEW YORK

• • •

WHAT ABOUT YOU?

It was not until late in our discussion that most women felt encouraged to talk about the possibility of their own victimization. Even when they were reporting the fears about themselves, they often expressed them in the terms of, "If something happens to me, how is this going to affect my family?"

Although the literature on women's fear of crime focuses primarily on rape, this research shows a wide gamut of emotions regarding different crimes, often linked to the objective reality of the individual woman. For example, Maria, a Latina who had been mugged at knifepoint in the subway, observed that her major fear is "to be cut" (with a knife). Patricia, a twenty-two-year-old African American woman who lives in a predominantly Black, lower-class area in Brooklyn, says:

I am very worried about gun problems. It is like, you know, Black on Black crime; it is very hard in my neighborhood.

Anything, any kind of dispute in my neighborhood, a gun is involved as the solution. And that worries me because it is not only consequences for two guys fighting and they are shooting or whatever, I may be passing from nowhere and I get a stray bullet.

In fact, Patricia said that her major fear was robbery, not rape:

Robbery, I really don't think about rape. . . . Because I live in the projects and, you know . . . it is the thing that happens a lot. Most of the time where I live the crime is robbery, stealing and pocketbook snatching . . . robbing girls for jewelry.

Ten of the Black women in the study mentioned their fear of hate crime. Lily and Marianne, two African American women in their twenties, expressed this fear:

I live in Bensonhurst, and there is a lot of racial tension in that area. You know the Yusuf Hawkins case [a Black teenager who was killed by a group of white teenagers]. I grew up there. White people and Black people are always fighting. You are hesitant to walk into the so-called white area after dark. If you walk over there and try to go to the store or something, they call you nigger.

When we were little I went to the elementary school in that area. We had a sniper who use to shoot at the Black students. And then, during and after the Yusuf Hawkins case, I was in my first year of high school. . . . Some of the white kids I went to school with, they were against the line calling us niggers and telling us to get out of their neighborhood. . . . They used to chase us from school.

Three African American teenagers living in the suburbs of northern New Jersey revealed similar feelings. They said they were especially worried now that they were in their junior year. "Next year we have to decide what school to go to," Betsy said. "And I am afraid because I have heard of so many racial incidents going on in college campuses."

White women, particularly young adult women, were more likely than Black or Latina women to express the fear of rape. Judith, a twenty-one-year-old middle-class white woman, reported being particularly afraid of rape by someone "breaking into the house or something, who will tie you up and rape and abuse you. . . ." Nancy, a thirty-year-old white woman, said that sometimes she felt afraid of being raped by someone following her, grabbing her, and pushing her into a car. Other women were also fearful that their daughters could be kidnapped and raped.

Why is it that the white women interviewed in this study reported being more concerned about rape than Black and Latina women? One plausible explanation is that the media tend to portray white women as victims of rape more often than Black or Latina women. Several recent cases of rape reported by the media, such as the Central Park jogger or the Greenwich Village victim whose body was found in a trailer, illustrate this important fact. Josephine, a twenty-three-year-old white woman, actually said to me:

> Just look at the newspaper; who are more likely to be raped? . . . I do not know if this is true or not, and it doesn't matter. It *seems that way* when you read the newspaper [italics added].

Josephine is right. Where fear of crime is concerned, what is more important than the actual statistics is our images of crime. These images are greatly influenced by the stories we hear from other people. They are also affected by the manner in which crime is presented and interpreted by the media, who benefit from our anxieties by increasing their newspaper sales and program ratings, and by politicians, who sensationalize certain crimes in order to promote their political agendas.

All the women who mentioned rape as their major concern feared being victims of strangers following them, breaking into their houses, lurking in the dark. These images are consistent with the prevalent ideologies of crime as presented by the media and politicians. Very few of the participants said they were afraid of being raped by their friends, boyfriends, or colleagues, who are the more likely aggressors. . . . [I]t is only in two special settings that women expressed fear of someone they knew: (1) several respondents said they were afraid of being sexually harassed by a coworker or a teacher; (2) a few respondents claimed to be afraid of becoming victims of domestic violence by their partners.

Participants in this study reported being victims of various crimes: burglary; purse snatching; pickpocketing; personal assault; sexual harassment at work, in school, and in the streets; mugging; rape; domestic violence; and the murder of a family member. The phrase "violence against women" normally refers to crimes such as sexual assault, domestic violence, rape, and other types of sex-related crimes. Yet many other crimes are committed against women simply because they are women. Two examples illustrate this point. Maria, a Latina woman, recounted:

> I have been assaulted several times. Once a man hit me in the eye with his fist and he also pulled my hair. I was leaving the train when this man pulled my hair and then he held me. When I was turning around, he took my purse.

Maria never reported this crime to the police: "What are they going to do? Nothing!" she claimed. This type of crime, however, was committed against Maria because she is a woman. Just imagine an offender pulling a man's hair—it does not happen. Furthermore, purse snatching is a very frequent crime against women in New York City. Some women are dragged, and some have been killed by cars while an aggressor is pulling on their purses. A large number of purse snatchings, however, are never reported to the police. Angela, a twenty-three-year-old African American study, reported another such victimization:

> I have been a victim of a crime three times. Exactly now is about two years ago. I live next to this beauty parlor where I go there occasionally. . . . I went to the store to purchase some beverages and when I came back there were four guys outside. . . . One of the guys came behind me with a gun. . . . I told the owner of the beauty parlor: if you don't let me in I will get shot. So she had to let me in. They took all my jewelry, my $495 watch, and my diamond ring that my boyfriend at the time had bought me.

Robberies around beauty parlors have been frequent in New York and New Jersey. Because women often go to these places to relax and have some time away from home, beauty parlors are seen as "easy targets" for robbers. Another example of a crime committed against a woman was that of Beth, a twenty-seven-year-old African American woman who reported having been hit by a taxi driver, over "one dollar and fifty cents." "I got bumped in my face," she said. Of course, "he thought that he could get away with it because I am a woman," she concluded.

Eight participants reported being burglarized. Other reported being in the midst of a shootout and being robbed at gunpoint. Three also reported that their children had been robbed at gunpoint. One Latina woman saw her brother being killed. Another was called by the police to identify her brother's body, which was found in a plastic bag. Although it is true that men are more likely to be the victims of murder, women have to live with the horror and trauma that accompanies it for the rest of their lives. "Every time I go out and I see the black trash bags, I feel a profound pain." These women were called to testify, to identify the bodies of their brothers. They were the ones who had to inform their families. In addition, they felt the moral responsibility of "holding the family together." One of them has been on antidepressants since her brother was killed. "That's the only way I can get out of bed,"

she said to me. Thus, in more than one way, these women were also victims of the murders.

Some places that seem to be especially frightening to women are those that limit the possibility of escape: trains, elevators, building entrances or hallways, and parking lots and garages were frequently mentioned. An empty street or a street or school hallway crowded with a group of male teenagers were also considered frightening situations:

> I freak when I walk down the street and I see a group of men walking toward me.
>
> Or when there is someone following you.
>
> My agony is the elevators.

Several of the women also said that they were afraid to enter their cars without checking the back seat. "Someone can be hiding there," said eighteen-year-old Kay. Having just received her driver's license and enjoying her relative freedom, she said it was "a drag to have to worry about things like that."

• • •

LIMITED LIVES

The fear of crime limits the lives of the women interviewed in this study in many ways, from the seemingly innocuous rituals of not walking out by themselves at night, to the most constraining: not taking a certain job, not attending night classes, or avoiding the streets completely. Some of these constraints seem also to be related to age and socioeconomic status. Teenagers and young adults were more likely to use expressions such as, "You cannot let fear stop you." Some of the older women's lives, on the other side, were highly constrained by their anxieties. Barbara, an elderly Latina woman, said, "I do not go out of my apartment; if I have to buy something, I ask my son to do it." Several senior and adult women mentioned that they do not go out at night.

The right to the use of public space is limited for women. Most participants avoid places such as parks and certain neighborhoods. As Gloria said, "I used to go and walk around the Village at night, forget it. . . . No, I wouldn't do it." Two women said that they were afraid to drive by themselves. According to Cecilia,

> Even when I drive, my biggest fear is running out of gas or having an emergency come up or falling asleep at the wheel of the car. I am afraid of driving alone.

At least five women said that they do not take the train and prefer the bus. Those who can afford to take a taxi at night do it. This means that fear of crime also has economic consequences for women, who have to pay more money for transportation in order to feel safe.

Several students said that they do not take night classes and base their choice of schools on the availability of safe transportation. Even when she wanted to attend another school, Sue, a twenty-two-year-old white student, concluded:

> That school was out of the question because it was too far from my home and I would feel very unsafe having to go there every day.

One twenty-nine-year-old Black woman of Caribbean descent, Louise, explained how she changed her class schedule because of the fear that some incidents provoked in her:

> I was followed from the #6 to the E and J trains, which is my usual route every single day because of sexual advances and other incidents that happen to me. I have decided since then not to take any classes after 6:00 p.m. because I had a horrible experience that haunts me until this day. . . . A man exposed himself to me and began to tell me horrible things. . . . It did happen a second time, again, the same person around the same time. . . . The most shocking thing to me is that it was a person of the same ethnic group as me. I still cannot understand because that is not part of my culture. I truly believe that it is something that he learned here in America. Everything changed since then because I expected that if something happen to me, my own ethnic group would protect me instead of scaring me to death.

I have never reported this to the police, but I think I should.

Two women reported having changed their jobs for fear of commuting by themselves. One of them said, "Your life is more important than a job." Two others said that they did not take certain jobs because they involved night shifts. Finally, one thirty-three-year-old woman, Emma, said that she did not take a job that she liked because it was located in a "bad neighborhood."

A group of Latina teenagers mentioned that they love to dance but don't go to clubs anymore because of what could happen there. Two of them said that even if they wanted to go, "My parents don't let me go." In general, Latina teenagers were more likely to report that their parents limited their behaviors for fear that something could happen to them, although a few of the white teenagers reported similar constraints. Finally, a twenty-five-year-old white woman, Rachel, said that she does not like to go shopping unless accompanied by her husband. She also mentioned being apprehensive when she goes with her small daughter because of "all the stories of kids being abducted at the malls." Rachel is not alone in her fears. According to a recent poll by America's Research Group (Charleston, South Carolina), a survey of 1,003 consumers found that 21.1 percent made fewer trips to stores in the past year because of the fear of crime (*Advertising Age* 1994).

Finally, for some women fear is so pervasive that it seriously interferes with the quality of their lives. Cora, a seventy-four-year old Puerto Rican woman, said that because she was old and "not as strong or as quick as I used to be," her life has been seriously limited:

> This fear has stopped me from enjoying my retirement in that I have to be careful of what I do.

A group of elderly white women said that the fear of crime limits their participation in church and community activities. "I am afraid of going to church meetings during the night," said Rosa. One of the elderly Latina women also mentioned that she does not recycle her cans, bottles, and newspapers because of her fear of going to the basement of her public housing apartment. "The dumpsters are in the basement and I don't go there by myself. I am not crazy," she concluded.

The fears that these women shared with me reflect that half of the U.S. population, who are supposedly guaranteed freedom, live imprisoned in the invisible cages of their worries, anxieties, concerns, and fears. Women's lives are highly constrained by the violence and harassment that they have to face daily in their homes, workplaces, schools, and on the streets. They are also affected by the images and representations depicted by an ideology of crime that supports and feeds such violence. Some women's fears have been influenced by previous victimization: mugging, rape, burglary, assault, purse snatching. Most feel vulnerable because being women almost guarantees that they will be harassed, stared at, followed, or molested in the streets. A few say that being victims of domestic violence has made them more afraid. Other women feel particularly vulnerable because of who they are: immigrant women who do not speak English, older women who depend on their welfare checks, African American women whose sons have been harassed by the police.

REFERENCE

Advertising Age. 1994. "Numbers for the 90s. Safer Shopping." 29 August, 3.

53

Who Benefits (and Who Doesn't)

CLARENCE LUSANE

*Things happen to people
who let things happen to my dope.*

WORDS OF A NEW YORK STREET DEALER

With billions of drug dollars floating around, one obvious question is who gets the largest share of these illicit profits. Money is made at all levels of the drug trade, from producers to chemists to distributors to street sellers, but the amount varies depending on class, race, and nationality. Trying to sort out who gets what part of the drug profits requires an analysis of the political economy of the drug trade and its relationship to the legal drug market.

In third world countries, peasants who grow opium or coca leaves receive more than they would for other crops but still receive only a fraction of the profits eventually derived from their labor. In Bolivia, a "carga" of leaves (45 kilograms) sells for about U.S. $55 to $100. This comes to about U.S. $1.20 to $2.20 per kilogram.[1] It is estimated that the markup from coca leaves to the final product of cocaine can be as high as 700 percent. It's even higher for crack cocaine. For heroin, the markup is estimated to be 2,000 percent.[2]

As the coca leaves transform into coca paste then into coca base, which is then made into cocaine hydrochloride (cocaine powder), the profits to be made at each stage take huge leaps. This is the major reason why many peasants have expanded from growing coca leaves into producing coca paste, which sells for about U.S. $250 per kilogram. It takes about 100 to 200 kilograms of coca leaves to make one kilogram of paste.[3]

Although profits are raked off unevenly, the illegal drug industry is the most equal opportunity employer in the world. There is room for corporate-like structures with thousands of employees that deal with transactions involving millions of dollars daily, as well as mom-and-pop operations that are run out of the back seat of a car. Individual entrepreneurs, limited partnerships, and financial syndicates all function side-by-side.

Many others may be tied by drug trafficking but never come in contact with any narcotics except, possibly, for personal use. Accountants, bankers, lawyers, pilots, arms merchants, real estate agents, jewelers, chemists, and others function as fronts for drug traffickers.

Finally, there are also those who benefit by the commercial and cultural exploitation of the drug crisis. Numerous products are sold that would otherwise sit on shelves or not exist at all if not for the spending trends of drug dealers, users, or the architects of the war on drugs. Expensive sneakers, gold chains, particular automobiles, and other extravagant commodities that are favored by drug dealers or their imitators rake in millions. Drug paraphernalia such as rolling paper, pipes, and weapons have also made fortunes. As for the drug war industry, the manufacturers of medical drugs used in drug treatment, such as methadone, or those involved in drug testing have made untold millions over the years.

THIRD WORLD CARTELS: THE FIRST TIER OF THE DRUG ECONOMY

Similar to other multi-nation businesses, the organizational control and the division of profits is determined through the prisms of race, national origin, and class position. Rivalry within the industry, attacks from without, and ever-changing market conditions mean that a constant flow of new players and power plays make it difficult to determine exactly who gets what. Despite these factors, drug trafficking becomes better organized and more sophisticated daily.

For many regions and nations around the globe, drug production and trafficking is not just a major component, but the center of their economy. In Southeast Asia, for example, opium growing and trafficking were the major factors in the growth of the economy of that region. In the 1950s, after mainland China became communist, and Chinese criminal syndicates scattered throughout the region, every area from Hong Kong to Thailand to Burma became engrossed in the opium trade. Very little distinction could be made between the public and private sectors when it came to the drug industry, as many government officials facilitated or participated directly in trafficking.

Corruption of political institutions is the normal method of operation for the drug lords, to the degree that elected officials and drug kingpins function as one. In Colombia, Bolivia, and Peru, for example, ties to the drug cartels exist at the highest levels of government. In Colombia, several presidents and, it's currently estimated, at least one-third of its members of Congress have been bought and paid for by the cartels.

• • •

Trafficking in illegal drugs has been, to say the least, very profitable for the top echelon of the trade. They engage in a vortex of obsessive profit-making that seems boundless. According to *Forbes,* cocaine sales created at least three billionaires among Colombia's Medellin drug cartel leaders: the late Gonzalo "the Mexican" Rodriguez Gacha, who was killed in a shootout with Colombian police in December 1989; Jorge Luis Ochoa Vasquez; and Pablo Escobar Gaviria.[4]

• • •

GANGS: THE SECOND TIER OF THE DRUG ECONOMY

At the next level are the national and regional suppliers and distributors, along with their managers and administrators. In the United States, the Justice Department estimates that there are hundreds of drug-trafficking organizations. This includes inner-city street "gangs" such as those affiliated with Los Angeles' Crips and Bloods, Chicago's Vice Lords and El Rukns, and Miami's Untouchables. These groups control a significant proportion of the cocaine, crack, and PCP distributed around the country.

No unity exists among law enforcement agencies on a definition of a gang, though the broadest interpretation seems to be on the West Coast and the narrowest on the East Coast. These differences make it difficult to assess more precisely what the so-called youth gang problem actually looks like. What would be called a gang in some cities, such as Los Angeles or Chicago, is not seen as a gang in other cities, such as Washington, D.C., or New York. According to an article in *Social Service Review*, officials have sometimes changed their definition of what constitutes a gang, as they did in Chicago, in order to reduce gang violence and gang homicide figures.[5]

Youth gangs often function as surrogate families. Gang members identify with the gang more so than with their real family, which is often broken apart and/or dysfunctional. Although an overwhelmingly male activity, women's participation is not uncommon, especially in situations where a female relative had been involved with the gang. Gang identity, collective and individual, is often shaped by the amount of violence and victimization that is exhibited.

Los Angeles, a city famous for its "gang problem," has about 450 gangs with about 45,000 members. But to refer to Los Angeles' renowned Crips and Bloods as gangs is misleading because they are not really gangs—that is, no central controlling group called Crips or Bloods exists. Both the Crips and Bloods are names that the city's prolific array of Black gangs have identified with. Police officials speculate that there are close to 200 smaller gangs, known as "sets," that identify with the name Crips, and about sixty-five to seventy sets that identify with the Bloods.[6] A Crip set is as likely to fight another Crip set as it is to fight Bloods, and the same holds true for Bloods. There are other Black and Latino Los Angeles gangs that don't identify with either.

Media sensationalism has propagated the erroneous and racist notion that all of the city's Black gangs traffic in drugs. The truth is that not all of Los Angeles' street gangs deal drugs and not all drug dealers are in gangs. Actually, Blacks make up only 39 percent of the Los Angeles County Sheriff's department gang records. Latinos are 59 percent and Asians account for about 2 percent. Supposedly, only seventy-two White gang members are known.[7] Local critics point out the discriminatory geographic patterns of police sweeps in Los Angeles and surrounding areas that register more Black and Latino youth as gang members.

One Black youth gang in Detroit, Young Boys, Inc., grew from twenty-four members to 300 members, with estimated weekly sales of $7.5 million in heroin and cocaine. In 1982, they grossed close to $400 million. According to Michigan State University criminologist Carl Taylor, who has written an authoritative book on Detroit's youth gangs titled *Dangerous Society*, the gang began with an investment of $80,000 that came from an insurance claim by two of the original members. By all accounts the group was extremely organized, highly disciplined, and unabashedly brutal. It imposed a "no drug use" policy on its members, and the penalty for violation was death.

Young Boys, Inc., was able to exploit the financial crisis that Detroit's Black community found itself facing in the Seventies. The gang functioned as a bank for many Black Detroiters by giving loans at usury rates for house purchases and starting small businesses, since Detroit's financial institutions had been chastised by the banking industry for not making loans to Blacks in the city. Gang members themselves bought homes in the suburbs and houses and apartments in the city to process and sell drugs. Some members opened legitimate businesses such as video and record stores.[8]

Although some of the money derived from drug trafficking was reinvested into the community, the structural and entrenched economic crisis confronting Detroit's Black poor, particularly that of males, went unaddressed. The social and economic crisis of the community was exacerbated by Young Boys, Inc.'s, care and feeding of Detroit's estimated 50,000 heroin addicts. Their activities accelerated the city's social deterioration and wrought uncalculated suffering.[9]

* * *

THIRD WORLD PEASANTS, IMMIGRANTS, AND AFRICAN AMERICANS: THE THIRD TIER OF THE DRUG ECONOMY

On the bottom of the drug economy pyramid are the growers and retailers. Hundreds of thousands of peasants around the world make up the front-line of drug production. Subject to eradication attacks, struggles between different cartels, and a volatile market, peasants often risk all to produce and sell the only profitable crop they can.

In the United States, the front line is occupied by street sellers, lookouts, couriers, workers in heroin "shooting galleries," and crack-house operators. These workers, many of whom are inner-city Black and Latino youth, are the first arrested and the first killed. Although profits can sometimes be high (a low-level lookout can earn $75.00 an hour on a good day), the risks are deadly and unforgiving.

Neither peasants nor inner-city dealers have any financial security, health insurance, or legal options that their middle- and upper-class employers have. They are the first and hardest hit when the market declines and they lack the skills and resources to invest and create long-term legitimate sources of income. The cycle of poverty and disempowerment that was for many the catalyst for entering the drug trade often remains the only way of life, other than death or imprisonment.

Many experts feel that while it is easy to enter the drug trade as a drug seller, particularly selling crack, few become rich. The Joint Center for Political and Economic Studies sponsored a roundtable discussion titled *Crimes, Drugs, and Urban Poverty*. During the discussion, Rutgers University sociologist Jeffrey Fagan said that, "Crack really is a deregulated market. It is totally

open . . . people have access to raw materials . . . and to selling locations." This view was echoed by Steve Rickman, of the Washington, D.C., Office of Criminal Justice, who said, "anyone can set up shop" on the unprofitable street corners that are not already under control.[10]

The big profits made from street sales are restricted to a few. According to Bruce Johnson of the Narcotics and Drug Research, Inc., "less than 20 percent of the people engaged in drug dealing have a net-worth cash return of as much as $1,000." This view appears to be backed up by a Rockefeller Foundation-funded study that reported that street-level dealers in Washington, D.C., made between $740 to $1,000 a week. Crack-house workers, as opposed to owners, appear to make very little. In one East Harlem neighborhood, studied by San Francisco State University's Phillippe Bourgois, crack-house workers made $50 to $70 for an eight-hour shift. This worked out to only 50¢ per vial of crack.[11]

The most significant conclusion from these discussions is that it is highly likely that low-level employees in the illegal drug industry, who profit little and face the most danger, may be won over to the legal labor market. Contrary to popular belief, winning these workers over does not mean dangling a large salary in front of them. It does mean making available meaningful, nonviolent, productive, and career-oriented work.

DOING THE LAUNDRY: WHITE PROFESSIONALS IN THE DRUG TRADE

Probably the biggest problem facing drug dealers is what to do with the mountains of $5, $10, and $20 bills that they accumulate. In these drug-conscious (and federally-observant) days, it takes considerable and creative financial acumen to turn drug dollars into "honest" money. As a pamphlet title *Money Laundering: A Banker's Guide to Avoiding Problems* notes: "For money laundering to be successful, there must be no 'paper trail.'"

The pamphlet, put out by the Supervision Policy/Research Office of the Comptroller of the Currency, notes a number of laundering schemes that banks knowingly participated in, including opening fictitious accounts and hiding behind phony real estate loans.[12]

A whole new cottage industry now exists of professional money launderers, many of whom work on one-time-only contracts and often for rival drug dealers or gangs simultaneously. These individuals, for the most part, are respectable White professionals with no criminal record. They include people like former Rep. Robert Hanrahan (R-IL) and Richard Silberman, former fundraiser for Jerry Brown in California, both of whom were convicted of laundering drug money. They include a former attorney general for the state of Kansas;[13] a consortium of Pennsylvania airline pilots, called the Air America Organization, who smuggled drugs and money for the Colombian cartels;[14] and three former DEA agents indicted in November 1988 for laundering more than $600,000 in drug money, to name a few.

Launderers range from cash-and-carry types, who pack suitcases and bags with cash, to more elaborate operators with electronic money-wiring facilities, telex machines, and automatic bill-counting equipment.

The most frequent vehicles for the laundering of drug money are banks in the United States, the Caribbean, and Hong Kong. Colombian traffickers prefer to use banks in Aruba, the Cayman Islands, Panama, and Uruguay. Chinese heroin deals prefer Hong Kong and Taiwan. And there are at least twenty-nine nations that have virtually no barriers to laundering money. This includes nations such as Austria, Bahrain, Barbados, Bermuda, Costa Rica, Grenada, Liberia, Liechtenstein, Monaco, Singapore, and Switzerland, as well as virtually unheard of islands such as Channel Islands, Caicos Islands, Isle of Man, Nauru, and Vanuatu.[15]

One indication of how profitable the illegal drug industry has been for certain areas of the United States is the growth in cash surplus at local banks. In the United States, a cash surplus results when all the banks in an area receive more money from a local Federal Reserve Board than they pay to it. In 1970, banks in Florida, a key drug entry point, had a currency surplus of $576 million. By 1976, this figure had grown to $1.5 billion. By 1982, the cash surplus in Jacksonville and Miami alone amounted to $5.2 billion. Similar trends exist in southern California, Texas, and New Orleans. Los Angeles had a surplus of $2.7 billion, while San Antonio's grew to $1.2 billion.[16]

• • •

NOTES

[1] Michael Stone, "Coke Inc.: Inside the Big Business of Drugs," *New York,* 16 July 1990, pp. 26, 28.

[2] "The General and the Cocaleros," *The Economist,* 9 December 1989, p. 40.

[3] James Cook, "The Paradox of Anti-Drug Enforcement," *Forbes,* 13 November 1989, p. 110.

[4] "The World's Billionaires," *Forbes,* 24 July 1989, p. 123; Douglas Farah, "Drug Lord's Surrender a Victory for Colombia," *Washington Post,* 18 January 1991, p. A14.

[5] Jill Walker, "Los Angeles Isn't Alone in Problems With Gangs," *Washington Post,* 3 July 1990, p. A10.

[6] *Drug Trafficking: A Report to the President,* Office of the Attorney General, Department of Justice, Washington D.C., 1989, p. 33.

[7] Ibid; Walker.

[8] Katherine McFate, "Black Males and the Drug Trade," *Focus,* May 1990, pp. 5–6.

[9] Carl S. Taylor, *Dangerous Society,* Michigan State University Press, East Lansing, MI, 1989, p. 99.

[10] Ibid; McFate.

[11] Ibid.

[12] *Money Laundering: A Banker's Guide to Avoiding Problems,* Office of the Comptroller of the Currency, Washington, D.C., December 1989, p. 2.

[13] Jefferson Morley, "Contradictions of Cocaine Capitalism," *Nation,* 2 October 1989, p. 346; Stephen Labaton, "The Cost of Drug Abuse: $60 Billion a Year," *New York Times,* 3 December 1989, p. D6.

[14] Ibid; *Drug Trafficking,* p. 37.

[15] Stephen Brookes, "Drug Money Soils Cleanest Hands," *Insight,* 21 August 1989, p. 15.

[16] Ibid; Morley, pp. 342–343.

54

A Year in Corporate Crime

ROBERT SHERRILL

The Wall Street Journal is the main reporter in our country of corporate crime. The Wall Street Journal has so much information on corporate crime it should be named The Crime Street Journal.

RALPH NADER

Taking Nader at his word, we decided the best way to survey the corporate crime wave in this country was to read a year's supply of *The Wall Street Journal*. What follows is a representative sample of the reports of big-business malfeasance published in the *WSJ* during 1996. And keep in mind that the paper's talented reporters could cover only a fraction of the misdeeds that occurred during the year. The stories, despite their relative restraint and sober factuality, display a panorama of civil and criminal corruption, sleaze, unhinged greed and other antisocial antics of some of the businesses that shape our lives.

Please note the punishments handed out to the malefactors. Although three executives were sentenced to spend a few months in prison (for fraud resulting in ten deaths) and a couple of others seem destined to land in jail eventually, the kind manner in which most erring business chieftains were treated solidly underscores the fact that in the United States a prison sentence is rarely looked upon as the proper fate of corporate villains.

Frequently, where criminal charges could have been brought the offenders were tapped with lesser counts of civil misconduct, and they agreed to make a settlement before trial, precluding the possibility of much heavier financial punishment. Don't be impressed by the seemingly hefty fines and restitution paid by some corporations; they were a minor inconvenience, for 1996 was a banner year for corporate profits.

Perhaps the best lesson to be drawn from this kind of survey was laid out years ago by Gilbert Geis, who frequently wrote on white-collar crime:

> The social consciousness of the corporate offender often seems to resemble that of the small-town thief, portrayed by W. C. Fields, who was about to rob a sleeping cowboy. He changed his mind, however, when he discovered that the cowboy was wearing a revolver. "It would be dishonest," he remarked virtuously as he tiptoed away. The moral is clear: since the public cannot be armed adequately to protect itself against corporate crime, those law enforcement agencies acting on its behalf should take measures sufficient to protect it. High on the list of such measures should be an insistence upon criminal definition and criminal prosecution for acts which seriously harm, deprive, or otherwise injure the public.

The trouble is, even if they wished to get tough—and the overriding trend is toward less toughness—most of the federal agencies with jurisdiction over the business world are woefully outmanned by the huge array of corporate lawyers. It isn't a David and Goliath situation—it's more like Shirley Temple versus King Kong.

JIM LEHRER'S FALLEN ANGEL

The Archer Daniels Midland company, super-briber to the political world, was involved in the most publicized corporate crime of 1996. Caught in a sting by Justice Department investigators, A.D.M., the planet's largest grain processor, pleaded guilty to charges of conspiring to fix prices for two products: lysine, a feed supplement for livestock, and citric acid, used in soft drinks and detergents. A.D.M., which was fined $100 million, and its Asian co-conspirators also agreed to pay more than $100 million to settle civil lawsuits brought by shareholders and customers. More lawsuits lie ahead.

Ironically, Archer Daniels' stock price actually jumped, because Wall Street judged the settlements and fines to be bargains. Any why not? After all, analysts figure that A.D.M. cheated its lysine customers alone out of more than $170 million. Although the $100 million fine is seven times larger than the largest antitrust penalty ever before levied by the Justice Department, it's (you'll pardon the expression) chicken feed for A.D.M., which in the last fiscal year had revenues of $13.6 billion from its agricultural products, most of which are heavily subsidized by the U.S. taxpayer.

Along with a bargain fine, Archer Daniels got a sweetheart deal: In exchange for pleading guilty and promising to help the Justice Department in its expanding investigation, A.D.M. was granted immunity against charges of price-fixing in the sale of high-fructose corn syrup, which, along with the corn-derived fuel ethanol, is A.D.M.'s leading product.

Also part of the deal was Justice's promise that its investigators wouldn't even bother to interview Dwayne Andreas, 78, who as chairman and chief executive has for decades treated A.D.M. as his personal fiefdom. His 47-year-old son, executive vice president and former heir apparent Michael Andreas, having been secretly taped in a price-fixing conversation with an Asian "competitor," was indicted for conspiracy; at the very least, his career at A.D.M. is over. But old man Andreas once again proved himself to be a masterful escape artist. Could this talent possibly be explained by

the more than $4 million he and his family and A.D.M. have given to Washington politicians since the seventies, most notably Kansas Republican Bob Dole (in return for billions of dollars in subsidies)? It also didn't hurt the elder Andreas's chances of getting special treatment that he personally donated $155,000 to the Democratic Party in 1993 and $100,000 in 1994, and was co-chairman of a dinner that raised $3.5 million for Clinton's presidential campaign in 1992.

Over the years his generosity has sometimes been suspect. He was acquitted of giving Hubert Humphrey an illegal $100,000 contribution in 1968; he slipped a thousand $100 bills into the Nixon White House in 1972, the year in which the term "money-laundering" entered the nation's vocabulary; and in that same season, a $25,000 check from Andreas somehow sneaked into the bank account of a Watergate burglar. But the old man has never tripped badly enough to earn a criminal record—except in 1993, when he and his wife paid an $8,000 fine for exceeding federal limits on political contributions.

Considering that it operates in one of the world's most piratical industries, A.D.M. has, like Andreas himself, led a charmed life, suffering only a couple of legal wounds: a 1978 conviction for fixing prices on grain sold to the Food for Peace program and a no-contest plea in 1976 to the charge of short-weighting and misgrading corn for export.

CLOSE RACE FOR 'CROOKEDEST'

Among the insurance industry's biggest companies, competition for the title of "crookedest" heated up in 1996. Three years ago, Metropolitan Life was fined $20 million for cheating its customers. Last year the nation's largest insurance company, Prudential, easily outdistanced Metropolitan by being fined $35 million. Also, Prudential is paying more than $1 billion in restitution to fleeced policy-holders.

After an eighteen-month investigation, a task force of insurance regulators from thirty states concluded that for thirteen years Prudential salespeople coast to coast had practiced a deception called "churning," often with the knowledge and sometimes approval of officials up to at least regional vice presidents. Indeed, some sleazy salespeople seem to have been promoted to managerial positions because of their success at duping customers.

"Churning" was a racket in which as many as 10 million customers were sweet-talked into using the cash value of their old insurance policies to pay the premiums on new, more expensive policies. They were not warned that the upgrading could be so costly that it would eat up their equity, leaving them with premiums they couldn't afford—and therefore no coverage.

Arthur Ryan, Prudential's chairman since late 1994, admitted that the charges were accurate and fired several salespeople and managers and a senior vice president. But some of the discarded employees transformed themselves from con artists into whistleblowers, providing investigators with sordid inside details about Prudential's operations. Potentially most damaging—and certain to be used by the army of plaintiffs' lawyers already assaulting the Rock—were sworn statements from some of the former employees claiming that Prudential officials had ordered them to destroy any documents that might reveal unsavory marketing practices.

As an embattled defendant, Prudential was far from lonely. Numerous other insurance companies were sued, or facing regulatory penalties, for identical or similar misconduct. When Mutual of New York paid $12.5 million to a mob of unhappy Alabama consumers, for example, *The Wall Street Journal* called the settlement "the latest in a series involving alleged deceptive sale practices at many of the nation's biggest insurers."

PIGGISH BANKS
AND LAUNDROMATS

Money-laundering, a highly profitable crime, is routinely committed by U.S. banks, but the laws against it are vague; it is hard to prove and thus rarely prosecuted. Justice Department investigators floundered through 1996 in an effort to

determine if Citibank had helped Raúl Salinas, brother of Mexico's former president, Carlos Salinas, launder at least $100 million in dirty money by transferring it to bank accounts abroad.

Raúl Salinas is now in prison in Mexico on charges of "inexplicable enrichment." But Citibank apparently found nothing inexplicable in the fact that Salinas, who never earned more than $190,000 per year working for the government, had the bank handle $100 million for him. Citibank officials say their internal investigation turned up no improprieties.

Some law-enforcement authorities thought it odd that Citibank refused to let its best and most experienced money-laundering compliance officer, Jane Wexton, a Citibank vice president and senior attorney, participate in the investigation. Allegedly Wexton was told she couldn't even ask questions about the case, although up to that point, says the *Journal,* she had been "regularly consulted by bank officials and outside regulators when any suspicious activities came up" because she was considered "a 'cop's cop' . . . a straight shooter who would be likely to tell regulators about something she considered wrong." Wexton, who had been at Citibank seven years, quit.

The *Journal* says "people inside Citibank" think the Justice Department is "developing an indictment against the bank."

DRUGS, DOGS AND QUACKS

Slowly, slowly, the Justice Department continues to try to catch up with the crooks who, according to the General Accounting Office, defraud the government out of $100 billion a year. There are reportedly 1,000 current investigations into health care fraud.

Like insurance companies, medical supply companies seemed to be competing for the title of crookedest. In September the *Journal* reported that SmithKline Beecham's clinical laboratory unit was "close" to an agreement to pay more than $300 million to the government to settle charges that it had bilked Medicare for unneeded blood tests. That settlement would still be less than the record $379 million paid by National

Medical Enterprises in 1994 for alleged fraud in psychiatric services.

Far back in the field, but closing fast, is Corning Inc., which paid $6.8 million to settle allegations that its Bioran Medical Laboratory in Cambridge, Massachusetts, regularly billed Medicare for blood tests that doctors hadn't asked for. That was in February. Eight months later Corning was forced to pay up again, this time a whopping $119 million to settle charges of fraudulent billing by one of its subsidiaries, Damon Clinical Laboratories. That must have been a proud moment for Corning—the penalty pushed it past the $110 million in civil and criminal penalties paid for similar misconduct by a San Diego firm, National Health Laboratories. The *WSJ* indicates that Corning is an old hand in this game. It paid $39.8 million in 1993 and $8.6 million in 1995 to settle federal charges of Medicare fraud.

In 1996 the medical industry provided the *Journal* with plenty of misconduct to write about, some that was criminal and some that was just plain corporate sleaze—such as Eli Lilly & Co.'s use of homeless, often alcoholic, men to test the safety of experimental drugs, paying them the lowest rate per diem in the human guinea pig business. The Food and Drug Administration's deputy director of drug evaluation told the *Journal* that using homeless drunks violates the agency's rule that drugs can be tested only on people who are able to make a "truly voluntary and uncoerced decision" to participate. Other test experts told the newspaper that using alcoholics could distort the evaluation of an experimental drug's safety. That may not worry Eli Lilly; after all, it's the company that put Oraflex on the U.S. market without telling the F.D.A. that the same drug had already killed several people in Europe.

The *Journal* medical story that takes the prize for describing pure gutter-level greed and callousness is the one about the British drug maker Boots and its main product, Synthroid. About 8 million Americans spend $600 million a year on drugs to control hypothyroidism, and Synthroid gets 84 percent of their money. It has been around since 1958 and was the first synthetic thyroid drug. When it came on the market, the F.D.A. approved it without asking for trial data.

That oversight has made it extremely difficult for rival thyroid drug manufacturers. Since there is no benchmark data on Synthroid, how could they persuade doctors that their products are just as good and are absorbed into the blood the same way Synthroid is?

In recent years the rivals began making some inroads in Boots's virtual monopoly by getting on state drug-approved lists. To beat them back, Boots hired a University of California research team to perform extensive tests comparing Synthroid with three rival—and much cheaper—thyroid drugs. If Boots thought the $250,000 contract would persuade the scientists to tout the superiority of Synthroid, it was in for a big surprise. In fact, the researchers found that the four drugs were essentially interchangeable and that F.D.A. support of the cheaper drugs would save thyroid patients $356 million a year. Furthermore, the researchers were going to say so in a paper that the *Journal of the American Medical Association* was ready to publish.

Curses! At that very moment Boots was about to sell itself to Germany's giant BASF, for $1.4 billion, and the research findings might queer the deal. But the solution was simple. The contract the university team had signed forbade publication of their findings without Boots's approval. It didn't approve, of course, and that was that. Synthroid still sits on top of the market, and millions of thyroid patients will pay for its being there.

Odds and Ends

Claiming that maybe as many as 25 million consumers were overcharged $600 million between 1989 and 1994, attorneys general in twenty-two states sued the biggest contact lens companies—Bausch & Lomb, Johnson & Johnson Vision Products and Ciba Vision Corporation—as well as several optometry trade organizations for price conspiracy. . . . The world's largest drug companies paid $350 million to appease thousands of retail druggists who claimed they were being overcharged. But the feud isn't over, and the Federal Trade Commission is investigating for signs of a conspiracy. . . . The Florida nursing home industry takes in $3.2 billion a year. But state inspectors give the lowest possible rating to one out of every twenty-six homes, and $2.5 million in fines have been levied for poor care. Trouble is, only one violator (poetically named Ambrosia Home) has paid up, a trifling $19,500. The big chains just appeal the fines and drag out the cases forever. . . . In one of the stiffest warnings issued in recent years, the F.D.A. told Pfizer that it must immediately stop making misleading promotional claims for Zoloft, a popular antidepressant (worldwide sales of $1 billion a year). . . . Sick or dead, you're a good target for fraud. The F.T.C. requires the nation's 20,000 funeral homes to give itemized price lists to prospective clients. But in a test run the commission found that as many as 8,000 undertakers are ignoring the rule. The F.T.C., timid as usual, has threatened only twenty of them.

SURPRISE! ARMS MERCHANTS CHEAT

When one of the bolts holding a 500-pound Maverick missile to the wing of a military plane breaks, and the missile is left dangling, attached by only the remaining bolt—and when that sort of accident begins to happen rather often—you'd think the Pentagon's sleuths would quickly find out what the hell was going on. But it took the military years to figure out that United Telecontrol Electronics was knowingly using defective bolts to hold the launchers in place. Phoney computer-controlled measurements made it look like the parts had passed inspection. Some company officials pleaded guilty to fraud; although their crime disrupted operations at air bases around the world—costing taxpayers big bucks—and put lives at risk, the penalties were far less than, say, a marijuana dealer would get. A former U.T.E. vice president got the maximum, an easy twenty-one months in prison and a $40,000 fine.

The Navy blames defective gearboxes on Navy F-18 fighters for seventy-one emergency landings and several in-flight fires, as well as the loss of an F-18 during the Gulf War. (They came to that conclusion after inspecting 150 gearboxes and finding that all—100 percent—were defective.) The outfit that made the lousy gearboxes, Lucas Industries, pleaded guilty to falsifying quality records and was penalized $106 million (not so large if you consider that Lucas has estimated yearly revenues of $6.7 billion).

Tec-Air Services and two of its execs pleaded guilty to fraud for failure to service properly aircraft emergency gear such as the oxygen apparatus that is supposed to be activated in case of accidental cabin depressurization. A big deal? Some would think so, seeing as how these systems are supplied to (among other aircraft) the Boeing jets used by the President and Vice President.

Does the Pentagon get angry about that sort of dangerous cheating? Nah. Maybe because it has become so commonplace. As the *Journal* says, "In courtrooms across the country, similar scenarios are playing out. Huge and small defense contractors are facing charges of manufacturing faulty products." Yes, and facing a host of other charges as well. Like fraud. And influence-peddling. And deceptive bookkeeping.

Alliant Techsystems was reportedly under investigation for overcharging the Pentagon by "tens of millions of dollars on various missile-production contracts." Alliant bought its space-propulsion contractors from Hercules, which itself is the target of a massive privately filed False Claims Act suit seeking hundreds of millions of dollars in damages for falsifying quality-control records, doctoring the books, etc. At first, Hercules denied the charges; then it claimed the government had known about the "multiple, recurring errors and procedural deviations" in the production of rocket motors, and therefore, being aware, it couldn't have been defrauded.

Arms merchants try to wear down accusers, and often succeed. But not always. In 1986 two women supervisors in a Hughes Aircraft plant alerted federal officials that the company was falsifying quality-control tests. Because of their whistleblowing, they say, they were harassed and were eventually hounded out. They sued Hughes under the False Claims Act—and a decade later they were finally victorious, winning $891,000 for themselves, $450,000 for their lawyers and $3.1 million for the government. This is nothing new for Hughes. Four years earlier it was convicted of criminal failure to test properly equipment vital to jet fighters, tanks and other U.S. weaponry and fined a piddling $3.5 million—and the Pentagon continued to do business with the company.

In this industry, it's constantly "déjà vu all over again." *Journal* reporter Andy Pasztor reminds us that "in the mid-1980s, Litton admitted that it had methodically defrauded the Pentagon for almost a decade by inflating prices, charging twice for some raw materials and failing to disclose rebates from vendors on dozens of Navy electronics contracts." Litton is still very much in the military business, and apparently some suspect it's still up to its old tricks, for federal agents raided its offices in Los Angeles to investigate alleged overcharges going back a decade.

If these rascals delay long enough, they are let off with cream-puff penalties. Lockheed Martin, the huge military contractor, closed the longest-running influence-peddling scandal—dating back to the eighties—by paying a mere $5.3 million on behalf of Martin Marietta Corporation, with which it merged in 1995. It was bad enough that such a trifling penalty could settle a suit accusing Martin Marietta of a $30 million overcharge. Worse, the cheap settlement wiped clean the slate of a company that had snagged its disputed contract with the help of an influence peddler and a crooked Navy department official. The contract was to build a supersonic low-altitude target to test the Navy's missile defense. The project was finally junked. Cost overruns—coming to almost 100 percent and due mostly to the contractor's faulty bid—helped kill it. But Martin Marietta walked away with $192 million.

McDonnell Douglas got off with a $500,000 fine for misleading the Pentagon on its $6.6 billion contract to build the C-17 cargo jet. The company kept telling the government it would break even, though its own estimates showed it would lose at least $1 billion—which it did. Guess who paid the difference.

SCREWING THE WORKERS

Of course, misdeeds of the sort mentioned screw the U.S. working class in a general way. But sometimes the screwing gets up close and personal, right in their own workplace. Three Kentucky coal executives were sentenced to prison in the deaths of ten workers, but it took seven years to get the villains behind bars. In 1989 methane gas exploded in a Pyro Mining Company shaft. In the resulting investigation, it was found that prior to the explosion the executives had lied to federal inspectors about the existence of the mine's hazardous conditions. For helping to kill ten men, the executives drew sentences ranging from five months (and a fine of $375) to eighteen months (and a $3,000 fine—or $300 per victim).

Using data from the federal Mine Safety and Health administration, the *Journal* reported that these sentences were among the longest ever handed down. "Most criminal complaints filed on behalf of the agency result in no prison time.... The agency estimates that only about 40 people have gone to jail since 1991 because of criminal safety violations at mines." With such light penalties, who's going to worry about obeying regulations?

Following up on a union complaint that Caterpillar's parts factory in York, Pennsylvania, had workplace health hazards, the National Institute for Occupational Safety and Health tried to inspect the place. Caterpillar said no. So NIOSH got a warrant from a federal judge ordering the company to admit the inspectors. Caterpillar still said no—the first time in fifteen years that any company had defied a federal warrant—and it was held in civil contempt. The company finally gave in. It could have been fined heavily, but of course it wasn't.

The minimum-wage law means nothing to many employers. They simply ignore it. The *WSJ* quotes Princeton labor economist Alan Krueger's estimate that as many as 3 million workers are paid less than the minimum wage and adds, "Violating the minimum-wage law has a certain economic logic to it because an employer, if caught, usually has to pay only the back wages that were due. Penalties are generally levied only on repeat or extreme violators."

A survey of garment makers, among the worst outlaws, found 43 percent paying illegally low wages. But trucking companies, eateries and construction firms are top criminals, too.

Here's one place where the corporate whine about "big" government becomes an especially bad joke. The Labor Department's team of inspectors has shrunk 15 percent in the past six years, and the remaining 500 inspectors are supposed to police 6 million workplaces for minimum-wage and overtime violations, child labor abuses, and other corporate tackiness.

Failure to pay for overtime work is especially widespread. The *Journal* says, "Violations are so common that the Employer Policy Foundation, an *employer-supported* think tank in Washington, estimates that workers would get an additional $19 billion a year if the rules were observed." [Emphasis in original.] In an analysis of more than 74,000 cases handled by the Labor Department over a four-year period, the paper concluded that one out of every fifty workers had been illegally denied overtime pay.

With Texaco leading the way, discrimination cases (involving age, disability, whistleblowing, race and sex) became boilerplate headlines in 1996. Texaco has had lots of company: Astra USA, WMX Technologies, Monsanto and Mitsubishi Motors, among others. Discriminators paid out many millions of dollars in penalties, but the corporate world was not visibly repentant or shaken. It's hard to make big corporations feel pain for their sins by merely fining them. Referring to one of Texaco's offers—$176 million—to settle a race discrimination case, the *Journal* made the point: Yes, if Texaco's offer were accepted, it would be the largest cash settlement ever to resolve such a case, but so what? After all, "it's not a princely sum to dole out over five years for a corporation with revenue of more than $30 billion."

THE ANTITRUST FARCE

When is a crime not officially a crime? Simple: When the laws against that activity are not enforced. Antitrust laws—those almost mythical beings so revered by populists—have been on the books for several generations and are supposed to be used to prevent the kind of concentration of market power that leads to price-fixing and the death of competition. That's what the Archer Daniels Midland case was all about. That's why the Federal Trade Commission used the antitrust laws to block Rite Aid's proposed $1.8 billion purchase of Revco, a combination that would have created the nation's largest drugstore chain. Antitrust laws were used (in addition to the instances previously cited in this article) to pry fines from Reader's Digest ($40 million) and U.S. Healthcare ($1.2 million).

But don't be misled. If the antitrust laws aren't dead, they're showing few vital signs. Concentrations of market power are coming on like a flood, and little is being done to stop the trend. In its yearly wrap-up, under the headline "Gorillas in Our Midst: Megadeals Smash Records as Firms Take Advantage of Favorable Climates," the *Journal* noted happily that "mergers, acquisitions and spinoffs totalled $659 billion in 1996, up 27 percent from $519 billion in 1995." The outlook was "as good as it gets," because "looser regulation is changing the competitive landscapes in telecommunications, utilities and broadcast, among other industries"—meaning government cops at the Justice Department and the Federal Trade Commission have all but surrendered.

Nowhere was this clearer than in the aerospace and military industry, where, in one of the largest mergers in U.S. history, Boeing bought McDonnell Douglas to become the only—yes, only—manufacturer of commercial jets in the United States, catapulting it ahead of Lockheed Martin, the number-one military contractor, as the world's largest aerospace company. This pairing-off received the same enthusiastic government support that a series of multibillion-dollar military industry mergers have received from the Clinton Administration over the past four years.

Other industries have been promised the same support. Last June the F.T.C. proposed that antitrust enforcers let cost savings justify mergers that would otherwise be considered illegal because they were anticompetitive. What's more, the commission adopted new rules that would speed mergers and acquisitions by radically shortening the time the F.T.C. takes to consider cases alleging anticompetitive conduct or consumer fraud.

Among those benefiting from the new atmosphere, the *Journal* predicted, would be airlines.

Their unauthorized meetings with foreign carriers to fix prices, which violates current antitrust laws, "will soon be routine—with the full approval of the U.S. government."

Crime marches on.

55

Pay Now, Pay Later

States Impose Prison Peonage

CHRISTIAN PARENTI

Kenneth Stewart Sr. is distressed. His son, Kenny, is on Virginia's death row and owes the state $57,756.20 for the cost of his jury trial—not his defense, but his whole trial. In Virginia, a jury trial is still a right, but one that losing defendants must literally pay for.

"Last week I sent $50 up to Kenny's account and there wasn't but six cents left in it after the authorities took their share," Stewart says.

Complicating matters are Kenny Stewart's abscessing teeth. "They charge fifty bucks to pull teeth, and Kenny needs most of his taken out," says the senior Stewart, who receives $850 a month in social security and wages as a part-time security guard. Every month he sends $200 to his son Kenny's prison account in the hopes that Kenny's teeth will be worked on.

On top of the $50 "co-payments" for medieval-style dental extractions, Kenny—like prisoners throughout the nation—must buy his own toiletries, underwear, socks, cigarettes, and stationery. Due to the notoriously bad prison food, most prisoners are forced to supplement the official diet with the occasional can of tuna fish or

soup mix from the commissary. But with a $57,000 debt around his neck, and a mouth full of dental bills, Kenny Stewart goes without the basics as he awaits execution.

Kenny Stewart's predicament is emblematic of a trend in American prisons to shift the costs of incarceration onto prisoners. Throughout the country, state corrections departments are squeezing whatever funds possible from the 1.5 million incarcerated Americans. Prisons extract money from their inmates by charging for court costs, imposing medical co-payments, seizing prisoners' assets, garnishing prisoners' wages, and pursuing former prisoners for the cost of their incarceration.

In Virginia, the accounting of justice is a picayune affair. Stewart and the 1,684 other Virginia prisoners who are paying the costs of their trials are, upon their defeat in court, slapped with an itemized bill that includes these standard charges: $100 forensics fee, $200 sentencing fee, $2 courthouse-maintenance fee, $65 court-reporter's fee, $12 sheriff's fee, $32.50 clerk's fee. $265 felony fee. And the most bizarre charge of all, as Stewart's lawyer David Baugh points out, is

From Christian Parenti, "Pay Now, Pay Later: States Impose Prison Peonage," *The Progressive,* July, 1996, pp. 26–29. Copyright © by The Progressive. Reprinted by permission.

the "jury lunch fee." "If they get beanie weenies, you get a break. If they get shrimp, you're screwed," says Baugh.

The minimum cost to a losing defendant is $720 for the first day of trial. After that the cost drops to $360 per day, or $30 per juror. Last year, Virginia collected $36 million in court fees—and that was for trial costs, not fines.

This arcane Virginia law, recently resurrected by get-tough legislators, is coming under increasing legal pressure. Baugh and the American Civil Liberties Union (ACLU) are planning to sue, arguing that charging for a jury trial violates the Constitution's Sixth Amendment (the right to a jury trial) and Fourteenth Amendment (equal protection under the law). No other state charges for criminal jury trials, though the National Center for State Courts reports that twenty-six states charge court fees in civil cases. Nonetheless, the Virginia attorney general's office defends its position. "Just because we're the only one that does it doesn't mean it's wrong," said spokesperson Mark Miner in a recent news report.

To help recoup a small portion of booming prison-construction costs, and to throw a bone to a vengeful public, many states enforce a pay-to-stay policy for inmates.

From small county jails to maximum-security state penitentiaries, inmates are being forced to pay room and board while they're locked away. According to an ACLU National Prison Project survey, twenty-one of the nation's state prison systems require at least some inmates to make payments toward room and board. Often the rule applies only to inmates employed in prison industries or by private firms that use prison labor. Such jobs pay between fifty cents and $4.25 an hour, with between 25 and 50 percent of the inmate's wages going toward room and board. However, many states and counties are beginning to charge regardless of a prisoner's ability to pay.

Michigan has one of the more established pay-to-stay programs. Relying on a previously dormant, sixty-year-old law known as the Prisoners Reimbursement Act, the Michigan attorney general's office has launched an aggressive campaign of lawsuits against inmates it deems able to pay the $40-a-day cost of their incarceration.

In 1994 the state collected $400,000 from prison inmates' bank accounts and pensions. And every year the Department of Corrections collects up to $1 million in rent from prisoners in halfway houses and prison work camps. "Taxpayers like the idea that we don't allow prisoners to profit from their crimes," says Attorney General Frank Kelley.

But not everyone agrees. Louise Grable contended that the state was unfair when it took 90 percent of her sixty-three-year-old, incarcerated husband's $398 monthly General Motors pension. He's currently serving three to sixteen years for sexually molesting a young girl, and he left his wife with no income other than the pension.

"I know what my husband did was wrong and he's paying for it, but I don't think that it's right for the state to walk right in and take everything," says Louise Grable, who is sixty-two. After a lengthy court case, she managed to have the pension deductions reduced.

In 1994, Michigan passed a similar law that will now charge up to $60 a day for room and board in county jails. Prisoners will also start paying for doctor visits and will soon face a "utility tax" on TVs, radios, and the other electrical appliances they use in their cells.

A bill in the California state assembly seeks to charge inmates $15 a day for room and board.

In Nevada, inmates, upon sentencing, are forced to disclose personal assets and to agree to pay a portion of the cost of their incarceration. Nevada prisoners who work pay from 20 to 50 percent of their wages for room and board. Mental-health services and emergency procedures can also drain a prisoner's account. If a prisoner incurred a few large expenses at the beginning of incarceration, "a guy could be stuck without any money for his whole stay," says Glen Whorton of the Nevada Department of Prisons. That, of course, means years without soap, socks, cigarettes, or stationery. "Hey, stay out of jail," says Whorton.

Increasingly, individual wardens and county sheriffs are devising such programs on their own, without approval from state governments.

In April 1994, the warden of Berks County prison in Pennsylvania began charging his prisoners for room and board. In the first six months, the prison collected $85,000 from 2,000 inmates; today the total is $295,000. Berks County prison now charges $10 per day, or 25 percent of costs. By mid-1995, other prisons in Pennsylvania were following suit.

In some cases, the prisons are even pursuing former prisoners for the cost of their incarceration.

Upon release from Berks County prison, former inmates have thirty days to set up a payment plan for their outstanding debts. When former inmates fail to pay, their cases are referred to Capax Credit Control. Between 1994 and 1995, 675 cases were forwarded to Capax.

Last year, Lancaster County jail joined the fray and started charging prisoners $10 a day for room and board. "It won't be a free ride any more," said Warden Vincent Guarini. Lancaster's policy even goes so far as to charge former inmates for pre-sentencing time.

As in other Pennsylvania institutions, former inmates of Lancaster are given thirty days to pay their debts, after which their case is handed to a private collection agency. So far, these Pennsylvania peonage policies have survived all legal challenges in circuit courts.

In Missouri, post-prison peonage is now state law. Last year, the Department of Corrections was authorized to take "10 percent of any wages, salary, benefits, or payments from any source." This lien on prisoners' assets and wages applies not only while a person is incarcerated but for "five years from the date that the offender is released."

In what passes for mercy among Missouri legislators, the law states: "Not more than 90 percent of the value of the assets of the offender may be used for the purposes of securing costs and reimbursement." The Missouri law will make failure to pay one's incarceration-related debts a violation of parole.

In a recent letter to *The Progressive,* prisoner Ricky Davis argued that the law will merely "ensure that re-entry into society is even more difficult for ex-cons" and that the wage attachments will lead to further poverty and thus more crime and recidivism. Davis also points out that former prisoners already pay for prisons—as does everyone else—with taxes.

One of the most inhumane policy innovations sweeping the corrections establishment is the introduction of co-payments for medical care, dental work, and mental-health services. These small (but, for prisoners, nonetheless expensive) fees are usually introduced with the stated aim of "deterring frivolous health complaints."

In October 1995, Allen County, Kentucky, started charging $10 for a doctor's visit. Not surprisingly, the average number of monthly doctor visits plunged from 1,125 to 225.

According to a 1994 report in *Corrections Forum* magazine, a similar co-payment program in Mobile, Alabama, showed a 50 percent reduction in inmate visits to the clinic. Between 1989 and 1993, the Mobile County jail saw its inmate population double while its total medical expenses dropped from $883,000 to $262,000.

The same story emerges throughout the nation, from the San Diego County jail to numerous state prison systems. Florida charges $3 a visit; Oklahoma $2, plus $2 per prescription; California $5 (soon to be boosted to $10), plus up to $200 for dentures and over $60 for eyeglasses; Nevada $4, with the costs of medication and prosthetics running much higher.

"No one is actually denied treatment due to lack of funds," says Whorton of the Nevada Department of Prisons. Instead, "the prisoner's account goes into negative balance," and, if someone is ultimately unable to pay, "the prisoner's welfare fund is charged." That fund is the repository of all prison canteen profits and is intended to be used for recreation and entertainment, such as buying basketballs, weights, and common-area televisions.

A similar practice holds true in many other states: Indigent inmates with no cash will have their accounts debited for medical costs and thus will be without any spending money until all debts are paid.

Colorado state prisons had charged $3 per visit, but the practice was found unconstitutional and amended. Now prisoners are charged only for a second opinion.

The Orleans Parish jail in Louisiana had been charging $3 to $5 for over-the-counter medication but this practice was likewise enjoined.

The ultimate in prison self-sufficiency is the prison as factory. This was the stated aim of Chief Justice Warren Burger, who in the 1970s urged Congress to turn prisons into "factories with fences." As part of the 1979 Justice Improvement Act, Congress established the Justice Department's Prison Industries Enhancement (PIE) program. Under PIE it again became legal for private corporations to employ prison labor. This practice had been outlawed by New Deal legislation—the Hawes-Cooper Act and the Ashurst-Sumner Act. The latter of these laws made transportation of prison-manufactured products across state lines a felony.

At first just a small-scale collection of pilot programs, PIE now licenses more than fifty "projects." Many states, most notably California, operate private prison-labor programs outside of PIE legislation and set their own rules for how firms can use prison labor.

California, Nevada, Oregon, and Washington lead the nation in leasing prison labor to private firms, as states are moving toward a partnership with private industry.

Nevada provides a glimpse of what may lie ahead on the road to self-sufficient prisons. Nevada prisoners already do the following:

- Build Bentley Nevada circuit boards for nuclear power plants (thirteen to fifteen inmates employed).

- Make waterbeds for Vinyl Products (sixty-five inmates employed).

- Restore cars for Imperial Palace (thirty to forty inmates employed).

- Hand-assemble Shelby Cobra road cars (fifteen inmates employed).

- And, in a perfect reflection of the dual economy in the United States, another thirty inmates build stretch limousines for Emerald Coach.

Other Nevada operations include making mattresses and fitting out ambulances for state government use.

On average, Nevada collects $800,000 to $1 million a year in room and board. Howard Skolnik, assistant director of the Nevada Department of Prisons, says: "The only thing stopping our expansion is that we've used all available space."

Despite his enthusiasm for the prison industries, Skolnik acknowledges that "it defrays some costs, but not much."

In Washington state, a company called Exmark uses a "flexible" pool of prison laborers to package everything from Microsoft Windows 95 to Starbucks Coffee products, JanSport gear, and literature for telecommunications giant US WEST, according to an article in *The Stranger.*

In Louisiana, Corrections Corporation of America—the largest private prison operator in the country, with, according to *The Washington Post,* profits of $4 million in 1993—has teamed up with the work-clothes manufacturer Company Apparel Safety Items (CASI). This is the first partnership between one of the nation's eighty-eight private prisons and a private manufacturer. The twenty-eight prisoners employed by CASI receive the minimum wage minus a 30 percent deduction for their room and board.

In Arizona, the press reports that even the 109 residents of death row are now pulling their own weight on a prison-run vegetable farm. One of the positive side-effects of employing the condemned—as far as the governor's press secretary is concerned—is that the inmates will now be too busy to file "frivolous lawsuits in attempts to circumvent their death sentences."

But other states have overcome numerous legal challenges. In Maryland, *Prison Legal News* reports, a prisoner named Jerome Johnson, who earned eighty-five cents a day and suffered from severe asthma, challenged the state's right to charge $4 to $10 for doctor visits and inhalers.

The court threw out Johnson's case, holding that the inmate's inability to pay for the medication necessary for him to breathe freely was not cruel and unusual punishment.

One particularly harrowing incident occurred in Comanche County, Oklahoma, where a jail simply refused to pay for a pretrial inmate's cancer treatments. Only after the prisoner filed a federal lawsuit were the treatments allowed.

Prison activists and critics of medical co-payment point to the long-term irrationality of such policies. "At first it seems to be a rational cost-cutting policy," says Jenni Gainsborough of the ACLU's National Prison Project. "But eventually prison administrators will be faced with widespread health crises. Even deadly contagious diseases such as HIV and tuberculosis often start out as a rash or a cough. If something like tuberculosis is not treated promptly—especially in a prison—epidemics are inevitable."

In purely financial terms, attempts to make prisons self-sufficient will, and do, fail. Prisoners are drawn from the poorest sections of the population and even if they are gouged for every penny in prison and hounded for all of their post-release lives by the debts of incarceration, they will never cough up enough money to pay for the overpriced, high-tech island of concrete in which they are kept. The main effect of squeezing funds from prisoners is to justify the gargantuan costs of America's gulag, which grows larger every day.

56

Japanese Say No to Crime

Tough Methods, at a Price

NICHOLAS D. KRISTOF

MITO, JAPAN

To understand why Japan is probably the safest industrialized country in the world, it helps to talk to a short, squat man with rumpled clothes, a gentle smile and big, leathery hands that once picked up a hammer and crushed his neighbor's head.

The murderer, who killed his neighbor after stealing money from him, is now 62, out of prison and on parole in this small city in eastern Japan. But his experience is testimony to the tough public attitude against crime.

No one in the murderer's family ever visited him in the 15 years and 3 months that he was in prison. His wife and son have met him since his release, but have told him never to return to the village where they live.

His three daughters, now all married, have refused to see him. "I have four grandchildren, I think," he said, but he has never met them or seen their pictures. The murderer, like other criminals interviewed, asked that he not be identified.

More than any other developed country, Japan has resolved to "just say no" to crime. Japanese

From Nicholas D. Kristof, "Japanese Say No to Crime: Tough Methods, at a Price," *The New York Times,* May 1, 1995, Copyright © 1995 by the New York Times Co. Reprinted by permission.

In Prison

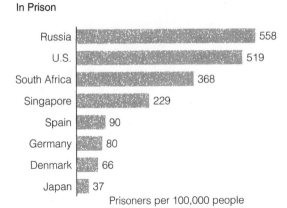

Source: Human Rights Watch and Sentencing Project

society ostracizes offenders and demands that they not just be caught but that they also confess and show remorse.

Japan and the United States have vastly different societies, but at a time when Americans say in some surveys that their No. 1 concern is crime, when Congress and the White House are searching for new ways to make streets safer, Japan may offer some lessons about ways to achieve security—and the price of that safety.

Handguns are banned in Japan, and the police and the courts have powers that would make American civil libertarians weep. Prison sentences are short, but prisons are grim places where some inmates are not allowed to talk to each other.

All this, coupled with strong families, a sense of social cohesiveness and an egalitarian distribution of wealth, has created what may be the real Japanese Miracle: not just one of the richest countries in the world but also just about the safest.

The poison gas attack on the Tokyo subway system in March killed 12 people, injured more than 5,500, frightened people across Japan and captured headlines around the world. But it was a fluke.

Until that terrorist attack, there had never been a killing recorded in Tokyo's subways, which carry 75 percent more people each day than New York City's. In contrast, an average of 18 people have been killed in New York's subways in each of the last five years.

Japan has been traumatized by the terrorist attack, but that is partly because people here are so accustomed to safety. Japanese have rarely had to worry about the killings and muggings that Americans fret about even in the best of times.

Last year 38 people were shot to death in all of Japan. In the United States, an average of 44 people are shot to death each day—in all, more than 16,000 gun homicides a year.

One measure of how life is more innocent in Japan is that the easiest place to steal money from is a police station. Anyone can go to a police box and, without showing identification, borrow the equivalent of $10 or $20 for transportation home.

"We can't ask them for I.D., because usually the reason they need to borrow cash is that they've lost their wallet," said Tokio Kunichika, chief of the police for the Tokyo rail system. "Basically we trust the people to return the money. And since it might not be convenient to come back here to return it, they can pay the loan back at their local police station."

The trust is perhaps understandable, in that Japan has only one-eighth as many thefts and burglaries as America. Japan, with 48 percent of the population of the United States, has one-twentieth as many killings, one-seventieth as many arson cases, and one-three-hundredth as many robberies.

Crime statistics are, of course, notoriously unreliable, but academic studies suggest that underreporting is less a problem in Japan than in the United States. And if anything, the figures understate the differences: a typical Japanese crime is a pick-pocketing, not a mugging, so the difference in the level of fear is even more stark than the statistics would suggest.

Japan's crime problems are now growing, but they are not spiraling out of control. For most of the post-war period, Japanese crime rates fell steadily. They bottomed in the 1970's and have been rising for about the last 15 years, but even now they are lower than they were in the late 1940's and the 1950's.

"There are special things about Japan, but we can learn from it," said Prof. David H. Bayley, an American criminologist at the State University of

New York and author of a book about the police in Japan.

"The lack of crime doesn't have to do with eating with chopsticks or taking your shoes off when you enter a house," Professor Bayley added. Instead, he said, it has to do with Japan's gun policy, police system and emphasis on social propriety.

PREVENTION

Stopping Trouble Before It Starts

One expert on Japan's criminal justice system is a lanky 19-year-old man with a round face, a shock of black hair and a long record of stealing from cars and homes. The young man lives in a halfway house in Mito, where he earlier spent a year in a juvenile detention home.

"I began to get in trouble in elementary school, when I was stealing money from my folks," the burglar said as he relaxed in a chair on the ground floor of his halfway house. "I left home, and I was hanging out and didn't have money. So I began stealing money from houses."

Every country has juvenile offenders, but in Japan they may have a greater chance of going straight.

In part this is because few young Japanese have drug problems. Japan vigorously cracks down on narcotics—as Paul McCartney found in 1980 when he arrived in Japan for a concert tour and was arrested for possession of marijuana—and illegal drugs are only a marginal problem in society.

Japan is also more likely than the United States to catch criminals. The proportion of crimes solved has been tumbling in Japan in recent years, but it is still 37 percent, compared with about 20 percent in the United States.

Burglary is also unattractive because it is not very lucrative. Japanese do not much like to buy second-hand goods, and the police keep a close eye on pawn shops. Anyone selling used goods in quantity gets a lot of police attention, and so there is not much of a secondary market for stolen goods.

In 1588 a warlord who pacified Japan, Hideyoshi Toyotomi, confiscated all swords and guns from ordinary citizens. Ever since then, ordinary Japanese have been pretty much barred from owning guns.

Outside the police, there are only 49 legal handguns in Japan, all belonging to marksmen. Even they cannot keep their guns at home but must deposit them at a shooting range.

Shotguns and a limited number of hunting rifles are allowed, but the owner must pass a rigorous licensing procedure and exam. This license must be renewed every three years—another onerous procedure—and so hunters do not lightly put up with the nuisance of owning a rifle.

As a result, there are about 425,000 guns in private hands in Japan, overwhelmingly shotguns and air rifles, and the number is dropping. In the United States, experts say, there may be 200 million guns.

Criminal gangs in Japan do have guns, smuggled from abroad, and they sometimes use them on one another in turf fights. But ordinary street criminals do not normally have access to guns, and a person caught with a loaded pistol can be imprisoned for 15 years, so a petty criminal would not be likely to carry one.

THE SOCIETY

Intact Families, Common Identity

Whatever the debate about whether gun control saves lives—and most Japanese think it does—Japan's experience underscores the importance of intact families and a relatively egalitarian society.

Criminologists say Japan's small income gaps, low unemployment and common social identity (85 percent of Japanese define themselves as middle-class) contribute to a shared antipathy toward crime.

Among Japanese offenders, the same kinds of risk factors seem to be associated with crime as in the West—particularly broken families. When a group of counselors was asked what they noticed

Crime in Japan . . . or the Lack of it

Japan and the U.S.

A comparison of Japanese and U.S. crime statistics in rates per 100,000 inhabitants:

	Japan	U.S.
Murder	0.98	9.5
Robbery	1.75	255.8
Arson	1.13	45.9
Burglary	186.9	1,099.0
Aggravated assault	5.4	440.0

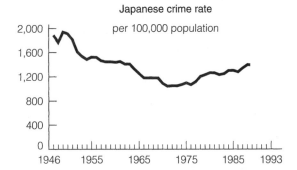

Japanese crime rate
per 100,000 population

Number murdered by firearms:

Japan (1994)	U.S. (1993)
38	16,189*

*1994 unavailable

Proportion of total murders committed with guns:

Japan (1994)	U.S. (1993)
3%	70%*

*1994 unavailable

Sources: F.B.I.; Japanese National Police Agency; Human Rights Watch and Sentencing Project.

---

about the juvenile offenders they worked with, they pointed to family backgrounds.

"It's common among those who come here that they have family problems when they grow up," said Yai Horie, a volunteer at a halfway house in Mito. "They don't have a mother, or a father, or they have only a stepmother, and they eventually become demoralized."

To the extent that broken families are connected with crime, Japan has a major advantage over the United States. By Western standards, Japanese families may often be dysfunctional—because the father spends very little time at home with children—but they are not broken.

Only 1.1 percent of Japanese children are born to unmarried mothers, for instance, while in the United States the figure is 30.1 percent.

What Americans call "family values" are rigorously inculcated in Japan. The value of good behavior, of fitting into a common society, is drummed into children from the moment they set off to first grade in identical school uniforms.

The version of "Goldilocks and the Three Bears" that is repeated in Japan, for instance, does not end with Goldilocks fleeing after she has

been discovered eating the bears' porridge and sleeping in their beds. Instead, she apologizes deeply, and she and the bears have a friendly lunch together.

THE SANCTION

Making Criminals Feel Guilty

The Japanese criminal justice system, like China's, places great emphasis on making criminals feel remorse. The idea is to create an environment where everyone agrees that crime is intolerable.

There is no plea bargaining as such in Japan, but criminals know that if they confess and seem penitent, then the police may let them off lightly. Likewise, prisoners are asked to write statements of remorse, and they can improve a convict's prospects of early release.

A 16-year-old gang member in Mito, for instance, had a clear record until one night two years ago, when gang members grew suspicious that another of their group was a spy for a rival gang. So they beat the victim and stomped on him, apparently intending to teach him a lesson. Instead, they killed him.

The 16-year-old was regarded by the police as genuinely horrified by what he had done, and they thought that he was so shaken by the experience that he would stay out of trouble. So he was sent to a juvenile detention center for 18 months.

"While in detention, I learned about my crime and about my shortcomings," the gang member wrote in the "expression of remorse" in his file.

The authorities released the young man from the detention center at the end of his sentence, but with conditions: that he live with relatives in a new neighborhood, that he work steadily at a job that parole workers found for him in a machine tool factory and that he have no contact with gang members.

"As long as he has no relationship with the bad guys from his past, he'll be 100 percent O.K.," said Shinji Yamada, his counselor.

Of course, Japanese criminals are as shrewd as Americans in knowing when it pays to display remorse that they do not feel. But since the whole system is geared to emphasize that crime is wrong, there may be less room in Japan for an alternative culture in which crime is "cool." Japanese social workers, and criminals, say that even among burglars, theft is considered deviant behavior.

CONFORMITY

Peer Pressure And Safe Streets

Professor Bayley, author of the book about the Japanese police, "Forces of Order," said some criminologists believe that a sense of orderliness and propriety is important to create an environment where crime is not tolerated.

"That's Japan," he added. "It's wall-to-wall propriety."

In Japan, for instance, even college students dress neatly, and the police take their shoes off at the entrances when conducting raids on private homes. The police installed a video monitor at the

door of the home of a top Government official, and so the official's wife now feels compelled to put on her makeup before taking out the garbage.

The social pressure to do things right creates a burden on those who fail to meet these standards, and it may explain Japan's high suicide rates—45 percent higher than America's. But the same social pressures may also help explain Japan's low crime rates.

While the West will never go as far as Japanese in creating a sense of propriety, Professor Bayley suggests that some Americans find the idea of greater orderliness attractive.

Permissiveness appears to be less fashionable in the United States these days, and school uniforms have made a small comeback in some areas. Likewise, the American enthusiasm for sending juvenile offenders to boot camp so they will learn to get up at dawn and make their beds reflects a Japanese-style interest in building discipline and orderliness.

THE POLICE

Broad Authority And Harsh Jails

In some respects, the differences between the law enforcement systems in Japan and the United States is not that it is harsher but that it is more authoritarian. Prison sentences are shorter than in America, for instance, and the prison population is minute by world standards.

For every 100,000 people, Japan has 37 prisoners, compared with 80 in Germany, 229 in Singapore, 519 in the United States and 558 in Russia. Escapes are almost nonexistent, and there appear to be few problems with prison gangs or rapes or assaults among prisoners.

The reason is that prisoners are sometimes kept in tiny, individual cells, and at times they are barred from talking to one another or even looking at one another.

"Prisoners can't hold a private conversation, even when they're working," said Nisaburo Sakata, who runs a counseling program in Mito for paroled prisoners. "If they just look at any women who happen to visit the prison, they're punished. So it's impossible to organize gangs in prisons."

Human Rights Watch, which is based in New York, published a report this year concluding that "prisoners in Japan experience routine violations of human rights." The report cited the case of a former prisoner who spent 12 years in prison entirely in isolation; even when she was taken out of her cell, the hallway would be cleared first.

To be sure, Japanese prisons seem to be less successful than the rest of the criminal justice system. The recidivism rate is higher in Japan than in the United States, meaning that if a Japanese adult does become a prisoner, he is more likely to return within five years than an American counterpart.

This may be because ex-convicts—like the murderer who defrauded and then bludgeoned his neighbor—are sometimes rejected even by their families, so they have no one to return to. In addition, many criminals are members of the underworld known as the Yakuza who have already made a career choice to become gangsters.

Not just prison guards but also the police in Japan have unusual authority by American standards. They can detain suspects for up to 21 days before bringing them to a judge, and in that period the police often conduct daily interrogations in which enormous pressure is put on detainees to confess.

Human Rights Watch and other critics say the anonymous interrogators sometimes rough up suspects and deprive them of food and water, as well as force them to stand in fixed positions for long periods. Partly because of these pressures, Japanese defendants are far more likely than Americans to admit crimes and plead guilty.

The police are also given much greater leeway than in America to question people on the street and search their pockets or bags. Japanese courts tend to defer to police judgments; in any case evidence can be admitted in court even if it turns out to have been seized illegally.

There is no question that these practices help in the fight against crime, but some people question whether they are worth it.

"I think that there is a lot that Japan has sacrificed for safety," said Koichi Miyazawa, a law professor at Keio University in Tokyo. "Safety has been achieved at the expense of freedom."

Kazue Akita, a criminal lawyer, acknowledged some ambivalence on the issue. "I'm afraid that people in Japan give too much power to the police," she said.

But she added that safety from crime is also necessary to maintain human rights. "It's really important for women to be able to walk on the street at night," she said. "There are problems of sexual discrimination in Japan, but at least women here can walk on the street at night."

12

World Health

I always look at the children of the rich . . .
their children are so healthy, so full of life.
They're so much more active than our children are.
I say to myself, "Could it be that we Indians are
idiots?" But no, I think their children are smarter
because they're better fed and better educated,
not because they were born that way.

ELVIA ALVARADO

The issue of world health dramatically illustrates the contradictions of the modern world. On the one hand, people today enjoy a life expectancy and quality of life unprecedented in history. Modern medical science has given us seeming miracles—people who could not conceive can now bear children, the dying are revived with organ transplants, and our youth are being vaccinated against diseases that once would have spelled an early death. Health care is being further improved, as practitioners of scientific medicine increasingly incorporate the wisdom of ancient medical practices of the world's cultures, bringing the best of the old and the new together in an effort to prevent illness and relieve pain and suffering. On the other hand, in many places in the world, children still face death at an early age from curable illnesses, and millions of people, young and old, are being ravaged by old and new epidemics—cholera, malaria, dengue fever, AIDS. Probably the most important health challenge facing humanity is the impact of inequalities of wealth and power on access to the best available health care. As Paul Starr put it in his work, *The Social Transformation of American Medicine,* "the dream of reason did not take power into account. The dream was that reason, in the form of the arts and sciences, would liberate humanity from scarcity and the caprices of nature, ignorance and superstition, tyranny, and

not least of all, the diseases of the mind and spirit. . . . Modern medicine is one of those extraordinary works of reason: an elaborate system of specialized knowledge, technical procedures, and rules of behavior. . . . But medicine is also, unmistakably, a world of power, where some are more likely to receive the rewards of reason than others" (Starr, 1982, pp. 3–4). Due to this situation, global health problems have frequently been examined from the conflict perspective.

In the poor nations, the health situation is devastating. People suffer conditions almost unimaginable to those of us living in the wealthy nations. Poverty, malnutrition, lack of sanitation, pollution, and poorly funded medical systems mean that illness is part of everyday life. In the United States and other wealthy nations, we expect to be healthy, and view illness as a deviation from the "normal" condition of health. One of the great structural-functionalist theorists of the twentieth century, Talcott Parsons, is remembered for his analysis of the "sick role." The sick are allowed to be deviant—that is, they are excused from their everyday role responsibilities, provided they conform to the dictates of the sick role—to follow the doctor's orders, and to try to get well. We expect the sick to return to the normal state of health, we expect our children to survive their infancy and early childhood, and we expect that adults will live into their 70s and 80s. Not so in the poor nations, where illness is routine, infant and child mortality are extremely high, and life expectancies are low. In "Death Without Weeping," Nancy Scheper-Hughes examines the tragic human costs of a 30–40 percent infant mortality rate among shantytown dwellers in Northeast Brazil. Mothers' seeming indifference to their infants' illnesses and deaths can be understood as a pragmatic survival response to the awareness that many of their children will not survive, no matter how much they love and care for them.

The inequality of power between wealthy and poor nations has an impact on health conditions. Although the life expectancy and quality of life in the poor nations have improved due to the spread of medical techniques developed in the wealthy nations, the power exercised by multinational corporations and foreign governments has had a variety of negative health consequences. In "Essential Drugs in Bangladesh," Zafrullah Chowdhury with Andrew Chetley expose the intervention by multinational pharmaceutical corporations, the U.S. government, and the World Bank to undermine a program of the Bangladesh government to make prescription drugs more available and affordable to its citizens.

Many of the wealthy industrialized nations have created national health insurance systems that seek to eliminate unequal access to health care. The United States is the only wealthy industrialized nation that lacks a comprehensive system of health insurance for its citizens. Although people in the wealthy nations are the healthiest worldwide, the chances for a healthy life are not equally distributed in these nations, even those with national health insurance. Lower-income workers, the poor, women, and members of racial or ethnic minority groups are often afflicted with illnesses caused or exacerbated by poverty. The existing medical services may also treat people unequally. For example, even in England, which has a national health care system intended to make health care available to all, black women migrants experience many difficulties getting equal treatment (Torkington, 1995).

A global health issue that has received much media attention in recent years is the spread of infectious diseases. This includes not only the dramatic spread of HIV/AIDS, which medical science can treat but not cure, but also the reappearance of diseases such as tuberculosis once thought to be under control with the discovery of antibiotics. In addition, we are now confronting deadly and frightening viruses such as Ebola and Hanta. Very little is known about these diseases. Anne Platt's "The Resurgence of Infectious Diseases" examines the underlying social causes of global epidemics and discusses the close connection between the spread of disease and the profitable exploitation of remote wilderness by vested interests that lack awareness of the health hazards of bringing humans into contact with viruses never before encountered.

The serious challenges to humanity in the area of health care have led activists to rethink the very definitions of *health* and *health care.* Notably, they have been moving from a narrower defintion of the problems to a broader, more social definition. In the past, health problems were often seen as individual problems, or as simply the problems of the single institution of medicine. The new definitions see health problems as related to the functioning of entire communities, nations, and the world. For example, it has been argued that the AIDS epidemic cannot be solved without redefining health as a *global human rights issue* (Mann, 1995). Empowerment of the poor is also an important concern of many health activists today. In "Sense and Sanitation," Sheela Patel describes a grassroots effort of the poorest urban dwellers in India. In Bombay—a huge city in which less than half of the residences are linked to sewers—the poor have begun a movement to improve health conditions by constructing public toilets.

WORKS CITED

Mann, Jonathan. 1995. "Health and Human Rights." *UNESCO Courier.* June 1995, pp. 26–31.

Starr, Paul. 1982. *The Social Transformation of American Medicine.* New York: Basic Books, Inc.

Torkington, Ntombenhle Protasia Khoti. 1995. "Black Migrant Women and Health." *Women's Studies International Forum.* 18(2) (March–April, 1995), pp. 153–158.

QUESTIONS FOR DISCUSSION

1. What are some of the conditions and institutions in Brazil that contribute to a high infant mortality rate? What does Scheper-Hughes mean by the statement that "under circumstances of high childhood mortality, patterns of selective neglect and passive infanticide may be seen as active survival strategies"? What do you believe to be the *most* important factor causing the deaths of so many infants?

2. After reading "Essential Drugs in Bangladesh," make a list of four things discussed in the reading that interested or surprised you. What do you expect would happen in the United States if Congress voted in a policy on essential drugs? Do you believe this is something that could happen in the United States? Explain.

3. What is meant by the "resurgence of infectious diseases"? What are some of the diseases involved? Why, according to Platt, is this resurgence occurring? What do you believe are the best strategies for solution to this problem?

4. Compare and contrast the efforts to improve health as discussed in "Essential Drugs in Bangladesh" and "Sense and Sanitation." Which do you believe will be more effective over the long run, and why?

INFOTRAC COLLEGE EDITION: EXERCISE

Health care reform has been a hotly debated issue in the United States, as people search for solutions to the growing crisis in health care. Some dimensions of the problem of health care in the United States that are generally acknowledged include the large numbers of people without health insurance, the high costs of health care, and the growing imbalance of power between the large Health Maintenance Organizations (HMOs) on the one side, and health care workers and consumers on the other. Learn more about some of these aspects by using InfoTrac College Edition. Look up one of the following articles:

"Health Care Reform: The International Way." Joseph White. *Issues in Science and Technology.* Fall 1995, v12n1p34. (*Hint:* Enter the search terms Medical Care and Canada, using the Subject Guide.)

"Health Care Workers Confront Managed Care." Marc Breslow. *Dollars and Sense.* January–February 1999, i221p18. (*Hint:* Enter the search terms Marc Breslow.)

"Community-Based Breast Cancer Intervention Program for Older African-American Women in Beauty Salons." *Public Health Reports.* March–April 1995, v110n2p179. (*Hint:* Enter the search terms African-American and Women, go to Subdivisions: Health Aspects.)

After reading the articles, think about solutions to the problems discussed. What should be included in a plan for health care reform in the United States, in order to deal with these problems?

FOR ADDITIONAL RESEARCH

Books

Charmaz, Kathy, and Debora A. Paterniti, eds. 1999. *Health, Illness and Healing.* Los Angeles, CA: Roxbury Publishing Company.

Kim, Jim Yong, Joyce Millen, John Gershman, and Alec Irwin, eds. 1998. *Dying for Growth: Global Inequality and the Health of the Poor.* Boston, MA: Common Courage Press.

McKenzie, Nancy F., ed. 1994. *Beyond Crisis: Confronting Health Care in the United States.* New York: Meridian.

Organizations

ACT UP (Aids Coalition to Unleash Power)
332 Bleecker Street, PMB 5
New York, NY 10014
(212) 966-4873
www.actupny.org

Doctors Without Borders, USA, Inc.
6 East 39th Street, 8th Floor
New York, NY 10016
(212) 679-6800
www.dwb.org

Partners in Health
113 River Street
Cambridge, MA 02139
(617) 661-4564
pih@igc.org
www.pih.org

Physicians for Social Responsibility
1101 14th Street NW, Suite 700
Washington, DC 20005
 (202) 898-0150
www.psr.org

ACTION PROJECTS

1. Sexually transmitted diseases are on the rise among youths in the United States today. Not only HIV/AIDS, but STDs such as syphilis, gonorrhea, herpes, and venereal warts are spreading rapidly. What accounts for this? Interview people in your area who are knowledgeable about youth culture—young people themselves, teachers, youth workers, counselors, and others—to find out. Use the sociological perspective—don't just get people's opinions, but try to ascertain what actual behavior exists that is helping to spread STD. Which cultural beliefs, values, and norms among today's youths promote this behavior and thus the spread of STDs? Does behavior differ among youths depending on their social class, race/ethnicity, gender, and sexual orientation? How and why? What programs are in place to help prevent the spread of STDs among young people? Are they effective? Why or why not?

2. Conduct an exploratory study into the relationship between power and access to health care in your own town, city, or region. Interview medical personnel, patients, and community health activists; pay visits to local hospitals and clinics, and observe what goes on. Where do low-income people go for medical care? If a person has Medicaid coverage, which doctors in the area will see them? Which will not? Why? What quality of medical care is available to low-income people?

How does it compare to the care available to the more affluent? What are some of the more serious illnesses common in the low-income community? Are there any special programs aimed at prevention and treatment? Based on your research, what solutions would you suggest to the problems of access to health care in your area?

3. Why is the United States the only advanced industrial nation without a national health insurance program? An excellent way to get more information on this question is to do some research into what happened to the Clinton administration's 1992 health care reform plan. Search your library's databases for magazine and newspaper articles on this reform plan. Identify professors on your campus, political leaders, or community people knowledgeable about the issue, and interview them for additional insights. What are the major features of the Clinton plan? Who opposed the plan, and why? Who was in favor of it, and why? Why did the plan never become a reality? Are there other health care reform proposals currently before Congress? Are these different from the Clinton proposal, and if so, how and why? What did your research tell you about why the United States lacks a national health insurance program? What (if anything), in your analysis, could be done in order to change this situation?

57

Death Without Weeping

Has Poverty Ravaged Mother Love in the Shantytowns of Brazil?

NANCY SCHEPER-HUGHES

• • •

I have seen death without weeping
The destiny of the Northeast is death
Cattle they kill
To the people they do something worse

ANONYMOUS BRAZILIAN SINGER (1965)

"Why do the church bells ring so often?" I asked Nailza de Arruda soon after I moved into a corner of her tiny mud-walled hut near the top of the shantytown called the Alto do Cruzeiro (Crucifix Hill). I was then a Peace Corps volunteer and a community development/health worker. It was the dry and blazing hot summer of 1965, the months following the military coup in Brazil, and save for the rusty, clanging bells of N.S. das Dores Church, an eerie quiet had settled over the market town that I call Bom Jesus da Mata. Beneath the quiet, however, there was chaos and panic. "It's nothing," replied Nailza, "just another little angel gone to heaven."

Nailza had sent more than her share of little angels to heaven, and sometimes at night I could hear her engaged in a muffled but passionate discourse with one of them, two-year-old Joana. Joana's photograph, taken as she lay propped up in her tiny cardboard coffin, her eyes open, hung on a wall next to one of Nailza and Ze Antonio taken on the day they eloped.

Nailza could barely remember the other infants and babies who came and went in close succession. Most had died unnamed and were hastily baptized in their coffins. Few lived more than a month or two. Only Joana, properly baptized in church at the close of her first year and placed under the protection of a powerful saint, Joan of Arc, had been expected to live. And Nailza had dangerously allowed herself to love the little girl.

In addressing the dead child, Nailza's voice would range from tearful imploring to angry recrimination: "Why did you leave me? Was your patron saint so greedy that she could not allow me one child on this earth?" Ze Antonion advised me to ignore Nailza's odd behavior, which he understood as a kind of madness that, like the birth and death of children, came and went. Indeed, the premature birth of a stillborn son some months later "cured" Nailza of her "inappropriate" grief, and the day came when she removed Joana's photo and carefully packed it away.

More than fifteen years elapsed before I returned to the Alto do Cruzeiro, and it was anthropology that provided the vehicle of my return. Since 1982 I have returned several times in order to pursue a problem that first attracted my attention in the 1960s. My involvement with the people of the Alto do Cruzeiro now spans a quarter of a century and three generations of parenting in a community where mothers and daughters are often simultaneously pregnant.

The Alto do Cruzeiro is one of three shantytowns surrounding the large market town of Bom

Jesus in the sugar plantation zone of Pernambuco in Northeast Brazil, one of the many zones of neglect that have emerged in the shadow of the now tarnished economic miracle of Brazil. For the women and children of the Alto do Cruzeiro the only miracle is that some of them have managed to stay alive at all.

The northeast is a region of vast proportions (approximately twice the size of Texas) and of equally vast social and developmental problems. The nine states that make up the region are the poorest in the country and are representative of the Third World within a dynamic and rapidly industrializing nation. Despite waves of migrations from the interior to the teeming shantytowns of coastal cities, the majority still live in rural areas on farms and ranches, sugar plantations and mills.

Life expectancy in the Northeast is only forty years, largely because of the appallingly high rate of infant and child mortality. Approximately one million children in Brazil under the age of five die each year. The children of the Northeast, especially those born in shantytowns on the periphery of urban life, are at a very high risk of death. In these areas, children are born without the traditional protection of breast-feeding, subsistence gardens, stable marriages, and multiple adult caretakers that exists in the interior. In the hillside shantytowns that spring up around cities or, in this case, interior market towns, marriages are brittle, single parenting is the norm, and women are frequently forced into the shadow economy of domestic work in the homes of the rich or into unprotected and oftentimes "scab" wage labor on the surrounding sugar plantations, where they clear land for planting and weed for a pittance, sometimes less than a dollar a day. The women of the Alto may not bring their babies with them into the homes of the wealthy, where the often-sick infants are considered sources of contamination, and they cannot carry the little ones to the riverbanks where they wash clothes because the river is heavily infested with schistosomes and other deadly parasites. Nor can they carry their young children to the plantations, which are often several miles away. At wages of a dollar a day, the

women of the Alto cannot hire baby sitters. Older children who are not in school will sometimes serve as somewhat indifferent caretakers. But any child not in school is also expected to find wage work. In most cases, babies are simply left at home alone, the door securely fastened. And so many also die alone and unattended.

Bom Jesus da Mata, centrally located in the plantation zone of Pernambuco, is within commuting distance of several sugar plantations and mills. Consequently, Bom Jesus has been a magnet for rural workers forced off their small subsistence plots by large landowners wanting to use every available piece of land for sugar cultivation. Initially, the rural migrants to Bom Jesus were squatters who were given tacit approval by the mayor to put up temporary straw huts on each of the three hills overlooking the town. The Alto do Cruzeiro is the oldest, the largest, and the poorest of the shantytowns. Over the past three decades many of the original migrants have become permanent residents, and the primitive and temporary straw huts have been replaced by small homes (usually of two rooms) made of wattle and daub, sometimes covered with plaster. The more affluent residents use bricks and tiles. In most Alto homes, dangerous kerosene lamps have been replaced by light bulbs. The once tattered rural garb, often fashioned from used sugar sacking, has likewise been replaced by store-bought clothes, often castoffs from a wealthy *patrão* (boss). The trappings are modern, but the hunger, sickness, and death that they conceal are traditional, deeply rooted in a history of feudalism, exploitation, and institutionalized dependency.

My research agenda never wavered. The questions I addressed first crystallized during a veritable "die-off" of Alto babies during a severe drought in 1965. The food and water shortages and the political and economic chaos occasioned by the military coup were reflected in the handwritten entries of births and deaths in the dusty, yellowed pages of the ledger books kept at the public registry office in Bom Jesus. More than 350 babies died in the Alto during 1965 alone—this from a shantytown population of little more than 5,000. But that wasn't what surprised me.

There were reasons enough for the deaths in the miserable conditions of shantytown life. What puzzled me was the seeming indifference of Alto women to the death of their infants, and their willingness to attribute to their own tiny offspring an aversion to life that made their death seem wholly natural, indeed all but anticipated.

Although I found that it was possible, and hardly difficult, to rescue infants and toddlers from death by diarrhea and dehydration with a simple sugar, salt, and water solution (even bottled Coca-Cola worked fine), it was more difficult to enlist a mother herself in the rescue of a child she perceived as ill-fated for life or better off dead, or to convince her to take back into her threatened and besieged home a baby she had already come to think of as an angel rather than as a son or daughter.

I learned the high expectancy of death, and the ability to face child death with stoicism and equanimity, produced patterns of nurturing that differentiated between those infants thought of as thrivers and survivors and those thought of as born already "wanting to die." The survivors were nurtured, while stigmatized, doomed infants were left to die, as mothers say, *a mingua,* "of neglect." Mothers stepped back and allowed nature to take its course. This pattern, which I call mortal selective neglect, is called passive infanticide by anthropologist Marvin Harris. The Alto situation, although culturally specific in the form that it takes, is not unique to Third World shantytown communities and may have its correlates in our own impoverished urban communities in some cases of "failure to thrive" infants.

I use as an example the story of Zezinho, the thirteen-month-old toddler of one of my neighbors, Lourdes. I became involved with Zezinho when I was called in to help Lourdes in the delivery of another child, this one a fair and robust little tyke with a lusty cry. I noted that while Lourdes showed great interest in the newborn, she totally ignored Zezinho who, wasted and severely malnourished, was curled up in a fetal position on a piece of urine- and feces-soaked cardboard placed under his mother's hammock. Eyes open and vacant, mouth slack, the little boy seemed doomed.

When I carried Zezinho up to the community day-care center at the top of the hill, the Alto women who took turns caring for one another's children (in order to free themselves for part-time work in the cane fields or washing clothes) laughed at my efforts to save Ze, agreeing with Lourdes that here was a baby without a ghost of a chance. Leave him alone, they cautioned. It makes no sense to fight with death. But I did do battle with Ze, and after several weeks of force-feeding (malnourished babies lose their interest in food), Ze began to succumb to my ministrations. He acquired some flesh across his taut chest bones, learned to sit up, and even tried to smile. When he seemed well enough, I returned him to Lourdes in her miserable scrap-material lean-to, but not without guilt about what I had done. I wondered whether returning Ze was at all fair to Lourdes and to his little brother. But I was busy and washed my hands of the matter. And Lourdes did seem more interested in Ze now that he was looking more human.

When I returned in 1982, there was Lourdes among the women who formed my sample of Alto mothers—still struggling to put together some semblance of life for a now grown Ze and her five other surviving children. Much was made of my reunion with Ze in 1982, and everyone enjoyed retelling the story of Ze's rescue and of how his mother had given him up for dead. Ze would laugh the loudest when told how I had had to force-feed him like a fiesta turkey. There was no hint of guilt on the part of Lourdes and no resentment on the part of Ze. In fact, when questioned in private as to who was the best friend he ever had in life, Ze took a long drag on his cigarette and answered without a trace of irony, "Why my mother, of course." "But of course," I replied.

Part of learning how to mother in the Alto do Cruzeiro is learning when to let go of a child who shows that it "wants" to die or that it has no "knack" or no "taste" for life. Another part is learning when it is safe to let oneself love a child. Frequent child death remains a powerful shaper of maternal thinking and practice. In the absence of firm expectation that a child will survive, mother love as we conceptualize it (whether in popular terms or in the psychobiological notion of maternal bonding) is attenuated and delayed with consequences for infant survival. In an environment already precarious to young life, the emotional detachment of mothers toward some of their babies contributes even further to the spiral of high mortality—high fertility in a kind of macabre lock-step dance of death.

The average woman of the Alto experiences 9.5 pregnancies, 3.5 child deaths, and 1.5 stillbirths. Seventy percent of all child deaths in the Alto occur in the first six months of life, and 82 percent by the end of the first year. Of all deaths in the community each year, about 45 percent are of children under the age of five.

Women in the Alto distinguish between child deaths understood as natural (caused by diarrhea and communicable diseases) and those resulting from sorcery, the evil eye, or other magical or supernatural afflictions. They also recognize a large category of infant deaths seen as fated and inevitable. These hopeless cases are classified by mothers under the folk terminology "child sickness" or "child attack." Women say that there are at least fourteen different types of hopeless child sickness, but most can be subsumed under two categories—chronic and acute. The chronic cases refer to infants who are born small and wasted. They are deathly pale, mothers say, as well as weak and passive. They demonstrate no vital force, no liveliness. They do not suck vigorously; they hardly cry. Such babies can be this way at birth or they can be born sound but soon show no resistance, no "fight" against the common crises of infancy: diarrhea, respiratory infections, tropical fevers.

The acute cases are those doomed infants who die suddenly and violently. They are taken by stealth overnight, often following convulsions that bring on head banging, shaking, grimacing, and shrieking. Women say it is horrible to look at such a baby. If the infant begins to foam at the mouth or gnash its teeth or go rigid with its eyes turned back inside its head, there is absolutely no hope. The infant is "put aside"—left alone—often on the floor in a back room, and allowed to die. These symptoms (which accompany high fevers, dehydration, third-stage malnutrition, and

encephalitis) are equated by Alto women with madness, epilepsy, and worst of all, rabies, which is greatly feared and highly stigmatized.

Most of the infants presented to me as suffering from chronic child sickness were tiny, wasted famine victims, while those labeled as victims of acute child attack seemed to be infants suffering from the deliriums of high fever or the convulsions that can accompany electrolyte imbalance in dehydrated babies.

Local midwives and traditional healers, praying women, as they are called, advise Alto women on when to allow a baby to die. One midwife explained: "If I can see that a baby was born unfortuitously, I tell the mother that she need not wash the infant or give it a cleansing tea. I tell her just to dust the infant with baby powder and wait for it to die." Allowing nature to take its course is not seen as sinful by these often very devout Catholic women. Rather, it is understood as cooperating with God's plan.

Often I have been asked how consciously women of the Alto behave in this regard. I would have to say that consciousness is always shifting between allowed and disallowed levels of awareness. For example, I was awakened early one morning in 1987 by two neighborhood children who had been sent to fetch me to a hastily organized wake for a two-month-old infant whose mother I had unsuccessfully urged to breast-feed. The infant was being sustained on sugar water, which the mother referred to as *soro* (serum), using a medical term of the infant's starvation regime in light of his chronic diarrhea. I had cautioned the mother that an infant could not live on *soro* forever.

The two girls urged me to console the young mother by telling her that it was "too bad" that her infant was so weak that Jesus had to take him. They were coaching me in proper Alto etiquette. I agreed, of course, but asked, "And what do *you* think?" Xoxa, the eleven-year-old, looked down at her dusty flip-flops and blurted out, "Oh, Dona Nanci, that baby never got enough to eat, but you must never say that!" And so the death of hungry babies remains one of the best kept secrets of life in Bom Jesus da Mata.

Most victims are waked quickly and with a minimum of ceremony. No tears are shed, and the neighborhood children form a tiny procession, carrying the baby to the town graveyard where it will join a multitude of others. Although a few fresh flowers may be scattered over the tiny grave, no stone or wooden cross will mark the place, and the same spot will be reused within a few months' time. The mother will never visit the grave, which soon becomes an anonymous one.

What, then, can be said of these women? What emotions, what sentiments motivate them? How are they able to do what, in fact, must be done? What does mother love mean in this inhospitable context? Are grief, mourning, and melancholia present, although deeply repressed? If so, where shall we look for them? And if not, how are we to understand the moral visions and moral sensibilities that guide their actions?

I have been criticized more than once for presenting an unflattering portrait of poor Brazilian women, women who are, after all, themselves the victims of severe social and institutional neglect. I have described these women as allowing some of their children to die, as if this were an unnatural and inhuman act rather than, as I would assert, the way any one of us might act, reasonably and rationally, under similarly desperate conditions. Perhaps I have not emphasized enough the real pathogens in this environment of high risk: poverty, deprivation, sexism, chronic hunger, and economic exploitation. If mother love is, as many psychologists and some feminists believe, a seemingly natural and universal maternal script, what does it mean to women for whom scarcity, loss, sickness, and deprivation have made that love frantic and robbed them of their grief, seeming to turn their hearts to stone?

Throughout much of human history—as in a great deal of the impoverished Third World today—women have had to give birth and to nurture children under ecological conditions and social arrangements hostile to child survival, as well as to their own well-being. Under circumstances of high childhood mortality, patterns of selective neglect and passive infanticide may be seen as active survival strategies.

They also seem to be fairly common practices historically and across cultures. In societies characterized by high childhood mortality and by a correspondingly high (replacement) fertility, cultural practices of infant and child care tend to be organized primarily around survival goals. But what this means is a pragmatic recognition that not all of one's children can be expected to live. The nervousness about child survival in areas of northeast Brazil, northern India, or Bangladesh, where a 30 percent or 40 percent mortality rate in the first years of life is common, can lead to forms of delayed attachment and a casual or benign neglect that serves to weed out the worst bets so as to enhance the life chances of healthier siblings, including those yet to be born. Practices similar to those that I am describing have been recorded for parts of Africa, India, and Central America.

Life in the Alto do Cruzeiro resembles nothing so much as a battlefield or an emergency room in an overcrowded inner-city public hospital. Consequently, morality is guided by a kind of "lifeboat ethics," the morality of triage. The seemingly studied indifference toward the suffering of some of their infants, conveyed in such sayings as "little critters have no feelings," is understandable in light of these women's obligation to carry on with their reproductive and nurturing lives.

In their slowness to anthropomorphize and personalize their infants, everything is mobilized so as to prevent maternal overattachment and, therefore, grief at death. The bereaved mother is told not to cry, that her tears will dampen the wings of her little angel so that she cannot fly up to her heavenly home. Grief at the death of an angel is not only inappropriate, it is a symptom of madness and of a profound lack of faith.

Infant death becomes routine in an environment in which death is anticipated and bets are hedged. While the routinization of death in the context of shantytown life is not hard to understand, its routinization in the formal institutions of public life in Bom Jesus is not as easy to accept uncritically. Here the social production of indifference takes on a different, even a malevolent, cast.

In a society where triplicates of every form are required for the most banal events (registering a car, for example), the registration of infant and child death is informal, incomplete, and rapid. It requires no documentation, takes less than five minutes, and demands no witnesses other than office clerks. No questions are asked concerning the circumstances of the death, and the cause of death is left blank, unquestioned and unexamined. A neighbor, grandmother, older sibling, or common-law husband may register the death. Since most infants die at home, there is no question of a medical record.

From the registry office, the parent proceeds to the town hall, where the mayor will give him or her a voucher for a free baby coffin. The full-time municipal coffinmaker cannot tell you exactly how many .baby coffins are dispatched each week. It varies, he says, with the seasons. There are more needed during the drought months and during the big festivals of Carnaval and Christmas and São Joao's Day because people are too busy, he supposes, to take their babies to the clinic. Record keeping is sloppy.

Similarly, there is a failure on the part of city-employed doctors working at two free clinics to recognize the malnutrition of babies who are weighed, measured, and immunized without comment and as if they were not, in fact, anemic, stunted, fussy, and irritated starvation babies. At best the mothers are told to pick up free vitamins or a health "tonic" at the municipal chambers. At worst, clinic personnel will give tranquilizers and sleeping pills to quiet the hungry cries of "sick-to-death" Alto babies.

The church, too, contributes to the routinization of, and indifference toward, child death. Traditionally, the local Catholic church taught patience and resignation to domestic tragedies that were said to reveal the imponderable workings of God's will. If an infant died suddenly, it was because a particular saint had claimed the child. The infant would be an angel in the services of his or her heavenly patron. It would be wrong, a sign of a lack of faith, to weep for a child with such good fortune. The infant funeral was, in the past, an event celebrated with joy. Today, however, under the new regime of "liberation theology," the bells of N. S. das Dores parish church no longer peal for the death of Alto babies, and no priest accompanies the procession of angels to the cemetery where their bodies are disposed of casually and without ceremony.

Children bury children in Bom Jesus da Mata. In this most Catholic of communities, the coffin is handed to the disabled and irritable municipal gravedigger, who often chides the children for one reason or another. It may be that the coffin is larger than expected and the gravedigger can find no appropriate space. The children do not wait for the gravedigger to complete his task. No prayers are recited and no sign of the cross made as the tiny coffin goes into its shallow grave.

When I asked the local priest, Padre Marcos, about the lack of church ceremony surrounding infant and childhood death today in Bom Jesus, he replied:"In the old days, child death was richly celebrated. But those were the baroque customs of a conservative church that wallowed in death and misery. The new church is a church of hope and joy. We no longer celebrate the death of child angels. We try to tell mothers that Jesus doesn't want all the dead babies they send him." Similarly, the new church has changed its baptismal customs, now often refusing to baptize dying babies brought to the back door of a church or rectory. The mothers are scolded by the church attendants and told to go home and take care of their sick babies. Baptism, they are told, is for the living; it is not to be confused with the sacrament of extreme unction, which is the anointing of the dying. And so it appears to the women of the Alto that even the church has turned away from them, denying the traditional comfort of folk Catholicism.

The contemporary Catholic church is caught in the clutches of a double bind. The new theology of liberation imagines a kingdom of God on earth based on justice and equality, a world without hunger, sickness, or childhood mortality. At the same time, the church has not changed its official position on sexuality and reproduction, including its sanctions against birth control, abortion, and sterilization. The padre of Bom Jesus da Mata recognizes this contradiction intuitively, although he shies away from discussions on the topic, saying that he prefers to leave questions of family planning to the discretion and the "good consciences" of his impoverished parishioners. But this, of course, side-steps the extent to which those good consciences have been shaped by traditonal church teachings in Bom Jesus, especially by his recent predecessors.

Hence, we can begin to see that the seeming indifference of Alto mothers toward the death of some of their infants is but a pale reflection of the official indifference of church and state to the plight of poor women and children.

Nonetheless, the women of Bom Jesus are survivors. One women, Biu, told me her life history, returning again and again to the themes of child death, her first husband's suicide, abandonment by her father and later by her second husband, and all the other losses and disappointments she had suffered in her long forty-five years. She concluded with great force, reflecting on the day of Carnaval '88 that were fast approaching:

> No, Dona Nanci, I won't cry, and I won't waste my life thinking about it from morning to night. . . . Can I argue with God for the state that I'm in? No! And so

I'll dance and I'll jump and I'll play Carnaval! And yes, I'll laugh and people will wonder at a *pobre* like me who can have such a good time.

And no one did blame Biu for dancing in the streets during the four days of Carnaval—not even on Ash Wednesday, the day following Carnaval '88 when we all assembled hurriedly to assist in the burial of Mercea, Biu's beloved *casula,* her last-born daughter who had died at home of pneumonia during the festivities. The rest of the family barely had time to change out of their costumes. Severino, the child's uncle and godfather, sprinkled holy water over the little angel while he prayed: "Mercea, I don't know whether you were called, taken, or thrown out of this world. But look down at us from your heavenly home with tenderness, with pity, and with mercy." So be it.

58

Essential Drugs in Bangladesh
The Ups and Downs of a Policy

ZAFRULLAH CHOWDHURY
WITH ANDREW CHETLEY

In 1982, the Bangladesh government introduced a radical and far-reaching National Drug Policy to curtail the abuses of pharmaceutical companies and ensure the availability of high-quality medicines at a price poor Bangladeshis could afford. It was the first time a country had drawn up such a policy in such a short period of time in one document. Since then, the policy has been opposed by many actors, including the national and international pharmaceutical industry, some governments in the North, and some sectors of the Bangladesh government itself. Yet despite constant efforts to undermine it, the policy has brought significant benefits.

When Bangladesh became an independent country on 16 December 1971, essential medicines were highly-priced and not easily available. According to the government's first Five-Year Plan:

> Many so-called manufacturers are engaged in bottling drugs imported in bulk,

From Zafrullah Chowdhury with Andrew Chetley, "Essential Drugs in Bangladesh: The Ups and Downs of a Policy," *The Ecologist,* vol. 26, no. 1 (January/February, 1996), pp. 27–33. Copyright © 1996 by The Ecologist. Reprinted by permission.

acting indirectly as the sales agents of foreign firms. Quality control of drugs is insufficient and spurious drugs are quite common.

It was not until 1982, however, that significant steps were taken to tackle this. On 27 April 1982, just one month after a military coup led by General Hussain Muhammad Ershad, the military government appointed an eight-member Expert Committee to review the drug situation in Bangladesh and make recommendations for a National Drug Policy consistent with the health needs of the country.

A NATIONAL DRUG POLICY

When the committee was appointed, 166 pharmaceutical manufacturers were licensed to produce medicines in Bangladesh, while 122 foreign companies from 23 countries were exporting drugs to the country—29 of them were British, 12 US, 11 Indian, 10 Swiss and 9 German. The total number of registered products, both locally produced and imported, was 4,340.

At its first meeting, the Expert Committee calculated that most medical practitioners usually prescribed from around a total of 50 of these thousands of drugs. They decided therefore to draw up a list of drugs they considered essential for Bangladesh's health needs in line with the advice given in a WHO publication, *The Selection of Essential Drugs,* and a list of drugs to be banned.

Sixteen criteria were drawn up by which to evaluate these and future drugs. Of these, 11 were based solely on scientific reasoning, concerning in particular effectiveness and therapeutic value; others were based on political and economic considerations or were intended to encourage the national pharmaceutical industry. For instance, multinationals were not to be allowed to manufacture antacids or oral vitamins in the hope that they would concentrate instead on antibiotics and life-saving drugs which were not so easily produced by smaller national companies.

The Expert Committee identified 1,742 non-essential or ineffective drugs to be banned and placed them in three categories. Production of drugs in the first category was to be stopped immediately, and stocks collected from pharmacies and destroyed within three months of the acceptance of the Expert Committee's report. Drugs in the second category were to be reformulated within six months on the basis of the guidelines suggested by the committee. A maximum of nine months was allowed in which to use up stocks of drugs in the third category. The Committee prohibited the import of raw materials from drugs in the first and second categories.

All pharmaceutical manufacturers in Bangladesh were affected by these recommendations, national companies the worst. Of the banned drugs that were manufactured in Bangladesh, 949 of them were made by national companies and only 176 by TNCs. The total number of imported banned drugs was 617, most of them originating (in descending order) from Britain, Switzerland, Germany and the US.

Other major recommendations were to establish a list of 150 essential drugs and a supplementary list of 100 specialized drugs; use generic (not brand) names for the manufacture and sale of drugs intended for primary health care; eliminate product patents and limit the use of process patents; draw up a national health policy; and revise the 1940 Drugs Act to control drug manufacturing quality, labeling and advertising, prices, prescriptions, technology transfer and licensing agreements, and pharmacies.

The NDP was to be applied consistently to both the private and public sectors, and to traditional and modern medical systems.

NO STORM UNLEASHED

As soon as they learned of the National Drug Police (NDP), which was approved by the Council of Ministers on 29 May 1982, TNCs, some of the larger national pharmaceutical companies, the Bangladesh Medical Association and representatives of many Western governments immediately began efforts to undermine it.

The TNCs' main objection to the policy was not because they had a particularly large market in Bangladesh; they were concerned that unless the NDP was nipped in the bud, the Bangladesh example would be copied by other countries.

Within hours of hearing about the policy, the US ambassador to Bangladesh, Jane Coon, called without prior appointment on General Ershad to convince him that the policy should not be implemented, or at the very least should be postponed, because it was unacceptable to the US.

Over the next few days, Ambassador Coon met with the editor of the most widely-circulated, pro-US Bengali daily newspaper, *Ittefaq,* the managing directors of US transnational companies, and others to devise a strategy to prevent implementation of the NDP. She informed offices of the US Agency for International Development (USAID) in neighbouring countries about the dangers of a policy such as the Bangladeshi one.

British, Dutch and West German ambassadors also called on General Ershad to express their dismay at the proposed drug policy. The West German ambassador was particularly angry, saying that it would deter German investment in the country. He said that the German multinational, Hoechst, had intended to expand in Bangladesh but was now reluctant to do so.

The British threat was more subtle. Foreign Secretary Douglas Hurd said in the House of Commons:

> We are keen that the Bangladesh government should use its scarce resources wisely. We are also keen that they should succeed in their policy of encouraging foreign investment to help them with the development of an industrial economy . . . It is important that in trying to achieve the aims of the pharmaceutical policy, [the Bangladesh government does] not discriminate against foreign-owned manufacturing companies in Bangladesh and [does] not frighten off prospective foreign investors.

Transnational companies started mobilizing the Bangladesh Medical Association and elite public opinion. Many directors and administrators of TNCs were retired army officers who were able to lobby the Martial Law Administration. Drug manufacturers met with senior officials of the army and the civil administration and with members of the governing Council of Ministers to persuade them that the NDP had no scientific basis. The finance and industry minister sided with the TNCs.

These pressures led General Ershad to call a public hearing on 6 June 1982 which lasted for over three hours. Rather than debate the scientific rationale of the NDP, however, TNC representatives argued that generic drugs policies had failed all over the world and that Bangladesh was attempting to ban drugs widely available elsewhere. Echoing the US ambassador, they called for the implementation of the NDP to be delayed, pending a review by an independent body. Arguing that the policy would deter foreign investors, they made a veiled threat to withdraw from Bangladesh.

Nevertheless, government officials stood their ground, realising that the TNCs' claims about the banned drugs and drug policies were false and that the companies were simply trying to defend their commercial interests. Thus on 7 June 1982, the health minister made the NDP document public, the first step towards giving it a legal framework, and on 11 June, General Ershad signed the Drug (Control) Ordinance 1982 which brought the National Drug Policy into effect.

AN "INDEPENDENT" US REVIEW

The day after the policy was made public, General Ershad received an unexpected invitation from US President Ronald Reagan to visit the United States. Just ten days later, General Ershad met President Reagan on 18 June in the US. Later he met Vice-President George Bush and, in response to Bush's suggestions, said that he would welcome a visit to Bangladesh by independent US scientists to advise him on the policy.

What he did not realise was that George Bush was a director of the US-based Eli Lilly, one of the world's top 10 pharmaceutical companies, and had substantial shares in other drug companies. Nor was he aware that only a few days earlier, Bush had told senior pharmaceutical industry executives from 22 countries that the Reagan administration planned to limit regulation of the pharmaceutical industry and end the "adverse relationship" between government and industry.

After high-level manoeuvrings between the US Pharmaceutical Manufacturers Association, the US State Department and the US Embassy in Bangladesh, Ambassador Coon told Health Minister Dr Shamsul Huq that a four-member "independent scientific committee" would be arriving in Dhaka on 20 July 1982 to examine the drug policy and advise the Bangladesh government.

When Dr Huq asked for the *curriculum vitae* (CV) of each of these experts, the US ambassador angrily implied that it was impertinent for a minister of a Third World country to ask for the CVs of experts from the US and discourteously left the minister without having tea which had already been laid on the table.

The health minister persisted with his request for the CVs until they were made available. Not one of the members of the "scientific committee" was an independent scientist: two were senior executives from Wyeth Laboratories, one was a Squibb vice-president, and the fourth was a director with the US Pharmaceutical Manufacturers Association.

General Ershad later told the health minister that he had expected the independent advisory committee to consist of officials from the US Food and Drug Administration and members of the National Academy of Sciences and other academics, definitely not representatives from the pharmaceutical industry.

The US "independent scientific committee" met executives from drug companies and *Bangladesh Aushadh Shilpa Shamity* (BASS, Bangladesh Association of Pharmaceutical Industries) and had a few meetings with the Bangladesh Medical Association. They did not wish to meet members of the Expert Committee and their findings were never made public.

US CITIZENS CHALLENGE THEIR GOVERNMENT

Although Bangladesh's drug policy was opposed by the US government, it received unexpected support from sectors of the US public. Well-known consumer rights activist Ralph Nader, concerned US citizens, consumer bodies, various church groups and human rights activists were among those who raised questions with their senators, congressional representatives and the State Department about US opposition to the policy. Many of them highlighted the US government's support for WHO's programme on essential drugs at the 35th World Health Assembly in May 1982.

The Public Citizen Health Research Group in Washington wrote to the Bangladesh Health Minister on 9 August that many of the drugs now banned from the Bangladesh market:

> were removed from the US market during the past 12 years on the recommendation of eminent medical scientists; many of the drug mixtures you are removing were deemed to be "irrational combinations" . . . and their use was "not recommended."

The Group also wrote to the US Secretary of State on 18 August that "to pressure the Bangladesh government to delay the withdrawal from the market of dangerous, ineffective, useless or unnecessarily expensive drugs is unconscionable."

The State Department responded that the US Embassy in Dhaka had simply "facilitated a dialogue" between the industry and the Bangladesh government and expressed the "hope" that implementation of the NDP might be delayed while this dialogue took place. Dr Sidney Wolfe of the Group replied:

> Imagine the outrage of the US public if a foreign government asked us to delay implementing a health-protecting decision of our Food and Drug Administration

WHO's Peculiar Silence

Although the National Drug Policy followed guidelines of the World Health Organization (WHO) on essential drugs and has been welcomed and publicized internationally by the WHO Action Programme on Essential Drugs, within Bangladesh itself, WHO's support appears inconsistent and slow to materialize. This is because the public face of the organization in any country is its country representative.

In Bangladesh, one WHO country representative was slow to respond to a request from GK for permission to translate the WHO essential drug list into Bangla. In 1992, the WHO representative even stated that the policy encouraged the smuggling of drugs. There are several possible explanations for WHO's varied support for national drug policies and rational drug use:

- WHO may be reluctant to upset the United States, which provides about 25 percent of WHO's regular budget. The US consistently baulks at supporting WHO policies which could be seen to interfere with the free trade interests of US companies.

- WHO's staff are primarily doctors who are skilled in the scientific diagnosis and treatment of diseases but who rarely show concern for social and environmental factors. Most fail to appreciate that a disease such as tuberculosis is an indicator of social inequality, and that malnutrition and insanitary conditions contribute significantly to the incidence of the most common diseases of the Third World such as diarrhoea or acute respiratory infections.

- Political and economic viewpoints are rarely considered as the organization attempts to maintain an "apolitical and neutral" stance which precludes "interference." Thus colonialism in its past and present forms, all of which have played an important role in the cause and spread of diseases, is ignored. As the British Medical Journal has pointed out, "WHO should be doing more to tackle the root cause of most diseases—poverty—and doing more to improve infrastructure of health care in the developing world."

- The action that WHO has taken on drugs is typical of a technological approach to health with which WHO medical officers feel familiar and comfortable. Doctors know about prescription and usage, efficacy and quality, but they do not know how to get drugs to the people who need them most. They do not realise that the class character and political will of the government determine the affordability and provision of essential drugs.

- WHO is known for its "fixation on medical technology—vaccines, drugs and doctors—[and] its unwillingness to grapple with the practicalities of delivering health care." But disease is not merely the consequence of poor health services; the provision of primary health care alone does not bring better health. To produce documents on drug policy but not to defend them actively or publicly, and to publish materials and then not to distribute them widely, as WHO has done in Bangladesh reflect a political decision *not* to act.

or the Environmental Protection Agency! Moreover, it is rather naive to ignore that Bangladesh is a US aid recipient and that a "hope" expressed by our State Department is perceived as a threat, veiled or unexpressed though it may be.

ANOTHER REVIEW

Despite such support, further pressure had built up on General Ershad in Bangladesh from TNCs, the elite, politicians and newspapers. Claims were made that the NDP would ruin the national drug industry and that doctors' clinical freedom was being infringed. Intellectuals were confused by the sudden decision to implement this kind of policy which was unexpected from a military government.

So General Ershad set up a Review Committee of six army doctors on 6 July 1982. Family members of two of the committee members had high-level jobs with national pharmaceutical companies which had licensing agreements with TNCs.

This Review Committee interviewed representatives of the national drug industry, TNCs, BASS, the Chemists and Druggists Association, the chamber of commerce and the Bangladesh Medical Association. It met separately with the former president of the Bangladesh Medical Association, Professor Firoza Begum, who was a director of Pfizer and owned shares in three other drug companies.

The BMA's statement was submitted to the Review Committee by a BMA former general secretary, Dr Sarwar Ali, who was the Assistant Medical Director of Pfizer (Bangladesh). The BMA said that many of the drugs that had been banned were safe, effective and necessary, and that adverse reactions and side effects were rare, only being caused by inappropriate self-medication. The BMA concluded that the problem lay not in the drugs themselves, but in the unrestricted availability of these drugs without prescription.

The Review Committee submitted its report direct to General Ershad on 12 August—a report which was diametrically opposed to that of the Expert Committee setting out the NDP. Following a suggestion of the Expert Committee, General Ershad brought the two committees face-to-face on 7 September to establish the scientific validity of their reports.

The meeting resembled a courtroom battle umpired by General Ershad. After a five-hour debate, General Ershad was convinced of the scientific validity of the NDP.

However, amendments were made to the NDP. The time allowed for banned drugs in the second category to be reformulated was extended from six months to one year and for drugs in the third category to be used up from nine to 18 months. Six products in the first category were taken off the proscribed list. Some 33 drugs produced under third-party licence were permitted until the expiry of the contracts; 55 other drugs manufactured by 52 small national companies were placed in a new category and allowed to manufacture for two years. After these adjustments, the final list of banned drugs stood at 1,666.

SIMMERING OPPOSITION

Since its inception in 1982, opposition to the National Drug Policy has continued, flaring up every now and then with calls to review the policy.

The drug industry has tried hard to discredit the public health group *Sabar Janya Shasthaya* (Health for All) which has publicized the content and ideas behind the NDP, exposed the misuse of medicines and the reasons for banning certain medicines.

The industry also whipped up attacks against *Gonoshasthaya Kendra* (the People's Health Centre), set up by the author in 1972 and Gonoshasthaya Pharmaceuticals Limited (GPL), established in 1981 to manufacture essential drugs of high quality and at low cost.

Simmering opposition came to a head in 1990. At the beginning of the year, President Ershad outlined to Parliament a National Health Policy. Although the 1982 Expert Committee had recommended that the drugs policy be part of a national health policy, it was only in 1987 that Ershad had appointed a committee to draw it up and in 1988 that it was ready.

As soon as the bill was introduced in Parliament in mid-1990, the Bangladesh Medical Association called for all doctors to strike in protest at the "anti-people health policy." While opposition political parties admitted in private that it was a good policy, they used the opportunity of the strike to discredit the government.

On 27 October 1990, the office, store and vehicles of GPL were again attacked and burned.

Tension and protests escalated after a prominent doctor and high-ranking BMA official was killed in November 1990, and finally the Ershad government fell on 6 December 1990. The National Health Policy was cancelled by the acting President and for a while it seemed as if the drugs policy, also considered "anti-people" by the BMA, would be withdrawn as well.

SAFE FOR NOW . . .

Political stability began to return to Bangladesh following elections in February 1991. Begum Khaleda Zia was sworn as Prime Minister in March and Parliament was convened a month later. On 30 April 1991, a cyclone devastated the southern part of Bangladesh; and in the midst of this disaster, the NDP survived one of its most rigorous tests since the introduction.

A consultative meeting, presided over by the new prime minister, with NGOs and foreign diplomats was held to address the provision of effective relief to cyclone victims. Swiss and US diplomats said that their airplanes were loaded with medicine and baby food, ready to fly into Bangladesh, but that the country's drug policy created an obstacle.

All the ambassadors and other participants were stunned by Khaleda Zia's prompt reply. She said that she greatly appreciated the offer to provide assistance, but that any medicine brought into Bangladesh must conform with the essential drugs list. The policy was safe.

. . . BUT UNDER ATTACK AGAIN

It was, however, a short-lived respite. In the midst of negotiations with the World Bank over loans for industrial development, the new government faced continued pressure from the Bangladesh Medical Association for the withdrawal of the drug policy. The Foreign Investors Chamber of Commerce and Industry (FICCI), led by Managing Director of Pfizer (Bangladesh) Ltd, S.H. Kabir, were also demanding a review of the NDP.

In response to such pressure, the government announced a 15-member Review Committee to revise the drug policy by 30 April 1992. The committee was chaired by the secretary of the Ministry of Health and Family Welfare; other members were medical teachers and from the pharmaceutical industry, the army, and the Chemists and Druggists Association. The deputy secretary of the health ministry, the director of the Drug Administration, the president of the Bangladesh Medical Association and two MPs were also members.

Meanwhile, in April 1992, the World Bank's head of its industry and energy unit in Bangladesh, Abid Hassan, met government officials to discuss an industrial sector loan. He followed up the meeting with a letter to the Joint Secretary of the government's economic relations division, Ayub Quadri, in which he made five specific "recommendations" pertaining to pharmaceuticals—"recommendations" which in the context of a loan negotiation were, in effect, "directives." These were:

- to allow the introduction of new products by using free sales certificates;
- to lift all controls on prices;
- to remove control over advertising from the drug licensing authority;
- to remove restrictions on foreign firms as to what they can produce;
- to abolish controls on the import of pharmaceutical raw materials.

At about the same time, the FICCI presented the Health Minister with a set of proposed amendments to the NDP which simply echoed the recommendations of the World Bank, with one small exception. The FICCI was not concerned about the removal of controls over advertising as drug companies were already printing and distributing product showcards giving misleading or untrue information.

Concern over these developments led the WHO and UNICEF country representatives to write to the head of the World Bank in Bangladesh, Christopher Willoughby, calling on

Achievements of the National Drug Policy

Substantial gains have been derived from Bangladesh's National Drug Policy which are evident when prices, production figures and quality indicators at the time when the policy was introduced are compared with those of a decade later. Between 1982 and 1992:

- essential drugs increased from 30 to 80 per cent of local production;

- drug prices stabilized, increasing by only 20 per cent, compared to an increase of 179 per cent in the consumer price index;

- Bangladesh companies increased their share of local production from 35 to more than 60 per cent. Overall, local production increased by 217 per cent.

- less dependency on imports and prioritization of useful products saved the country approximately US$600 million;

- the quality of products improved: the proportion of drugs tested which were found to be substandard fell from 36 per cent to under 7 per cent.

Availability of essential drugs increased dramatically with the increase in national production, which led to a decrease in imported drugs. Eighty per cent of locally-produced drugs are for primary and secondary health care.

The procurement of raw materials at the most competitive prices led to a sharp drop in such prices and, in turn, a fall in the maximum retail price of finished drugs. The drop in price in real terms made drugs more affordable for consumers. The smallest price decrease was 12 per cent, the largest 97 per cent. The highest price increase was for aspirin which went up from Taka 0.10 to Taka 0.55. As drug prices are low, the incidence of counterfeiting is also low. High-priced drugs, whether imported or produced locally, are most subject to counterfeiting.

The NDP eliminated most transfer pricing and over-invoicing for imports of capital machinery, raw materials and packaging materials, which were common practices before 1982.

The quality of locally-manufactured drugs improved through greater vigilance; testing became easier because, according to the Expert Committee's criteria for drug assessment, most drugs contain only one active ingredient. In 1981, 327 products were tested of which 36 per cent were substandard; in 1992, more than 2,600 samples were taken with less than seven per cent found to be of substandard quality. Substandard drugs are produced mainly by small companies; if all the companies producing substandard drugs are taken together, they account for about one per cent of the market.

The number of drugs per prescription issued by doctors has decreased from five or six drugs per prescription in 1982 to between two and four in 1990. This was due in part to the elimination from the market of many useless and combination products and to increased awareness among doctors of the need to prescribe more wisely. Such awareness was stimulated by public discussions about the National Drug Policy.

Despite these achievements, misbranded and spurious drugs, mostly smuggled from India, banned under the national drug policy continue to be prescribed by unqualified village practitioners who are often the only health providers in rural areas.

the World Bank to ensure that gains secured by the NDP were not undermined, in particular "low-cost and price-controlled basic essential drugs, affordable by the majority of the population, ensuring cost-effective process in primary health care."

Abid Hasan was ordered to send another letter to Ayub Quadri, clarifying the World Bank's position on the NDP. His cleverly-written letter of 8 June 1992 began with the statement, "First and most important, we fully support the drug policy objective of increasing the supply of essential drugs (EDs) and making these available at affordable prices."

But it went on to argue that the present controls could be replaced by more liberal policies or procedures "without sacrificing the objective of increasing availability of good quality and affordable EDs." The five recommendations were repeated, albeit in a modified form, and the final paragraph contained a veiled threat:

> We would appreciate if these views are brought to the attention of the drug policy review committee urgently, specially since one aspect (import controls) of the above is germane to [Industrial Sector Adjustment Credit-II] negotiations.

A NEW DRUG POLICY?

Amidst confusion and disagreement among the Review Committee, its term was extended by six months to 31 October 1992. Just two days before that date, the drug review sub-committee agreed on a draft National Drug Policy 1992, which amounted to little more than a camouflaged and elaborate version of the recommendations made by the World Bank representative, Abid Hasan.

The new policy provided for the creation of a Drug Registration Advisory Committee (DRAC) in place of the Drug Control Committee (DCC), comprising experts from medicine, pharmacology and pharmacy, and representatives from the manufacturers and other trades and professional groups. The term "professional groups" effectively excluded consumer groups but would open the door to groups directly or indirectly related to the industry.

The DRAC would be authorized to approve the safety, efficacy and quality of drugs and to determine patterns of disease prevalence and therapeutic need.

The sub-committee prepared a list of over-the-counter drugs to be sold without prescription for the short-term relief of symptoms considered not to require medical advice or accurate diagnosis.

Other recommendations were to encourage foreign companies to set up manufacturing plants in Bangladesh to produce "research-based" new drugs; to remove the requirement of prior approval for imports of raw and packaging materials; manufacturers to set each drug's Maximum Retail Price; and separate administration of traditional medicines "which should not be amalgamated with [pharmaceutical] drugs at any level, be it manufacturers or dispensaries."

Before this version of the NDP could get off the ground, in November 1992, the association of the traditional medical systems sued the Ministry of Health and Family Welfare, claiming that the Review Committee was invalid as it did not include any traditional practitioners. The court issued a stay order, preventing the government from giving any further consideration to the report.

A LULL IN THE STORM

For almost a year, it appeared that the main elements of the original NDP would survive. But in 1993, another major threat came, this time from Bangladesh's finance minister, M Saifur Rahman, who believed that the NDP was a major impediment to foreign investment in Bangladesh. He said publicly that the National Drug Policy should be "immediately scrapped" because it was "ruining the blue-chip pharmaceutical industry," adding that the industry should be free of all controls.

About a week later, the health minister said that the government might review the drug policy, a comment which was sufficient warning for WHO and UNICEF representatives to go back to the head of the World Bank in Bangladesh, Christopher Willoughby.

The trio met with the finance minister and the health minister the following week. Finance Minister Saiful Rahman was visibly angry and asked whether he should comply with Abid Hasan's 1992 directive or theirs. Willoughby was forced to admit that Abid Hasan's letter had been sent without proper clearance.

ERODING THE NDP

Finance Minister Saifur Rahman thus changed his strategy to defeat the NDP. Instead of attempting to abandon its basic principle of a limited list of essential drugs, he encouraged the Drug Control Committee (DCC) to take a number of drugs off the banned list at each of its meetings.

The DCC had been reconstituted in early 1992 with an abundance of new members representing the drug industry. At its first meeting in March 1992, a nonessential combination cough rub, Vaporub, was approved, along with 12 single-ingredient products of no proven superiority over existing ones. Three of the drugs were not recorded in the British Pharmacopoeia, the US Pharmacopoeia or *Martindale: The Extra Pharmacopoeia.*

In late 1992, a DCC meeting attended mainly by members from the pharmaceutical industry, and with barely the minimum number of committee members present, permitted combination products such as antacids with Simethicone, multivitamins with minerals and vitamin B-complex syrup, with had been deemed since 1982 to be of little therapeutic value, to be sold. This violated the principles of the national drug policy. Since then, the number of new products registered for manufacture has grown rapidly.

On 12 January 1994, the DCC agreed to several price deregulation measures. Rather than removing all price control, which would provoke reactions from consumers and professionals, a list of 117 drugs which would be subject to prior control was drawn up. These were referred to as "listed drugs" to avoid definition problems associated with the concept of essential drugs and to bypass High Court restrictions on amending the essential drugs list. The list was an arbitrary one and the criteria for inclusion on it were not defined by the committee. High-selling common drugs as well as expensive newer drugs were not included.

Price control of imported raw and packaging materials was to be discontinued, and manufacturers of "unlisted" drugs were free to fix their own price.

An extensive press campaign by Health For All (HFA) managed to delay the implementation of these measures until June 1994. As expected, the prices of drugs which are not "listed" have started to shoot up.

UNWRITTEN CHAPTER

The Bangladesh National Drug Policy has been both praised and attacked since its inception in 1982. It has survived several onslaughts from vested interests at home and abroad which could have destroyed it completely.

Consumer organizations, health activists and a few journalists continue to struggle to retain the benefits achieved through the drug policy, while TNCs and local producers alike, backed by sections of the World Bank, press for further deregulation and liberalization.

Bangladesh's National Drug Policy has been far from perfect. Virtually every year since the policy was adopted, there has been prominent media coverage in Bangladesh of substandard, counterfeit and spurious drugs, the sale of expired drugs, the smuggling of banned drugs, the preponderance of unauthorized retail pharmacies—and inadequate provision to control these violations. Quality assurance, transparency concerning drug registration and misleading drug promotion are current controversial topics. But not one popular newspaper has called for the withdrawal or suspension of the 1982 policy.

In summarizing the Bangladesh National Drug Policy in 1992, pharmacologist Milton Silverman—the author of several books on medicine use in developing countries—and his colleagues wrote that "the end of the chapter is yet to be written."

The ending envisaged in 1982 when the National Drug Policy was drafted was one which puts people's health first, which delivers essential drugs, which encourages a strong national pharmaceutical sector that operates within reasonable regulations to ensure public health concerns are foremost. The struggle to achieve that goal continues.

FURTHER READING

"Making National Drug Policies a Development Priority: A Strategy Paper and Six Country Stories," *Development Dialogue,* 1995:1. Dag Hammarskjöld Foundation. Övre Slottsgatan 2, S-75310 Uppsala, SWEDEN.

59

The Resurgence of Infectious Diseases

ANNE PLATT

More than half a century after the discovery of antibiotics, infectious diseases are on the rise—with the most dangerous ones now carried by some 2 billion people. In the coming years the epidemics are only likely to get worse—unless public health considerations are brought into the future planning of all public works.

In May 1993, a physically fit 20-year-old Navajo Indian—a cross-country and track star—began gasping for air while driving to his wife's funeral near Gallup, New Mexico. For several hours, the man suffered from what seemed to be a severe but otherwise unremarkable case of the flu. Then, abruptly, his condition worsened. Blood filled his lungs. He was taken to an emergency room, where he died—drowning in his own serum. Reports confirm that around the same time, three other healthy Navajos in the Four Corners area (where New Mexico, Arizona, Colorado, and Utah meet) died from cases of flu or pneumonia gone suddenly awry. Clearly, something horrific was on the loose—but what?

Medical authorities from the state of New Mexico, the Indian Health Department, and the U.S. Centers for Disease Control and Prevention questioned families, relatives and friends, pored over medical records, and investigated possible links among the victims. They collected blood samples to analyze for viruses. Newspaper headlines warned cryptically of "Navajo flu," and "the mystery illness." As the death toll mounted, people in the affected region panicked, and some began to avoid anyone who was sick, for fear of contamination.

Investigation found that ground zero was a mouse—a rural deermouse, *Peromyscus maniculatus,* which is a native of most of North America, including the American southwest. Apparently, the deermouse harbored a strain of the Hanta virus (named after the Hanta River in Korea where it was originally discovered), which causes severe damage to the pulmonary tract and lungs. It was not immediately clear, however, why this normally reclusive animal had suddenly begun to appear—and leave its droppings—in kitchens and playgrounds.

The press treated the Hanta outbreak of 1993 as an anomaly, a disturbing but isolated incident. In fact, however, it was part of a larger pattern that involves a growing list of illnesses—and growing risks to hundreds of millions of people. At the end of a century in which infectious diseases were thought to be well controlled, disease-causing (pathogenic) microorganisms are breaking out all over the world. Some of these pathogens, such as the *Escherichia coli* O157:H7 bacterium, Hepatitis C, and Rift Valley Fever are new and unfamiliar. Others are old ones we thought had been beaten, such as the microbes that cause tuberculosis, malaria, the plague, and measles.

AN EPIDEMIC OF EPIDEMICS

Despite the fact that most such afflictions are curable—and despite the major advances that have been made in sanitation, medical care, and increased public awareness of diseases and health in this century—infectious diseases still kill more people than cancer, or car accidents, or war. They kill more than 16.5 million people each year, and global incidence is on the rise.

Worldwide, 3.3 million people are killed by tuberculosis each year, while another 2 million die from malaria, predominantly in tropical regions. And for every person who dies, more than one hundred are infected. One-third of the world's people—some 1.8 billion—now carry the tubercle bacillus, the bacterium that causes tuberculosis. More than 500 million people are infected with tropical diseases such as malaria, sleeping sickness, river blindness and schistosomiasis.

As the 20th century approaches its end, changes of unprecedented magnitude and speed are taking place in the planet's physical and social environment, with the cumulative effect of allowing infections to spread far faster than anyone has been able to spread the means of preventing and treating them. The killing of forests, contamination of water, destabilization of climate, and explosion of urban population have all con-tributed to the weakening of public health protections. As a result, transmissions of infectious diseases through all media—air, water, insects, rats, and the human body itself—are on the rise.

Haunting these changes is an omnipresent biological pattern. As evolutionary biologists E. O. Wilson and others have observed, rapid disruption inevitably seems to favor some life forms over others.

This is a pattern seen at all levels, from microbes to mammals and from algae to trees. When a forest is burned, opportunistic, short-lived species spring up ahead of stable, long-term ones. When a building is bulldozed to an empty lot, weeds spring up before the return (if ever) of whatever species were there before. When coastal wetlands are disrupted to make room for resort hotels, algal blooms spread over the fragmented wetlands and choke them.

Among microbes, as among larger life forms, there are opportunistic varieties: bacteria or viruses, that invade human blood or cells. Just as weeds exploit disturbance more quickly than slower-growing trees of a stable forest ecosystem, or insect pests exploit the reduced biodiversity of a crop, infectious agents can adapt fast enough to overwhelm societies whose "natural" environments are disturbed.

An undisturbed ecosystem imposes a set of checks on the growth of microbes, but during severe disruptions, the balance may be skewed in their favor. The more disruption there is in the human habitat, the bigger the biological risks are for the people.

BIOLOGICAL MIXING

The environment in which the human species evolved and developed its basic defenses against disease was one that remained—despite natural disturbances—basically stable for thousands of years. In the past 100 years, it has changed radically—in forest cover, air and water quality, diet, and most recently in patterns of weather. The changes have boosted the spread of microbes, thereby increasing human vulnerability to infection.

The planet has become not only more vulnerable, but also—in effect—smaller and easier for small organisms to move around on. Surging growth in global tourism, migration, and trade has done the trick; distances that were once rarely covered within the lifespan of a microbe are now covered routinely and easily, not just thanks to increased air travel but thanks to new roads cutting through wilderness areas. Microbes, which can hitch rides on anything from the boots of travelers to the wheel wells of planes, are extending their reach and coming into contact with more people.

One result is that viruses which previously remained hidden in remote rainforests suddenly have access to large human populations. Rapid settlement of the Amazon basin, for example, contributed to the spread of malaria-carrying mosquitos. The paving of the Kinshasa Highway across central Africa gave a fateful boost to the outbreak of AIDS. And the sheer volume of traffic between different ecosystems has brought about a process of planetwide biological mixing. This mixing goes in both directions: mobility brings microbes to human populations, but human expansion also carries diseases into new areas. The bestselling book *The Hot Zone,* by Richard Preston, recounts how the deadly Ebola viruses were released from a remote African rainforest, as though the rainforest were a kind of Pandora's Box. According to Princeton University ecologist Andrew Dobson, however, what more often happens is that humans bring new vectors into the rainforest.

One reason the spread has been hard to stem is that these microbes are invisible stowaways, carried unwittingly by their hosts—and we have no way of knowing where the next outbreak will occur. Some changes in the environment may reduce the risk of disease transmission, while other factors will increase it. Changing agricultural practices can create new jobs and increase crop output, for example, but the changes may also invite new species to colonize. In any alteration of the landscape, there is potential for unforeseen consequences. After the Indira Gandhi Canal was built in Rajasthan, to irrigate

desert-like areas of India, farmers switched from cultivating traditional crops of jowar and bajra to more commercially profitable wheat and cotton, which require large amounts of water. Large numbers of people came to the area in search of work. Then the monsoons came.

As it turned out, the main canal—445 kilometers long, from Masitanwali to Ramgarh—served as an ideal breeding site for mosquitos. Instead of high crop productivity and prosperity, the heavy rains brought the farmers tragedy and death in the form of cerebral malaria, a parasite which is carried by *Anopheles* mosquitos. Excess rain, combined with water logging behind the canal and inadequate drainage in the fields, created an epidemic of malaria that quickly spread through extensive canal areas.

"The ignition wire of construction-related stagnant water, and the gunpowder of immigrant labor, created an explosion of malaria," reads a World Bank-commissioned report on India's Sardar Sarovar Dam Project. Malaria and waterborne diseases are common during the monsoon season, but the canals carried the malaria epidemic to a much larger area, exposing workers and farmers who then transmitted the disease to friends and families.

The dangers of irrigation projects are no worse than those of hydroelectric projects, road-building, logging, mining, and even urbanization. By displacing wild populations and habitats, development activities such as logging often deprive microbes of their usual hosts—in effect forcing them to find new ones. Essentially, this is what happened in 1975 with the Borrelia bacterium that causes a relapsing fever known as Lyme disease. The bacterium lives in four different tick species which spread it in their saliva—usually to deer and raccoons. But suburban development in Old Lyme, Connecticut, tended to substitute people for these other mammals in the disease cycle. Borrelia bacterium has since spread throughout the United States, and Lyme disease is rampant in New England and the upper midwest states of Wisconsin and Minnesota.

Along with the 100 to 200 microbes known to be dangerous, another 1,000 or more may be "out there," according to Paul Ewald, a biologist at Amherst College in Massachusetts. "It's a lottery" as to whether or not a pathogen will be introduced into the population. Understanding the various links between environmental disruption, microbial outbreaks, and health may be nearly impossible, but recognizing some general patterns will allow for better prediction and disease prevention. Four syndromes—water contamination, climate change, human actions that magnify natural disasters, and increasing social disruption—have been associated with such "unpredictable" occurrences as the Hanta and Gandhi Canal incidents.

BAD WATER

Many of the outbreaks have been linked to the degradation of natural systems, particularly of water. Infectious illnesses are widespread in areas with overburdened sanitation facilities and unsafe drinking water. Even in regions where water quality is considered fairly good, that impression is belied by the high levels of human waste and sewage carried in prominent rivers—the Danube, the Volga, and the Mekong, among them—from which disease-carrying bacteria and viruses make their way into drinking supplies.

Waterborne diseases—including diarrheal diseases caused by *E coli,* salmonella, vibrio cholera, and viral diseases such as hepatitis A and dysentery—cause hundreds of times more illness worldwide than chemical contamination of drinking water does. In 1992, 68 countries reported 461,783 cases of cholera and 8,072 deaths. In Russia, the very rivers that Russians depend on for life—the Volga, the Dvina, the Ob—are now hazardous to public health. The rivers harbor strains of cholera, typhoid, dysentery and viral hepatitis that spread through water systems and contaminate drinking supplies.

"Epidemic diarrheal diseases are both preventable and curable," says Dr. Ronald Waldman, former coordinator of the World Health Organization's Global Task Force on Cholera Control. "With a rapid and effective response,

case-fatality rates from cholera, for example, can be kept to less than 1 percent. As health-care professionals, we cannot allow a lack of preparedness to be responsible for an unnecessary loss of life."

One method for predicting cholera may be detecting and monitoring plankton blooms. Rita Colwell, the University of Maryland biologist who is now President of the American Association for the Advancement of Science and Dr. Paul Epstein, of the Harvard School of Public Health, have studied the bacteria and viruses that catch a ride with plankton, and (with others) have shown that where there are blooms, there are corresponding outbreaks of diseases in humans, marine mammals and fish. In the late 1980s, toxic algal blooms and morbillivirus were associated with dolphin die-offs in the Mediterranean Sea.

Cholera, which comes from the Spanish words for anger, cólera, erupts seasonally, when the temperature, sunlight, nutrient levels, and acidity are right for it. (At other times of year, the bacteria become dormant and hibernate with their plankton hosts.) During blooms, and in areas that have a history of waterborne illness, medical providers can curtail outbreaks by teaching residents how to diagnose cholera and how to avoid infection—by boiling water for drinking and washing food thoroughly.

In 1905, a new type of cholera was identified in the corpses of Moslem pilgrims who died at the El Tor quarantine camp en route to Mecca. The El Tor strain was more virulent than the classic strain of cholera because it proved to be a hardy survivor outside the human body, and the infection lasted longer in victims. By 1982, El Tor was the dominant strain in Bangladesh.

Even where there is a well-trained medical community, however, a seemingly minor event can sometimes trigger an epidemic. In 1991, for example, bilge water from a Chinese freighter was responsible for releasing this Asian strain of cholera into Peruvian waters. Once the bacterium of *Vibrio cholerae O1,* biotype El Tor, was released, it quickly spread through the marine environment and into drinking water supplies where it infected people. The bacterium also infected fish, mollusks, and crustaceans—which heightened the public health threat, because *ceviche,* raw fish with lemon juice, is an important local food. By 1993, there were more than 500,000 cases of cholera throughout Latin America, with 200,000 in Peru alone.

Given worldwide population growth and increasing pressures on scarce water supplies in many areas, the incidence of waterborne infections can be expected to keep rising, unless water quality protection is made a top priority in the planning and management of all water-using activities—from irrigation and hydropower to the disposal of sewage.

BAD WEATHER

Second, many outbreaks seem to be related to ongoing and incipient changes in climate. Warmer weather can expand the range of vectors: in July 1994, for example, 24 cases of malaria were reported in Houston, Texas, where warm weather had attracted malaria-infected mosquitos from Latin America.

Such conditions are likely to worsen before they improve. The Netherlands-based institute, Research for Man and the Environment (RIVM), reported last year that a global mean temperature increase of three degrees Celsius in the year 2100 increases the epidemic potential of mosquito populations in tropical regions two-fold and in temperate regions more than ten-fold. The RIVM model estimates an increase in malaria cases of several millions in the year 2100. And RIVM calculates that more than 1 million people could die each year as a result of "the impact of a human-induced climate change on malaria transmission" during the next 60 years—as many people as were killed in both World Wars. Yet, that would be the toll of just one disease among many. To address these threats, governments need to begin tracking the broad connections between their energy and transportation policies. Because these policies affect the rate of climate change, they can profoundly affect the long-term risks to public health.

HUMAN AMPLIFICATION

A third syndrome is that infectious outbreaks appear to follow on the heels of human activity that in some way magnifies the effects of natural disturbances—whether of floods, storms, or earthquakes. For example, some experts believe the outbreak of plague in Surat, India, in September 1994 was connected to the flooding of the Tapti River that summer, and an earthquake a year earlier. The quake had left the landscape devastated by ruin, and thousands of people homeless. Emergency aid and medical supplies were flown in for the survivors, but the effort was so successful that excess food had to be stored in warehouses, where rodents crawled in and feasted. The rodents reproduced quickly, allowing the pneumonic plague bacterium—harbored in the fleas that infest the rodents' fur—to greatly extend its range.

During the summer, the monsoon flooded the Tapti River and inundated the poorest districts of Surat with three meters of water. Again, people were forced to leave their homes. The rodents, too, were forced to seek shelter on drier land. Rats and people crowded together on the same high ground, increasing people's exposure to the plague bacterium. Although India was medically prepared to deal with water-borne diseases such as gastroenteritis, cholera and dengue fever, it had no plans for plague. The disease had not been seen in more than 40 years.

When the first cases were reported, more than one-third of the Surat's private physicians left town. Panic erupted as plague cases proliferated, overwhelming the medical system with fearful patients. Video footage showed victims being carted off in wheelbarrows, while families packed onto trains and roads to flee. All over the world, news reports warned travelers to stay away from India.

At Surat, a combination of weather patterns and earthquake and flood damage was exacerbated by social factors: shantytowns, squalid living conditions, warehoused food, inadequate health care, and panic fanned into hysteria by media coverage that focused on the disaster without explaining its causes. What could have been prevented or controlled at an early stage became a financial as well as social disaster, as international airline flights to and from India were canceled and trade came temporarily to a halt.

SOCIAL DISRUPTION

Finally, the vulnerabilities brought about by increasing environmental disruption and exposure are, in many ways, further amplified by the world's growing problem of internecine social disruption. With about 30 civil wars now taking place in the world, the systems needed for prevention and treatment of disease have been repeatedly shattered—often opening the way for infections to spread unchecked.

In Russia, health conditions are worsening because of the combination of unstable political conditions, deteriorating infrastructure, and the transitional economy. According to the head of the Russian Academy of Medical Sciences, as quoted in the 1992 *Russian State Report on the State of Health of the Population of the Russian Federation,* "we [Russia] have already doomed ourselves for the next 25 years."

Poor hygiene and diet, compounded by inadequate food supplies and high levels of pollution have brought an onslaught of ecological and human health problems to this part of the world. In the early months of 1994, Russians were hit by 22 percent more cases of tuberculosis than in the same months of 1993. Measles rose by 260 percent and mumps by 10 percent during the same period. In 1976, diphtheria had all but disappeared from the former Soviet Union thanks to childhood immunizations. But it surged back in the 1990s, rising from 1,200 cases in 1990 to 15,210 in 1993. The disease had come back everywhere, including Moscow and St. Petersburg.

Although it is a half-day's plane ride from Moscow, Russia's Pacific coast of Khabarovsk is not immune to the wave of infectious diseases that have swept across Russia since the collapse of the Soviet Union. People travelling to this mountainous coastal region have brought diphtheria with them. Meanwhile the mass exodus of refugees from southern regions of Russia has made it nearly

impossible to stop the transmission of diseases that move directly from person to person.

Last summer, after recording nearly 1,000 cases of cholera, health officials in Dagestan, approximately 120 kilometers west of the Caspian Sea, imposed a quarantine to stop people from leaving the area. The quarantine was not successful at controlling the epidemic, since by then most people in the area were harboring the cholera bacterium whether they showed signs of it or not. In nearby Chechnya, all attempts at public health control have failed because of the war.

What has befallen Russia is echoed, in varying degrees, in India, in Latin America, and in the Four Corners area of the American Southwest. The paradigm is relentless: disrupted environments increase biological stresses on humans everywhere; mobility and population expansion increase their exposure to opportunistic microbes; and political or economic disruption prevent the application of known preventions or cures.

WRITING THE PRESCRIPTION

How this cycle can be broken is not something that can be determined by medical research alone. Enough research has already been done to yield the knowledge needed to control many of the epidemics spreading around the world: tuberculosis and malaria, for example. The problem is that the knowledge is not widely applied, and the diseases continue to spread.

Stopping the world's growing "epidemic of epidemics" may not be possible, in fact, until considerations of human health are integrated into all major human activities—including the planning of irrigation and dam projects, road building and transportation systems, agricultural practices, and extractive industries such as mining and logging.

In the future, along with keeping ecosystems intact and minimizing habitat alterations, communities should require planners to prepare for the unanticipated consequences of development. And they would do well to provide ongoing health education for their populations, especially in areas that are particularly vulnerable to environmental disruption.

Since infections do not stop at the borders of communities and countries, neither should control efforts. Individual nations need to coordinate with the World Health Organization and with each other to establish a reliable global surveillance system which would provide early warning, monitor incidence, and coordinate the response. Adequate medical supplies and complete treatment therapies can then be targeted to at-risk populations. Individual communities will need to adapt these programs to local conditions. In China, for example, the government set up a program to control schistosomiasis 40 years ago. At the national level, public health, water resources, agriculture, planning, and finance officials draft laws, plan programs and monitor progress. But it is up to leaders at the county and city levels—who know the local weather and terrain—to find and exterminate snails, educate their people, and treat the afflicted. The program is working.

When we begin routinely to take health impacts of our industries and societies into account, the outbreaks of disease that now shock us won't seem so puzzling. When the Hanta virus broke out in Arizona, for example, it was not as mystifying to some of the Navajo medicine men, whose traditions had taught them to see the interconnectedness of all living things, as it was to the medical specialists and the media.

The scientists, alerted that deer mice were carrying the problem, tested the mice and were able to identify the exact culprit. But they failed to notice that it was the environment that was changing, not the pathogens.

The Navajo medicine men meanwhile observed that prior to the outbreak, snow melt cascading down to the valley desert below, combined with a spring of heavy rains, had reminded some of their elders of the years 1918 and 1933, when there had been similarly unpredictable weather. In each of those years, there had been a disease. In each of those years, pinon trees produced an abundance of pine nuts. Mice had descended on the extraordinary harvest and reproduced ten-fold in one season. The rains had then forced the mice out of their flooded burrows to scurry about above ground, looking for food and shelter and increasing their exposure to humans. Disease was what happened when the balance of life was upset.

"When there is disharmony in the world, death follows," said one of the medicine men. He understood that when strange symptoms appear, they are not anomalies.

60

Sense and Sanitation

SHEELA PATEL

Bombay is the financial capital of India with some of the highest property values in the world. Half of its ten million people pay incredible prices for homes. The other half live in informal settlements; more than a million of them on pavements in makeshift structures of bamboo, plastic, cloth, wood and tin. These people pay a high price too, though the currency they hand over is not rupees but their own health, living as they often do without water or sanitation of any kind.

Yet the Bombay Municipal Corporation (BMC) is ambivalent: "We can't give toilets to slum dwellers; this will encourage people to migrate to the city!" they say, or: "The slums along the highway should have sanitation so foreign visitors don't have to see this embarrassing sight of squatting people with umbrellas along the road." As if people were flocking to Bombay to enjoy the luxury of public toilet blocks, or the psychological comfort of tourists should be the primary motivation of municipal sanitation programmes!

Or else they say: "Don't the poor deserve the same as everyone else—an individual toilet?" or: "Since poor communities don't maintain public toilets, let's give them toilets in their own home so they will be forced to keep things clean."

Everyone, it seems, has an opinion about how to solve the problem of sanitation in informal settlements. But what do the residents who live there, the people on the footpaths and in the slums, believe is a workable solution?

In 1984, I and 12 other people formed the Society for the Promotion of Area Resource Centres (SPARC). We sought to create an organization which would make space for poor communities to focus on issues which concern them, to understand why they face certain problems and then to reflect on the solutions. Over the last ten years, through our alliance with Mahila Milan—a national network of women's collectives—and the National Slum Dwellers Federation (NSDF), we have used this approach to address many issues, including land tenure, shelter, employment and credit.

The way women living on pavements in the Byculla area of Bombay formulated their opinion on toilets illustrates our approach. We visited slums with and without public toilets, and the few government-constructed tenement blocks in which each dwelling has an individual toilet. Through these site visits and numerous discussions the women arrived at an assessment of the status quo.

Less than half of Bombay is linked to sewers. In most slums the residents either defecate in the open or—in the few locations where they exist—use community latrines. Municipal maintenance is infrequent and poor, and the number of users

per toilet is far too high. The toilets are dirty, uncared-for, overflowing and often unusable.

In slums without toilets people created makeshift arrangements which emphasized privacy, but not the disposal of feces. In slums with toilets the number of users could be as high as 500 people per seat. Little children never got a chance to use the facilities when adult men were lined up waiting. In the government-constructed, multi-storey tenements, women were very unhappy to have an individual toilet inside their homes. In all the areas visited women had taken the drastic step of blocking it up. Many slums have low water pressure; toilets begin to stink, and since the tenement is only one room a dirty toilet next to the cooking area presents a serious health hazard. "If we have to cope with a dirty toilet," the women said, "it is better that it is outside the house—we have other uses for that space."

Having completed the rounds of other informal settlements, SPARC, Mahila Milan and NSDF began to develop their own views. They agreed a preference for community toilets with a ratio of one toilet for every 25 people—a toilet block of four or five seats could be shared by 20–25 households who would jointly manage them. The blocks would include separate seats for men and women, an outside open channel over which the children could squat and a flushing mechanism which would draw their waste into the main collection pit.

When we enter into dialogue with the authorities this is now the basic formula we present to them as the people's solution. It is not perfect. It is not ideal. And it is not permanent. But it represents a pragmatic solution which will make basic sanitation available to all the poor people in the city and establish a partnership between city authorities and communities.

The collective, hard-earned experience of SPARC, Mahila Milan and NSDF suggests that proactive dialogue must be properly prepared for by the participants. Each group must make a substantial investment so that it can come to the negotiating table with a clear sense of what is important to them and what is not, what contribution they can make towards the solution and what concessions would be acceptable.

In 1989–90 we surveyed slums in ten cities. We helped local city federations identify a core team of community leaders (men and women) who visited all their informal settlements. Almost invariably sanitation was identified as one of the most persistent and serious problems they faced. For example, in Kanpur, a city in the state of Uttar Pradesh in the north of India, the slum dwellers surveyed their area and found they needed 500 toilets. They suggested the municipal officials study their proposal and, if it were acceptable, construct a number of toilet blocks which the communities would then maintain.

We began training slum dwellers in other cities to organize themselves and to enter into dialogue with the municipal administrations. We also ensured that federations were able to visit each other's settlements to gain ideas and confidence. These types of support provided both capacity-building experience and tangible evidence—assets which helped their participation in the city's decision-making process.

We are not participating in a project in Bombay which will provide 20,000 toilet seats for one million people living in the city's slums. According to our data collection there are at present just 3,000 toilet seats for these people; 80 per cent of the toilets are not fully operational and need to be torn down or repaired. Negotiations to explore how communities can be assisted to take on construction, maintenance and management of toilets are in progress.

We and the communities with which we work have come a long way in the ten years since women pavement dwellers first began to discuss the problem of toilets in their ideal settlement. As more and more communities, women and city officials cooperate they become more capable of refining the solution, adding new dimensions and adapting to different contexts. This improves the material condition of people living in informal urban settlements. But, more importantly, it is a process of empowerment and involvement. Once people starting talking about toilets other things follow.

13

Prospects for the Future

For every complex problem, there is an easy answer. And it is wrong.

H.L. MENCKEN

Previous chapters have introduced the reader to a variety of global problems and to a variety of efforts under way to respond to these problems. The purpose of this final chapter is to take a more in-depth look at the nature of efforts to solve the world's social problems, the challenges they face, and their prospects for success. In this chapter, as in earlier chapters, we will pay special attention to the role of "bottom up" movements. These movements are increasingly influential on the world scene, in many ways setting the agenda for discussion and debate on world issues.

What is the relationship between "bottom up" movements and the "top down" strategies of corporations and governments, as they work to find solutions to global problems? As discussed by Marvin E. Olsen in Chapter 1, this relationship may be one of "confrontation" (conflict) or "involvement" (cooperation). From the order perspective, we might expect that the relationship will often be one of cooperation. The first reading in this chapter, "The Civil Society Sector" by Lester M. Salamon and Helmut K. Anheier, supports this idea. The authors discuss their research into the scope, structure, and financing of the multitude of "nongovernmental" organizations that have arisen in the twentieth century. This study is part of a discussion that has flourished in recent years over the nature of "*civil society*"—that is, all of the various organizations and institutions in society that lie outside of the government and capitalist structures. Salamon and Anheier argue that in many cases the relationship between nongovernmental organizations, government, and

business is conflictual, but that in just as many cases, the relationship is one of "interdependence and mutual support." A noteworthy finding of their research is that governments provide a large percentage of the funding of nongovernmental organizations around the world.

From the conflict perspective, the solutions to social problems will in large part be determined by the outcome of conflict between the "bottom up" and "top down" actors in the world scene. The world system perspective suggests that many of the problems examined in previous chapters can be seen as social-structural problems within the world capitalist economy and the competitive inter-state system. In this perspective, "top down" approaches to solving social problems often fall short, or create new sets of problems. This is because corporations, government elites, and wealthy classes typically stand to benefit in myriad ways from the perpetuation of social problems. This perspective characterizes "bottom up" movements as *antisystemic movements* that have arisen in response to the structural problems of the world system. Although these movements have great diversity, they have in common that they seek to transform the inegalitarian nature of the system, and to increase the material well-being, power, and esteem of those who are disadvantaged under the system (Arrighi, Hopkins, and Wallerstein, 1989). Such antisystemic movements abound in the world today—movements of women, racial and ethnic minorities, indigenous peoples, consumers, workers, gays/lesbians; movements for peace, environmental preservation, human rights, democracy, children's rights—and these movements are growing. Even the labor movement, often depicted as declining, is growing, if we look at the movement on a worldwide basis, rather than in individual regions such as the United States, where the labor movement has suffered serious setbacks (Silver, 1996). Torry D. Dickinson provides insight into the nature of antisystemic movements from the world system perspective in "Global Women's Movements: A World System Perspective." Dickinson points out that the expansion of the world economy is negatively affecting women in many places, leading to the growth of movements. However, the world economy is also developing unevenly, creating a different set of challenges for women in different regions. As a result, the women's movements around the world have developed a variety of different goals and strategies.

A major challenge faced by movements aimed at solving social problems is the mobilization of resources toward their goals. "Bottom up" movements in particular typically lack a wealth of resources—money, advanced technology, access to the mass media—with which to organize effectively. Their prospects for success depend on maximizing other assets—strength in numbers, a unifying ideology, a sense of purpose. As we saw in Chapter 3, "Democracy and Human Rights," many "bottom up" movements around the world today are being unified and energized by the concepts of *human rights* and *democracy*. For example, the effectiveness of women's movements worldwide, despite their diversity, has clearly been enhanced by the emergence of the principle of "women's rights as human rights" (Lederman and Chow, 1996). The principle of democracy has also unified and energized movements in China, Hong Kong, and Taiwan, and in the United States (So, 1992; Marable, 1997).

Another challenge is that of determining where to intervene in the world system in an effort to change it. Movements may be more or less effective, depending on the goals they set. In "Places to Intervene in a System," Donella H. Meadows discusses various approaches to change. She argues that the most effective approaches are to work on changing the *paradigm*, or the fundamental ideas and mind-set of a system, and the fundamental *organization* (structure) of the system. The ideas of human rights and democracy may be seen as efforts to change the paradigm of the mod-

ern world system. Antisystemic movements may be seen as efforts to change the basic structure as well as the paradigm of the world system. The reader might go back to some of the movements discussed in previous chapters and reflect on how they fit into Meadows' analysis: Which involve a paradigm shift? Which seek to restructure the system? Which seem to have the more limited goals that are lower down on Meadows' list of priorities? Frank Lindenfeld analyzes the advantages and disadvantages of a limited approach to change in "Possibilities and Limits of U.S. Microenterprise Development for Creating Good Jobs and Increasing the Incomes of the Poor."

The chapter continues with a hopeful statement about a challenge that has haunted the "bottom up" movements in the modern world. This is the issue of *oligarchy*, analyzed by Robert Michels in 1915 as "the *iron law of oligarchy*." Michels argued that movements inevitably change from democratic to undemocratic in the course of time. Thus, movements to transform the world are ultimately defeated, as they come to resemble the undemocratic structures they sought to replace. The reality that many movements have succumbed to oligarchy has been grounds for despair and turned many people away from involvement in social change. But is oligarchy inevitable? In "Redefining Leadership," Linda Stout writes about the experiences of one grassroots organization in the United States that has fought to establish and maintain a democratic character. The Piedmont Peace Project has "built a model based on sharing power and created a structure for broad-based leadership." Research indicates that many nongovernmental movements around the world today are successfully confronting the issue of oligarchy (Fisher, 1994).

Finally, what do sociology, sociologists, and students of sociology have to do with solving the world's social problems? The chapter concludes with one sociologist's reflections on this question. In "What Are You Going to Do for Us?" MaryBe McMillan poses some of the ethical and intellectual dilemmas faced by those in this field who seek to make their work relevant to solving problems in "the real world."

WORKS CITED

Arrighi, Giovanni, Terence K. Hopkins, and Immanuel Wallerstein. 1989. *Antisystemic Movements*. New York: Verso.

Fisher, Julie. 1994. "Is the Iron Law of Oligarchy Rusting Away in the World?" *World Development,* 22(2): 129–142.

Lederman, Joanne, and Esther Ngan-ling Chow. 1996. "Gender-Based Violence and International Human Rights: Women Claim Their Humanity." Paper presented at the American Sociological Association Meetings, New York, August 16–20, 1996.

Marable, Manning. 1997. "Black (Community) Power!" *The Nation,* December 22, 1997, 21–24.

Silver, Beverly J. 1996. "The Decline of Labor Militancy: A Core-Centric Myth?" Paper presented at the American Sociological Association Meetings, New York, August 16–20, 1996.

So, Alvin Y. 1992. "Democracy as an Antisystemic Movement in Taiwan, Hong Kong and China: A World Systems Analysis." *Sociological Perspectives*, 35(2): 385–404.

QUESTIONS FOR DISCUSSION

1.   Can "bottom up" movements find solutions to world problems without coming into conflict with governments, corporations, and other powerful organizations? Discuss this after reading "The Civil Society Sector," "Global Women's Movements: A World System Perspective," and reviewing "Strategies for Changing Society" in Chapter 1.

2.   What is the most effective place to intervene in the world in an effort to solve the problems discussed in this reader? Do you agree with Meadows' argument in "Places to Intervene in a System"? Explain.

3.   Is oligarchy inevitable? What is your reaction to the ideas presented by Stout in "Redefining Leadership"?

4.   What is *your* answer to the questions in the last six lines of McMillan's poem, "What Are You Going to Do for Us?"?

 INFOTRAC COLLEGE EDITION: EXERCISE

The journal *The Futurist* is devoted to discussion of prospects for the future of the world. Although not a journal of sociological research, it deals with many interesting social topics. Explore the journal, using InfoTrac College Edition. InfoTrac contains the last few years of the journal, and you can view a list of all the articles published in it. Click on the PowerTrac icon, and under Choose Search Index, select Journal Name (in) and enter the search terms The Futurist.

Review the list of articles. What are some of the social problems covered? After looking over the list, select and read one or two of the articles that interest you. What is your reaction to what you read? Reflect on what you think your reaction might have been *before* you took this course. How has your thinking changed?

FOR ADDITIONAL RESEARCH

Books

Benjamin, Medea, and Andrea Freedman. *Bridging the Global Gap: A Handbook to Linking Citizens in the First and Third Worlds.* 1989. Cabin John, MD: Seven Locks Press.

Olson, Annette. 1994. *Alternatives to the Peace Corps: A Directory of Third World and the U.S. Volunteer Opportunities.* 6th ed. Oakland, CA: Food First Books.

Walls, David. 1993. *The Activist's Almanac: The Concerned Citizen's Guide to the Leading Advocacy Organizations in America.* New York: Simon and Schuster.

Organizations

Center for Campus Organizing
P.O. Box 748
Cambridge, MA 02142
(617) 725-2886
ucp@igc.apc.org

Center for Third World Organizing
1218 East 21st Street
Oakland, CA 94606
(510) 533-7583
www.ctwo.org

Global Exchange
2017 Mission Street, Suite 303
San Francisco, CA 94110
(415) 255-7296
www.globalexchange.org

Labor Education and Research Project (Labor Notes)
7435 Michigan Avenue
Detroit, MI 48210
(313) 842-6262
www.labornotes.org

ACTION PROJECTS

1.   If your library has *Hope* magazine, look through issues from the last few years. What did your exploration teach you about the scope and diversity of efforts to provide solutions to social problems? Identify what you feel are the *top three* hopeful movements covered by the magazine. What interested or inspired you about these movements?

2.   Organize a film festival on your campus to raise students' awareness about "bottom up" movements around the world. Your college librarian should have catalogs from video distributors that are good sources for such videos—for example, California Newsreel, New Day, Appalshop. A good source for such videos is also the American Friends Service Committee Video and Film Library, 2161 Massachusetts Avenue, North Cambridge, MA 02140, (617) 497-5273. Reflect on what you learned about these movements from the videos you showed and the audience response to them.

3.   How do people involved in "bottom up" movements deal with the challenges they face? This chapter discussed some challenges, such as developing a unifying ideology, deciding upon goals, and the issue of oligarchy. Interview someone who is active in a "bottom up" movement in your region to learn about how the movement has dealt with the challenges it has faced. In addition to challenges discussed in this chapter, another important issue to cover is how the movement survives opposition—that is, the backlash against it by its opponents. What forms of opposition has this movement faced? What has it done to survive in the face of this opposition?

61

The Civil Society Sector

LESTER M. SALAMON AND
HELMUT K. ANHEIER

The twentieth century has been a time of immense social innovation. Paradoxically, however, one of the social innovations for which the twentieth century may deserve to be best known is still largely hidden from view, obscured by a set of concepts that cloud its existence and by statistical systems that consequently fail to take it into account.

That innovation is the civil society sector, the plethora of private nonprofit, and nongovernmental organizations that have emerged in recent decades in virtually every corner of the world to provide vehicles through which citizens can exercise individual initiative in the private pursuit of public purposes. If representative government was the great social invention of the eighteenth century, and bureaucracy—both public and private—of the nineteenth, it is organized, private, voluntary activity, the proliferation of civil society organizations, that may turn out, despite earlier origins, to represent the greatest social innovation of the twentieth century.

That this is so, however, may be one of the best kept secrets of modern social development. Despite its importance, the nonprofit, or civil society, sector has remained an uncharted subcontinent on the social landscape of modern society. One reason for this is doubtless empirical. The data systems used to gather information about the structure of economic life have systematically overlooked this sector or essentially defined it away. This is certainly true of the UN System of National Accounts, the major guidance system for the world's statistical agencies, which defines out of the nonprofit sector any organization that receives half or more of its income from services fees or government grants.

Beyond these empirical problems lies an even more important conceptual, and even ideological, one. So diverse is the array of organizations that falls within the ambit of this "sector" that we have tended to focus on the differences among these organizations rather than the things they have in common. This tendency has been encouraged by political ideologies on both the political Left and the political Right that have systematically led observers to overlook this sector or downplay its role—the Left because too positive a view of the capabilities of nonprofit institutions could undermine support for the expansion of modern governmental protections against the risks of poverty and distress; and the Right because acknowledgment of the continued presence—indeed massive growth—of nonprofit organizations could undercut conservative complaints that state expansion displaces nonprofit groups. The result has been a curious conspiracy of silence about one of the most striking features of modern social life and an inability to move beyond the prevailing two-sector model of "market" and "state," of public and private sectors, that has long dominated our images of modern society and kept the nonprofit sector largely hidden from view.

The purpose of this article is to bring this largely invisible subcontinent on the landscape of twentieth century social reality into better focus conceptually as well as empirically. To do so, we draw on the results of a major inquiry we have just completed into the scope, structure,

From Lester M. Salamon and Helmut K. Anheier, "The Civil Society Sector," *Society*, Jan/Feb., 1997, pp. 60–65. Copyright © by 1997 Transaction Publishers; all rights reserved. Reprinted by permission.

financing and role of the "nonprofit sector" in a broad cross-section of countries around the world (the U.S., the UK, France, Germany, Italy, Sweden, Hungary, Japan, Brazil, Egypt, Ghana, Thailand, and India). These countries were selected to present different levels of development, different levels of government social welfare spending, and different religious and cultural traditions. We focus on the key institutional component of civil society: the private voluntary, or nonprofit, sector, which we define as a set of entities that is (i) organized, (ii) private, (iii) non-profit-distributing, (iv) self-governing, and (v) voluntary to some meaningful extent. Included here are membership associations, local community groups, clubs, health care providers, educational institutions, social service agencies, advocacy groups, foundations, self-help groups, and many more. While this does not capture all components of civil society, such as a free press, civil rights protections, religious life, or the participation of individuals in public life outside organizational settings, it does nonetheless embrace the key institutional embodiment of this concept.

Four major findings emerge from our effort to chart the nonprofit sector at the international level. Taken together, they fundamentally revise common conceptions about the scope, character, role and operations of this sector and challenge some of the central theories in the field. In the process, they point to a new way to think about the whole concept of "civil society."

MAJOR ECONOMIC AND SOCIAL FORCE

In the first place, despite the rise of the modern welfare state, the civil society sector turns out to be a major social and economic force. It accounts for a far larger share of national employment and recent employment growth than is widely assumed, and has become a pervasive mechanism through which individuals and societies pursue a wide assortment of public and private purposes.

Employment

Looking first at employment, in the eight major countries for which we collected complete empirical data (the U.S., the UK, France, Germany, Italy, Sweden, Hungary, and Japan), the civil society sector employed as of 1990 the equivalent of 11.9 million full-time paid workers. This represents close to one out of every 20 jobs overall, and an even higher one out of every 8 jobs in the fast-growing service sector, which is where the civil society sector operates. In addition to this paid employment, moreover, nonprofit organizations in these countries employ the equivalent of 4.8 million full time employees as volunteers. Adding these to the total would boost nonprofit employment by another 40 percent.

Expenditures

In addition to a sizable employment base, the civil society sector also boasts substantial operating expenditures. In the eight countries for which we were able to assemble complete operating expenditure data, the civil society sector, as we have defined it, had expenditures in 1990/2 of $614 billion. This is the equivalent of almost 5 percent of the combined gross domestic product of these countries and is 20 percent larger than the gross national product of Spain.

Contribution to Employment Growth

Not only is the civil society sector in these seven countries an immense economic presence, moreover, it has also been a growing presence, particularly in recent years. In fact, it has been a more potent source of job growth than most other segments of the economy. In Germany, France, and the U.S., for example, the civil society sector, with 6 percent of total employment, nevertheless accounted for nearly 13 percent of the job *growth* during the decade of the 1980s. Put somewhat differently, one of every 8 new jobs created in these countries during the 1980s was created within the civil society sector.

Activities

The real evidence of nonprofit presence is not, of course, the expenditures these organizations make or the employment for which they account, but the activities in which they engage, and here as well the presence is considerable. Thus, for example, nonprofit organizations accounted as of 1990 for:

- four out of every 10 hospital patient days and virtually all sport facilities in Germany;
- one-third of all child day care and 55 percent of all residential care in France;
- over half of all hospital beds and half of all universities in the United States;
- over 75 percent of all universities in Japan;
- over 20 percent of all primary and elementary education in the UK;
- over 40 percent of all residential care facilities in Italy; and
- 40 percent of all dwelling units constructed or rehabilitated in Sweden.

WIDESPREAD PRESENCE

Not only is the civil society sector a major force overall in these countries, but also its presence is far more widespread than in commonly assumed. An extensive civil society sector, it turns out, is hardly an exclusively American phenomenon. While nonprofit employment as a share of total employment, at 6.8 percent, is higher in the U.S. than elsewhere, the degree of variation among the major developed countries is considerably less pronounced than might be expected. Except for Italy, Japan, and Sweden, which are at the low end of the range, the remaining countries are all in the range of 4 percent of employment. And when only the service sector is considered, even Japan falls in the relatively high range of 9–10 percent.

What is more, the number of associations has increased substantially in recent years. In France, over 60,000 associations were created in 1990 alone, compared to less than 18,000 in 1961. Similarly, in Germany, the number of associations

per 100,000 population nearly tripled from 160 in 1960 to 475 in 1990. Even Hungary, within two years of the fall of communist rule, boasted over 13,000 associations. And Sweden, often regarded as the prototypical welfare state, displays some of the highest participation rates in civil society worldwide: most Swedes belong to one or more of the country's close to 200,000 membership associations, creating a dense social network of 2,300 associations per 100,000 population.

Nor is the civil society sector restricted to the advanced industrial nations. Although detailed employment and expenditure figures on the civil society sector are much more difficult to obtain, such organizations also appear to be increasingly widespread in the developing countries as well. Some 45,000 nonprofit organizations are functioning in Sao Paolo, Brazil, for example, and another 16,000 in Rio de Janeiro. In Brazil as a whole, the number of nongovernmental organizations is close to 200,000. In Egypt, a survey of nonprofit organizations found some 20,000 such organizations in existence as of the early 1990s. In Thailand, Bangkok alone boasts approximately 2,200 nonprofit organizations, and close to 11,000 have been identified countrywide. India has one of the most complex and diverse civil society sectors in the world, including a rich tapestry of Gandhian rural development organizations as well as numerous "empowerment-oriented" associations tied to various political, religious and ethnic movements.

VARIED PATTERNS

If the civil society sector is surprisingly ubiquitous, it nevertheless often takes quite different forms in different places. The extent of the variation is somewhat "constrained," however. Specifically, four components seem to dominate the sector almost everywhere. In particular, education and research, health, social service, and culture and recreation organizations account overall for nearly 80 percent of sector expenditures. What is more, these four components account for at least 75 percent of sector expenditures in seven of the eight countries, and in the only exception (Sweden) they account for well over half.

Nevertheless, considerable variation is also apparent within this overall pattern. Thus, for example, culture and recreation falls out of the top four fields of nonprofit activity in three of the eight countries (Germany, Japan, and the United States); education falls out in two (Germany and Hungary); and health falls out in three (Hungary, Sweden, and the UK). Indeed, in terms of expenditure dominance, it is possible to detect at least four distinct patterns of non-profit composition among the countries we have examined, although significant variations can still exist within these patterns both in terms of the specific subtypes of organizations that are responsible for the pattern and in terms of the rest of the composition of the sector in each country. More specifically, higher education dominates nonprofit expenditures in Japan and the UK, health in the United States and Germany, social services in Italy and France, and culture and recreation in Hungary and Sweden.

A rather different structure of civil society organizations is evident in the developing world, where *development and housing organizations* play a much larger role. Accounting for only 6 percent of nonprofit expenditures on average in the eight developed countries we examined, this class of organizations was far more prevalent in the developing countries, where far less specialization characterizes civil society organizations, so that the principal function of a health organization may be to develop a new water supply, or of an educational agency to sponsor a training enterprise. In Egypt, for example, a quarter of all non-profit organizations we surveyed turned out to fall into this category.

The overall picture of the composition of non-profit activity presented here does not change substantially when account is taken of the input of volunteers. Based on surveys of volunteer activity in four project countries—France, Germany, Italy, and the U.S.—the four major fields of nonprofit action identified above remain the principal fields even after taking account of volunteer inputs. At the same time, the inclusion of volunteer effort does have some impact on the results.

LIMITED ROLE OF PRIVATE GIVING

A fourth key finding that emerges from the work reported here is the relatively limited role that private giving plays in nonprofit finance.

According to conventional wisdom, what sets nonprofit organizations apart from their counter-parts in the public and business sectors is their reliance on private charitable giving, as opposed to tax revenues or service fees, as their principal source of support. In point of fact, however, our data reveal that *private giving is not only not the major source of nonprofit income in our eight countries. It is not even the second most important.* Rather, the average share of total nonprofit income originating from private philanthropic giving in the eight countries for which we collected comparable data was only 10 percent, and this includes individual giving, foundation giving, and corporate giving combined. By contrast, almost half (49 percent) of all nonprofit income in these eight countries comes on average from fees and sales, and 41 percent from government.

This pattern of nonprofit finance is remark-ably common among the countries we examined, moreover. Although private giving is higher in the U.S. than elsewhere, even there it represents no more than 19 percent of total income, com-pared to 51 percent from fees and dues and another 30 percent from government. In only two countries (France and Germany) does fee income lose its primary position as a source of nonprofit income, but it loses it not to private giving but to government. Thus the public sector accounts for 68 percent of the income of German nonprofit organizations and 59 percent of the income of French nonprofit organizations.

UBIQUITY OF CIVIL SOCIETY ORGANIZATIONS

What are the implications of these findings? In the first place, it appears that the nonprofit sector is a far more ubiquitous presence in societies throughout the world than previous thinking had suggested.

The common presumption of American exceptionalism so far as private, nonprofit institutions are concerned, a presumption stimulated by the work of Alexis de Tocqueville more than 150 years ago, turns out to be wide of the mark. A significant range of organized private activity outside both the market and the state seems increasingly to be a basic feature of advanced, democratic societies regardless of other social or historical features. What is more, such organizations are also increasingly evident in developing societies as well. Whether this is a product of the increased diversity of demands for "public goods" that accompanies development, or some other set of factors, is difficult to determine. It seems likely, however, that the recent worldwide loss of confidence in the capability of state institutions on their own to promote development, provide for social welfare, and protect the environment is playing a crucial role, as is the growth of new communications technologies and the emergence of educated middle class elements to take advantage of them. The upshot is a set of conditions unusually propitious for the blossoming of voluntary organizations of many sorts. We seem, in short, to be witnessing a "global associational revolution" of extraordinary scope and dimensions. Given recent research suggesting the contributions such institutions make to "social capital," to the norms of reciprocity and trust that help to sustain democracy and foster economic growth, this is a hopeful finding indeed.

While the apparent ubiquity of civil society organizations is encouraging, however, it is also clear that these organizations are not equally evident in all places. As we have seen, the nonprofit share of total employment ranged from a high of 6.8 percent in the U.S. to a low of 0.8 percent in Hungary and 1.3 percent in Italy. What is more, considerable variations exist in the composition of this sector from place to place.

What this suggests is that while conditions may be propitious for the emergence of civil society organizations throughout the world at the present time, the ability of different societies to take advantage of these conditions may nevertheless vary considerably. The nonprofit sector, it appears, requires what sociologist John Hall has

termed "social moorings," i.e. suitable social, economic, and political circumstances, and these are hardly equally present everywhere. To the contrary, the long-term viability of these organizations may still hang very much in the balance in a significant number of places. One reflection of this is the ambiguous legal status these organizations often occupy. In Japan, for example, the formation of such organizations is still treated legally as a privilege granted by particular government ministries rather than a right that is available to all citizens. France denies most nonprofit organizations the right to acquire endowments and in much of the developing world authoritarian regimes often seek to hold the nonprofit sector at bay through restrictive legal provisions and even more restrictive implementing procedures. All of this should make us cautious about overly romantic hopes that such institutions will inevitably blossom everywhere without close attention to the conditions required for their growth.

CIVIL SOCIETY AND THE STATE

A third implication of the findings reported here is to cast significant doubt on some of the prevailing theories for explaining the presence or absence of nonprofit organizations, particularly the dominant "market failure/government failure" theory and the conservative concept of an inherent conflict between the nonprofit sector and the state that is a natural corollary of it. Viewing the nonprofit sector as one response to the market's inherent limitations in generating an adequate supply of "collective goods," this theory suggests that nonprofit organizations should be most plentiful where cultural or religious diversity makes it difficult for a population to come to agreement about the level of public goods that should be provided, thus limiting the ability to generate the majorities needed to get government to supply them. According to this theory, the civil society sector thus essentially "fills in" for government where political controversy, rooted in population diversity, makes it impossible for government to respond to collective needs. From this it follows that a zero-sum relationship should exist between the nonprofit sector and the state, with the nonprofit sector most in evidence where state activity is most constrained, and the nonprofit sector most constrained where government is most active.

This line of thinking has found reflection as well in conservative political ideology, which faults the state, among other things, for displacing voluntary groups and undermining the spirit of community to which they give rise. According to this view, which is articulated particularly forcefully in Robert Nisbet's influential book *Community and Power,* the state is a bureaucratic monolith inherently hostile to alternative centers of power. As the state expands, it therefore renders voluntary organizations functionally irrelevant, thereby contributing to their decline and undermining the spirit of community which they sustain.

In fact, however, this zero-sum, conflictual image of the relationship between the civil society sector and the state finds only limited support in the data examined here. While there are clearly circumstances where the civil society sector is at odds with the state, there are at least as many where the relationship is one of interdependence and mutual support. This is clearly evident in the striking record of governmental support for the nonprofit sector reported above. As we have seen, such support is four times larger, on average, than the support these organizations receive from private charitable sources. The state has thus emerged in the modern era not as a displacer of nonprofit activity, but as perhaps the major "philanthropist," underwriting nonprofit activity and significantly extending its reach.

Behind these numbers, moreover, lies a rich history of inter-relationships that has developed over decades and that often has deep roots in social traditions and political realities. The massive pattern of government reliance on nonprofit organizations and nonprofit reliance on the state that we found in Germany, for example, reflects the hold of the doctrine of "subsidiarity" that grew out of Catholic social thought in the 1930s and that was fastened on German social policy in the postwar era, obliging the state to rely on nonprofit organizations to deliver the lion's share of state-financed welfare services. Similar patterns are evident, however, albeit for different reasons in the United States, France, and Italy. In the United States, long-standing conservative hostility to federal welfare protections made it impossible for Great Society reformers in the 1960s to expand government social welfare protections except through an elaborate system of what Salamon has termed "third-party government" in which the national government financed a sizable array of welfare services but turned to other institutions, among them private, nonprofit groups, to deliver the resulting services. In this way, it was possible to expand America's relatively meager social welfare protections without expanding the size of the state bureaucracy. France's left-wing government pursued a similar course in the 1980s. Rather than shying away from the civil society sector as an expression of right-wing idealism, the Mitterand government seized on this sector as a mechanism for permitting the extension of welfare-state services without the need to expand the increasingly mistrusted apparatus of the modern welfare state. In Italy as well, a left-leaning government, aware of shortcomings in the country's vocational training system, moved in the late 1970s to decentralize control of this system and encourage the creation of nonprofit training centers for technical occupations, often in close partnership with local businesses.

These and other examples thus confirm the suggestion of John Hall that "[t]he notion that groups, albeit of the right type, should balance the state is subtly wrong. This manner of conceptualizing state-society relations leaves much to be desired because it tends to see the state exclusively as a threat." In fact, state support can empower civil society organizations as well as constrain them. While conservative political thinkers may argue that an inherent conflict exists between the civil society sector and the state, and that state activity must be cut back to permit room for the expansion of voluntary effort, the more common relationship, at least in the developed nations, appears to be a pervasive partnership instead.

At the same time, reliance on the state can be taken too far. This is the message that emerges from the recent history of Japan, where nonprofit organizations have been actively enlisted in the provision of state-financed services, but always on terms defined mostly, indeed almost exclusively, by the state. The upshot has been to convert nonprofit organizations into mere "agents" of the state, rather than true "partners" with it. Thus, for example, Japan's social welfare corporations (*Shakaifukushi hojin*) are so restricted to performing tasks assigned to them by central and local governments that they think of themselves more as quasi-governmental organizations than private nonprofit organizations, even though this is their legal form. This is the concern that Habermas evidently had in mind when he complained in *The Legitimation Crisis* about the tendency of the Keynesian welfare state to delegate an ever increasing array of functions to "politically opportune" private actors that thereby surrender their role as critics of state action. To avoid this, the civil society sector must balance its receipt of state support with access to other sources of support. In part this can be fee and service income. But equally important is a meaningful level of private philanthropic support. From the evidence we have assembled, it appears that as little as 10 percent of total income needs to come from this source in order to achieve a reasonable degree of autonomy. Clearly, however, the 1 percent level evident in Japan is not enough.

RETHINKING CIVIL SOCIETY

Taken together, finally, these findings seem to suggest the need for a new way to conceptualize the notion of "civil society." Of late, this term has been used to depict a particular class of social institutions, a class that lies outside both the market and the state. In a sense, the term has thus been appropriated by a particular sector. The rationale for this has been the argument that this set of institutions, in all its diversity of interest groups, savings associations, church choirs, sports clubs, charities, and philanthropic foundations, is uniquely engaged in creating the networks of civic engagement that produce and enforce communal values and notions of trust so necessary for cooperation and civil life.

While there is validity in this line of argument, however, it has the unfortunate effect of relegating the other sectors to the status of being "uncivil." More importantly, it overlooks the extent to which the "civil society sector" relies on the other sectors to survive. Indeed, it might be argued that a true "civil society" is not one where one or the other of these sectors is in the ascendance, but rather one in which there are three more or less distinct sectors—government, business, and nonprofit—that nevertheless find ways to work together in responding to public needs. So conceived, the term "civil society" would not apply to a particular sector, but to a *relationship* among the sectors, one in which a high level of cooperation and mutual support prevailed. This is the conclusion that seems to emerge from our data. Generally speaking, the nonprofit sector is large and civil society advanced where such partnerships among the sectors developed; both are underdeveloped where such cooperation is absent. What this suggests is that developing mutually supportive relationships between the nonprofit sector and the state, and with the business community as well, may be one of the highest priorities for the promotion of democracy as well as economic growth throughout the world.

SUGGESTED FURTHER READING

Fukuyama, Francis. *Trust: The Social Virtues and the Creation of Prosperity.* New York: The Free Press, 1995.

Halls, John (ed.). *Civil Society: Theory, History, Comparison.* Cambridge, UK: Polity Press, 1995.

Powell, Walter W. *The Nonprofit Sector: A Research Handbook.* New Haven: Yale University Press, 1987.

Putnam, Robert D. *Making Democracy Work: Civil Traditions in Modern Italy.* Princeton: Princeton University Press, 1993.

Salamon, Lester M. *Partners in Public Service: Government-Nonprofit Relations in the Modern Welfare State.* Baltimore: Johns Hopkins University Press, 1995.

Salamon, Lester M. and Helmut K. Anheier. *The Emerging Nonprofit Section—An Overview.* Manchester: Manchester University Press, 1996.

• • •

62

Global Women's Movements

A World System Perspective

TORRY D. DICKINSON

IN GLOBAL MOVEMENTS, WOMEN LEAD

Female heroes in regional and global movements are taking lead in subverting the system of global domination. Today women can be found at the heart of almost all major social movements that address both general social problems and gender-related issues. Although many people often assume that feminists form narrow "interest groups," feminists address broad social needs. Woman-centered movements, which are becoming more interconnected at the global level, create new building blocks that establish more democratic ways of living. In these movements you'll find that, next to activist women, there are some male feminists, who have learned that sexism—like racism and ageism—doesn't really work for anyone. First-hand knowledge about the central social location of the gender hierarchy often seems to enable feminists to see global social problems and social alternatives in comprehensive, visionary ways.

Since 1975, most women around the world have found that their work load has increased, their real wages and job-related benefits have declined, their state subsidies have decreased, and their work options have become more limited. Throughout the world, more women now head households and largely support their families. Many women see their daughters' options becoming more limited, their female and male children working for low wages or even being held in captivity as bonded laborers or sex workers. As these global changes have become institu-tionalized in different ways in poorer areas of the South and in richer areas of the North, there is a diversity of global women's movements trying to redefine the world of work and eliminate male domination in society.

WIDESPREAD EVIDENCE OF WOMEN'S ORGANIZING

Female garment workers and electronic workers go on strike in Mexico and the Philippines, demanding that transnational corporations pay higher wages and provide safer working conditions. Strikers tell their side of the story to news reporters and feminist filmmakers, and to people living on the other side of the world, who buy the jeans and computers made on the global assembly line. Likewise, U.S. and European women, who have seen their manufacturing jobs go to low-wage areas in the South, begin to recognize that their lives are interconnected with women and teen girls in export enclaves. Through activities like factory strikes, fair trade organizing, consumer boycotts and the development of local markets, women living on opposite sides of the globe are exploring different ways to fight transnational managers who run world-wide industries.

As U.S. and European women demand that the state reinstate welfare benefits and mothers' pensions, women in sub-Saharan Africa and Latin America lead anti-austerity protests against state governments that cut subsidies for working people. Just as welfare cuts in the North bring new family hardships, reductions in state subsidies for food, fuel and transportation in the South mean drastic declines in family resources. This is especially the case as prices increase and rampant inflation ensues. Women have waged a battle against many of these more than 100 subsidy reduction and structural adjustment programs, which allow the North's transnational corporations to continue dominating economic life.

Like other women who confront state policies that hurt working-class families, Indian women in Ecuador played a prominent role in an August, 1997 protest against austerity measures. They confronted the military, piled logs, burned tires, and stopped traffic on a highway leading to Quito. Many protests against austerity measures, which have erupted since the mid-1980s in both Third World and former communist countries, are organized by women's groups.

Women fight growing homelessness and underemployment by organizing urban squatting movements in the North and in the South. For example, low-income women in a suburb of Philadelphia, Pennsylvania—many who had relied on welfare payments to provide temporary stability—began a campaign in the late 1980s to seize and renovate abandoned apartment buildings and to set up a tent city on a vacant lot. This group of women from Kensington (who had witnessed the closing of the town's textile factories and the disappearance of jobs) repeatedly confronted the police and the court system, as have other urban squatters in many countries, including the United States, France, and South Africa. As work options become more limited, growing numbers of squatters have settled around the outer rim of large Third World cities and have built improvised homes made out of reused cardboard, metal, wood and cement blocks. Feminist activists can often be found fighting for a place to live in the city and inventing ways to keep living costs low.

Women form an integral part of environmental, land seizure and community self-reliance movements in Latin America. These movements try to reclaim working people's land, which has been illegally expropriated through false legal means or through the expansion of lodging, mining, and agricultural production. In some cases, rural women have helped organize land use movements when landowners suddenly began forbidding the use of rent-free housing, tillable land for subsistence gardening, and community land that had been (by custom) a community commons for the working people in the area.

In the Philippines, women began protesting the expansion of cash crop production when land owners began denying access to the forest, which had served as a source of fish, wild animals and edible plants. Rather than seizing land for communal use, these women began establishing their own producing and market system, where they made and bought goods from each other (and not from capitalist producers). As they began embracing indigenous and non-commercial foods, these women started to share information about cultivating native plants that provided better nutrition and reflected their cultures. As in India, Bangladesh, and Brazil, women in the Philippines have become very concerned about maintaining their control over fertile seeds for native plants, which today's multinational corporations often claim as their private property through the patenting process.

WOMEN'S WORK AND THE GLOBAL PRODUCTION SYSTEM

The global profit-making system rests on a two-sided system of exploiting and sustaining labor. *Direct exploitation* is carried out in the wage labor environment, where work is organized by and benefits employers. For example, transnational corporate managers engage in direct exploitation when they pay workers less than their work is worth.

Indirect exploitation takes place in the less visible realm of non-wage work, which is organized by laboring households and inter-household networks (whose members do wage work at various points in their lives). When laboring people do their own work and create their own goods (instead of buying goods and services from the capitalist market), systemic indirect exploitation is taking place. For example, when household members take care of children and elders, prepare their own food, sell goods in informal markets, and take in boarders, they are doing two things at the same time. By supplementing their wage incomes (and holding the family together), they are helping themselves. But, when they enable themselves to survive on meager wages, they are also being subjected to indirect exploitation.

Because they can continue to pay low wages, company managers become rich when workers do a lot of home-based work to supplement inadequate wages. Employers count on households to do this home-based survival work, just as much as household members have come to count on each other for survival. Both indirect and direct exploitation are integral processes of the world-economy.

The capitalist world-economy has developed to the extent that it has because it draws on value created at home. Contrary to many feminists' arguments, there has been no clear separation between public and private in the world-economy. There has been no clear line drawn between value that is created in wage labor and value that is created in the household. Low wages force households to organize their own working day; labor's non-wage work, in turn, has sustained unequal global wage levels. The payment of low wages, on the one hand, places families in situations where they need to do lots of home-based work, like cultivating vegetables, raising chickens and pigs, and building shanties out of recycled discards. All this self-organized work allows laboring families to get by with low wages. On the other hand, the real value of this home-based work, which supplements wage incomes, enables employers to keep paying inadequate wages.

Feminists in the world-economy have had to deal with this Catch-22 dilemma of survival, where laboring families help to sustain the system when they do supplemental work. Some feminists have focused on the need to change wage labor relations (by improving wages, benefits and health conditions at work). Others have focused on how wage labor and non-wage work are connected at the household level; they think about how to change both work realms together (by increasing women's wages, increasing men's work at home, and trying to get the state to guarantee a basic living wage and social services). And still others have tried to figure out how to sever home-based work from formal labor relations, and how to totally redefine work relations from this vantage point (by disengaging from wage labor, and by naming family and community work as the center of a new and completely different set of work and marketing relations).

BENDING THE GLOBAL SYSTEM: FEMINIST STRUGGLES AGAINST INEQUALITY

Confronting Direct Exploitation

Over the last century, many labor movements—including women's movements—have been aimed at primarily direct exploitation. They have tried to increase workers' wage levels and consumer options. And many still do. By putting pressure on employers and the state to increase wages and subsidies, female and male activists have both challenged the way the system functions and demanded that a redistribution of income take place. Because there are limits to how much income can be redistributed in the unequal world-economy, this is an anti-systemic act. It's especially anti-systemic when women demand equality because the entire system has historically rested on the super-exploitation of female wage laborers.

The economic despair that has set in since 1975 may be diminishing the expectation that the system can deliver a better way of life. If many simultaneous movements try to increase wages and state subsidies around the world, the system may begin to split apart. After all, the world-economy is based on unequal levels of work and remuneration. Ultimately, the highly unequal capitalist world-economy cannot meet the demand for equality. And equality is one ideal that bourgeois democracy has promised to deliver.

Confronting Indirect Exploitation

Because workers in the South are paid lower wages than those in the North, households in the South have to generate probably 50 percent of their own monetary and in-kind income through the process of indirect exploitation. As a result, woman-centered non-wage struggles that attempt to subvert and reroute indirect exploitation have been highly visible in the South during the 1980s and 1990s. Feminist movements in the South, in particular, have seized self-organized, "women's" work as a political wedge. They are beginning to decide how cooperative laboring groups can reorganize and directly benefit from self-organized work, and how this non-wage work can simultaneously weaken global capitalism.

At the same time, through their critique of the global economy, feminists in the North began to see that the modern world is based on the indirect exploitation of households, where non-wage work by loving families was subverted and used to sustain the entire world-economy. As a result, women's movements in the 1980s and 1990s began building bases of resistance in households and communities, where indirect exploitation occurred. The idea became to attack the system from its invisible, taken-for-granted base, which has been enabling the entire system to operate.

Thus, various groups of feminists in both the North and South developed diverse, but often common approaches for promoting social change in work relations. Struggles directed against capital and the state, which try to get businesses and the state to address needs that arise through the process of direct exploitation, are complemented by "non-wage, community-centered movements" that focus on problems of indirect exploitation. These community-centered efforts are carried out at the same time that women are fighting to get businesses and the state to increase pay, state subsidies and state-organizing services like education. A common strategy of non-wage, community-centered struggles is to build on organizational strengths that emerge through the process of indirect exploitation. Many feminist-inspired move-

ments, for example, promote environmental change, women's micro-enterprise and marketing networks, and rebirth of the commons. A key goal is to develop cooperative, women-centered, non-capitalist relationships that may eventually undermine and even replace capitalist relationships. In this way, activists have begun using non-wage work as an avenue to become independent from global production and consumption.

Community-centered activists seem to be saying, "We don't want to be more fully integrated into your system. We'll take our non-wage work, households and community relationships, and we'll use them to begin building our own local and regional economies." At this time, it's not clear how many women's global labor movements are consciously trying to extract non-wage work from the world-economy's orbit. Although these movements are anti-systemic, we don't know if woman-centered, regional production and cooperative arrangements will provide a long-term replacement for capitalist relations. By valuing women and their community-based social and work relationships, this approach offers a chance to subvert the gender hierarchy, as well as a creator of that hierarchy: the world-economy.

Confronting Male Dominance

Along with institutional and community-based feminisms, another set of feminist movements primarily confronts male domination. Typically these movements push for change within the global system and in non-capitalist community relationships. These "anti-patriarchal movements" complement, and sometimes even overlap with, the two other main types of feminist movements. The struggle against domestic violence, for example, seeks involvement from the state (through the court system and through state funding), and simultaneously develops an independent base of female power. Many feminists have struggled to end rape and torture during civil, regional and world wars. And women fight to end the traffic in women and to stop the global sex trade by working with the state and international organizations. Activists form their own anti-trafficking networks and encourage women and girls to develop themselves and their political power. Anti-patriarchal struggles are anti-systemic because they try to undermine patriarchal relationships, which sustain the world-economy, prop up household inequality, and undergird dominant cultures.

ANTI-SYSTEMIC GLOBAL FEMINISMS AND OUR FUTURE

As institutional feminists work to expose how limited income security and democracy is for the majority of laborers, community-based feminists try to redefine non-wage work and pull it out of the world-economy. Anti-patriarchal struggles fight to create more egalitarian and democratic cultures. These approaches, which are adopted by many women's global labor movements, are anti-systemic in the long run. They all offer hope to diverse groups of women around the world, who have been struggling for equality, democracy and a better way of living in a shared world.

Feminist movements may start to have multiple impact if they link institutional, community-based and anti-patriarchal struggles together. Perhaps some women and men are already crossing over between these three types of movements, and others as well. Perhaps new types of feminist "crossover" movements have already started to emerge. We will only know if we take the history of social change seriously. And this just may mean engaging in it.

63

Places to Intervene in a System

In Increasing Order of Effectiveness

DONELLA H. MEADOWS

Folks who do systems analysis have a great belief in "leverage points." These are places within a complex system (a corporation, an economy, a living body, a city, an ecosystem) where a small shift in one thing can produce big changes in everything.

The systems community has a lot of lore about leverage points. Those of us who were trained by the great Jay Forrester at MIT have absorbed one of his favorite stories. "People know intuitively where leverage points are. Time after time I've done an analysis of a company, and I've figured out a leverage point. Then I've gone to the company and discovered that everyone is pushing it **in the wrong direction!**"

The classic example of that backward intuition was Forrester's first world model. Asked by the Club of Rome to show how major global problems—poverty and hunger, environmental destruction, resource depletion, urban deterioration, unemployment—are related and how they might be solved. Forrester came out with a clear

From Donella H. Meadows, "Places to Intervene in a System," *Whole Earth,* Winter, 1997, pp. 78–84.

9. Numbers (subsidies, taxes, standards).

8. Material stocks and flows.

7. Regulating negative feedback loops.

6. Driving positive feedback loops.

5. Information flows.

4. The rules of the system (incentives, punishments, constraints).

3. The power of self-organization.

2. The goals of the system.

1. The mindset or paradigm out of which the goals, rules, feedback structure arise.

leverage point: Growth. Both population and economic growth. Growth has costs—among which are poverty and hunger, environmental destruction—the whole list of problems we are trying to solve with growth!

The world's leaders are correctly fixated on economic growth as the answer to virtually all problems, but they're pushing with all their might in the wrong direction.

Counterintuitive. That's Forrester's word to describe complex systems. The systems analysts I know have come up with no quick or easy formulas for finding leverage points. Our counter-intuitions aren't that well developed. Give us a few months or years and we'll model the system and figure it out. We know from bitter experience that when we do discover the system's leverage points, hardly anybody will believe us.

Very frustrating. So one day I was sitting in a meeting about the new global trade regime, NAFTA and GATT and the World Trade Organization. The more I listened, the more I began to simmer inside. "This is a **HUGE NEW SYSTEM** people are inventing!" I said to myself. "They haven't the *slightest idea* how it will behave," myself said back to me. "It's cranking the system in the wrong direction—growth, growth at any price!! And the control measures these nice folks are talking about—small parameter adjustments, weak negative feedback loops—are **PUNY!**"

Suddenly, without quite knowing what was happening, I got up, marched to the flip chart,

tossed over a clean page, and wrote: **"Places to Intervene in a System,"** followed by nine items (see left). Everyone in the meeting blinked in surprise, including me. "That's brilliant!" someone breathed. "Huh?" said someone else.

I realized that I had a lot of explaining to do.

In a minute I'll go through the list, translate the jargon, give examples and exceptions. First I want to place the list in a context of humility. What bubbled up in me that day was distilled from decades of rigorous analysis of many different kinds of systems done by many smart people. But complex systems are, well, complex. It's dangerous to generalize about them. What you are about to read is not a recipe for finding leverage points. Rather it's an invitation to think more broadly about system change.

That's why leverage points are not intuitive. I've included a brief overview of systems theory (see box on next page) to provide the context for the list that follows.

9. NUMBERS.

Numbers ("parameters" in systems jargon) determine how much of a discrepancy turns which faucet how fast. Maybe the faucet turns hard, so its takes a while to get the water flowing. Maybe the drain is blocked and can allow only a small flow, no matter how open it is. Maybe the faucet can deliver with the force of a fire hose. These considerations are a matter of numbers, some of which are physically locked in, but most of which are popular intervention points.

Consider the national debt. It's a negative bathtub, a money hole. The rate at which it sinks is the annual deficit. Tax income makes it rise, government expenditures make it fall. Congress and the president argue endlessly about the many parameters that open and close tax faucets and spending drains. Since those faucets and drains are connected to the voters, these are politically charged parameters. But, despite all the fireworks, and no matter which party is in charge, the money hole goes on sinking, just at different rates.

The amount of land we set aside for conservation. The minimum wage. How much we spend

Systems Theory

To explain numbers, stocks, delays, flows, feedback, and so forth, I need to start with a basic diagram.

The "state of the system" is whatever standing stock is of importance—amount of water behind the dam, harvestable wood in the forest, people in the population, money in the bank, whatever. System states are usually physical stocks, but they could be nonmaterial ones as well—self-confidence, trust in public officials, perceived safety of a neighborhood.

There are usually inflows that increase the stock and outflows that decrease it. Deposits increase the money in the bank; withdrawals decrease it. River inflow and rain raise the water behind the dam; evaporation and discharge through the spillway lower it. Political corruption decreases trust in public officials; experience of a well-functioning government increases it.

Insofar as this part of the system consists of physical stocks and flows—and they are the bedrock of any system—it obeys laws of conservation and accumulation. You can understand its dynamics readily, if you can understand a bathtub with some water in it (the state of the system) and an inflowing faucet and outflowing drain. If the inflow rate is higher than the outflow rate, the stock gradually rises. If the outflow rate is higher than the inflow, the stock goes down. The sluggish response of the water level to what could be sudden twists of input and output valves is typical—it takes time for flows to accumulate.

The rest of the diagram is information that causes the flow to change, which then cause the stock to change. If you're about to take a bath, you have a desired water level in mind. You plug the drain, turn on the faucet and watch until the water rises to where you want it (until the discrepancy between the desired and the actual state of the system is zero). Then you turn the water off.

If you start to get in the bath and discover that you've underestimated your volume and are about to produce an overflow, you can open the drain, until the water goes down to your desired level.

Those are two negative feedback loops, correcting loops, one controlling the inflow, one controlling the outflow, either or both of which you can use to bring the water level to your goal. Notice that the goal and the feedback connections are not visible. If you were an extraterrestrial trying to figure out why the tub fills and empties, it would take a while to realize that there's a goal and a discrepancy-measuring process going on within the creature manipulating the faucets. But if you watched long enough, you could figure that out. Now let's take into account that you have two taps, a hot and a cold, and that you're also adjusting for another system state—temperature. Suppose the hot inflow is connected to a boiler way down in the basement, four floors below, so it doesn't respond quickly. And the inflow pipe is connected to a reservoir somewhere, which is connected to the planetary hydrological cycle. The system begins to get complex and interesting.

Mentally change the bathtub into your checking account. Write checks, make deposits, add a faucet that dribbles in a little interest and a special drain that sucks your balance even drier if it ever goes dry. Attach your account to a thousand others and let the bank create loans as a function of their combined deposits, link a thousand banks into a federal reserve system—and you begin to see how simple stocks and flows, plumbed together, make up systems way too complex to figure out.

on AIDS research or Stealth bombers. The service charge the bank extracts from your account. All these are numbers, adjustments to faucets. So, by the way, is firing people and getting new ones. Putting different hands on the faucets may change the rate at which they turn, but if they're the same old faucets, plumbed into the same system, turned according to the same information and rules and goals, the system isn't going to change much. Bill Clinton is different from George Bush, but not all that different.

Numbers are last on my list of leverage points. Diddling with details, arranging the deck chairs on the *Titanic*. Probably ninety-five percent of our attention goes to numbers, but there's not a lot of power in them.

Not that parameters aren't important—they can be, especially in the short term and to the individual who's standing directly in the flow. But they **RARELY CHANGE BEHAVIOR.** If the system is chronically stagnant, parameter changes rarely kick-start it. If it's wildly variable, they don't usually stabilize it. If it's growing out of control, they don't brake it.

Whatever cap we put on campaign contributions, it doesn't clean up politics. The Feds fiddling with the interest rate haven't made business cycles go away. (We always forget that during upturns, and are shocked, shocked by the downturns.) Spending more on police doesn't make crime go away.

However, there are critical exceptions. Numbers become leverage points when they go into ranges that kick off one of the items higher on this list. Interest rates or birth rates control the gains around positive feedback loops. System goals are parameters that can make big differences. Sometimes a system gets onto a chaotic edge, where the tiniest change in a number can drive it from order to what appears to be wild disorder.

Probably the most common kind of critical number is the length of delay in a feedback loop. Remember that bathtub on the fourth floor I mentioned, with the water heater in the basement? I actually experienced one of those once, in an old hotel in London. It wasn't even a bathtub with buffering capacity; it was a shower. The water temperature took at least a minute to respond to my faucet twists. Guess what my shower was like. Right, oscillations from hot to cold and back to hot, punctuated with expletives. Delays in negative feedback loops cause oscillations. If you're trying to adjust a system state to your goal, but you only receive delayed information about what the system state is, you will overshoot and undershoot.

Same if your information is timely, but your response isn't. For example, it takes several years to build an electric power plant, and then that plant lasts, say, thirty years. Those delays make it impossible to build exactly the right number of plants to supply a rapidly changing demand. Even with immense effort at forecasting, almost every electricity industry in the world experiences long oscillations between overcapacity and undercapacity. A system just can't respond to short-term changes when it has long-term delays. That's why a massive central-planning system, such as the Soviet Union or General Motors, necessarily functions poorly.

A delay in a feedback process is critical **RELATIVE TO RATES OF CHANGE** (growth, fluctuation, decay) **IN THE SYSTEM STATE THAT THE FEEDBACK LOOP IS TRYING TO CONTROL.** Delays that are too short cause overreaction, oscillations amplified by the jumpiness of the response. Delays that are too long cause damped, sustained, or exploding oscillations, depending on how much too long. At the extreme they cause chaos. Delays in a system with a threshold, a danger point, a range past which irreversible damage can occur, cause overshoot and collapse.

Delay length would be a high leverage point, except for the fact that delays are not often easily changeable. Things take as long as they take. You can't do a lot about the construction time of a major piece of capital, or the maturation time of a child, or the growth rate of a forest. It's usually easier to slow down the change rate (positive feedback loops, higher on this list), so feedback delays won't cause so much trouble. Critical numbers are not nearly as common as people seem to think they are. Most systems have evolved or are designed to stay out of sensitive parameter ranges. Mostly, the numbers are not worth the sweat put into them.

8. MATERIAL STOCKS AND FLOWS.

The plumbing structure, the stocks and flows and their physical arrangement, can have an enormous effect on how a system operates.

When the Hungarian road system was laid out so all traffic from one side of the nation to the other had to pass through central Budapest, that determined a lot about air pollution and commuting delays that are not easily fixed by pollution control devices, traffic lights, or speed limits. The only way to fix a system that is laid out wrong is to rebuild it, if you can.

Often you can't, because physical building is a slow and expensive kind of change. Some stock-and-flow structures are just plain unchangeable.

The baby-boom swell in the US population first caused pressure on the elementary school system, then high schools and colleges, then jobs and housing, and now we're looking forward to supporting its retirement. Not much to do about it, because five-year-olds become six-year-olds, and sixty-four-year-olds become sixty-five-year-olds predictably and unstoppably. The same can be said for the lifetime of destructive CFC molecules in the ozone layer, for the rate at which contaminants get washed out of aquifers, for the fact that an inefficient car fleet takes ten to twenty years to turn over.

The possible exceptional leverage point here is in the size of stocks, or buffers. Consider a huge bathtub with slow in and outflows. Now think about a small one with fast flows. That's the difference between a lake and a river. You hear about catastrophic river floods much more often than catastrophic lake floods, because stocks that are big, relative to their flows, are more stable than small ones. A big, stabilizing stock is a buffer.

The stabilizing power of buffers is why you keep money in the bank rather than living from the flow of change through your pocket. It's why stores hold inventory instead of calling for new stock just as customers carry the old stock out the door. It's why we need to maintain more than the minimum breeding population of an endangered species. Soils in the eastern US are more sensitive to acid rain than soils in the west, because they haven't got big buffers of calcium to neutralize acid. You can often stabilize a system by increasing the capacity of a buffer. But if a buffer is too big, the system gets inflexible. It reacts too slowly. Businesses invented just-in-time inventories, because occasional vulnerability to fluctuations or screw-ups is cheaper than certain, constant inventory costs—and because small-to-vanishing inventories allow more flexible response to shifting demand.

There's leverage, sometimes magical, in changing the size of buffers. But buffers are usually physical entities, not easy to change.

The acid absorption capacity of eastern soils is not a leverage point for alleviating acid rain damage. The storage capacity of a dam is literally cast in concrete. Physical structure is crucial in a system, but the leverage point is in proper design in the first place. After the structure is built, the leverage is in understanding its limitations and bottlenecks and refraining from fluctuations or expansions that strain its capacity.

7. REGULATING NEGATIVE FEEDBACK LOOPS.

Now we're beginning to move from the physical part of the system to the information and control parts, where more leverage can be found. Nature evolves negative feedback loops and humans invent them to keep system states within safe bounds.

A thermostat loop is the classic example. Its purpose is to keep the system state called "room temperature" fairly constant at a desired level. Any negative feedback loop needs a goal (the thermostat setting), a monitoring and signaling device to detect excursions from the goal (the thermostat), and a response mechanism (the furnace and/or air conditioner, fans, heat pipes, fuel, etc.).

A complex system usually has numerous negative feedback loops it can bring into play, so it can self-correct under different conditions and impacts. Some of those loops may be inactive much of the time—like the emergency cooling system in a nuclear power plant, or your ability to sweat or shiver to maintain your body temperature. One of the big mistakes we make is to strip away these emergency response mechanisms because they aren't often used and they appear to be costly. In the short term we see no effect from doing this. In the long term, we narrow the range of conditions over which the system can survive.

One of the most heartbreaking ways we do this is in encroaching on the habitats of endangered species. Another is in encroaching on our own time for rest, recreation, socialization, and meditation.

The "strength" of a negative loop—its ability to keep its appointed stock at or near its goal—depends on the combination of all its parameters and links—the accuracy and rapidity of monitoring the quickness and power of response, the directness and size of corrective flows.

There can be leverage points here. Take markets, for example, the negative feedback systems that are all but worshiped by economists—and they can indeed be marvels of self-correction, as prices vary to keep supply and demand in balance. The more the price—the central signal to both producers and consumers—is kept clear, unambiguous, timely, and truthful, the more smoothly markets will operate. Prices that reflect full costs will tell consumers how much they can actually afford and will reward efficient producers. Companies and governments are fatally attracted to the price leverage point, of course, all of them pushing in the wrong direction with subsidies, fixes, externalities, taxes, and other forms of confusion. The REAL leverage here is to keep them from doing it. Hence anti-trust laws, truth-in-advertising laws, attempts to internalize costs (such as pollution taxes), the removal of perverse subsidies, and other ways of leveling market playing fields.

The strength of a negative feedback loop is important **RELATIVE TO THE IMPACT IT IS DESIGNED TO CORRECT.** If the impact increases in strength, the feedbacks have to be strengthened too.

A thermostat system may work fine on a cold winter day—but open all the windows and its corrective power will fail. Democracy worked better before the advent of the brainwashing power of centralized mass communications. Traditional controls on fishing were sufficient until radar spotting and drift nets and other technologies made it possible for a few actors to wipe out the fish. The power of big industry calls for the power of big government to hold it in check; a global economy makes necessary a global government.

Here are some other examples of strengthening negative feedback controls to improve a system's self-correcting abilities: preventive medicine, exercise, and good nutrition to bolster the body's ability to fight disease, integrated pest management to encourage natural predators of crop pests, the Freedom of Information Act to reduce government secrecy, protection for whistle blowers, impact fees, pollution taxes, and performance bonds to recapture the externalized public costs of private benefits.

6. DRIVING POSITIVE FEEDBACK LOOPS.

A positive feedback loop is self-reinforcing. The more it works, the more it gains power to work some more.

The more people catch the flu, the more they infect other people. The more babies are born, the more people grow up to have babies. The more money you have in the bank, the more interest you earn, the more money you have in the bank. The more the soil erodes, the less vegetation it can support, the fewer roots and leaves to soften rain and runoff, the more soil erodes. The more high-energy neutrons in the critical mass, the more they knock into nuclei and generate more.

Positive feedback loops drive growth, explosion, erosion, and collapse in systems. A system with an unchecked positive loop ultimately will destroy itself. That's why there are so few of them.

Usually a negative loop kicks in sooner or later. The epidemic runs out of infectable people—or people take increasingly strong steps to avoid being infected. The death rate rises to equal the birth rate—or people see the consequences of unchecked population growth and have fewer babies. The soil erodes away to bedrock, and after a million years the bedrock crumbles into new soil—or people put up check dams and plant trees.

In those examples, the first outcome is what happens if the positive loop runs its course, the second is what happens if there's an intervention to reduce its power.

Reducing the gain around a positive loop—slowing the growth—is usually a more powerful leverage point in systems than strengthening negative loops, and much preferable to letting the positive loop run.

Population and economic growth rates in the world model are leverage points, because slowing them gives the many negative loops, through technology and markets and other forms of adaptation, time to function. It's the same as slowing the car when you're driving too fast, rather than calling for more responsive brakes or technical advances in steering.

The most interesting behavior that rapidly turning positive loops can trigger is chaos. This wild, unpredictable, unreplicable, and yet bounded behavior happens when a system starts changing much, much faster than its negative loops can react to it.

For example, if you keep raising the capital growth rate in the world model, eventually you get to a point where one tiny increase more will shift the economy from exponential growth to oscillation. Another nudge upward gives the oscillation a double beat. And just the tiniest further nudge sends it into chaos.

I don't expect the world economy to turn chaotic any time soon (not for that reason, anyway). That behavior occurs only in unrealistic parameter ranges, equivalent to doubling the size of the economy within a year. Real-world systems do turn chaotic, however, if something in them can grow or decline very fast. Fast-replicating bacteria or insect populations, very infectious epidemics, wild speculative bubbles in money systems, neutron fluxes in the guts of nuclear power plants. These systems are hard to control, and control must involve slowing down the positive feedbacks.

In more ordinary systems, look for leverage points around birth rates, interest rates, erosion rates, "success to the successful" loops, any place where the more you have of something, the more you have the possibility of having more.

5. INFORMATION FLOWS.

There was this subdivision of identical houses, the story goes, except that the electric meter in some of the houses was installed in the basement and in others it was installed in the front hall, where the residents could see it constantly, going round faster or slower as they used more or less electric-

ity. Electricity consumption was 30 percent lower in the houses where the meter was in the front hall.

Systems-heads love that story because it's an example of a high leverage point in the information structure of the system. It's not a parameter adjustment, not a strengthening or weakening of an existing loop. It's a **NEW LOOP**, delivering feedback to a place where it wasn't going before.

In 1986 the US government required that every factory releasing hazardous air pollutants report those emissions publicly. Suddenly everyone could find out precisely what was coming out of the smokestacks in town. There was no law against those emissions, no fines, no determination of "safe" levels, just information. But by 1990 emissions dropped 40 percent. One chemical company that found itself on the Top Ten Polluters list reduced its emissions by 90 percent, just to "get off that list."

Missing feedback is a common cause of system malfunction. Adding or rerouting information can be a powerful intervention, usually easier and cheaper than rebuilding physical structure.

The tragedy of the commons that is exhausting the world's commercial fisheries occurs because there is no feedback from the state of the fish population to the decision to invest in fishing vessels. (Contrary to economic opinion, the price of fish doesn't provide that feedback. As the fish get more scarce and hence more expensive, it becomes all the more profitable to go out and catch them. That's a perverse feedback, a positive loop that leads to collapse.)

It's important that the missing feedback be restored to the right place and in compelling form. It's not enough to inform all the users of an aquifer that the groundwater level is dropping. That could trigger a race to the bottom. It would be more effective to set a water price that rises steeply as the pumping rate exceeds the recharge rate.

Suppose taxpayers got to specify on their return forms what government services their tax payments must be spent on. (Radical democracy!) Suppose any town or company that puts a water intake pipe in a river had to put it immediately DOWNSTREAM from its own outflow pipe.

Suppose any public or private official who made the decision to invest in a nuclear power plant got the waste from that plant stored on his/her lawn.

There is a systematic tendency on the part of human beings to avoid accountability for their own decisions. That's why there are so many missing feedback loops—and why this kind of leverage point is so often popular with the masses, unpopular with the powers that be, and effective, if you can get the powers that be to permit it to happen or go around them and make it happen anyway.

4. THE RULES OF THE SYSTEM (INCENTIVES, PUNISHMENTS, CONSTRAINTS).

The rules of the system define its scope, boundaries, degrees of freedom. Thou shalt not kill. Everyone has the right of free speech. Contracts are to be honored. The president serves four-year terms and cannot serve more than two of them. Nine people on a team, you have to touch every base, three strikes and you're out. If you get caught robbing a bank, you go to jail.

Mikhail Gorbachev came to power in the USSR and opened information flows (*glasnost*) and changed the economic rules (*perestroika*), and look what happened.

Constitutions are strong social rules. Physical laws such as the second law of thermodynamics are absolute rules, if we understand them correctly. Laws, punishments, incentives, and informal social agreements are progressively weaker rules.

To demonstrate the power of rules, I ask my students to imagine different ones for a college. Suppose the students graded the teachers. Suppose you come to college when you want to learn something, and you leave when you've learned it. Suppose professors were hired according to their ability to solve real-world problems, rather than to publish academic papers. Suppose a class got graded as a group, instead of as individuals.

Rules change behavior. Power over rules is real power.

That's why lobbyists congregate when Congress writes laws, and why the Supreme Court, which interprets and delineates the Constitution—the rules for writing the rules—has even more power than Congress.

If you want to understand the deepest malfunctions of systems, pay attention to the rules, and to who has power over them.

That's why my systems intuition was sending off alarm bells as the new world trade system was explained to me. It is a system with rules designed by corporations, run by corporations, for the benefit of corporations. Its rules exclude almost any feedback from other sectors of society. Most of its meetings are closed to the press (no information, no feedback). It forces nations into positive loops, competing with each other to weaken environmental and social safeguards in order to attract corporate investment. It's a recipe for unleashing "success to the successful" loops.

3. THE POWER OF SELF-ORGANIZATION.

The most stunning thing living systems can do is to change themselves utterly by creating whole new structures and behaviors. In biological systems that power is called evolution. In human economies it's called technical advance or social revolution. In systems lingo it's called self-organization.

Self-organization means changing any aspect of a system lower on this list—adding or deleting new physical structure, adding or deleting negative or positive loops or information flows or rules. The ability to self-organize is the strongest form of system resilience, the ability to survive change by changing.

The human immune system can develop responses to (some kinds of) insults it has never before encountered. The human brain can take in new information and pop out completely new thoughts.

Self-organization seems so wondrous that we tend to regard it as mysterious, miraculous. Economists often model technology as literal manna from heaven—coming from nowhere, costing nothing, increasing the productivity of an economy by some steady percent each year. For centuries people have regarded the spectacular variety of nature with the same awe. Only a divine creator could bring forth such a creation.

In fact the divine creator does not have to produce miracles. He, she, or it just has to write clever **RULES FOR SELF-ORGANIZATION.** These rules govern how, where, and what the system can add onto or subtract from itself under what conditions.

Self-organizing computer models demonstrate that delightful, mind-boggling patterns can evolve from simple evolutionary algorithms. (That need not mean that real-world algorithms are simple, only that they can be.) The genetic code that is the basis of all biological evolution contains just four letters, combined into words of three letters each. That code, and the rules for replicating and rearranging it, has spewed out an unimaginable variety of creatures.

Self-organization is basically a matter of evolutionary raw material—a stock of information from which to select possible patterns—and a means for testing them. For biological evolution the raw material is DNA, one source of variety is spontaneous mutation, and the testing mechanism is something like punctuated Darwinian selection. For technology the raw material is the body of understanding science has accumulated. The source of variety is human creativity (whatever THAT is) and the selection mechanism is whatever the market will reward or whatever governments and foundations will fund or whatever tickles the fancy of crazy inventors.

When you understand the power of self-organization, you begin to understand why biologists worship biodiversity even more than economists worship technology. The wildly varied stock of DNA, evolved and accumulated over billions of years, is the source of evolutionary potential, just as science libraries and labs and scientists are the source of technology potential. Allowing species to go extinct is a systems crime, just as randomly eliminating all copies of particular science journals, or particular kinds of scientists, would be.

The same could be said of human cultures, which are the store of behavioral repertoires accumulated over not billions, but hundreds of thousands of years. They are a stock out of which social evolution can arise. Unfortunately, people appreciate the evolutionary potential of cultures even less than they understand the potential of every genetic variation in ground squirrels. I guess that's because one aspect of almost every culture is a belief in the utter superiority of that culture.

Any system, biological, economic, or social, that scorns experimentation and wipes out the raw material of innovation is doomed over the long term on this highly variable planet.

The intervention point here is obvious but unpopular. Encouraging diversity means losing control. Let a thousand flowers bloom and **ANYTHING** could happen!

Who wants that?

2. THE GOALS OF THE SYSTEM.

Right there, the push for control, is an example of why the goal of a system is even more of a leverage point than the self-organizing ability of a system.

If the goal is to bring more and more of the world under the control of one central planning system (the empire of Genghis Khan, the world of Islam, the People's Republic of China, Wal-Mart, Disney), then everything further down the list, even self-organizing behavior, will be pressured or weakened to conform to that goal.

That's why I can't get into arguments about whether genetic engineering is a good or a bad thing. Like all technologies, it depends upon who is wielding it, with what goal. The only thing one can say is that if corporations wield it for the purpose of generating marketable products, that is a very different goal, a different direction for evolution than anything the planet has seen so far.

There is a hierarchy of goals in systems. Most negative feedback loops have their own goals—to keep the bath water at the right level, to keep the room temperature comfortable, to keep inventories stocked at sufficient levels. They are small leverage points. The big leverage points are the goals of entire systems.

People within systems don't often recognize what whole-system goal they are serving. To make profits, most corporations would say, but that's just a rule, a necessary condition to stay in the game. What is the point of the game? To grow, to increase market share, to bring the world (customers, suppliers, regulators) more under the control of the corporation, so that its operations become even more shielded from uncertainty. That's the goal of a cancer cell too and of every living population. It's only a bad one when it isn't countered by higher-level negative feedback loops with goals of keeping the system in balance. The goal of keeping the market competitive has to trump the goal of each corporation to eliminate its competitors. The goal of keeping populations in balance and evolving has to trump the goal of each population to commandeer all resources into its own metabolism.

I said a while back that changing the players in a system is a low-level intervention, as long as the players fit into the same old system. The exception to that rule is at the top, if a single player can change the system's goal.

I have watched in wonder as—only very occasionally—a new leader in an organization, from Dartmouth College to Nazi Germany, comes in, enunciates a new goal, and single-handedly changes the behavior of hundreds or thousands of millions of perfectly rational people.

That's what Ronald Reagan did. Not long before he came to office, a president could say, "Ask not what government can do for you, ask what you can do for the government," and no one even laughed. Reagan said the goal is not to get the people to help the government and not to get government to help the people, but to get the government off our backs. One can argue, and I would, that larger system changes let him get away with that. But the thoroughness with which behavior in the US and even the world has been changed since Reagan is testimony to the high leverage of articulating, repeating, standing for, insisting upon new system goals.

1. THE MINDSET OR PARADIGM OUT OF WHICH THE SYSTEM ARISES.

Another of Jay Forrester's systems sayings goes: It doesn't matter how the tax law of a country is written. There is a shared idea in the minds of the society about what a "fair" distribution of the tax load is. Whatever the rules say, by fair means or foul, by complications, cheating, exemptions or deductions, by constant sniping at the rules, the actual distribution of taxes will push right up against the accepted idea of "fairness."

The shared idea in the minds of society, the great unstated assumptions—unstated because unnecessary to state; everyone knows them— constitute that society's deepest set of beliefs about how the world works. There is a difference between nouns and verbs. People who are paid less are worth less. Growth is good. Nature is a stock of resources to be converted to human purposes. Evolution stopped with the emergence of *Homo sapiens*. One can "own" land. Those are just a few of the paradigmatic assumptions of our culture, all of which utterly dumfound people of other cultures.

Paradigms are the sources of systems. From them come goals, information flows, feedbacks, stock, flows.

The ancient Egyptians built pyramids because they believed in an afterlife. We build skyscrapers, because we believe that space in downtown cities is enormously valuable. (Except for blighted spaces, often near the skyscrapers, which we believe are worthless.) Whether it was Copernicus and Kepler showing that the earth is not the center of the universe, or Einstein hypothesizing that matter and energy are interchangeable, or Adam Smith postulating that the selfish actions of individual players in markets wonderfully accumulate to the common good.

People who manage to intervene in systems at the level of paradigm hit a leverage point that totally transforms systems.

You could say paradigms are harder to change than anything else about a system, and therefore this item should be lowest on the list, not the highest. But there's nothing physical or expensive or even slow about paradigm change. In a single individual it can happen in a millisecond. All it takes is a click in the mind, a new way of seeing. Of course individuals and societies do resist challenges to their paradigm harder than they resist any other kind of change.

So how do you change paradigms? Thomas Kuhn, who wrote the seminal book about the great paradigm shifts of science, has a lot to say about that. In a nutshell, you keep pointing at the anomalies and failures in the old paradigm, you come yourself, loudly, with assurance, from the new one, you insert people with the new paradigm in places of public visibility and power. You don't waste time with reactionaries; rather you work with active change agents and with the vast middle ground of people who are open-minded.

Systems folks would say one way to change a paradigm is to model a system, which takes you outside the system and forces you to see it whole. We say that because our own paradigms have been changed that way.

0. THE POWER TO TRANSCEND PARADIGMS.

Sorry, but to be truthful and complete, I have to add this kicker.

The highest leverage of all is to keep oneself unattached in the arena of paradigms, to realize that NO paradigm is "true," that even the one that sweetly shapes one's comfortable worldview is a tremendously limited understanding of an immense and amazing universe.

It is to "get" at a gut level the paradigm that there are paradigms, and to see that itself is a paradigm, and to regard that whole realization as devastatingly funny. It is to let go into Not Knowing.

People who cling to paradigms (just about all of us) take one look at the spacious possibility that everything we think is guaranteed to be nonsense and pedal rapidly in the opposite direction. Surely there is no power, no control, not even a reason for being, much less acting, in the experience that there is no certainty in any worldview. But everyone who has managed to entertain that idea, for a moment or for a lifetime, has found it a basis for radical empowerment. If no paradigm is right, you can choose one that will help achieve your purpose. If you have no idea where to get a purpose, you can listen to the universe (or put in the name of your favorite deity here) and do his, her, its will, which is a lot better informed than your will.

It is in the space of mastery over paradigms that people throw off addictions, live in constant joy, bring down empires, get locked up or burned at the stake or crucified or shot, and have impacts that last for millennia.

Back from the sublime to the ridiculous, from enlightenment to caveats. There is so much that has to be said to qualify this list. It is tentative and its order is slithery. There are exceptions to every item on it. Having the list percolating in my subconscious for years has not transformed me into a Superwoman. I seem to spend my time running up and down the list, trying out leverage points wherever I can find them. The higher the leverage point, the more the system resists changing it—what's why societies rub out truly enlightened beings.

I don't think there are cheap tickets to system change. You have to work at it, whether that means rigorously analyzing a system or rigorously casting off paradigms. In the end, it seems that leverage has less to do with pushing levers than it does with disciplined thinking combined with strategically, profoundly, madly letting go.

64

Possibilities and Limits
of U.S. Microenterprise Development
for Creating Good Jobs and Increasing
the Incomes of the Poor

FRANK LINDENFELD

• • •

REFLEXIVE STATEMENT

It is surely no news to this audience that there is a high and growing level of poverty and economic inequality in the United States. Politicians of both major parties have shredded many of the last remains of welfare, leaving over 36 million people to live below the official poverty level without a safety net (U.S. Bureau of the Census 1997). In 1996, 13.6% of the workforce was jobless including the unemployed, the so-called discouraged workers, and those who find only part time hours but who want to work full time.[1] Many of those who do have jobs don't earn enough to get out of poverty even if they work full time, because the minimum wage rates are so low. There are just not enough well-paying jobs to go around.

There are many possible solutions. The conservative prescription for increasing employment—keeping wages low, cutting taxes of the rich, and forcing welfare recipients into low-paying jobs—has only increased the gap between rich and poor. Alternative reforms, however, could substantially reduce poverty and unemployment. For example:

- A guaranteed minimum income of $20,000 per year, regardless of a person's employment status

- An increase in the minimum wage to $10/hour

- A reduction of the standard work week to 30 hours

- More government funded jobs (e.g., doubling the number of teachers in the public schools)

- Free tuition at public colleges and universities

- Higher progressive taxes on wealth, corporate and personal income, especially for those earning over $150,000 per year

- Taxes on transfer of investment capital

- Greater public accountability of large private corporations including revocation of charters in cases of fraud

- Banning imports of goods or services produced under exploitative labor conditions

- Promoting an increase of worker cooperative networks

- Increasing the extent of employee ownership of stocks of the companies where they work

- Strengthening local communities and creating jobs by promoting the development of small businesses and microenterprises, emphasizing local production for local and regional consumption.

Here I will focus on the last alternative—strengthening local community economies, and creating and maintaining jobs by starting or expanding small businesses and microenterprises (small, often home-based firms that provide work for their owners and up to four employees). By itself, this strategy cannot eliminate unemployment and poverty, but it can provide work and higher incomes for some people. I have witnessed the success of this approach first-hand.

My concern with issues of poverty and jobs led me to become interested in microenterprise development in North Central Pennsylvania. Several years ago, I began working with my colleague at Bloomsburg University, Dr. Pamela Wynn, on a microenterprise project funded by the Center for Rural Pennsylvania.[2] This collaboration resulted in the development of the Local Enterprise Assistance Program (LEAP). LEAP's objectives are to create jobs by promoting self-employment—especially for low and moderate income households—and to further community economic development in the rural area around Bloomsburg, Pennsylvania. LEAP is a partnership between Bloomsburg University and the Rural Enterprise Development Corporation, a local non-profit community group. Despite a relatively low level of funding, from 1994–1998 LEAP provided business training and technical assistance for over 80 persons and made three loans from its own resolving loan fund.

AN OVERVIEW
OF MICROENTERPRISE

I will briefly survey some of the various types of microenterprises. Then, I will review the potential and limitations of microenterprise development programs in the United States and offer suggestions that may help overcome some of these limitations.

There are some 10–20 million businesses in this country, depending on who is counting. According to Dun & Bradstreet, almost two-thirds are microenterprises.[3] Microentrepreneurs may include: self-employed doctors, lawyers, accountants, management consultants, computer consultants, owner-operators of small machine shops and auto repair shops, craftspeople, building contractors, restauranteurs, mom & pop grocery store owners, beauticians, small retail shop owners, child care providers, house cleaners, and street vendors.

Microenterprises vary in many ways. While self-employment may provide some with six-figure incomes, many microentrepreneurs earn only a few thousand dollars per year. The major dimensions along which microenterprises vary are: the *number of jobs* they provide—from the part-time employment of the business owner up to full-time jobs for four employees; the *gross annual sales* of goods and services—from a few hundred dollars to millions; and the *percent of sales "exported"* to other regions or countries vs. sales to local customers. Most of the businesses assisted by microenterprise development programs such as LEAP have gross annual sales of under $30,000, sell primarily within the local community, and provide employment solely for their owners.[4]

MICROENTERPRISE
DEVELOPMENT PROGRAMS
IN THE UNITED STATES

Microenterprise development programs try to increase the chance of business success for their participants by providing some combination of business training, technical assistance and access to credit. They offer business training classes ranging from a few hours to 60 hours or more, often with the goal of having each participant complete a business plan. They provide participants with technical assistance using consultants and community volunteers to help them with questions about market research, pricing, marketing, management, accounting, insurance and finance. Many provide credit through contacts with banks and community development loan funds; some also operate their own loan funds.

As of 1998, there were an estimated 300 or more private, non-profit microenterprise development intermediary organizations in the United States. Most began during the late 1980s and early 1990s and most are affiliated with the Association for Enterprise Opportunity, a national trade organization. In addition, numerous statewide associations have sprung up to promote knowledge-sharing among member groups to increase access to state funding and to influence public policies to encourage microenterprise development. In Pennsylvania, for example, the PA Microenterprise Coalition has 15 members, including LEAP.

Microenterprise organizations promote small business development but can go far beyond this in their social concerns and practical business support for low income as well as women and minority participants. The goals of such microenterprise programs include not only job creation and retention, but broader purposes. These include increasing the income of low and moderate income families, promoting personal growth among participants, and furthering civic development of the communities they live in.

The Self Employment Learning Project profiled seven successful microenterprise development groups, one of which began in 1983 and the rest between 1987 and 1989.[5] By 1994 these organizations had served over 13,000 clients and had extended business loans totaling more than $8 million. Of their clients, 73% were women; 60% were from a minority racial or ethnic group; and 43% had incomes below the poverty level prior to receiving assistance (Edgcomb, Klein and Clark 1996). These authors divide microenterprise programs into three categories: those that are primarily individual loan programs, such as PEPP/MICRO in Arizona; those that focus on group lending, such as the solidarity circles of the Los Angeles-based Coalition for Womens Economic Development; and those like Women Venture in St. Paul, Minnesota, that use training strategies where access to credit is an option.

The best programs in my opinion have two characteristics. First, they operate on a scale large enough to make a difference in reducing poverty in their communities. An example is Working Capital, with branches in New England, Delaware and Florida. Second, they offer comprehensive development help that includes business training, technical assistance, access to business credit, and encourage the formation of peer enterprise associations for mutual aid.

OUTCOMES

Personal Outcomes

Participants in microenterprise training programs increase their business and personal skills and they report increases in self-confidence and self-esteem. Many program participants show increases in income, either from self-employment or working for others.

An analysis of self-employment assistance demonstration programs for the unemployed in Washington state and Massachusetts compared unemployed persons randomly assigned to a self-employment assistance "treatment" group with those assigned to a control group. Some participants in both groups created jobs for themselves through self-employment. The study showed that self-employment programs can help some of the unemployed by increasing their chances of success in running a business. The demonstration projects in both states increased the likelihood of participants' self-employment and total time in employment, and reduced the length of unemployment. Both showed increases in income for those in the self-employment group. In Massachusetts, the difference in annual income between the self-employment "treatment" group and the control group was almost $6,000 (Benus, Johnson, Grover, and Shen 1994).

A study of self-employment among welfare recipients showed that "there is a pool of current welfare recipients with self-employment experience who could benefit from microenterprise training and loan programs, if earnings from self-employment are not expected to result in economic self-sufficiency" (Spalter-Roth, Soto, Zandniapour and Braunstein 1994, 3). Welfare recipients who participate in self-employment assistance programs have increased their incomes, where microenterprise income is part of an income package that includes some government assistance. National sample surveys conducted in the 1980s showed family incomes of welfare recipients who were also self-employed was $12,254, compared with $11,753 for those on welfare who also worked for others, and $8,547 for non-employed welfare recipients (Spalter-Roth et al. 1994).

Three of the programs tracked by the Self Employment Learning Project—Good Faith Fund, Womens Self Employment Project, and Institute for Social & Economic Development—actively target welfare clients to help them increase their incomes by developing a microenterprise, enabling some to leave welfare. Typically, participants patch together a total income from various sources, including microenterprise and welfare. A combined analysis of 302 participants in five of the programs tracked by the project shows a median annual business income (salary plus owner's draw) of $13,200, with 23% earning over $24,000, and 26% under $6,000 (Clark and Huston 1993).

Business Outcomes

Each year, hundreds of thousands of new businesses are begun in the United States, while other hundreds of thousands fail. Census data illustrate the volatility of U.S. businesses. During the period 1989–1992, for example, about 6 million workers lost jobs in retail and service establishments through the demise of the companies that employed them, while over 8 million found jobs in newly established retail and service businesses.[6] According to the U.S. Small Business Administration, three-fourths of all United States firms existing in 1992 survived four or more years, at least until 1996. Of the firms with no employees in 1992, 72/4% were still in business four years later; of those with 1–4 employees, almost 90% survived four years.[7]

How effective are microenterprise development programs in helping their clients beat the odds, and not only to survive but to thrive as well? Existing evidence indicates that they do in fact help businesses survive and grow. A comprehensive longitudinal study of 405 microenterprises conducted by the Self Employment Learning Project found significant positive outcomes of self-employment. One-third reported positive changes in net worth. A large majority of poor microentrepreneurs showed significant income gains over time (62%) and 23% had income gains large enough to move out of poverty. In this panel study, 73% of businesses operated by low income respondents (and 78% of the total sample) that took part in the first wave interviews were still in operation when re-interviewed three years later (Self Employment Learning Project 1996). This shows a survival rate similar to the overall U.S. rate for small businesses.

Data based on self-reports from nine members of the PA Microenterprise Coalition for July 1995–June 1997 indicates that they served a total of over 2,300 participants, 70% of whom had low or moderate incomes when they began the program. The aggregate data for these programs show a survival rate of 86% for the 668 small businesses they assisted during this two-year period.[8] About half of the businesses provided full-time work for their owners. In addition, these businesses hired a total of 417 full-time and 100 part-time employees.[9]

LIMITATIONS OF MICROENTERPRISE

Self-employment appeals to people because they are in charge of their work lives and because it enables them to find ways to make money by providing products or services through their own unique talents and skills. However, there are a number of limitations of microenterprise development as a strategy for helping poor and moderate income people. In the first place, self-employment can help only a small fraction of them. Only some of us are prepared to put in the long hours and intense effort to run our own businesses, and have the combination of talents, background and personal characteristics that make for success in self-employment. Given the choice, many of us would prefer to collect a steady paycheck, letting someone else take the risks and worry about the hundreds of details involved in running a business. How many low and moderate income people would like to run their own businesses? It's hard to say, though this is a dream shared by millions of people in this country from all walks of life. I would guess that those seriously interested in running their own business and obtaining any necessary additional training to do so to be somewhere between 5%–20% of low and moderate income people. A study of unemployment insurance claimants in Washington State, for example, reports that of the over 42,000 new unemployment insurance claimants notified of a meeting to explain a self-employment demonstration program, 7.5% showed up. Given the millions in the United States whose family incomes barely reach $20,000 per year today, even 5% represents a significantly large pool of people potentially interested in self-employment.

A second major limitation of many microenterprises is that while they may provide jobs for the owners and their employees, their earnings are often relatively low. To begin with, only some entrepreneurs are able to successfully expand their business into full-time employment. To earn enough for their families, microenterprise owners may hold down a full-time or part-time job working for someone else while running their business part-time. After several years of struggle to get established, some microentrepreneurs are able to earn a decent income and to provide high-paying jobs with benefits for their employees. All too often, however, the income of the self-employed and those they hire barely reaches existing minimum wage levels.

Third, a major problem for many is that they just don't have enough customers and sales. Microenterprises operate in a marketplace where it is difficult for a business to survive if competitors can offer significantly lower prices because their labor costs are lower or they can purchase items by the truckload. Small business owners are often forced to exploit themselves and their employees just to stay in business.

Fourth, even if they want to run a business, people of modest means run into the almost insuperable obstacle of lack of equity capital. *The defining characteristics of the poor is that they lack money!* For example, many don't have enough to pay for basic subsistence needs, let alone cash for a down payment on a truck or the first two months of payments for motor vehicle insurance.

A final issue that deserves mention is that there are thousands of low income people that run part-time low profile businesses "under the table." They do not report their income on tax returns, don't carry the necessary insurance, lack business licenses, etc. Many small entrepreneurs do not dare to declare their income for fear that they can't afford the self-employment taxes, or that it will interfere with their eligibility for government benefits. In addition, many microentrepreneurs find the cost of legally hiring another person and paying for their social security and workmens' compensation taxes prohibitive. When they do provide jobs for others, it is often on an unofficial cash basis.

SUGGESTIONS FOR OVERCOMING SOME OF THESE LIMITATIONS

Not all of these problems can necessarily be overcome, though there are some potential solutions worth considering. Part of the answer for the low incomes generated by many microenterprises may be community programs to raise wage levels. Sustainable wage campaigns have worked for increases in minimum wage levels in a number of United States cities with some success. A doubling of the current U.S. minimum wage would be even better.

Another answer is to supplement dollars with a locally traded alternative currency as part of a community barter network. Ithaca Hours in Ithaca, New York, for example, has enabled thousands of community members to trade goods and services with each other in exchange for a local scrip, the Ithaca Hour, worth the equivalent of $10 U.S. Ithaca Hours have helped push local wages upward. Some farmers, for example, offer agricultural workers a choice of working for hourly wages of U.S. $5–$6, or one Ithaca Hour. The alternative currency is accepted by hundreds of local businesses for partial payment of goods and services. A few accept it as full payment, without requiring U.S. dollars. As of Spring, 1997, the 5,700 Ithaca Hours then in circulation had generated over $1.5 million in local transactions.[10]

One of the solutions for the low sales volume that microenterprises often experience is to increase their "exports" to markets beyond their locality and region. Intermediary organizations can help with this by publishing catalogs and brochures for participating businesses, and helping distribute them to potential customers in other states or countries. "Export" sales may generate much-needed local income, though dependence on such sales may make the local producers vulnerable to fluctuations in outside demand.

The other side of this coin is the destructive impact of cheap "imports" from outside the community, offered for example through Walmarts and other retail chains. Growing food locally, and providing services and manufacturing through local efforts, is vital to the economic health of a community or a region. Even when local products or services cost a little more, it makes sense to buy from local merchants and service providers to help them in business, keeping the local economy healthy by supporting local jobs.

On the most general level, the problem of stagnant wages and large-scale unemployment is a result of the process of globalization carried out by transnational corporations over the last several decades. Globalization undercuts U.S. labor and environmental standards and eliminates domestic jobs. The best cure for this is to move away from globalization, toward an economy based on local production for local and regional consumption (see the various articles in Mander and Goldsmith 1996). A conscious policy of import substitution through small business and microenterprise production could be part of a renewal of local com-

munities and help them keep local wealth from being siphoned out to metropolitan centers.

The lack of equity capital is another critical weakness of microenterprise. There are ways around this, including sweat equity and group savings clubs that provide a pool of capital distributed periodically to each member in turn. One of the most promising solutions is the individual development account (IDA). IDAs are savings accounts set up by poor people that can be used to pay for education, down payment on a home, or capital to start a business. Typically, anti-poverty agencies encourage the poor to save by matching their IDA savings, with the help of capital contributed by foundations or government.[11] The accounts would be provided through local non-profit community development organizations.

The initial capital requirements for many businesses are quite substantial, far beyond the $10,000 that would be set aside in savings at the end of four years using IDA accounts of $500 in savings per year, with a 4:1 match. In these cases, it makes sense to consider cooperative ventures. Twenty persons pooling $10,000 each would be able to make an equity investment of $200,000 to start a cooperative business enterprise and to leverage additional loan capital.

POLITICAL OBJECTIONS

In addition to the various practical limitations of microenterprise development as a strategy for helping to better the lot of working people, there is also a political objection. To put it simply, the capitalist system itself fosters low wages, unemployment, ill treatment of workers, etc. Entrepreneurial development programs generally carry an ideological message that favors private enterprise and business interests. The capitalist system, and especially its embodiment in the major transnational corporations, militates against the interest of low and moderate income people, for example by keeping their wages low and cutting job benefits. Yet the message of microenterprise development programs—sometime hidden but often plainly evident—is that capitalism is good in that it enables everyone to have a chance at the American dream of owning a business and eventually becoming wealthy through entrepreneurial endeavor.

Most of us see this as a self-serving ideology that promotes the interests of the rich by making working people, especially those who are poor, believe the system is legitimate and that they themselves could achieve social mobility within it by the entrepreneurial route. Admittedly, it is all too easy for microenterprise developers to accept this myth without challenge, letting low and moderate income people who participate in their programs believe that with enough hard work they, too, could become rich. I don't think this is inevitably the case; we can raise the consciousness of participants in business training programs by helping them understand the structural barriers to success for microenterprise, and how the system is rigged in favor of the large corporations. Even if they won't become millionaires through private enterprise, they may nevertheless be able to raise their incomes and provide a better life for their families.

We also need to help those we work with to understand the desirability of providing adequate wages and benefits, and good working conditions for those that they themselves hire. It is not so much private enterprise as its embodiment in large impersonal corporations that is a major problem with capitalism, together with the undue influence of corporations and the wealthy on the political process. This needs to be made clear in our educational work.[12]

MICROENTERPRISE AND PUBLIC POLICY

Public policy should encourage the formation and stabilization of microenterprise development organizations, as one means of furthering community economic development and helping to create jobs. Though not *the* solution to poverty, microenterprise development groups can help generate more good jobs for the self-employed

and those they hire, and increase the incomes of program participants.

The quality of life for many low income people would be improved if they were provided with help that resulted in their successful self-employment. Providing such help will cost money for staff; training programs that work with low income participants will generally be more expensive than those designed exclusively for middle-class persons who have had the benefit of better education. With economies of scale, however, microenterprise development programs may provide training, technical assistance and business loans for as low as $2000 per participant. Assuming that approximately 1000 non-profit agencies throughout the United States work with an average of 500 participants per year (a potential of 500,000 clients per year), the total cost would be only $1 billion. If only one-third of these participants actually start or expand businesses, this would result in adding or stabilizing as many as 200,000 jobs per year, 165,000 though self-employment and at least 35,000 others hired by them. This is a benefit that would be definitely worth many times more than its dollar cost.

During the 1980s and 1990s, state and federal agencies have provided a major part of the funding for non-profit microenterprise development organizations. They have done so through such avenues as the Small Business Administration's microenterprise program, Community Development Block Grants, and refugee resettlement programs. More federal and state funding for microenterprise programs would be an inexpensive investment that can pay large dividends by helping to create jobs and reduce unemployment and poverty.

FUTURE RESEARCH

Better outcome studies are needed to measure the effects of microenterprise development programs, and to compare businesses assisted with those that did not receive agency help. There are many other questions that still need to be answered by future research. For example, are some aspects of business development programs more useful than others in promoting positive outcomes? Under what conditions might microenterprise programs be even more effective in creating and retaining jobs and promoting local economic development? Under what conditions can microenterprises successfully compete with big business?

SUMMARY

By itself, microenterprise development is hardly a cure for mass poverty and unemployment. It is, however, potentially *part* of the answer. Preliminary evidence indicates that many clients of microenterprise development programs have succeeded in increasing their business sales as well as their family incomes. In addition, many have benefitted from personal growth and an increase in their self-esteem and self-confidence. Even when they have not pursued their own businesses, program participants seem to have an easier time finding jobs than they did before taking the business training classes.

An investment of $1 billion in operational funding for microenterprise development in the U.S. could lead to an increase in the number of development agencies and build the capacity of existing ones, enabling them to help create or sustain 200,000 jobs or more per year for microenterprise owners and those they hire. This would make a positive contribution to reducing unemployment and increasing income for the poor and disenfranchised of our society. Promoting microenterprises and small businesses will also have a beneficial effect on government revenue, reducing welfare expenditures and turning some former welfare recipients into taxpaying contributors. Microenterprise development, as part of a combined package of reforms that emphasize local economic development—including higher minimum wages, a guaranteed income, a shorter work week, a ban on imports made by exploited workers, and tax increases for the rich—could help eliminate much of the unemployment and poverty in this country.

NOTES

[1] In 1996 the U.S. labor force was 133,943,000 with 7,236,000 officially unemployed. An additional 5,451,000 were counted as not in the labor force but wanted a job. Also, 6,293,000 persons in the labor force worked less than 14 hours (U.S. Bureau of the Census, 1997).

[2] The Center for Rural Pennsylvania is an agency of the PA state legislature which provides funding for rural research and action projects.

[3] The Statistical Abstract of the U.S. shows 16,154,000 non-farm proprietorships and 1,494,000 partnerships for 1994. Corporate data for 1994 were not available, but in 1993 there were 3,965,000 corporations (U.S. Bureau of the Census, 1997). The Dun and Bradstreet Census, which probably omits many of the smallest enterprises, lists 10,382,000 businesses for 1997. Fully 65% of the D&B listed companies had 0–4 employees!

[4] According to the Self Employment Learning Project's study of 405 microentrepreneurs tracked over 5 years, 47% have gross sales under $1,000 per month, 21% sales of $1,000–2499 per month, and 25% sales of $2500 per month or more. Two thirds employ the owner-operator alone, while 34% also provide jobs for others. (Self Employment Learning Project 1966).

[5] They are Women Venture (St. Paul, Minnesota) which traces its history to 1983, Coalition for Womens Economic Development (Los Angeles), Good Faith Fund (Pine Bluff, Arkansas), Institute for Social and Economic Development (Iowa City), Coalition PEPP/MICRO (Tucson), Rural Economic Development Center (Raleigh, NC), and Womens Self Employment Project (Chicago).

[6] "Job growth and destruction in the U.S., 1989–1992," U.S. Census Bureau.

[7] SBA Office of Advocacy, "Characteristics of Small Business Employees and Owners, 1997," table 5.8. (www.sba.gov/ADVO/stats/ch_emp_o). The SBA data exclude both the very largest companies as well as all small businesses with gross annual receipts of less than $500 in 1992.

[8] Of the total businesses that were counted as still in existence, an estimated half had their initial contact with a microenterprise program during the second 12 months.

[9] Impact study data collected for presentation to legislative colloquium, available from PA Microenterprise Coalition, c/o MEDA, 2501 Oregon Pike, Ste. 2, Lancaster, PA 17601. For other discussions of

microenterprise in Pennsylvania, see Webb (1998) and Kraybill and Nolt (1995).

[10]For more on alternative currency, see the spring 1997 special issue of *Yes!* Krimerman (1995) and Meeker-Lowry in Mander and Goldsmith (1996).

[11]The fall 1997 issue of *Assets,* a newsletter published by the Washington, DC Corporation for Enterprise Development describes a nationwide policy demonstration of IDAs for poor people, involving 13 agencies in as many states with 2,000 IDA participants. In Ithaca, NY, for example, the Alternatives Federal Credit Union is setting up 100 IDAs in which poor people save at least $20 per month, and match participant savings by a contribution of 3:1 for four years.

[12]My own ideal would be a mixed economy, with a large cooperative or worker-owned sector, a small business and microenterprise sector, and a public sector. There would be room for corporations, but they would be more closely regulated, and those with over $1 billion in assets would be subject to strict measures of public accountability.

REFERENCES

Benus, Jacob M. , Terry R. Johnson, Michell Wood, Neelima Grover, and Theodore Shen. 1994. *Self Employment Programs: A New Re-Employment Strategy.* Unemployment Insurance Occasional Paper 95-4. Washington, DC: U.S. Department of Labor.

Clark, Peggy, and Tracy Huston. 1993. *Assisting the Smallest Businesses: Assessing Microenterprise Development as a Strategy for Boosting Poor Communities.* Washington, DC: The Aspen Institute.

Edgcomb, Elaine, Joyce Klein, and Peggy Clark. 1996. *The Practice of Microenterprise in the U.S.* Washington, DC: The Aspen Institute.

Kraybill, Donald B., and Steven M. Nolt. 1995. *Mish Enterprise: From Plows to Profits.* Baltimore: Johns Hopkins University Press.

Krimerman, Len. 1995. "Local Currencies: Their Pitfalls & Potential." *Grassroots Economic Organizing Newsletter,* 19.

Mander, Jerry, and Edward Goldsmith (eds.). 1996. *The Case Against the Global Economy, and for a Turn Toward the Local.* San Francisco: Sierra Club Books.

Self Employment Learning Project. 1996. "Microenterprise Assistance: What Are We Learning About Results?" Paper presented at the 1996 annual meetings of the Association for Enterprise Opportunity, Providence, Rhode Island.

Spalter-Roth, Roberta, Enrique Soto, Lily Zandniapour, and Jill Braunstein. 1994. *Micro-Enterprise and Women: The Viability of Self-Employment as a Strategy for Alleviating Poverty.* Washington, DC: Institute for Women's Policy Research.

U.S. Bureau of the Census. 1997. *Statistical Abstract of the U.S., 1977.* Washington, D.C.

Webb, Ronald J. 1998. *Microenterprise—An Economic Development Strategy: Lessons Learned in Pennsylvania.* Washington, D.C.: Research Institute for Small & Emerging Business.

65

Redefining Leadership

LINDA STOUT

When I first started to talk about peace issues and the concerns of my community, I was advised by middle-class people that I wasn't the right spokesperson. But I found that my words and my life experiences connected to other people's experience. I knew I didn't fit the "traditional" definition of a leader— the white educated male who talked facts and statistics, not feelings and experiences. I was a low-income woman who used the words of my community to share my feelings and experiences. When I spoke truthfully and honestly about what I saw and how things needed to change, people in my community wanted to join in. I began to realize that people in my community were looking to me as a leader. Because of my own experiences, I realized that there could be, and in fact, had to be, a different definition of a leader. As a result, when I founded PPP, one of the most important things I promised was that we would work to make everyone feel that she or he could be a leader. Instead of using the traditional structure of leadership, with one person at the top, we set out to define a new vision of leaders and leadership.

Redefining leadership is really a survival issue for people of color, women, and low-income communities. Traditional leadership has historically excluded us for the most part. Most of the exceptions are people who have been recognized because the traditional leaders in the power structure have identified them as representatives of our communities. Often these people have ended up being co-opted (they have taken on the identity and values of those in power); sometimes they have been chosen to begin with because they looked or acted more like people in the power structure than people in low-income communities. You can see that this is the opposite of low-income people finding our voices. To preserve our communities and the integrity of our lives, we must find leaders outside the traditional power structure.

Even within progressive organizations we need to redefine leadership. The most common leadership model does not encourage people to become leaders. Organizations usually have a single leader, and even if that person is very positive and in touch with the community, everyone else becomes dependent on that one individual. If something happens to that person the organization will likely fall apart. At PPP, we have worked toward building a model of shared leadership.

Out of an organization that develops many leaders come many strategies, styles, and ideas for making change. This variety makes an organization more vital and more likely to reach lots of different people. In contrast, if one leader defines an organization, then only the people who can identify with that single person will feel included. If one leader makes the decisions on how things should be done, the strategies become limited to a single personal style. In the early years of PPP, I used to think that I needed to let other people do things their way so that they would learn from their own mistakes. What I learned, though, is that there are a lot of *different* ways to successfully reach a goal. I would think I knew the "best way," only to find out that a totally different approach was even more successful. Through the shared leadership model at PPP, I've learned from others new ways to do our work. And I have also learned how important it is to have a range of different approaches to draw upon in achieving our goals.

The most powerful leadership occurs when other members identify with the leader and feel that they could also work in that role. Leaders are more effective when other people can say, "Oh, I can do that" or "That's just how I feel too"—instead of focusing on what a brilliant or exceptional person the leader is. Traditionally leaders are seen as separate from their communities. They're regarded as smarter, more talented, more powerful people whose individual personalities set them apart. But if a leader truly shares an identity with other people, then the people see they have something in common with a leader and together, they move forward.

It is often said that there are too many leaders and not enough followers. It is also said that leaders are born, not made. Both of these sayings come out of a traditional model of leadership that honors hierarchical power and class status. In economic systems like capitalism, where only a few people can be at the top (those who own the companies) and most people have to be at the bottom, a system with few leaders "works." You don't want too many leaders or shared leadership because the structure requires lots of followers. People have come to think that this system is "normal" or natural because it is so promoted and accepted in the United States.

If progressive groups are to begin to help bridge the divisions that currently exist in the United States, our view of leadership must reflect a different vision of society—one rooted in real economic and social democracy. This new view of what constitutes a leader and leadership will require us to honor several principles. First, the idea that, while leaders have traditionally been white, male and middle-class people, we are committed to building an organization of diverse leaders and to making it possible for anyone to become a leader. Second, that *all* people *can* learn to be leaders through support and training in which people get to recognize themselves as leaders and learn the skills of leadership. And third, that shared leadership is the result of shared decision making. We must believe in many leaders.

In order to put into practice these principles, PPP has built a model based on sharing power and created a structure for broad-based leadership. Through our organizing work, our staffing, and our workshops, we have consciously challenged the ways that traditional ideas about leadership have kept our members, and people like us, from recognizing their own power and their own potential to be leaders.

At PPP, our training workshops help our members look at who the traditional leaders in this country are. We ask people to name those who are in power and list the skills that they feel those people have as leaders. The list usually includes things like being able to speak in public, writing skills, and being able to deal with media. Then we look within our own group to see which of these skills we have *as a group*. We may not have one person who possesses all these different skills, but almost always together we possess all of the different skills that people have listed. Then we ask *why* the people in power have power. Often, we discover that the reason isn't just the skills we've listed but some additional reason: the person has money or is white or has a certain educational level or is male. You might think this process would discourage members, but it actually helps them analyze what parts of our idea of leadership are based on skills and what parts are based on privilege. It is the first step to beginning to think of *ourselves* as leaders.

The next step we take is to redefine how a person becomes a leader. At PPP we believe that leadership can be taught. For instance, since deal-ing with the media is an important leadership skill in our organization, we find the person within our group who can begin to teach that skill to everyone else. This is a very different perspective from the traditional one that sees leaders as "born, not made." We are committed to following a pattern that allows many new leaders to be "made."

When traditional leaders make speeches, to take another example, they tend to use all sorts of facts and quote statistics. When we teach public speaking at PPP, we emphasize speaking from the heart, speaking from experience and telling our own stories. We also work to recognize that we do not have to speak in a particular way. In fact, trying to talk like people think a leader should talk, like using "standard" grammar, can actually stop us from being articulate and from speaking with our own power.

In teaching public speaking, we work in teams and allow people to choose their teammates. Usually, teams consist of people whose skills or talents differ. Someone with a lot of enthusiasm and good ideas but who doesn't write well may work with another person who can write. Someone who is afraid or shy will team up with someone who is more outgoing. The pairs work up a joint presentation: the more outgoing person might speak first, but the other person will present also. In this way, both people have input into creating the final presentation. We also talk about how to deal with our feelings, especially nervousness. We tell people not to try to pretend it is not there, but to acknowledge it—to say "This is the first time I have ever spoken in front of a group like this and I am very nervous about it" as a way to be more comfortable.

If we are presenting to an "outside" audience, the final step in our training workshop is to get our folks to "translate" their speeches. We ask our members to think what needs to be translated for middle-class folks to understand us. The outcome is still powerful because it comes from the heart, and the process of translating is done in an empowering way. Even though the final words may not be the way we speak at home, folks totally understand why we're changing them and that they are the "expert" translators themselves.

In addition to building leadership through workshops, we promote new leadership within our organizing work. New leadership sometimes emerges in very unexpected places. For example, a very quiet, shy, older woman, whom I'll call "Doris," came to PPP's first open house. She wanted to volunteer. We had a list of tasks from stuffing envelopes to making phone calls to helping lobby to going to events, but Doris would only volunteer to clean the office and do yard work. We let her clean, but we also talked about her as a staff. When we see a volunteer who will only do domestic tasks, our goal is to give that person an opportunity to move beyond that role. Doris had a lot of potential, but she was nervous and afraid. We decided to ask her to come to a "Get Out the Vote" workshop. We asked our trainer to pay special attention to Doris and to try to get her involved without making her feel overwhelmed. During the workshop, our trainer finally got Doris to agree to do a role play in which she asked other people to volunteer to work. He stayed with her and supported her, and she did it. Afterwards, Doris was incredibly excited and so happy and proud of herself. The following fall, one of our leaders ended up in the hospital on "Get Out the Vote" day, and Doris took on the job. She organized her whole community by going door to door, recruiting volunteers, and setting up rides to get out the vote. Doris had found her voice. Her story is just one example of the ways I have seen leadership develop in people when they are given the chance to break out of old roles. It's important to recognize that Doris not only became a leader at PPP—the experience changed her life in many ways.

Another way that we try to break the traditional leadership mold in our organizing is by having the people who have become leaders become the trainers for the new leaders. We also ask volunteers to begin to teach others organizing skills they have learned. For example, once I was speaking with a woman who had consistently turned out ten volunteers. I said, "OK, you've been getting out ten people for 'Get Out the Vote' day. Now it's time to get out a hundred." She said, "Forget that." And we just made a joke

about it. Then I said, "No, in all seriousness, here is what I think we could do. What if you were able to get your ten folks to each do what you are doing? Then you don't make any more phone calls than before. Instead of getting them to get out the vote, your job is to get each of them to get ten other people to do what they've been doing." We discussed how she might ask them to volunteer and what kind of training they might need. She agreed to try to get her ten volunteers to come to a training workshop, and she was very excited to think her ten calls might eventually bring in one hundred people.

The rewards of traditional leadership are reward to the individual ego—getting strokes for "my" idea, "my program," "my total dollars raised." At PPP, we see it differently—the reward of leadership lies in giving what you've learned to others. People are recognized as leaders at PPP not for any single event they've organized, but for the new people they have motivated to join the organization through their leadership example.

Recognizing new leaders is an important process. The first principle is to not make any assumptions about who will or will not be a leader. Recognizing new leadership means being very observant and being ready to react when a person seems ready to take a next step toward becoming a leader. When people say no, it is often not because they don't want to be involved. They are often afraid, or have other issues that keep them from believing they can make a difference. My assumption is always that people want to take on leadership responsibilities and want to move forward to make things happen. That is what I mean when I say we must look at everyone as a potential leader.

A good example of this willingness to believe that people want to be leaders is the way in which our organizer George Friday handled our first organizing meeting in Asheville, North Carolina. Only three of the expected sixteen people showed up for that workshop. Each of the three was somewhat quiet and made references to past failed attempts to organize their community. George could have spent the time waiting and disrespected the three who came by focusing on those who hadn't showed up. Instead, she took the time to appreciate each person and their commitment to being there. She spent several hours going through the original agenda including discussing who leaders could be and encouraging folks to share their visions for their community. She then brought up planning the next meeting and developing a list of who should be invited. All three of the folks were now enthusiastic and saw themselves as important and felt confident that they could bring others. When the next meeting happened, twenty-five people came and the three original folks were there as "leaders."

Just as important as recognizing new leadership is encouraging new leaders along the way by acknowledging and valuing their accomplishments. Most low-income folks and many women have a difficult time understanding that their actions have real results. Even when they see the results, it is hard for them to accept that they were responsible. It is important to help people see their own impact in order for them to begin to develop as leaders. For example, after "Doris" organized the get-out-the-vote efforts in her community, we told her, "You made this happen. If you had not got volunteers to go to every door and ask people to vote, probably only half of that number would have voted. And, if you had not got volunteers to drive people to the polls, many who really wanted to vote would not have had the opportunity. *You* made this happen." Telling people explicitly what they did and the result of their efforts allows them to believe they can make a difference. Sometimes, just asking the question "What do you think would have happened had you not done that?" helps people see what they were able to accomplish. Calling their attention to their contribution is extremely important to creating leadership. In our society we often assume that if someone does a job well, they're just doing their job—there is no need to let them know that they are doing valuable work. We are only too willing to criticize—but not to praise or respect people's efforts. At PPP we always try to acknowledge people's efforts. We all need that.

The definition of leadership we put into practice at PPP is not without some problems. Sometimes the most difficult work we do is supporting people along their way to becoming leaders. It is tempting to give a task to a volunteer and just say "do it," but to build leadership, we all need to share in the task by giving each other ongoing support. Support may mean giving encouragement and appreciation even when a task gets only half done. It may mean pitching in and sharing the task if another person seems overwhelmed or too scared to continue. Frequently we must call and check in with volunteers who have agreed to take on a particular task. If it is not happening, we don't assume that the person is too busy or just doesn't care. We assume that there is a way we need to support them. They really want to do it; otherwise they wouldn't have volunteered. A lot of people want to make those ten phone calls but then they get scared. Sometimes it just means a phone call to talk them through their feelings. This kind of encouragement takes time and attention away from other tasks that seem more important, and all of us at PPP have failed to check in with volunteers and then gotten upset when the work didn't get done. We have to constantly remind ourselves that every member and every volunteer is a potential leader.

As I've said before, one of the issues we have struggled with in building a new style of leadership at PPP is sexism. The tendency to think of leaders only as white middle-class men is as true within progressive organizations as it is in other areas of society. It remains true even though there are more women working in progressive organizations and in the overall workforce now than in the past. I'd like to describe the kind of leadership in fighting sexism that was shown by women at a conference on voter registration that PPP staff and members attended. The male participants at this conference wouldn't let the women in attendance speak. They interrupted us and spoke out of turn, but the male facilitator would let the men talk as long as they liked. I saw all the women around me just shut down. Some of us talked about it and decided that we should do something. We passed a note around to all the women during the middle of the session to tell them to meet in a side room before we went to lunch.

When we broke for lunch, all of the women gathered in the room. We talked about what we were feeling. Several women were very articulate and objected to specific actions as sexist. Others admitted feeling confused or angry but hadn't seen the behavior of the men as sexism. We decided on an action to take: any time someone said something sexist or acted sexist, we would all stand up. A white middle-class woman asked, "How will we know if it is sexist? And what if I think it is sexist and you don't?" I remembered attending other gatherings where the women's caucuses were very divided and worried that we'd fail to act because we couldn't agree, when a PPP member said, "Well, we all have to stick together on this. If any one person thinks it's sexist, whether I see it or not, we all stand up together. It doesn't matter, if one woman stands up, we all stand up." And that was what we did. When we all stood up, one person explained what we were doing. Some of the men got very angry, and others acknowledged and supported us by standing with us.

As a result of our leadership, one of the men who was to lead a workshop challenged us to use his time for whatever we wanted to do. We seized the opportunity to do a workshop on sexism. We did an exercise in which the men listened to how the women felt when they were treated as we had been in the conference. Then we gave them a chance to respond. Several men were angry, and a couple even left the workshop early. But many men said they learned from the experience. One even said it changed his life and would change the way he worked with women in the future.

We take on the issue of sexism in our own PPP workshops when helping our members learn new leadership models. Our experience in the communities PPP works in has taught us that, while women in these communities may actually do most of the volunteer work for a specific PPP campaign or project, these same women often do not see themselves as leaders. Again and again, they will choose men as their spokespersons or as members of the PPP steering committee for their community. Sometimes they will choose men who do none of the door-to-door, or telephone, or other community work and who participate only in the sense of expressing their views at meetings. Their views are often treated by the women present as authoritative, simply because men have expressed them.

At PPP, we began to see this pattern as a problem and to think how we might help our women members identify themselves as leaders. For example, when one community group needed to choose a steering committee, we asked all the members to choose the five *people* who did the most work for the group. After five women were named and stood up in the meeting, we asked the members whether any of these women should be on the steering committee. After the women agreed to serve on the steering committee and were approved by the members, we asked the members to choose the five *men* they thought did the most work for the group. These men were also approved as steering committee members. This approach allowed the PPP members to begin to see a different leadership model—one that tied leadership to work rather than gender. It was a revolutionary idea for folks, and exhilarated them in the way it released them from choosing leaders whom they sometimes did not respect nor feel were true representatives. It permitted them to empower themselves to become leaders.

Confronting the kinds of oppression—like sexism—that exist within progressive groups and organizations is difficult but also critically important. By being made aware of how they may be giving up their power, women and others may find new ways to understand the behaviors of people in their own organization and outside it. A new or renewed understanding of how sexism, homophobia, racism, and other forms of oppression are at work all around us is a necessary part of the PPP's redefinition of leadership. Leaders allow all voices to be heard.

When I talk to people who are struggling to develop their own leadership, I tell about my own struggle to believe I could be a leader. People see me now as a leader, though I keep reminding them that I am no different than other people in our community. I do not pretend that I have always had the skills that I have now. This is important to communicate to people who are beginning to develop their own leadership skills. I once found in tears two of our PPP staff who were writing their speeches to deliver at a fundraising event. They felt afraid that they could not do their speeches exactly the same way I give a speech now: without notes and without seeming to have any fear or nervousness about it. So what I said to them was, "I didn't start out this way. What I had to do when I first started was to write out word for word what I was going to say. I memorized it and then I took my cards with me. I mostly read my speeches in the beginning." I talked about how I felt then and how I would cry every time I had to do a speech. They had seen me speak after five years of regular speech-making and felt that they had to do the same. This was very disempowering and only through sharing my story were they able to feel they could then develop their own way of speaking. To model leadership, those of us who are seen as leaders must always remember how we got to where we are now, and we must constantly turn to others to teach them the first step.

Because of my traveling around and now writing this book, a lot of people want to see me as the "only" leader for PPP. They think that I must be the exception. I know this isn't true because of working with an entire staff of leaders. I know this isn't true because of all the leaders who are in every community PPP serves. "Shared leadership" is a real and essential part of the structure of PPP and continually renews our belief that all of us are leaders in our own way.

People often ask us at PPP if we have a problem with too many leaders. We have a series of joint management teams for the office, training, and staff; advisory committees for each community which share decision-making powers with community people; a committee for each event which includes staff, community leaders, and new people coming into the leadership; and finally, the overall structure of PPP is governed by a combination of community folks, supporters, and staff.

I suppose from a traditional perspective on power and leadership, that might look like too many leaders. But our experience at PPP has shown us that the more leaders we have, the more we share leadership, and the more new people we have taking on leadership roles, the stronger and more powerful we become. I think of leadership as being an expanding circle. The most the circle enlarges to include, the bigger and more powerful it grows, and the stronger the organization becomes.

66

What Are You Going to Do for Us?

MARYBE MCMILLAN

• • •

REFLEXIVE STATEMENT

When I first saw the theme for the annual meeting of the Southern Sociological Society, I immediately shook my head in dismay at its irony. First, the vocabulary with which the theme was stated, "Inciting Sociological Thought: Engaging Publics in Dialogue," was indicative of the larger problem of sociologists' ability to communicate with the general public. Second, even if sociologists make

From MaryBe McMillan, "What Are You Going to Do for Us?," *Humanity and Society*, vol. 22, no. 4 (November, 1998), pp. 422–428. Copyright © 1998 by MaryBe McMillan. Reprinted by permission.

their research findings intelligible for a general audience, they rarely present these findings to the working class, the poor, or other disadvantaged groups, whom their research directly impacts. Additionally, sociologists rarely solicit ideas from such groups about how research could benefit them. Sometimes we are so preoccupied with sharing our knowledge and "inciting sociological thought" that we miss opportunities to learn from the people and groups that we study.

Since the program chairs were soliciting submissions for the meeting related to this theme, I decided to reflect critically on the topics. I brought up such issues as ethnocentrism and the complexities of action research, and did so in a nontraditional way. Barriers to reaching a broad audience with our research include not only the scientific jargon that we use but also the formulaic and sometimes tedious style in which we write. A standard academic paper—with its introduction, literature review, methodology, data, and discussion—allows for little creativity and appeal to a popular audience. Thus, for the annual meeting, I decided to write a poem and present it as a poster. The poster included some explanatory text, which is below, and a collage of headlines and photographs related to contemporary social problems. My hope is that the poem generates questions and discussion about the potential and difficulty of action research.

"ENGAGING PUBLICS IN DIALOGUE" OR (IN PLAIN ENGLISH) TRYING TO TALK TO PEOPLE ABOUT SOCIAL ISSUES:

Can Sociologists Speak in a Way That Most Folks Understand and Is What They Have to Say Relevant Anyway?

With their study of social problems like inequality, racism, and crime, sociologists have many relevant potential insights to offer to policy makers and those people directly affected by policy. I say *potential* because there are barriers that stand in the way of sociologists sharing their research and engaging others in meaningful discourse. Most of these barriers are extensions of one central obstacle: the tendency to engage only those like ourselves.

I think it would be fair to say that most sociologists are well educated members of the middle class. Most of us would also like to think of ourselves as progressive and politically liberal. We share our research with like-minded colleagues in our departments or at professional meetings; we publish our results in journals that are read mostly by other sociologists. We may address legislators and other government elites about policy matters like welfare reform, but they are also much like ourselves.

How often do we address the people like welfare mothers whom policy will directly impact? If we do, would they understand what we were saying? I say this not to patronize the working class, but to point out that typically when we address people, we do so in a language particular to social scientists.

More important than simply addressing people whom our research and social policy directly impacts is talking with them about their lives. What kind of changes would help them? Too often, we devalue such experiential knowledge. Even as we venture into communities to do research that we hope will help some disenfranchised group with economic or social development, we often do not include this group as a full partner in the research.

Several experiences have taught me how isolated academics are from people of other classes and races, how such isolation contradicts the goal of reducing inequality, and how people outside of academe are keenly aware of these shortcomings and thus, highly suspect of any efforts at applied research. While I do not have the definitive answer to how we can overcome such barriers, I think that the first step is to recognize and develop an awareness of the jargon we use, of the people with whom we interact, and of the compatibility of our actions with our notions of social justice. The following poem is designed to make people think about the issues that arise when sociologists go outside academe into the "real world," where sociological concepts such as race and class are personified and made real.

"What Are You Going to Do for Us?"

I.

"You're white?"
One little boy says to me
As he grabs my hand.
The look in his eyes
And his continued attention
Make me realize that this is a question,
Not a statement.

Nothing I learned in my sociology classes
Has taught me as much about race,
About how isolated we are
From one another
As my experience
As a volunteer at Head Start.

"What is that?"
The little girl asks me.
I look down at my arms,
And notice that she is pointing
At the freckles and moles
That dot my skin.
As we sit on the floor,
The girls run their fingers
Through my hair
And try to braid it like theirs.

I realize that I am
Just as curious as they are.
I want to ask one of the teachers
How long her hair can stay braided
In its many beaded braids.
But I don't.
Unlike these kids,
I have been fully socialized
And have all the inhibitions
Such socialization brings.

The teachers think
That I am in college
Studying to be a schoolteacher.
They think that I come every week
Because I am required to.

After a year or so,
They realize that any requirement
Would long since be over.

"I come here just because,"
I tell them quietly,
Embarrassed at the luxury
Of having leisure time.
"Yes, I'm in college."
"No, I'm not going to be a teacher."
"I'm studying sociology."
The conversation ends.

II.

I'm not surprised
That they don't know what sociology is.
I do not explain
Because I am not sure what to say,
Maybe sociology is . . .

Correlates, variables, hypotheses,
To regress something on something else.
R squared, p<.05, t-test,
Statistically significant,
But is it socially significant?
Causally related to what?
I wonder . . .
Does our research have any effect?

The equations predict the log odds
That crime will follow either a convex or
    bimodal pattern.
Translation:
The less who understand,
The more important we feel.

Class, population, respondent,
Where is the person amidst the jargon?
Would a sociologist know a working class
    person if she saw her?
Would we engage a homeless person in
    conversation?
Or would we cast our eyes downward
And walk quickly past,
Embarrassed at our indifference?

III.

I am a long way
From the cafeteria at Head Start.
As we sit in the restaurant,

Eating our hummus, our tofu, and our black
  bean salsa,
Discussing sustainable agriculture,
Organic farming, free range chicken,
Detailing the hazards of pesticides and
  ground beef.

"You can get free range chicken at
  Wellspring,"
One of the group says,
As he bites into his grilled portobello
  sandwich,
"And they carry organic green beans and
  carrots."
"I think ground beef is the worst thing you
  can eat,"
Another one says, "You just never know
  about E Coli.
I'm afraid to eat hamburger any more."

We rejoice in aiding organic food
  production
And keeping the independent farmer going,
While eating healthy, too.
But have we stopped to ask,
What are poor people to eat?

No, we can't ponder that now because
It's another day,
And we're off to another meeting.
This one is a labor discussion group,
Relation to the means of production and
  such stuff.

We sit in the hotel,
Eating our $7.00 eggs and toast,
Wiping our mouths with linen napkins
While the African-American waiters keep
  our coffee cups filled.
Finally, we get around to the question,
What are the issues facing the working class?

While we contemplate this question,
Construction workers are next door at the
  local diner,
Sitting in the vinyl booths,
Eating their $1.99 breakfast specials.

We talk about false consciousness,
But do we know what the "true"
  consciousness is,

Of the person in the diner next door,
Whom we have never engaged in
  conversation,
Who lives in a part of town,
Far from where we live,
Whose kids don't go to school with ours,
Who doesn't shop in the stores we
  patronize?

IV.
We finally stop to think about such
  questions.
We decide to venture out,
Out into the unknown land
Of migrant farm workers,
And we find that there are only more
  questions,
Questions that make us twist uncomfortably
  in our chairs.

"What are you going to do for us?"
The labor organizer asks.
We have come to seek
His help in making contacts,
His input about our project,
His partnership in our research.
All the things
That we think action research should be.

"What are you going to do for us?"
He asks again.
"When you see the problems,
Do you have lawyers who can help us?"

Research should be . . .
But did we realize
How it would be?
Did we realize
The commitment such research requires?

The organizer tells us,
"You can't just show up one time
And come back three months later
And expect people to trust you.
You have to develop a relationship
With the people."

I had been patting myself on the back,
For venturing out of the ivory tower,
For engaging in the conversation.

But as I sit silently in my chair
I realize that conversation is simply not
  enough.

V.

"What are you going to do for us?"
The question echoes unanswered in the
  silence.
It is NOT a question
Simply asked by one labor organizer
To a few sociologists.
It is a question
Posed by the many social groups we study.
It is a question
Asked of all sociologists.

"But we do research,"
The sociologists reply,
"We are not supposed to do anything else.
That's not what sociology is about.
It's about objectivity, scientific research and
  all that."

Perhaps I should tell that to the teachers
As we stand on the playground at Head
  Start.
Perhaps I should tell that to the kids,
Who at four years old might not yet know
  the word,
"Racism,"
But who, because of it,
Are here at Head Start,
A program created
Because inequality persists in our society.

What will you say
When someone asks you,
What sociology is about?
What will you say
When they ask,
"What are you going to do for us?"